Lecture Notes in Electrical Engineering 168

For further volumes:
http://www.springer.com/series/7818

Ade Ogunsola and Andrea Mariscotti

Electromagnetic Compatibility in Railways

Analysis and Management

Springer

Authors
Ade Ogunsola
Parsons Group International
Abu Dhabi
UAE

Andrea Mariscotti
University of Genoa
Genoa
Italy

ISSN 1876-1100
e-ISSN 1876-1119
ISBN 978-3-642-30280-0
e-ISBN 978-3-642-30281-7
DOI 10.1007/978-3-642-30281-7
Springer Heidelberg New York Dordrecht London

Library of Congress Control Number: 2012937670

Printed on acid-free paper

Springer is part of Springer Science+Business Media (www.springer.com)

Preface

A railway is a complex distributed engineering system and thus the construction of a new railway or the modernisation of a existing railway requires a deep understanding of the constitutive components and their interaction - with themselves and with systems (or installations) external to the railway. This interaction requires a system-thinking approach, in which the overall functionality (or set of functions) can be demonstrated to have been achieved. The said functions are themselves requirements and the requirements covered in this book relate to electromagnetics. The railway electromagnetic environment is characterised by a complex mix of high power sources, sensitive safety critical systems, intentional transmitters in close proximity to each other and distributed over the entire length of the railway line. As a distributed complex system, it has the additional complexities of the propagated field being dependent, somewhat, on the position and current loading capability of the load (train) that is in motion. The railway electromagnetic environment is a complex superposition of electromagnetic environments, including those of light industrial/residential/commercial and heavy industrial applications. Electrified railways can require very high output electrical systems and power electronics, that are characterised by large excursion of voltage and current variables during commutations and throughout their operating modes, and that have a peculiar current distribution to the moving loads, leading to additional specific electromagnetic environment. A parallel can be made between a railway system and a printed circuit board: the physical dimensions and layout of the railway is analogous to that of printed circuit board - the physical dimensions of the printed circuit board is comparable with the larger geometry of the railway; furthermore, modern electronics systems consist of densely populated circuitry, compromising analog and digital functional blocks with a set of intricate inter-connects; there are of course differences in the frequency range and the physical dimensions of transmission lines and the current and voltage ranges of many phenomena. With this in mind, it is logical to assume that all the theory and practices used to design high speed circuit boards for minimum emissions and maximum immunity are applicable, in some way, to the design of railway system for system compatibility and safety.

Railway applications, particularly those relating to Signalling and Telecommunication, have not been immune to the adoption of advanced electronics; over the past 20 years or so, train control and supervisory functions have increasingly been performed by advanced communication and electronics technologies taking advantage of the growth in semiconductor and telecommunication industries. The use of Insulated Gate Bipolar Transistor, IGBT, in power electronics and the use of Global System for Mobile Communication, GSM, in Train Control applications are an example of the technology uptake in railway applications. These advances, and many more, in railway applications presents new challenges in understanding the electromagnetic behaviour of railway systems. Electromagnetic theory, therefore, forms an important foundation in modern railways system; however, it is the applied electromagnetic discipline, Electromagnetic Compatibility, that concerns the ability of an electrical/electronics systems to co-exist in an electromagnetic environment without causing neighbouring systems to mal-function, while at the same time having an adequate level of immunity to the electromagnetic disturbance that neighbouring system generate, that informs the aim of this book.

The book is organised into 8 Chapters; Chapter 2 presents an overview of railways system with particular focus on the railway traction system and the associated electromagnetic aspects. Chapter 3, covers signalling, communication and other associated systems. Chapter 4 deals with the subject matter of EMC Management and presents a summary of the processes, that need to be implemented for the proper management and delivery of Railway EMC. Chapter 5, follows on with the subject of EMC assurance as applied to railway project covering the relatively new subject of EMC and Functional Safety. Chapter 6 gives an overview of standards and the standardisation process, detailing applicable standards used for various systems within railway applications, as well as railway application standards. Chapter 7 presents the electromagnetic theory and formulation as applicable to railway systems, covering such subjects as antenna, shielding and cable-to-cable coupling, to name a few. Chapter 8 deals with earthing and bonding, lighnting protection system and stray current protection systems.

Ade Ogunsola
Abu Dhabi, UAE
Andrea Mariscotti
Genoa, Italy

Acknowledgements

The railway electromagnetic environment has become increasingly complex with the increasing use of information technology and advanced semiconductor components for a variety of applications. This continual growth in the application of such technologies makes it impossible to write a book on Electromagnetic Compatibility with specific focus on railway applications that is complete and completely satisfying. The field is vast and diverse, and it is difficult to decide what to include and what not to include. Our natural thought was to structure this book in terms of the major components of a railway system, without falling into the trap of writing a general book on railway engineering. Then we had to decide on the appropriate level of the electromagnetic content of the book, on the one hand assuming that the readers of this book are people who have a general understanding of electromagnetics, but not necessarily an understanding of the philosophy of electromagnetic compatibility and vice versa. We feel, we just about got the balance right, while recognising that more could be done or could be included. The content stems from our collective experience of working in the field of Electromagnetic Compatibility (EMC) Engineering as well as the application of the same on railway projects as well as academia.

A number of people have helped make this book possible. I, (Ade Ogunsola) am grateful to all those that gave me the opportunity to walk the path of an EMC Engineer. This includes Brian Sketchly, Dave Eardley and Dr Alwyn Finney, all of GPT Limited, who provided the opportunity and encouragement to explore. Further thanks goes to Tracy Best and Paul Thomas for giving me the opportunity to think "systems". To colleagues at London Underground, particularly John Bradshaw - The EMC Engineer and Melville Rodrigues - Tube Lines EMC Engineer, sounding boards for many aspects further explored in this book. To Andy Bourne for allowing me to work EMC magic for 8 years without interruption. To my academic colleagues of the Department of Electrical and Electronics Engineering, University of Lagos, Nigeria. Thanks to Prof Ike Mowete (Woked Fellow) for the numerous discussions on electromagnetic. To Leonardo Sandrolini and Ugo Reggiani of the University of Bologna, Italy for the wonderful research collaboration. To my EMC standardisation colleagues of IEC TC 77 and AFSEC TC 77 for the numerous intellectual discussions on EMC. To Parsons and colleagues at Parsons for the opportunity to turn numerous railway projects into my own personal laboratory. To Ibiyemi and Andrew Akoto, Kehinde and Taiwo Ogunsola for their encouragement and support. To my late parents, who saw this project start but not its completion. And finally without both the moral and emotional support of Folashade, Adeola and Jadeola, this book would not have been started or finished. Thank you.

Andrea Mariscotti would like to thank the many friends who encouraged and supported the development of various content during on-site activities and their preparation, meetings, participation to committees and the development of European Union projects in the field of Electromagnetic Compatibility and measurements applied to the railway field, that have been an interesting and grateful occasion of study and discussions. In particular, my grateful acknowledgement goes to my friend Marco Ruscelli, who is a living reference for track circuits and on-site testing, and his other colleagues at Ansaldo STS in Genoa; it's surprising how seeing several of my former students working ay Ansaldo STS doesn't make me feel old. Further thanks to friends in other companies for the interesting discussions, ideas and problems (Giancarlo Cantobelli at Alstom, Pina Imbesi at Bombardier, Markus Meyer at Emkamatik, Dieter Wuergler at Enotrac, Bernhard Hemmer at Siemens, Noel Haddad at SNCF, Giulio Depaz at Vossloh-Kiepe, to cite just a few). Not to forget the academic community, I would like to add some persons with whom I worked pleasantly and profitably, but not limiting the relationships to work: Paolo Pozzobon of my university with whom I have been working for years and introduced me to the railway world; for the joint development of the first models during the late '90s Sergio Brillante, Gianfrancesco D'Addio and Patrizia Ferrari, who formed with me the first group working on railway EMC, continuing later on with Maurizio Vanti. Francescaromana Maradei, Mauro D'Arco and Sergio Pignari from the Trenitalia commission back 10 years ago. The large group of the many friends and acquaintances that I have met on-site during tests and inspections, in numerous countries, working together, lending instruments as well as problems and solutions, eating something when it is possible, is uncountable.

Ade Ogunsola, PhD
Andrea Mariscotti, PhD

Contents

Chapter 1
Introduction

Numerous components and component technologies result in new opportunities for integration inside railway systems. The desire to address these important problems in real life and advances in information technologies motivates the development of new train control and system monitoring concepts. The goal is to develop train control and system monitoring systems which have low operational cost, reasonable investment cost, and high safety and comfort levels. This is true for rolling stock, where the running conditions are efficiently and promptly adapted to the required operating profile, responding to requirements related to comfort, energy efficiency and reduction of power consumption, minimization of wearing, mechanical stress and noise, etc., as well as for the infrastructure and also for station services, where increased customer safety and comfort, reduction of waiting times, improvement of awareness and of the information level are all attainable by means of controlled climatization and air conditioning, networked safety cameras, interactive terminals, visual and audio broadcasting, modern ticketing and access control, etc..

The higher level of informatization, networking and deployment of sensors and control devices, and the use of smarter electronic equipment with larger power concentrations and increased switched power require a systematic approach to design, certification, installation, verification and testing, in particular for the various aspects of electromagnetic interaction and possible interference, under the term "electromagnetic compatibility" or EMC in short.

EMC referred to a single equipment represents the ability of the equipment to operate as intended in the assigned electromagnetic environment, contemporarily not impairing the functions of other equipment located nearby. It is evident that this definition implies the electromagnetic characterization of the target environment, based on as widest as possible measurements aimed at establishing the behavior of the environment under the same category; furthermore, the knowledge of the propagation and coupling of electromagnetic emissions identifies the expected level of disturbance appearing on different

A. Ogunsola and A. Mariscotti: Electromagn. Compat. in Railways, LNEE 168, pp. 1–4.
springerlink.com

interfaces (called "ports"), all identified as relevant for coupling of the said disturbance to nearby victim equipment; the disturbance, then, can cause interference if too close to the susceptibility of electrical and electronic devices in terms of amplitude, frequency band, and more generally time-frequency behavior. The level of coupled disturbance depends highly on how the equipment is installed in the assigned environment and the degree of coupling of ports with respect to a wide set of electromagnetic phenomena, conducted and radiated, steady and transient; this coupling is determined by the electrical and electromagnetic characteristics of the equipment enclosure (the port for radiated phenomena), of connectors and connected cables (their length, location, routing, shielding, etc.) and of grounding (grounding efficiency in terms of the offered grounding impedance, presence of noise, etc.).

EMC Engineering as a discipline, has evolved in response to the immense growth in density, sensitivity, bandwidth and complexity of electronic systems over many years, often compactly packaged with high power electronics. Challenges that are fundamental to major technologies and applications have received due attention, harvesting structured methodologies that blended into their respective engineering: layout of high-speed and RF integrated circuits, integration of interference-handling into the wireless communications' signal management and processing, and others. However, each new integration or co-location of different circuits or systems, or new configuration or packaging, poses challenging risks of interference that need specific attention, even if the system concept is well established and the system project is considered low-risk.

At the upper level (the apparatus level) several equipment are located in the same cabinet or rack, and several factors and elements are shared among equipment, possibly supplied by different manufacturers. It is thus necessary to establish and operate a common framework of rules and criteria, to be followed by manufacturers and designers, in order to reach a satisfactory EMC performance, with the due margins.

At the system level, when different apparatus and equipment are interconnected in various forms, the problem of harmonizing rules and criteria is even more complex, taking into account the wider diversity of the same apparatus and equipment, and the intrinsic variability represented by cabling and interconnections. The distributed nature of signalling and the power network systems within railway applications adds to the complexity of the electromagnetic interactions and its coupling. As a railway system transverses a number of electromagnetic environments, the number and complexities of the interfaces, requires that system integration with respect to electromagnetic as an interface is well defined and specified. The functional requirements of the interface need to be adequately specified, with responsibilities allocated to the interfacing parties. Most importantly, the interface requirements needs to be subjected to a verification and validation process. There is a general change in the electromagnetic characteristics of a railway environment, mainly due to the implementation of modern technologies in power electronics, communication and electronics.

It is easy to see therefore that EMC has particular relevance for contractual and technical aspects, demonstrating that an equipment, apparatus or system is fit for purpose under a well defined framework of rules and criteria, coded in standards, that must be correctly assigned and enforced when performing the EMC management at the upper system level and adequately followed and implemented at the lower subsystem, apparatus and equipment level. The objective of achieving EMC is to reduce the spurious interactions within and between systems below the threshold of interference to the systems' functions and performance. This undertaking may take the form of a stand-alone project or of an adjunct project within the system-development project. The development of a generic EMC solution to a class of systems (e.g. design rules for integrated circuits) warrants a major stand-alone effort whose cost is justified by the savings thereafter. However, a recovery effort for an (operating) system that is already adversely impacted by electromagnetic disturbance, EMD, and requires mitigation in order to reach its full performance, is highly cost sensitive. The cost of such an effort is proportional to the system's complexity. It entails diagnosis, analysis, measurements, changes in architecture, signal design and routing, layout and packaging. In a railway system, such mal-operation could be costly (in terms of system delivery and safety)[1].

The present book is born from the awareness of the complexity of the EMC problem in a complex distributed system, such as a modern guideway transportation system, that includes traction power supply, rolling stock equipped with static power converters and auxiliaries, signalling systems, station auxiliaries for ventilation and air conditioning, lighting, passenger announcements and display of information, ticketing, escalators and elevators, etc. EMC related issues often require a multi-stage approach, where the issue must first be identified and its boundaries defined, in terms of relevant elements, subsystems, etc.. Then, a theoretical model is proposed, for example by means of equivalent circuits, closed form or empyrical expressions, finite element methods or other simulation tools, in order to evaluate the associated electromagnetic phenomena, such as the characterization of propagation and coupling, possible shielding and reflections, and so on, defining the sources and victims relationship. The results must be evaluated against a characterization of victims and sources, as complete as possible and as required by the problem; depending on the safety function of the involved system, this characterization may be complex and time consuming, for a complete and accurate result, reducing the uncertainty to a minimum. The overall process develops not only with a scientific basis, but more importantly with a well grounded a normative base, that is with the most complete and adequate

[1] System delivery in this context refers to time to market, industry acceptance, lost of market share etc, whereas system safety refers to injury or loss of life due to mal-operation of equipment.

reference to applicable standards; this means that the standard requirements and the coded practices must be part of the system specifications since the beginning and followed up and managed throughout the process. All this often is beyond the competence of a single person and in many cases it is difficult to identify an EMC expert within most manufacturers. The complexities of the subject matter and its application within railway systems informs that management process, based on systems engineering principles, be implemented in which the necessary components of EMC engineering is applied to achieve the desired level of compatibility and safety, within budget and schedule.

In this book, a general description of railway systems is given with particular emphasis on the elements that contribute to or influence the overall electromagnetic profile of the railway; this is followed by an overview of management and assurance processes that are generally applied to the railway applications, detailed review of standards and regulations applicable to railway applications and a detailed treatment of electromagnetic principles as applied to railway.

Chapter 2
Railway Traction Systems

This Chapter presents a general description of the elements that characterize a railway electric transportation systems: the supply system, the traction line, the rolling stock, signalling, and additional subsystems, that equip modern railway transportation systems, in particular urban and suburban railway transportation facilities. Throughout this chapter, basic definitions will be provided and refined. The chapter focuses on elements that are important in achieving systems and electromagnetic compatibility with emphasis on the electromagnetic compatibility aspects and thus some knowledge of railway engineering is assumed. The reader is referred to publications such as [644] for a detailed discipline specific treatment of the subject matter covered herein.

2.1 Traction Systems and Traction Supply

The general structure of a traction system is made of several electric supply substations (ESSs) connected to traction line (TL) sections, that may be electrically continuous or separated, depending on the type of electrification and power management conditions, and one or more trains moving with variable tractive conditions. Traction systems may differ for a variety of characteristics, that may be relevant depending on the application, the performed analysis and the considered phenomena. A classification follows, where an attempt is made to treat the most relevant characteristics of some traction lines categories, without the claim of being absolutely correct, because of the many variants and implementation deviations, due to the development of these systems in many countries over a long period of time with different local requirements and characteristics:

- heavy traction railways feature large gauge running rails, heavy locomotives or electric motor cars with absorbed power in the range of MW, moderate to high cruise speeds (e.g. above 100 km/h), large installed power at substations, ranging approximately from 1 MW/km to some MW/km, depending on the profile of the territory, slope of grades and on the headway

A. Ogunsola and A. Mariscotti: Electromagn. Compat. in Railways, LNEE 168, pp. 5–94.
springerlink.com

and operating profile of trains, that determine their number in a supply section; they include a vast range of railways, from the high speed lines supplied in ac at high voltage and featuring very high speeds and installed power per km, to conventional lines with trains of various lengths, smaller installed power and infrastructure particulars that do not permit high speed operations, down to commuter and low speed lines with medium size train consists, that don't always require heavy traction rails and powerful locomotives, and that are normally reserved for implementation in cities outskirts, rural areas and countries under development;
- light railways are implemented to serve medium to large size cities, or to serve the outskirts and bordering towns, replacing metros for medium distance connections; the supply voltage is usually dc at 750 or 1500 V, also for compatibility of equipment with existing metro lines; trains are normally fixed compositions of electric cars like in a metro, with larger headway but with smaller absorbed power, because of the size of the train consist and the reduced cruise speed; the installed power is thus lower, comparable to or smaller than metros, in the order of a fraction of MW/km;
- metros, with several characteristics in common with light railways, with slightly more frequent starts and stops and possibly steeper velocity profiles, always embedded in urban areas and for this reason feature a meshed line structure, with several intersecting power lines, and large short circuit currents; the headway may be very tight, depending on the implemented solutions, from several minutes down to 120 seconds for most modern lines in large populated cities, or even 90 seconds during peak hours. This in turn determines the installed power that may be larger than a normal light railway above 1 MW/km;
- trolley buses and tramways are low speed transportation systems, with a less demanding operating profile and with lighter and shorter vehicles and train consists. We could say that the supply voltage is always dc and the voltage level may take different values depending on the specific system and solution,with a range between about 500 and 750 V (e.g. 600, 650, 700 and 750 V nominal voltages are very common). The installed power barely corresponds to the rated power of one or few vehicles in a supply section and corresponds thus to one or few hundreds kW/km; moreover, the vehicles are often equipped with braking energy recovery systems (based for example on on-board batteries or super-capacitors) and for trolley buses with a combustion engine, that frees the vehicle from the overhead line supply (to move in urban traffic and in historical centres where the traction line infrastructure is reduced to a minimum or completely removed to reduce visual impact).

In the following more details are given for each type of traction system making use of existing examples. They represent the reference for the analysis of the infrastructure properties and characteristics that determine the electrical behavior of a traction system with respect to the conducted emissions of power supply substations and circulating rolling stock, their propagation and coupling with the rest of the network, and, in particular, signalling devices. For practical reasons the supply systems are identified with their most common voltage levels or their supply frequency, even if special configurations exist (such may be treated on a case by case basis). Of course, the voltage levels are nowadays standardized and the use of "15 kV system" or "25 kV system" expressions identify almost unambiguously the system general characteristics. Yet, through the years the voltage level as well as the supply frequency have been the results of a compromise between opposite requirements: for example a lower frequency, it is known, gets a proportionally lower line longitudinal reactance and line voltage drop, so that substations may be pushed farther apart; a lower voltage is not only profitable for the rating of electric machinery and conversion devices, but also in order to reduce clearance requirements, in particular if the electrification of an existing system within its right-of-way is concerned.

2.1.1 AC Railway Systems: 15 kV 16.7 Hz

2.1.1.1 General Description

This is a supply system that is in use in the German-speaking European countries (Germany, Austria and Switzerland) and with some differences in Sweden and Norway. At the turn of the century, systems were built around an inductive-resistive commutation pole shunt for traction units and then soon with series-wound commutator motor allowing higher supply voltage.

The reduced supply frequency (one third of the industrial frequency of 50 Hz in Europe) was motivated mainly by a correspondingly smaller longitudinal inductive reactance of the traction line and thus a smaller longitudinal voltage drop. The 16.7 Hz frequency (in reality equal to 50/3 Hz) can be obtained by a 50 Hz supply network by means of the rotating converters, that are still in use in many substations.

The most modern version of the rotating converters are the static converters, that transform and adapt the input frequency by means of semiconductor inverters or cycloconverters. These converters have a large rated power and a correspondingly low switching frequency; in order to reduce their harmonic emissions on the 16.7 Hz side they are equipped with harmonic filters of various types (low pass and resonant), that have proved in recent tests their effectiveness (static converters in operation on the Swiss network showed a negligible impact on the harmonic distortion of the line voltage[1]).

[1] These results to authors' knowledge have not yet been published and they were synthetically commented in a private communication.

The 16.7 Hz railway systems are equipped also with autonomous generation plants operating directly at 16.7 Hz: the energy is derived by hydrogenerators and combustion engine generators.

The type of traffic is both conventional and high speed, so ranging up to 250 km/h and slightly above; the installed power is that characteristic of heavy traction systems, and it is particularly large in high traffic areas, where local, regional and long distance trains are all absorbing power by the same network.

To ensure interoperability and the standardization of current collection mechanism, the absorbed traction current is drawn from an overhead contact wire system, the catenary, placed on the track axis above it at a standard height.

2.1.1.2 Traction Current Arrangement and Supply Architectures

The traction circuit arrangement is characterized by the way in which the return current is taken away from the rolling stock back to the supply substations. For 16.7 Hz systems we distinguish:

- Rail return (RR) systems, where the return current flowing through the axles and the running rails keeps flowing in the same running rails until the substation is reached; the rail potential may reach dangerous values and the rail current amplitude (in terms of the net differential mode component) may represent a significant disturbance coupled to signalling circuits;
- Return conductor (RC) systems, where the return current divides between the running rails and a purposely laid down conductor at the same potential, the return conductor; a percentage of the return current is thus driven away from the running rails, closer to 100% for larger mutual inductance between the catenary and the return conductor;
- Rail return with booster transformer (BTRR) systems, where the mutual inductance between the overhead and the running rails is increased, thus forcing almost equal currents in the two circuits, and in turn reducing to zero the earth current;
- Return conductor with booster transformer (BTRC) systems, where the mutual inductance between the overhead and the return circuits is increased, thus forcing almost equal currents in the two conductors, and in turn reducing to zero the current in the running rails.

The four systems are depicted in Figure 2.1, where the ideal circulation of the traction current is indicated by the arrows.

The operating principle of the booster transformer is based, as for the autotransformer described later on, on two magnetically coupled identical windings (i.e. a single-phase transformer with a 1:1 ratio): the parameters describing its operation are synthetically the self inductance of each of the two windings, the mutual inductance (or in other terms the magnetizing inductance) and thus the magnetic coupling factor. The larger the magnetic

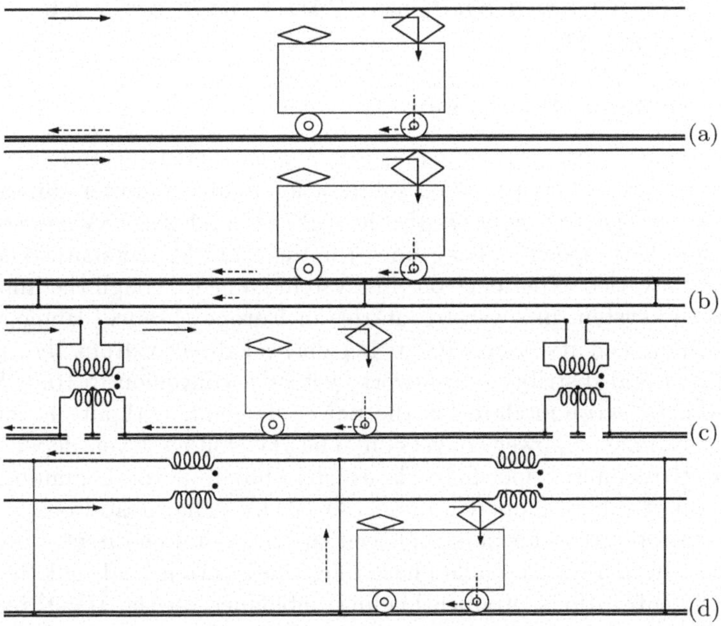

Fig. 2.1 Supply configurations for (a) RR, (b) RC, (c) BTRR, (d) BTRC systems

coupling and the magnetizing inductance, the lower the no-load current (i.e. the magnetizing current) and the better the correspondence between the currents in the two windings. We could say that for a large enough coupling close to unity the two windings are carrying identical currents; since one of the two windings is in series to the overhead supply conductors, the current in the other winding is forced to be equal, recalling the current from the connected conductor of the return circuit, that is the running rails or the return conductor. The mutual or magnetizing inductance is required to be large not only to reduce the no-load current, but from another viewpoint to operate in a source current way on the remaining traction circuit, recalling the said current with an impedance much larger than that of the traction circuit section itself. The effect of the booster transformer on the traction line impedance is to increase it by its stray (or leakage) reactance.

Bringing the return current as close as possible to the live conductors (that carry the traction power supply current) has several benefits: first, the mutual inductance between the two parts of the traction circuits is minimized and so is the inductive voltage drop, that is the major voltage drop term in AC railways; second, the magnetic field emissions and the induced voltage on nearby installations are minimized as well, being also proportional to the mutual inductance.

2.1.2 *AC Railway Systems: 25 kV and 2x25 kV 50/60 Hz*

2.1.2.1 General Description

There are many railway systems equipped with a 50 Hz supply (or 60 Hz supply); it represents a preferred solution for light commuter railways and normal heavy traction railways (both in 25 kV and 2x25 kV versions), as well as new high speed railway lines (in the 2x25 kV version). The main reasons are a varied experience on this system gained through the implementation of the system in several countries in Europe (France, Hungary, UK, to cite some), and its simplicity, being derived directly from the national transmission and distribution network, yet at a conveniently High Voltage (HV) level, to accommodate for the power demand, without the need for frequency change or power conversion. The HV voltage level, depending on the respective country specificity, is usually chosen at the beginning of the HV interval of values, namely around 130-170 kV. The substations are made simply by supply transformers that feed the traction line directly. However, being the traction line a single-phase line, the traction load and the power demand at substations produce loading unbalance on the HV three-phase lines, and voltage unbalance as a consequence. The general solution is that of connecting successive substations at different phase pairs on a dedicated HV line, so that the overall traction load is somehow balanced, but it is clear that even unbalances of some % may be of concern. In the likelihood that a railway system represents a significant load, such as a high speed railway line, and the resultant unbalance is a major concern, the architecture of the HV supply feeders is subdivided into a first lower HV voltage for a dedicated bus, connected to a series of substations on the territory, fed by a dedicated transformer or autotransformer at a higher HV level, connected on the primary side to the national transmission network.

In general, 50 Hz systems are managed with one-side fed isolated supply sections (SSs) to avoid unwanted short circuits between adjacent sections, since, first, the single phase Electrical Supply Substations (ESSs) are connected in turn to different phase conductors at HV level and thus have a relative phase jump of 120° electrical degrees, and, second, even in the case of two single phase sections with the same phase rotation, there is the possibility of a phase difference between two HV feeder lines because of their own resistive and reactance differences. The electrical separation of adjacent SSs is obtained by means of the so called phase separation sections (PSSs) [123]. In case of a failure to some component, that brings an ESS out of service, two SSs are reconfigured and brought together bypassing the interposed PSS, so that the supply of the otherwise isolated supply section is ensured.

2.1.2.2 Traction Current Arrangement and Supply Architectures

There are two different solutions, one based on a single polarity at +25 kV with respect to the running rails at 0 V (indicated as 1x25 configuration), the other based on a double polarity system at +25 kV and -25 kV respectively, again with respect to the running rails at 0 V (indicated as 2x25 configuration).

The operating principle of the 1x25 system is the same as a 16.7 Hz system: the same configurations (RR, RC, BTRR and BTRC systems) and the schemes shown in Figure 2.1 are equally applicable, with a more profitable sizing for the booster transformer, being the magnetizing current for the same transformer inductances three times smaller.

On the contrary the operation of a 2x25 system is based on Autotransformers (ATs), a coupled transformer similar to a booster transformer, but featuring much large nominal power and connected between the two polarities, where it applies balanced currents. In a standard 2x25 kV AT system the rolling stock is supplied through the overhead circuit at +25 kV (the catenary) and the traction return current flows from the rolling stock axles back to the supply through the rails; the rail current that reaches the ESS is largely reduced by the balancing action of ATs, that force almost equal currents in the +25 kV and –25 kV circuits (the latter representing the negative feeder or NF). The train return current flows through the rails and then to the NF of the neighboring ATs, that are the supply section (SS) boundaries. The reason for this configuration is a power balancing in the two live circuits, with a reduced overall line inductance (being smaller the area of the loop between positive and return current flows); an additional benefit is the reduction of the induced voltage on nearby circuits and installations as well. Due to limitations in AT power rating and stray inductance, some current keeps flowing into the rails and also remote ATs may be involved. An accurate treatment of the equations of single- and double-feed SS is presented in [312, 310].

Depending on the headway and on the power absorption of circulating rolling stock, that define the overall absorbed power and the installed power, there are solutions based on a variable number of ATs and ESSs and their reciprocal distance.

For high speed lines with a large traction load and power demand the preferred solution is that of large rated power ESSs (from 50 to 80 MVA) with usually one or two ATs in between (also of considerable power, normally between 12 and 20 MW). Of course ESS and AT position, and thus their separation distance, must cope with the territory characteristics and the identification of suitable locations, and for ESSs the demands extends also to the accessibility of a HV overhead line. This line arrangement has been adopted for new lines in many European countries.

For older lines in which the 2x25 kV scheme (or in some cases 1x25 kV) was adopted for conventional lines (e.g. in United Kingdom, Hungary, Slovakia,

Czech Republic), the position/location of ESSs and ATs may follow slightly different schemes, particularly due to i the smaller power demand.

Single-phase traction loads represent a load of important value even at the HV level; due to the uneven power absorption, they cause a relevant voltage unbalance on the feeding HV three-phase line and this can be limited by the following methods [476]:

- as already pointed out, substation transformers may be simply allocated to alternate phases of the supply system by dividing the traction line into separate feeding sections (also called supply sections); the unbalance tends to globally cancel out, even if it is to be noted that, in particular on low headway high-speed lines or on freight lines, the substations are not loaded simultaneously and evenly;
- by means of various transformer connections the unbalance compensation can be locally accomplished with some limitations: some of the most common solutions are briefly reviewed below and are treated in greater detail in the cited references, where also some experimental data are reported;
- a higher degree of load balancing is obtained if local compensating equipment, added to the substation, is used; static passive balancing is implemented by means of capacitors and reactors connected on the line side, as extensively shown in [476]; it is also possible to adopt rotating or static converters for balancing action, rated for a fraction of the total power of the traction system load;
- the problem of balancing is completely removed if the traction line is fed entirely by means of a static or rotating converter, that thus offers a balanced load to the HV three-phase line under all conditions.

Since the transformer connection solution to reduce the degree of the voltage unbalance is a good compromise and performs satisfactorily, particularly for networks with limited short-circuit power, it is treated in the following with some details. This type of solution finds application in eastern countries, such as Korea (2x25 kV 60 Hz) [488], Taiwan (2x26 kV 60 Hz with substation transformer rating of 30 MVA, composed of 10 and 15 MVA units), continental China [685], and Japan.

In [217, 476] a detailed description of the three most common connections is given: Vee connection, Scott connection and Le Blanc connection (shown in Figure 2.2). Further architectures, such as Woodbridge, various three-phase connections to single phase, Impedance Matching and Multi-Purpose Balanced transformers are also provided in [476, 685].

The Vee connection has two primary windings (with the same number of turns) regularly connected to two phases of the incoming HV line and the center of the two windings is connected to the third one. The secondary windings are normally coupled to the two HV windings with the same number of turns. Given the two absorbed apparent power terms S_1 and S_2 at the traction line interface, we may write the HV line currents and the related absorbed apparent powers as

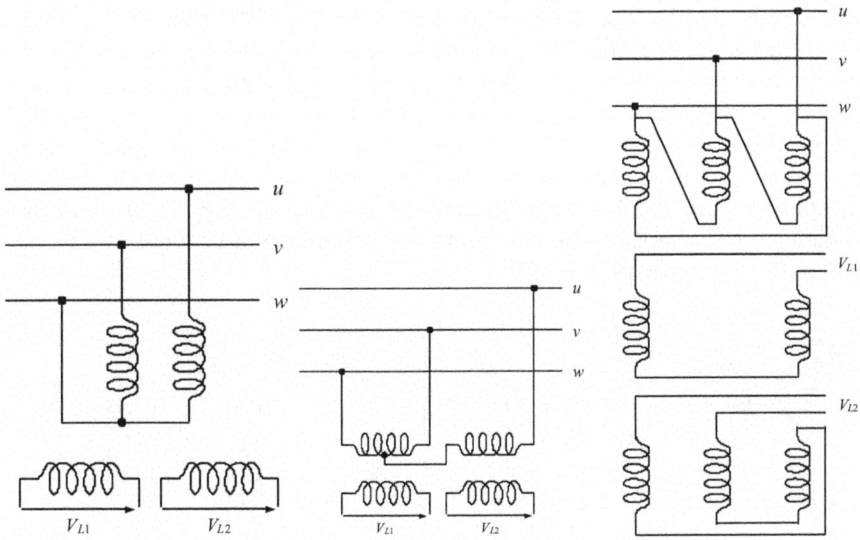

Fig. 2.2 Special transformer connections for AT traction supply: (a) Vee, (b) Scott and (c) LeBlanc connections

$$
\begin{aligned}
I_u^* &= \frac{S_1}{V_u - V_v} & S_u &= \frac{V_u}{V_u - V_w} S_1 \\
I_v^* &= \frac{S_2}{V_v - V_w} & S_v &= \frac{V_v}{V_v - V_w} S_2 \\
I_w^* &= -(I_u^* + I_v^*) & S_w &= -\frac{V_w}{V_u - V_w} S_1 - \frac{V_w}{V_v - V_w} S_2
\end{aligned}
\tag{2.1}
$$

Assuming three balanced voltages on the HV side 120° apart, the three absorbed power terms become

$$
\begin{aligned}
S_u &= \frac{S_1}{\sqrt{3}} \angle 30^\circ \\
S_v &= \frac{S_2}{\sqrt{3}} \angle - 30^\circ \\
S_w &= \frac{S_1}{\sqrt{3}} \angle - 30^\circ + \frac{S_2}{\sqrt{3}} \angle 30^\circ
\end{aligned}
\tag{2.2}
$$

The Scott connection[2] is more complex and it is made of two single-phase transformers with identical secondary turns N_2 connected to either sides of

[2] A Scott-T transformer (also called a Scott connection) is a type of circuit used to derive two-phase (2-φ) current from a three-phase (3-φ) source or vice-versa. The Scott connection evenly distributes a balanced load between the phases of the source.

the traction supply line across the neutral section; the primary windings on the contrary have different number of turns N_{11} and N_{12} in $\sqrt{3}/2$ ratio; the first primary winding is supplied by the full line-to-line voltage of the incoming HV 3-phase line, while the second primary winding is supplied between the mid point of the first one and the other line voltage. Given an identical power absorption on the two line sections with identical currents in amplitude and displacement factor, the loading on the HV 3-phase line is balanced and the three line currents are equal in amplitude and with the correct 120° phase displacement.

$$\begin{aligned} I_u^* &= -\frac{2}{\sqrt{3}}\frac{N_2}{N_1} I_1 & \qquad I_u^* &= \frac{2}{3}\frac{S_1}{V_u} \\ I_v^* &= -\frac{I_u}{2} - \frac{N_2}{N_1} I_2 = \frac{N_2}{N_1}\left(\frac{I_1}{\sqrt{3}} - I_2\right) & \qquad I_v^* &= -\frac{1}{3}\frac{S_1}{V_u} + j\frac{1}{\sqrt{3}}\frac{S_2}{V_u} \\ I_w^* &= -\frac{I_u}{2} + \frac{N_2}{N_1} I_2 = \frac{N_2}{N_1}\left(\frac{I_1}{\sqrt{3}} + I_2\right) & \qquad I_w^* &= -\frac{1}{3}\frac{S_1}{V_u} - j\frac{1}{\sqrt{3}}\frac{S_2}{V_u} \end{aligned} \tag{2.3}$$

Assuming three balanced voltages on the HV side 120° apart, the three absorbed power terms become

$$\begin{aligned} S_u &= \frac{2}{3} S_1 \\ S_v &= -\frac{1}{3} S_1 \angle 240° + j\frac{S_2}{\sqrt{3}} \angle 240° \\ S_w &= -\frac{1}{3} S_1 \angle 120° - j\frac{S_2}{\sqrt{3}} \angle 120° \end{aligned} \tag{2.4}$$

The LeBlanc connection is more complex than the Scott and requires a heavier and more expensive construction: the two secondary windings connected to the traction line circuits of opposite polarity are formed by several individual windings, with three different number of turns. However, in [217, 476] it is shown that after some numerical manipulations the resulting HV three-phase load is identical to the Scott connection.

Balancing and compensation are particularly considered in Japan where static converters for active control are added to the already implemented Scott configuration, as in the case shown by Akagi [1]: at the Shintakatsuki substation a 33 kV 48 MVA modular thyristor inverter was installed by the Central Japan Railway Company to further reduce voltage imbalance, voltage drops and negative-sequence component; the connection occurs through a HV/MV transformer bringing the output of the thyristor inverter to the 154 kV voltage level of the primary line. The compensating action results in a reduction of rms voltage variations from 9 kV on one phase to less than 3 kV and correspondingly the voltage imbalance factor from 3.6% to 1%, at the train passage.

2.1.3 Comparison of Autotransformer and Booster Solutions

In this section the two most widely applied solutions for ac systems are considered in more detail and compared to evaluate their convenience and performance with respect to the most important characteristics: required installed power, unitary cost, effectiveness in power balancing and rail voltage limitation, reduction of interfering induced voltage. The most common nominal power values and related cost are derived from examples of installations and commercial information available from the major manufacturers and railway operators.

Booster transformers are produced not only for 15 kV and 25 kV voltage levels at 16.7 and 50 Hz, but also for other lower voltage levels and for other operating frequencies, such as 25 and 60 Hz; the nominal power ranges from 100 to 1000 kVA and the winding current is usually limited to something less than 1 kA. In Europe, the two most prominent examples of BT usage are UK for 50 Hz systems (with a BT spacing of 3.2 km for both RR and RC systems) and Sweden for 16.7 Hz systems (with a BT spacing of 5 and 2.8 km for RC and RR configurations respectively).

Autotransformers are manufactured for modern 2x25 kV systems, normally operated at 50 Hz, but also 60 Hz versions exist, for those countries with a 60 Hz national grid. Their nominal power values are much larger than those of BTs and normally range between 10 and 20 MW.

The effect on the traction line of BT and AT is that of reducing the pantograph impedance (the line impedance perceived by moving trains), the line voltage drop (with a most cost effective sizing of the traction line) and the induced voltage on wayside installations, such as cables and longitudinal metallic parts.

Some mathematical modeling is needed to compare the performance of the two devices and to evaluate them with respect to Rail Return and Return Conductor reference systems. The effect and performance are evaluated against two main phenomena: the line voltage drop compensation and the rail voltage.

We begin by defining the impedance of the rail return circuit with respect to earth Z_{RR} and the impedance of the contact line with respect to the rail return (also called loop impedance) Z_{CR}; the rail-to-earth leakage conductance G_{RR} is the shunt term driving a portion of the return current away from the rails to the earth. For the central portion of a line section (neglecting end effects) we may define the rail current fraction $q = -Z_{CR}/Z_{RR}$ and the corresponding screening factor $k = 1 + q = 1 - Z_{CR}/Z_{RR}$, as well as the propagation constant related to the transmission line elements formed by the series impedance Z_C and the shunt conductance G_{RR}: $\gamma_R = \sqrt{Z_C G_{RR}}$; the real part of γ_R determines the attenuation of the propagation phenomenon and its inverse $\tau = 1/\gamma_R$ is called effective length. When considering a source current absorbed by a train or due to a fault, the loop impedance seen from the substation is given by $Z_C = Z_{CC} - Z_{CR}^2/Z_{RR}$, where Z_{CC} represents the impedance of the contact system with earth return, so excluding the return circuit (formed by the rails and possibly the return conductor).

Now the BT simply ensures that a large fraction of the return current is made to flow through the purposely chosen return conductor (the rails for a RR system or a trackside conductor, possibly at the same height of the contact wire to improve magnetic coupling and mutual inductance, for RC systems); from a circuit point of view this is accomplished by increasing the line impedance with the large magnetic coupling (and thus mutual inductance) of the two BT windings. With an idealised booster transformer the catenary current would be equal to the return-wire current to preserve the ampere-turn balance. In practice, magnetising current is required, and this will flow in the primary winding with an earth-return component. Another important departure from the ideal transformer case is the leakage reactance of the transformer. At each booster location a finite voltage drop will occur when a load current is flowing; this predominantly inductive voltage drop is algebraically summed with that due to the inductance of the catenary/return wire circuit and the general effect is to produce a system which has a comparatively high inductive source impedance.

The reduction of the induction effects onto external circuits may be evaluated considering simply the point screening factor or calculating the current-length integral. This index, since the line cross section remains the same across the various solutions, weights the return current distribution for the longitudinal length where it occurs, and may be led back to the net earth current leaving the return circuit through the rails and associated rail-to-earth conductance. For this reason a larger G_{RR} results in a larger current-length integral and a worse screening effect.

The Swedish Railways some years ago used the 16.7 Hz line between Kiruna and Svappavaara as a laboratory to test both solutions (BT and AT with a 5 and 13 km spacing respectively) with respect to the said performance indexes, external induction and rail voltage. The results, supported by additional calculations concerning the sensitivity with respect to element spacing and rail leakage conductance, confirmed that BT and AT solutions are not that different:

- the induced voltage with BT more tightly spaced is subject to oscillations moving longitudinally with maxima amid two adjacent BTs, and the maxima increases going farther from the substation feeding the line at one end only; for ATs there is a similar phenomenon in principle, but much less remarkable and the increase of maxima is more regular; the overall values are approximately the same with some local peak values higher by about 10-20% for BTs;
- for a RC system, when the return current from train axles reaches the trackside RC, the rail potential problem is eliminated with a general reduction of an order of magnitude, even if in the rail section between train location and the first rail-RC bonding the situation is identical to the RR

system.The rail potential is, of course, inversely proportional to the rail leakage conductance, but the phenomenon is far more evident with the AT system, in particular for the improvement obtained for leakage values above 0.1 S/km (approaching the EN 50122-2 limit of 1 S/km the rail voltage reduction is in the order of 30-50%).

On the contrary, the equivalent feeding impedance for the AT system is approximately one half that offered by BTRR and BTRC systems and so it is the longitudinal voltage drop, that allows a wider separation of the supply substations.

2.1.4 Other AC Systems

There are examples of ac railway systems operating at frequencies other than those previously considered above. The primary rationale for the implementation of a low frequency supply is to reduce the longitudinal inductive voltage drops and to increase the substations separation and to reduce costs. Hhistorically, several systems were designed and constructed at the beginning for low frequency and supplied by a non-standard voltage, and they have been maintained until the present day.

The line connecting Lancaster, Morecoinbe and Heysham in the UK supplied by an overhead line at 6.6 kV 25 Hz was commissioned in 1908, but upgraded to a 50 Hz system in the early '50s. In the United States there are old railway systems supplied at 15, 20 or 25 Hz; this type of supply may be derived by the national grid at 60 Hz by using rotating converters or cycloconverters. Examples are the New Jersey Coastline between Rahway and Matawan-Longbranch at 12 kV 25 Hz [32] and other systems in Eastern US and Canada at various voltages between about 12 and 20 kV; the use of the 25 Hz frequency for transportation and also for power transmission began in the US in the early '30s, and the reason was simply that of reducing inductive voltage drops on the transmission lines at the price of slightly larger electric machinery. All these systems are very old and in some cases no longer in service; the arrangement of the traction supply system is very simple, without booster transformers or autotransformers.

Examples of higher voltage systems are those for the ore carrying heavy haul trains in South Africa and United States: a 50 kV voltage level was adopted for both the Sishen - Saldahna railway line in South Africa running a total of 861 kilometres [243] and the Black Mesa - Lake Powell 125 km long railway line in the United States [42]. In both cases large power locomotives are needed to haul train consists of about eighty cars with maximum speeds of about 100-120 km/h: the solutions adopted many years ago were fof a high profile for performances, availability and robustness, but interference and electromagnetic compatibility in general were not a concern, since the line is singly owned and dedicated to a single type of traffic, without the need for high-performance signalling systems. Given the noteworthy absorbed power of this type of rolling stock, the 50 kV voltage level, though uncommon, was chosen to ensure a profitable deploy of substations and electric supply system elements on the territory. In particular for the Black Mesa - Lake Powell line, the exigency of a power upgrade that did not require the construction of a new high voltage line and supply substation led to an analysis and improvement of the existing line: being supplied from one side only, the line resonances excited by locomotives rectifier harmonics [536] caused significant overvoltages; the proposed and implemented changes were a split of the line in two section by interposing series capacitors and the compensation of harmonics and reactive power demand by the tuned filters installed mid-way between the substation and the other line end [42].

2.1.5 Short Circuit Current and Protections for AC Systems

Short circuit events are relevant to EMC for induced interference and for coordination of electrical safety requirements; to this aim the time duration of the short circuit is relevant and determined by the fault clearing time of the protection device (circuit breaker).

The calculation of short circuit current is beyond the scope of this book and reference may be made to applicable standards used for electric supply networks, such as IEC/EN 60909-0 [166], ANSI/IEEE Std. C37.010 [334] and ANSI/IEEE Std. 141 [362].

Circuit breakers for AC railways are very similar to those used in electrical distribution and transmission networks: they implement overcurrent protection with an inverse curve, where the clearing time depends on the current amplitude. A distance to fault feature is almost always implemented with more than one zone of intervention, that can cover the different protection characteristics when the traction line is reconfigured following an out-of-service substation. With a directional impedance measurement the zones can be defined as slightly less (let's say 80-90 %) of the substation separation and slightly less than two substation separation, with the possibility of a third zone slightly beyond two substation separation, partially overlapping with the protection zones of the other substations. Modern protection relays are equipped with a reverse reach characteristic for close-up faults on the same substation busbar. An additional problem with modern traction systems is that the current absorbed by several traction loads in the same supply section is so large that the need to distinguish it from a weak fault is usually satisfied with blind sectors in the polar plot of the measured impedance to supposed fault seen by the relay. Another solution for protecting weak faults (usually located at the end of the supply section) is to use a delayed overcurrent protection, set to a suitable threshold with respect to traction loads and with a time delay long enough to let the distance protection relays, if present, to trip first, as well as quick overcurrent curves, and to filter out large peak load currents caused by transient concurrent current absorptions. This of course has an impact on electrical safety and induced interference since the worst case clearing time becomes longer.

Even if the protection relay can be made to operate across booster transformers or autotransformers located within a supply section, in both cases the accuracy of the impedance measurement and the distance to fault determination is compromised: BTs cause step changes of the apparent line impedance, while with ATs there is a "fold back" causing ambiguity of the measured impedance value versus distance, as shown in Figure 2.3.

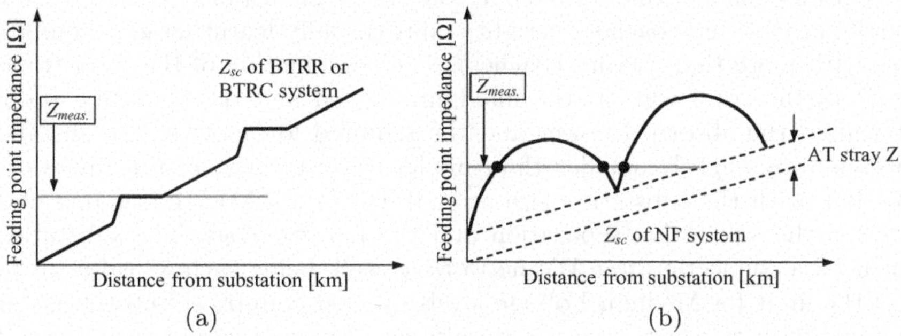

Fig. 2.3 Apparent line impedance: (a) BT step change, (b) AT fold back

A further complication comes from regenerative braking that both increases the total current flowing into the fault and modifies the fault impedance seen by the protection devices, thus hindering its operation. The most viable solution is to detect the line voltage on-board and prevent regenerative braking when it drops below a satisfactory threshold, such as during the vast majority of faults.

2.1.6 DC Railway Systems: 1.5 and 3 kV

2.1.6.1 General Description

This supply system is always used for heavy traction railways in its 3 kV version, while it can be used also for light railways at the lower voltage level of 1.5 kV (or 1500 V, as it is more commonly written). The 1500 V system was introduced in the southern and southwestern parts of France; the power limitation, at about 4-5 MW, for this type of system lead to the introduction of 3 kV systems in the late '20s. Several countries in Europe have conventional railways supplied at these two voltage levels: Italy, Spain, Poland, Belgium, Russia, part of the Czech Republic (at 3 kV), the Netherlands and France (at 1500 V) to cite some. Light railways at 1500 V, limited to a small region or around a populated area, may be found in United Kingdom, Denmark (S-railways in Copenhagen), Istanbul (under construction) and United States.

The main advantage of a dc supply is that the longitudinal voltage drop is only due to the longitudinal resistance and inductive reactance is always zero. Yet, there are some drawbacks regarding the generation of a large enough dc voltage (not that easy at the time when the systems were constructed), the interruption of a large dc current during short circuits (for which special purpose circuit breakers are designed and produced, called also High Speed Circuit Breakers) and the significant impact on corrosion produced by the resulting stray current.

The traffic on conventional dc lines travels at cruise speeds up to about 160-180 km/h, but the train composition may be largely variable: local commuter trains (made up of four to eight coaches), long distance passenger trains (usually around ten coaches), freight trains (usually featuring long compositions with more than twenty coaches). The combination of the total trailed weight of the train consist, the maximum speed and the operating profile determines the absorbed power and the required locomotive. The installed power is only slightly smaller than modern ac railways, in the range of 1 MW/km, with the substation size around five to ten MW, depending obviously on the substation separation and the line topology. The substations thus need a connection to a HV network as well, being such a load at or beyond the limit for Medium Voltage levels in most countries. Substations are equipped with rectifier groups in 6-pulse or 12-pulse configurations, usually one for each track, without electrical separation between SSs and possibly a larger number of rectifier groups at complex junctions. Substations are however equipped with feeder switches, to configure the supply scheme, by isolating a substation (by-pass), by choosing to supply the left or the right section, or to completely switch off a line section (e.g. for maintenance or in case of accidents or emergency).

At the very beginning of the Italian Alta Velocità[3] project the use of a dc supply was considered, laying down the main design elements for the hypothetical increase of the line voltage to twice and even four times the rated voltage of the conventional line, that is 6 and 12 kV: the semiconductor devices were at that time available for those voltage levels, and the circulating current would have been limited to the present values, in order to keep short circuit protection safe and affordable. At the end of this first exploration phase, the ac solution, already successfully implemented in other European countries, such as France, won the bet.

Conventional lines have been selected for the intercommunication with the new high speed lines (all with an ac supply system) in order to ensure high speed trains the access to conventional stations; in countries with a widespread use of dc supply systems (such as Italy, Belgium and the Netherlands) this has required complex technical solutions, to isolate the two power supply systems at the interconnection, to avoid any short circuit and current leakage during train passage and to avoid crossed interference onto the

[3] In Italian, "Alta Velocità" means "High Speed".

signalling systems. The current collection mechanism and the synchronization of pantograph operations during the handover between two supply systems separated by a neutral section is subject to careful verifications during interoperability assessment of a traction line and of a new rolling stock.

2.1.6.2 Traction Current Arrangement and Supply Architectures

The rectifiers are supplied with transformers that have a standard secondary voltage, that after the rectifying action produces exactly the rated no load dc voltage V_{d0} of 3600 V (or 1800 V): for a diode rectifier with 6-pulse reaction the relationship with the input line-to-line rms voltage is $V_{d0} = 1.35\,V_{ll}$. In the case of a 12-pulse reaction, the transformers have double secondary windings, one delta and one star, to accommodate for the required phase rotation of 30 electrical degrees at the network fundamental. The rectifying process has some side effects: it produces harmonics both in the dc output voltage and at the ac input, modelled as current harmonics, the rectifier being a distorting load, and it requires a transformer oversizing with respect to the rated active power. Under the assumption of an ideal sinusoidal balanced three-phase supply network, the harmonics on the dc side voltage are located at frequencies that are $6k$ multiples of the fundamental, with k integer, giving rise to 300, 600, 900, 1200 Hz harmonics in a 50 Hz system, and 360, 720, 1080, 1440 Hz terms in a 60 Hz system; for the ac side, the current harmonics are the result of the intermodulation with the same fundamental frequency and are of the type $6k \pm 1$ times the fundamental frequency, giving rise to 250, 350, 550, 650, 850, 950, 1150, 1250 Hz harmonics in a 50 Hz system, and 300, 420, 660, 780, 1020, 1140, 1380, 1500 Hz terms in a 60 Hz system. The reason for using 12-pulse reaction is the reduction of the said harmonic impact: the characteristic harmonics at both input and output sides produced by each 6-pulse group cancel each other and only those for k even "survive", showing less populated spectra with 600 and 1200 Hz harmonics (or 720 and 1440 Hz ones) on the dc side voltage, and 550, 650, 1150, 1250 Hz harmonics (or 660, 780, 1380 and 1500 Hz ones) on the ac side. Because of the harmonic distortion, however present also for 12-pulse realizations, dc substations are very often equipped with resonant filters at the output tuned to cancel out the main characteristic harmonics of the dc voltage ripple. It is thus evident that a dc substation is a complex non-linear electrical system that requires a sophisticated modeling for the accurate evaluation of both harmonic distortion (characteristic and non-characteristic) and the tracking of active and reactive power flows, naturally related to a single frequency ac network. A more detailed treatment may be found below in sec. 2.1.10.

2.1.7 DC Metro Systems: 600, 750 and 1500 V

Metro systems are characterized by a large installed power per km to accommodate for a large number of circulating trains, possibly as long as the station boardwalks, so up to seven or eight coaches, with highly variable operating conditions and significant power absorption and power flows to accommodate for the required headway and transit times. Electric substations are numerous and they are installed at a close distance between each other, in order to feed the power demand and keep the line voltage above a minimum threshold imposed by the requirement on traction effort and speed profile. The substation rectifiers are all in 12-pulse configuration for improved power sizing and to reduce harmonic distortion. Short circuit currents increase correspondingly and they are approximately as twice as larger than those of dc railways.

Since metro lines and substations are located largely in urban areas where the space for a HV substation and HV feeding lines is not available, and for economical reasons, the substations are supplied by Medium Voltage cable networks that follow the same line route and must be considered as a relevant source of inductive interference to other telecommunication and signalling cables, both in normal and short circuit conditions. Concerning the definition of the current spectrum assigned to these sources, it is remembered that the MV network is affected by a non-negligible harmonic distortion, as outlined below in sec. 2.1.10.2. Substations are connected with an in/out scheme, with each MV section terminating on two MV feeding points on either sides for redundancy: to avoid major network resonances and to ensure network stability, only one MV feeding point is normally connected at a time; the network is reconfigured in case of an out-of-service of a feeding point switching in and connecting to the next one (see sec. 3.3 for further details and considerations).

The traction supply is distributed to trains by means of either an overhead or a third rail system; at a lower and more common (especially for modern systems) rated voltage of 750 V, taking into account the longitudinal voltage drops that bring it to the lower threshold normally around 600 V or even less, the circulating current absorbed by the trains in a supply section may reach some 6-7 kA, with a significant impact on both current collection and supply conductors sizing. For the first issue it is not uncommon that the total power demand of a train consist is served by several current collection devices (pantographs or sliding shoes) in parallel, each belonging to a different coach and connected in parallel by the train bus at the line voltage often running on train roof. For the general sizing of current carrying conductors (either traction line conductors or feeding conductors, all at line voltage) some solutions are normally applied. Feeders are made of several cables in parallel to reduce the voltage drop between the substations and the traction line. The traction line conductors are mostly of standard size for compatibility with vehicles and so they are supported by parallel cables, with numerous bonding points, that increase the overall available cross section. Being that the substation separation distance is limited to a few km, it is also viable that the substation feeder is extended trackside and connected to an intermediate point along the traction line, rather than only directly at the longitudinal position facing the substation itself, thus bringing a feeding point of moderate short circuit power to intermediate positions that are affected by the largest voltage drops. This solution puts feeders at traction voltage in parallel in cable trays or cable trenches trackside, adding another source of induced interference onto trackside telecommunication and signalling cables.

The monitoring and control of the substations is usually performed by a dedicated SCADA (Supervisory Control And Data Acquisition system), interfaced with an ATO/ATC system. Signals from the field devices (located at switches and protections, transformers, converters, etc. to probe their status, operating conditions, temperature, etc.) are collected and conveyed through RTUs (Remote Terminal Units) to the SCADA modules at the upper hierarchy level; connection and data transmission may be implemented by means of copper cables (normally used for data networks), but fiber optic is normally adopted in the most dangerous zones, so close to HV equipment at substations.

2.1.8 DC Trolleybus and Tramways: 600 and 750 V

Given the intimate connection to urban structures, these systems are the most prone to custom solutions and to a vast variability related to various technical solutions, concerning substation sizing and location, arrangement of the feeding conductors and of the grounding of the return circuit; grounding and stray currents, in particular, are to be considered carefully, since the track is mostly buried in asphalt and routed in the urban environment.

The absorbed power is moderate since speed and tractive effort profiles are not so steep, vehicles are made of a single traction unit or by two articulated units, and the installed power is normally in the order of some hundreds kW maximum. The current circulating in the overhead system is limited to some hundreds of Amperes and it is supplied by various substations and feeding arrangements. In some cases few substations at some line locations (main junctions and terminal points) are chosen, with a significant power rating for each substation, re-using for the 750 V voltage level equipment and components for metro applications; on the other side there is a minimum impact solution with small and distributed substations, assembled and housed in watertight cabinets mounted on poles or located in small rooms and cabinets, since such a few hundreds kW substation may be fed by a dedicated LV line derived from urban MV/LV cabins nearby (size, clearances and creepage distances are small for LV applications and protection devices are widely available for such rated currents).

For this application the monitoring and control is performed by simpler apparatus that may connect to the various substations by Power Line Carrier (using a coded high frequency carrier on the same traction supply system) or telephone lines (widely available in urban areas).

2.1.9 Traction Line Frequency Behavior

All the traction supply circuits are designed and sized to carry the nominal current with acceptable voltage drops (with the traction supply voltage within the acceptable range all along the traction line) and self heating (at the prescribed ambient temperature for both conductors and connecting terminals). To this end, the design phase is preceded by electromechanical simulations to ascertain the worst case configuration for the rating of substations and traction line conductors. Sizing is performed at the traction supply fundamental and additional factors are normally included for the effects produced by the main harmonics, that is additional over-heating in electrical machinery and line conductors, that leads to a certain amount of over-sizing. Besides this, however, the design phase rarely takes into account the frequency behavior of the traction line, such as line resonances and anti-resonances, interaction with substation and on-board filters, dramatic changes of the return current distribution. This behavior is considered in detailed in section 7.6 with expressions for resonance and anti-resonance frequencies and the corresponding value of the pantograph impedance. The effect of the substation filters in dc traction lines is treated below in sec. 2.1.10.3. In the following, published results of the frequency behavior of dc and ac traction lines are presented, extending the analysis to include the determination of the relationships between various line variables, in particular absorbed traction current and rail currents or rail-to-rail voltage.

2.1.9.1 Traction Supply Line Impedance and Resonances

Historically, the modeling of the electrical behavior of the traction line has developed with the need for a more accurate and sophisticated analysis tool, while investigating new electrification solutions. In the '70s and '80s, several authors and research groups were investigating the possibility of complete characterization at the fundamental supply frequency of the traction line, including electric machinery and power intakes, by means of coupled impedances and impedance matrices; they were thus extending techniques used for electric power networks, and in particular overhead power line networks, to electric traction application, including and taking into account the geometrical position of conductors, the properties of materials and all the cross-connections and bonding conductors. As clearly stated later on in [478], "For load-flow and voltage profile studies along a railroad proposed to be electrified, the most important factor is the series impedances of the power distribution and return systems. For harmonic distortion, electrostatic interference, and lightning performance analysis, the system shunt impedances need to be taken into account together with the series impedances." It is true that resistive and inductive effects, both self and mutual, were enough to demonstrate the basic properties, such as line voltage drop and power and/or current flow, that are the ground for the electrification design, and for the evaluation of, at that time, pioneering solutions, such as autotransformers and booster transformers (see sec. 2.1.3). The principles and the basic equations for the determination of the self and mutual terms between conductors are reviewed in Chapter 7; the concepts of geometrical mean radius, skin effect, the effect of ground were introduced in the determination of self and mutual coupling terms, achieving greater accuracy and extending the frequency interval of validity. We could say that the '80s and early '90s saw a significant improvement and a sophisticated level of modeling, supported also by the first experimental findings, as exemplified by the works of Mellitt, Hill, and other pioneers: underground and buried cables [672, 651, 215, 480], tunnel effects on line p.u.l. parameters [497, 527], electrical characterization of running rails, sleepers and ballast [301, 307, 306, 305, 308], improved modeling of transformers to extend their frequency range of applicability [241, 240] are all examples of these achievements. Since then, the investigation of material and device electrical properties have been duly extended to cover a wider frequency range and to integrate traction line models with signaling and rolling stock models [232, 655].

Starting from the transmission line equations arranged to analyze the traction line behavior at various degrees of approximation, the following points may be outlined.

Modes of Propagation. Contrary to the general characteristics of a two wire transmission line, a set of traction line conductors exhibits more than one propagation mode, depending on their connections: in [234, 235] the propagation modes are theoretically considered up to a significantly high frequency, including the effect of the ground image currents also for low ground conductivity; these propagation modes are identified as Transmission Line (TL) and Fast Wave (FW), the former characterized by a smaller attenuation constant at high frequency [234]. TL and FW modes are practically coincident up to a critical frequency value; above it they start to differentiate and the attenuation constants approach asymptotically zero and infinity, respectively. Based on the consideration that the propagation along the line occurs with a finite attenuation at all frequencies, the TL is considered the dominant one, so that the authors proposed a simplified approach focusing only on the less attenuated modes. The proposed formulation is based on the solution of Sommerfeld integrals and leads to a better approximation of the TL attenuation constant than the Carson formulation. For distances between the line conductor and ground, typical of a railway system, the critical frequency is located between a fraction of MHz and some MHz, for increasing ground conductivity. Extending the considerations to a multiconductor case the following conclusions are reported in [235]:

- the first propagation constant of the system is related to the common mode of propagation, which refers to the conductor-to-ground path and the attenuation of the common-mode current is mainly affected by the ground losses; the attenuation constant of the common mode computed by the

proposed method is larger and smaller than that of Carson's expressions in the low and high frequency range respectively, identified as the frequency interval below and above the critical frequency value; on the contrary when the phase constant of the common mode of propagation is computed, the results are overlapping;
- for the other modes, all differential and related to the propagation for pairs of line conductors, the attenuation increases monotonically with frequency because they are mainly related to the losses in the conductors themselves and the agreement with the Carson approach is much better.

For problems concerning the propagation of harmonics and audiofrequency components Carson's expressions remain thus a satisfactory method and they coincide with the exact solution in all the cases considered by [234, 235].

When the interest is on low frequency components, such as the fundamental supply frequency and a few harmonics, for example for load flow and electromechanical simulations, the multiconductor representation of the traction line can be further simplified. In [515] an extreme simplification is proposed down to the fundamental modes, one for each traction line potential (that for a 2x25 kV line are +25 kV, 0 V and -25 kV), in order to study low frequency large scale problems, such as load flow, line voltage drops and line voltage stability; further examples of this type of applications can be found in [600, 246]. References [477] and [478] are two examples of calculations of the equivalent loop impedance for AT systems at 12 and 25 kV and 25 and 50 Hz supply frequency respectively, made considering a variable number of tracks from one to four.

The propagation modes are a theoretically sound basis to define what is normally observed in practice, while evaluating the frequency response measured at various terminal pairs in a real traction line:

- there are main resonances, appearing in all the frequency responses, determined by the main traction loops between the overhead supply line and the equivalent return circuit;
- there are additional resonances at higher frequency, not only justifiable with a main higher order resonance, but also caused by additional resonances between subsets of conductors;
- also the anti-resonances, the main one uniquely determined in the case of a locomotive positioned on a straight line, are in reality determined by various kinds of couplings between the traction line conductors.

For all these resonances and anti-resonances, the shape of the peaks (either in the positive or negative direction, depending on the type of the evaluated electrical variable, namely voltage, current, impedance or admittance) is influenced by the associated equivalent electrical parameters, that in turn depends on which conductors are involved, how they are coupled to each other for the specific resonance or anti-resonance and the amount of lumped elements that are added to an in principle distributed circuit, such as the traction line itself. It is clear that the analysed network is an articulated mixed distributed-lumped parameters network and that the identification or prediction of the propagation modes and related resonances and anti-resonances is a complex task.

Resonances and Anti-resonances. We have already said that the observed line resonances and anti-resonances may be different from the theoretical ones in amplitude and frequency; this happens for a series of factors: a variable number of tracks (changing the p.u.l. parameters of each track), characteristics of connected transformers, use of high voltage cables for connections and feeders (their p.u.l. shunt capacitance between core and shield is at least ten times larger than that of the line), circulating vehicles (loading the line with their on-board transformers and filters and in some cases with the roof cable connecting pantographs); also external lines, such as another traction line or an electric distribution or transmission high voltage line may have a weak influence on the traction line frequency behavior.

Since it is not possible to further analyse the general considerations above with a theoretical development without turning to the multiconductor transmission line modeling considered in sec. 7.6, the theoretical development is supported by a series of examples and published results available in peer review literature:

- in [526], at the end of sec. IV, the effect of the long high voltage cables used in the Loetschberg tunnel and of their stray capacitance is highlighted: lower impedance at high frequency, increased number of anti-resonances and resonances, with the latter that cannot be easily detected because of the general damping due to many shunting resistive elements (the other trains and the equivalent loss resistance of cables and transformers); direct measurements [541] show various degrees of oscillating overvoltages at switch-on, obtained during line construction (so without any traction load), particularly large when one of the substations was out-of-service (Frutigen, for example); even in the presence of long cables and related capacitance, SBB requires that the network resonance is located above the 5th harmonic (at 83.3 Hz) to avert major dangerous overvoltages sustained by vehicle emissions;
- even if in theory network resonance occurs at a fixed characteristic frequency, whatever the train position [314, 488, 507], network resonance as well as resistive losses (also named damping factor) are also determined by a complex combination of infrastructure elements, as pointed out in [687]; it is shown in [687]that the damping may vary by about 10% for an assigned position as a function of earth resistivity and about 20% increasing the number of simulated tracks from one to three and then to five (particularly at the higher resonance frequencies, around 16 kHz), but a huge variation of an order of magnitude or even more is caused by changing the longitudinal position, as expected;
- in [539] it is shown that the effect of the circulating vehicles on a example railway network is relevant, in particular if the vehicles are far from

substations; also the damping effect of artificially added Re 6/6 tap changer locomotives to the already existing more modern Re 465 is noticed and will be again recalled in the next paragraph.

Voltage Distortion Due to Vehicle Emissions and Pantograph Impedance. Vehicle emissions were modelled in the past as a parallel combination of current generators for each harmonic component and voltage distortion was calculated as the product of the traction line impedance seen at the pantograph and the said harmonic current sources. This holds true if the power source is stiff and the power supply impedance can be kept much lower at all frequencies; a more complete model requires the equivalent harmonic input impedance of the vehicle to be included in the Norton equivalent circuit of the current sources. This approach is normally followed when vehicle-line interaction is to be analyzed with a classical frequency domain approach; yet, vehicle emissions depend on vehicle operating conditions (acceleration, braking, coasting, and at a finer level of detail, total rms absorbed current, speed, etc.) and thus a frequency domain approach is surely inadequate to represent the time-varying topology of a switching converter [553].

Voltage distortion and traction network instability may be a very challenging problem, as it is well known considering the famous accident that occurred in April 1995 on the Swiss network (and, after a preliminary fixing of the first one, again in September of the same year) [553, 540]. Many trains stopped in the Zurich area with the locomotives on-board control shutting down the loco main switch for excessive harmonic content on the traction line. The network on those two Sundays was lightly loaded and the circulating rolling stock was all of the new type, Re450 and Re460, – as they discovered at a later analysis – equipped with high performance control systems and high frequency line-side converters, with improved power factor and lower losses, with respect to older locos. Yet, it was demonstrated that the lack of the power losses associated with the "dumb old trains" had led the network in a possibly unstable configuration, excited by the harmonic emissions of the new locomotives chasing the "perfect power factor correction". More specifically, the input pantograph impedance of a Re460 shows intervals on the frequency axis where the behavior is almost purely inductive with a small negative resistance value. On the contrary tap-changer locomotives, such as Re 6/6, show an inductive/negative-resistance behavior only when braking and around the fundamental. Also the bandwidth of the on-board control plays a role in determining network resonances. The old tap-changer locomotives could change the tap position on the input transformer in a fraction of a second, offering thus a few Hz bandwidth. Faster than them, phase controlled thyristor locomotives apply their control every half cycle, so with a maximum bandwidth of 33 Hz. Modern converters are able to react even faster, with a bandwidth in the order of some hundreds Hz, with complex emissions patterns, characterized by time-varying harmonics and a relocatable energy content, and have required a new method of analysis of network instability, abandoning the old assumption of a linear network connecting equivalent current sources representing the loads. The sources are moving and are not really current sources, since they react to traction line current and voltage changes as they perceive them (local measurement), the network - even if linear - exerts a phase displacement on the various harmonic components and finally time delay is a factor, including not only the time required to travel the line between two sources located apart (that would be fairly low), but also the time the estimation and control algorithms on-board require to determine the network state, to identify the best control action and to apply it to the converter devices through the modulator.

For the determination of the network frequency response at various location is a valuable mean of tackling possible network instabilities, there was proposed a software modification to the on-board control to transform the traction converter into a signal generator for a volt-amperometric measurement at the pantograph [542]. The swept frequency interval for the driving test signal is limited by the switching frequency, so for the used train is limited to 1900 Hz (a FLIRT with a non interlaced 2 kHz switching frequency of the single unit and an overall 4 kHz switching frequency per traction converter); the other converter units may be kept available to move the vehicle at a reduced speed along the line, to accommodate for several measurement locations. This measurement system was successfully used in the Netherlands on the Betuweroute line [543].

2.1.9.2 Distribution of the Return Current

The return current that leaves the wheels divides among the running rails (in either direction), other conductors at the ground potential and the earth itself; the reason for analyzing it is the evaluation of the amount of coupled disturbance onto signalling devices, as a function of frequency, rolling stock position, infrastructure conditions. The subject is of great relevance in deciding the limits of interference, the margins and the corresponding immunity requirements for connected equipment. Under the term "infrastructure conditions" there lie all the characteristics and peculiarities of a specific electric traction system, from the location, rating and equivalent circuit of electric machinery, to the number, position and interconnection of system conductors, and in particular of the return circuit, down to the properties of some materials and parameters, concerning the rails and ballast. As we will see below in sec. 3.1.1 the variables that are relevant for interference evaluation are the currents flowing in the running rails and the voltage between the rails, and these are the variables that EN 50238 is focused on, for the demonstration of compatibility of train detection systems. In [508, 514] the transfer functions[4] between the excitation of the system (the total return current, that in this

[4] The term "transfer function" is very common in the railway world, even if a little abused, these relationships being frequency responses expressed in terms of amplitude only.

case corresponds to the absorbed pantograph current) and the said rail variables were considered extensively, with the support of a sensitivity analysis with respect to the track parameters that are the most relevant and likely to modify their value.

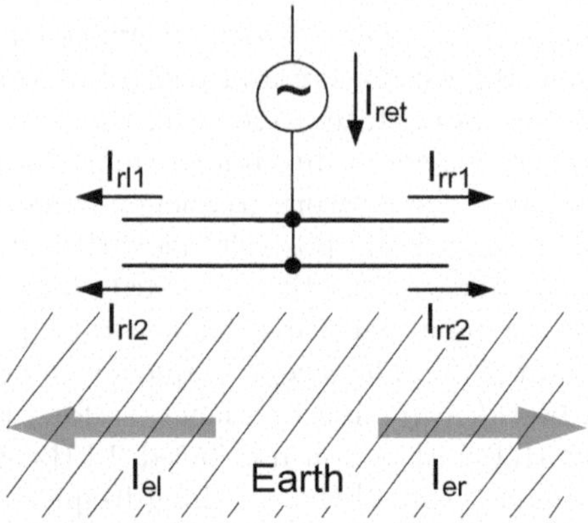

Fig. 2.4 Distribution of the return current leaving the vehicle axles

The earth current represents the amount of current leaving the rails and then reclosing far away at the substation. Its value is mainly influenced by the earth resistivity and the rail-to-earth conductance, that in turn determine also the loading of the free track on the track circuits themselves. For very high loading the shunting effect of rolling stock axles cannot be detected anymore and this poses an upper limit on the rail-to-earth conductance, which is typically around a fraction of 1 S/km (the assumed standard value was set to 0.2 S/km). This value is also compatible with the permissible values specified for dc systems against stray currents in EN 50122-2 [86], of 1 S/km for open air sections and 0.5 S/km in tunnels for heavy traction application. It is intuitive to state that insulated rail joints would reduce to zero the running rail current, but resonance effects and the shunting operated by the rail-to-earth conductance keep it far from this ideal condition.

The reference system for the analysis in [508] was a simple very long single-rail track circuit laid down on a straight uniform line supplied from the left end and terminated in either of two conditions on the right end, open circuit and short circuit; an open circuit will let all the return current flow to the left-hand side section, while a short circuit condition will recall a certain amount of current also in the right-hand side section. It is clear that this is an over-simplified system, with some extreme approximations, but it is valuable indeed for a clear visualization of the mixed effects of the various considered parameters. The results are shown up to 20 kHz and the frequency range is split into two intervals for clarity. A 1 A return current is applied at each frequency and the frequency spectra of the said rail variables are

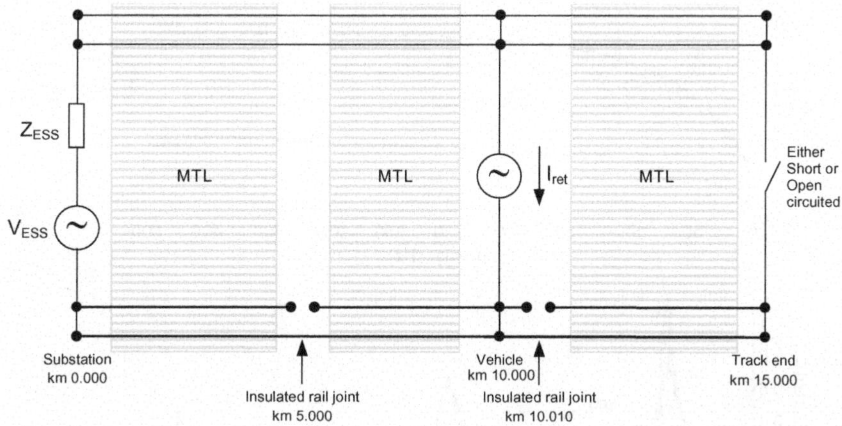

Fig. 2.5 Schematic of the railway traction line section used for the analysis

calculated all along the victim track circuit: the dashed curve refer to the current measured at the point closest to the injection point (z=9.800 km), the dotted curve to the current measured at the point most distant from the injection point (z=5.000 km)

As it was anticipated, the signal rail, insulated by IRJs located at each end, experiences a non-negligible current, with the occurrence of resonances in the higher part of the frequency range; the current level is larger for the open circuit condition, that makes more current flow in the left direction from the injection point.

In [508], the authors equally presented results for the dc system and from the results shown they do not differ greatly from the ones shown in Figure 2.6 and 2.7, except for a larger initial current at low frequency in the earthed rail and the effect of the low frequency resonance caused by the substation filter (in the reference the Italian 3 kV filter was assumed and the effect was a resonance followed by an anti-resonance in the 100-200 Hz frequency interval with a maximum deviation of 50% of the current amplitude from the average value).

The main features that can be noted are: the current in the signal rail is negligible at very low frequency close to the injection point and increases with frequency and with distance; at about 500 Hz the two rail currents are balanced and almost equal, so that their difference is of less concern; at larger distance from the injection point the influence of frequency is less remarkable; the signal and the earthed rail both experience the same system wide resonances for the two terminal conditions, indicating that even a resonance of the aerial conductors can exert some influence on the return circuit and vice-versa.

The rail-to-rail voltage was also considered observing major differences between the two terminal conditions and the two railway systems.

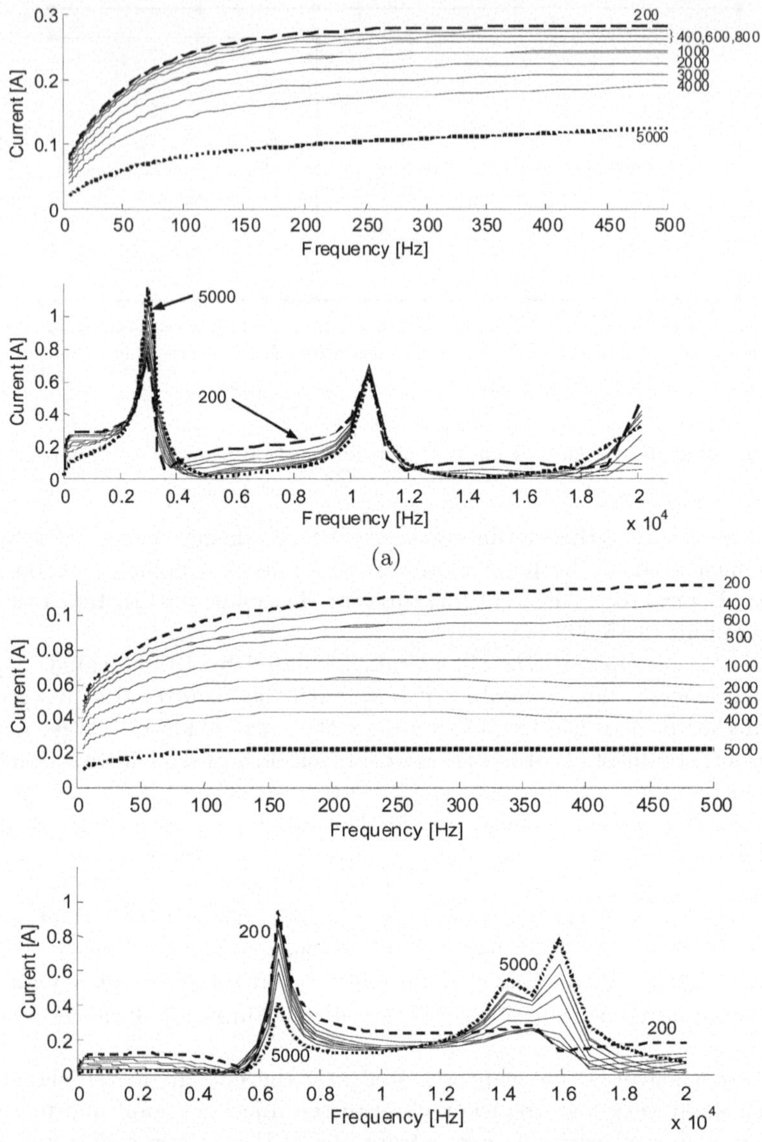

Fig. 2.6 Signal rail current for (a) open circuit and (b) short circuit conditions on an AC 16.7 Hz system

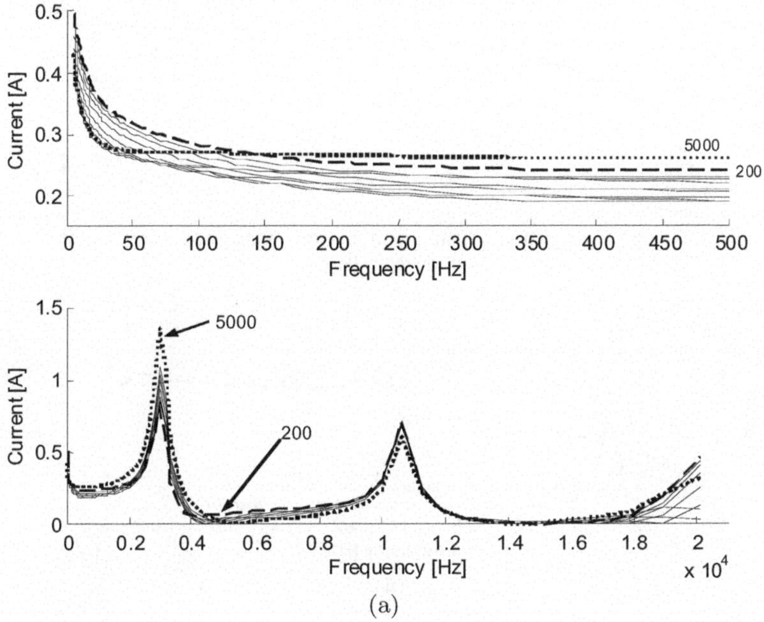

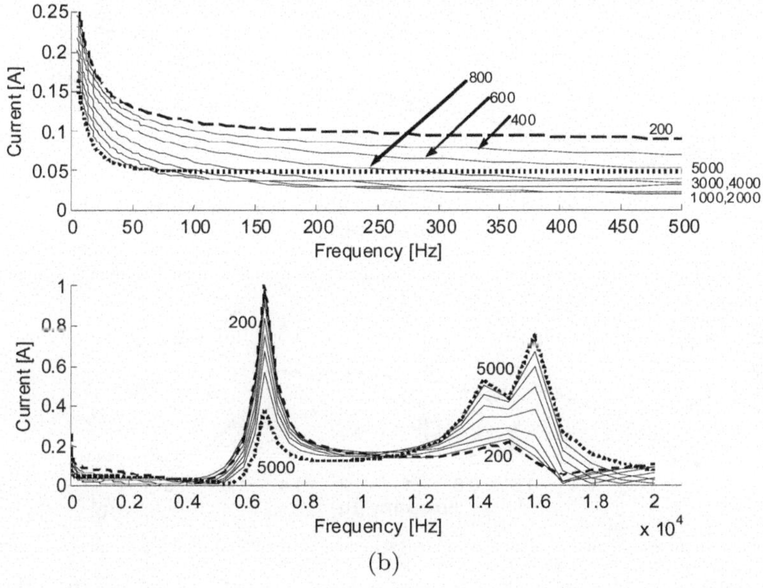

Fig. 2.7 Earthed rail current for (a) open circuit and (b) short circuit conditions on an AC 16.7 Hz system

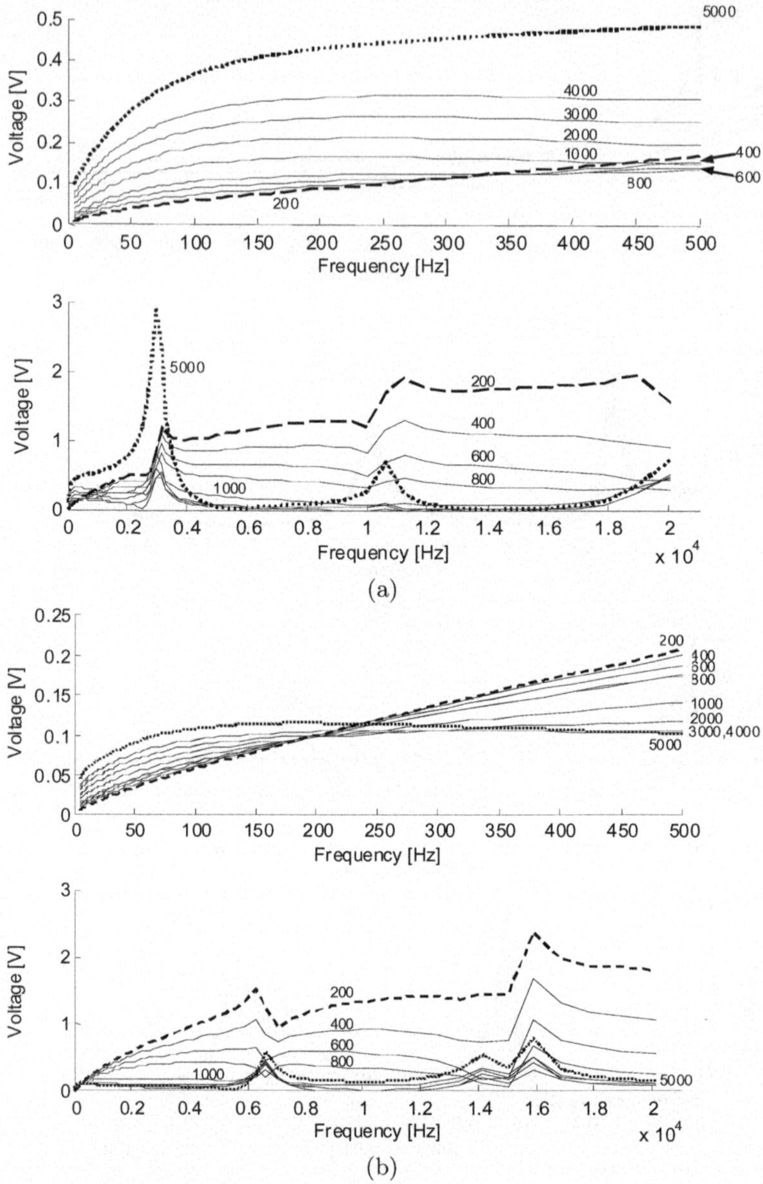

Fig. 2.8 Rail-to-rail voltage for (a) open circuit and (b) short circuit conditions on an AC 16.7 Hz system

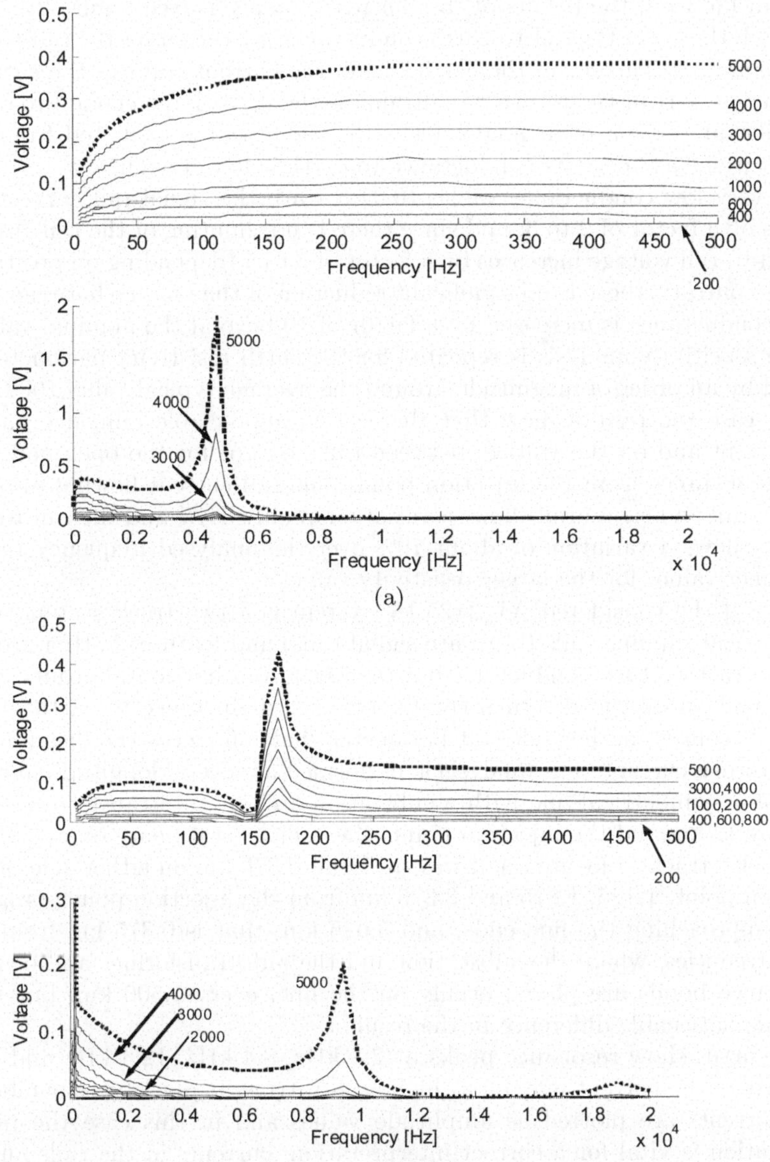

Fig. 2.9 Rail-to-rail voltage for (a) open circuit and (b) short circuit conditions on a DC system

Following from the results of the sensitivity analysis (see Figure 2.10), as expected, the lower the rail-to-earth conductance g_{re} the larger the rail-to-rail voltage and the current in the earthed rail; the current stays in the earthed rail, that is part of the return circuit, unless the g_{re} is large enough to drain a significant portion of it; analogously, the signal rail is insulated from the return circuit as far as its conductance to earth is low enough.

For very low conductance values (0.02 S/km), the signal rail current decreases by a factor of 2 to 3, and since there is no shunting of the rail circuit, the rail-to-rail voltage increases by a factor of 3 to 5 (depending on position). On the contrary, there is no significant reduction of the voltage between rails if the conductance is increased by a factor of 10 beyond the nominal value.

The sensitivity analysis is repeated for the earth resistivity parameter ρ_e (varied by an order of magnitude around the average typical value 100 Ωm); in this case the results show that there is no appreciable influence on the rail current and on the voltage between rails, except for the open end case at large distance from the injection point. The rail current flowing into the earthed rail and signal rail shows a variation of only 3-4% and the rail-to-rail voltage shows a variation of about 10% over the analysed frequency range, the higher values for the larger resistivity value.

In [514] the considered AT 2x25 kV system is a two track system, with symmetrical running rails (both are signal rails) and features both a ground through-earth return conductor (that realizes also a horizontal ground electrode) and an overhead return conductor; the conductance to earth of the former becomes another relevant parameter that influences the distribution of the return current. The analysis is performed at several longitudinal positions of a symmetrical line with a substation and an autotransformer separated by 12 km and the injection point (the vehicle, stationary on one of the two tracks, track 1) located half way at 5 km: 0.375 km on either side of the injection point, 1.125, 1.875 and 3.375 km from the injection point, progressively approaching the line ends, and 5.625 km, that is 0.375 km from the two extremities, where the substation and the autotransformer are located. Impedance bonds are placed evenly on the line, every 1.500 km, but they cause no noticeable difference in the results.

All curves show resonance peaks at 2.8 kHz, 8.4 kHz, 12.4 kHz and 17.9 kHz, where the current may be as high as 4-5 Arms. It must be remembered that currents are plotted as amplitude values and in this case the phase information is vital for a correct interpretation: currents in the rails and in the return conductors of the same (left or right) section of either track are opposite in phase and these conductors form loops. For frequencies other than system resonance the differential rail current is $<$2-5% of the traction return current for all positions, thanks to the high degree of symmetry of the supply system and the low impedance of the return conductors.

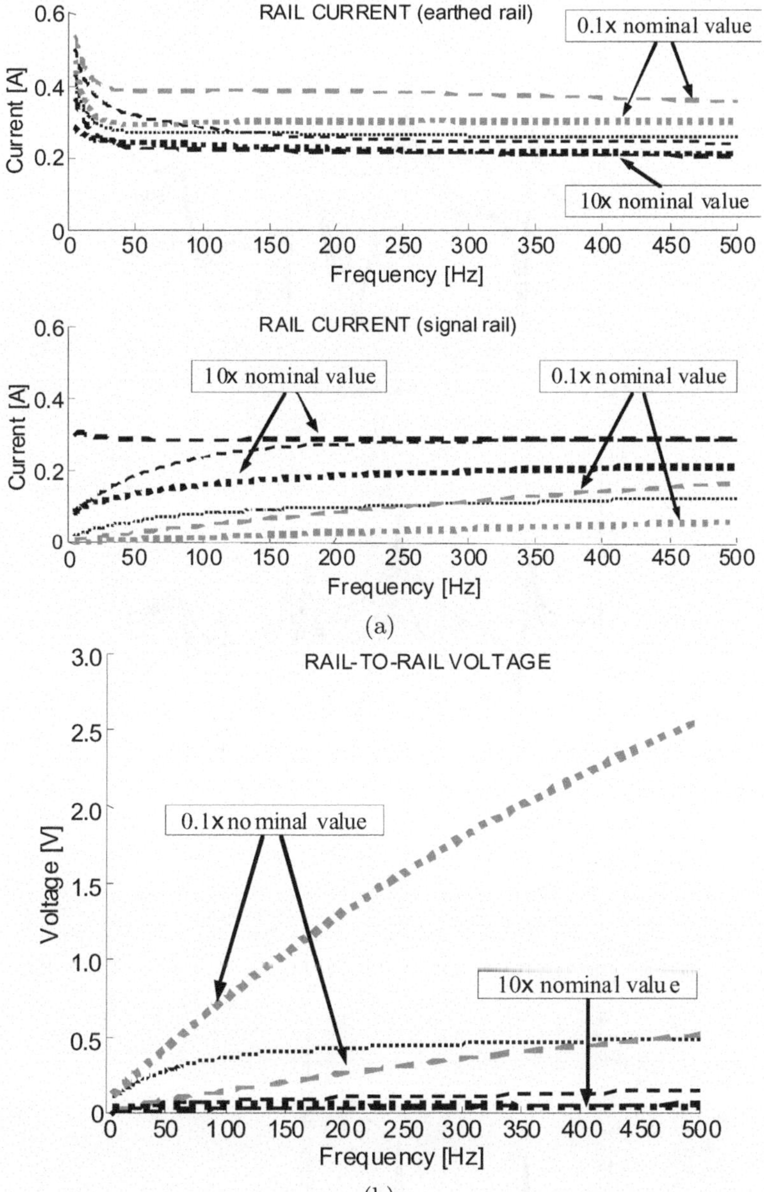

Fig. 2.10 Sensitivity analysis versus rail-to-earth conductance g_{re} (nominal value 0.2 S/km): (a) rail currents and (b) rail-to-rail voltage for open circuit condition on an AC system

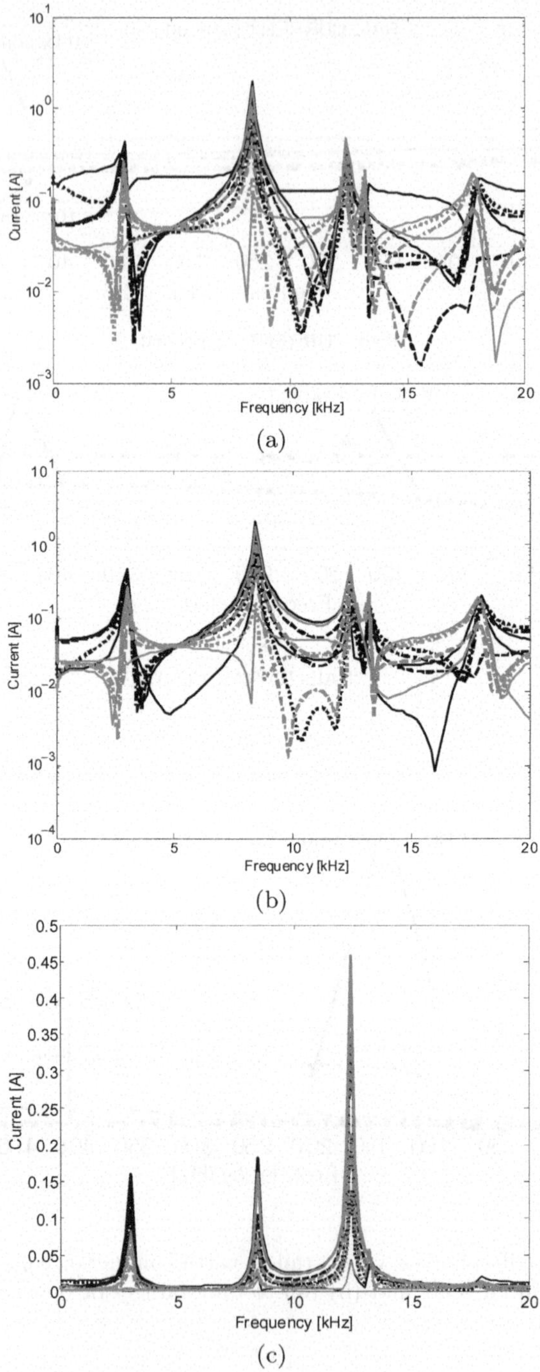

Fig. 2.11 Rail currents of both tracks of an AT 2x25 kV system: (a) track 1 rail, (b) track 2 rail, (c) d.m. rail current

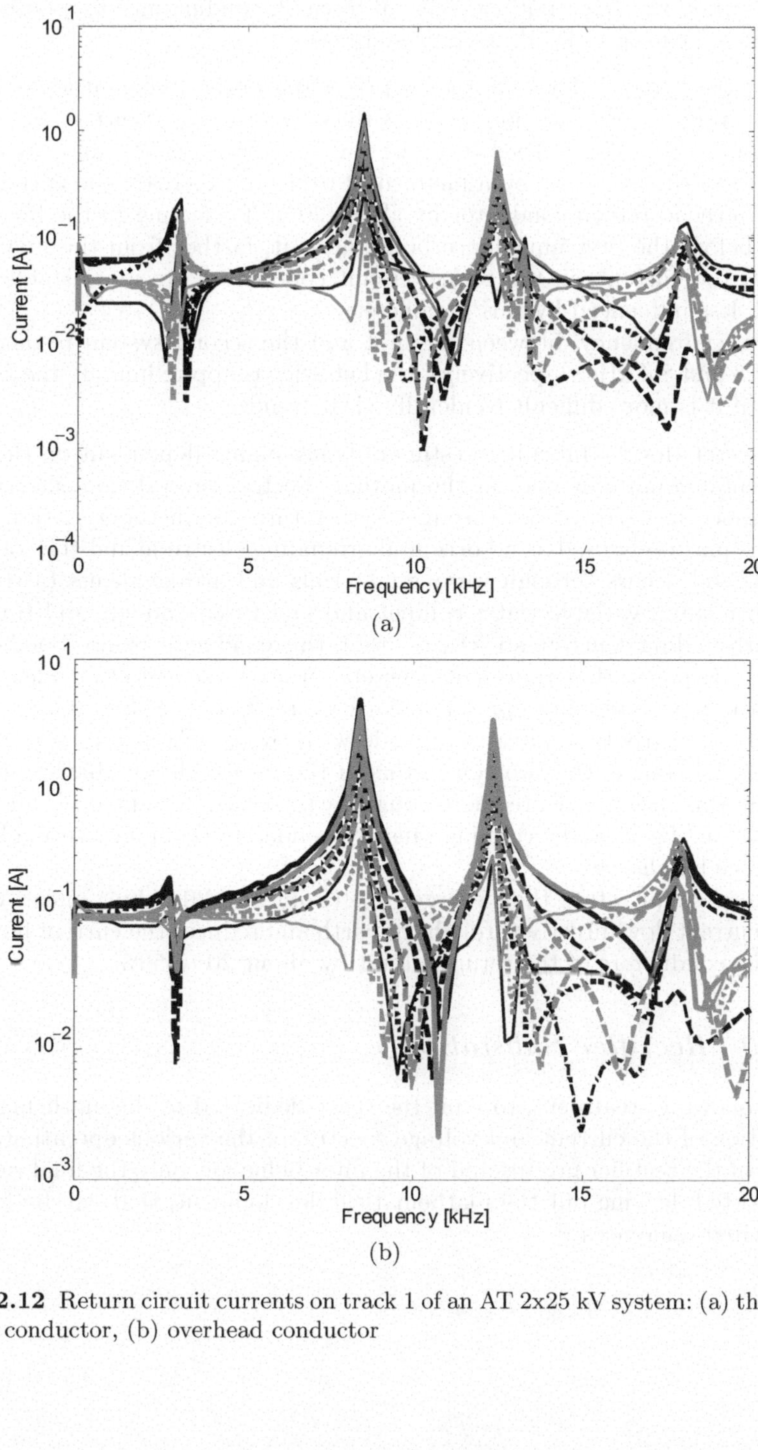

Fig. 2.12 Return circuit currents on track 1 of an AT 2x25 kV system: (a) through-earth conductor, (b) overhead conductor

Concerning the sensitivity to the rail-to-earth conductance g_{re}, based on the plots appearing in [514], we may state that:

- at low frequency, below about a few kHz, a one order of magnitude smaller conductance g_{re} (0.02 S/km) reduces the current in the return circuit conductors at the ground level (running rails by about 10-20% and through-earth return conductor by a factor of 2 to 5) and increase the current in the overhead return conductor by about 20%, if referring to the line section before the first impedance bond; after it, farther from the injection point, the currents in the return circuit conductors are more stable and much less influenced by this parameter;
- at higher frequency, between the first and the second system resonances (at 2.8 and 8.4 kHz respectively), the behavior is approximately the same, even if it is more difficult to identify clear trends.

A parameter that is difficult to estimate being highly dependent on the environmental conditions and on the mutual effect of several elements is the conductance to earth of the through-earth return conductor g_{Re}: this conductor is purposely used as a horizontal grounding electrode and the contact with the soil occurs through various materials and arrangements (a trench filled with sand, variable water content and moisture, viaducts and tunnels with surrounding concrete structures, etc.); this conductor is also bonded to the wayside poles, that represent concrete encased vertical electrodes; it is then customary assigning g_{Re} a standard value of some S/km and for this reason the sensitivity analysis was made with two bracketing values, 2 and 20 S/km. Of course, the two conductors of the return circuit that are influenced by this parameter are the through-earth return current itself and the overhead r.c., because the running rails are bonded to them only through the impedance bonds.

We may simply state that a larger g_{Re} (from 2 to 20 S/km) makes more return current flow directly through the earth and reduces the current flowing in all the conductors of the return circuit by about 20 to 50%.

2.1.10 Rectifier Substation

In the following treatment, to limit the space dedicated to the mathematical description of the current and voltage spectra, of the various operating conditions of the rectifier groups and of the filter behavior, only the final results are reported, leaving out the mathematical development, that can be found in the cited references.

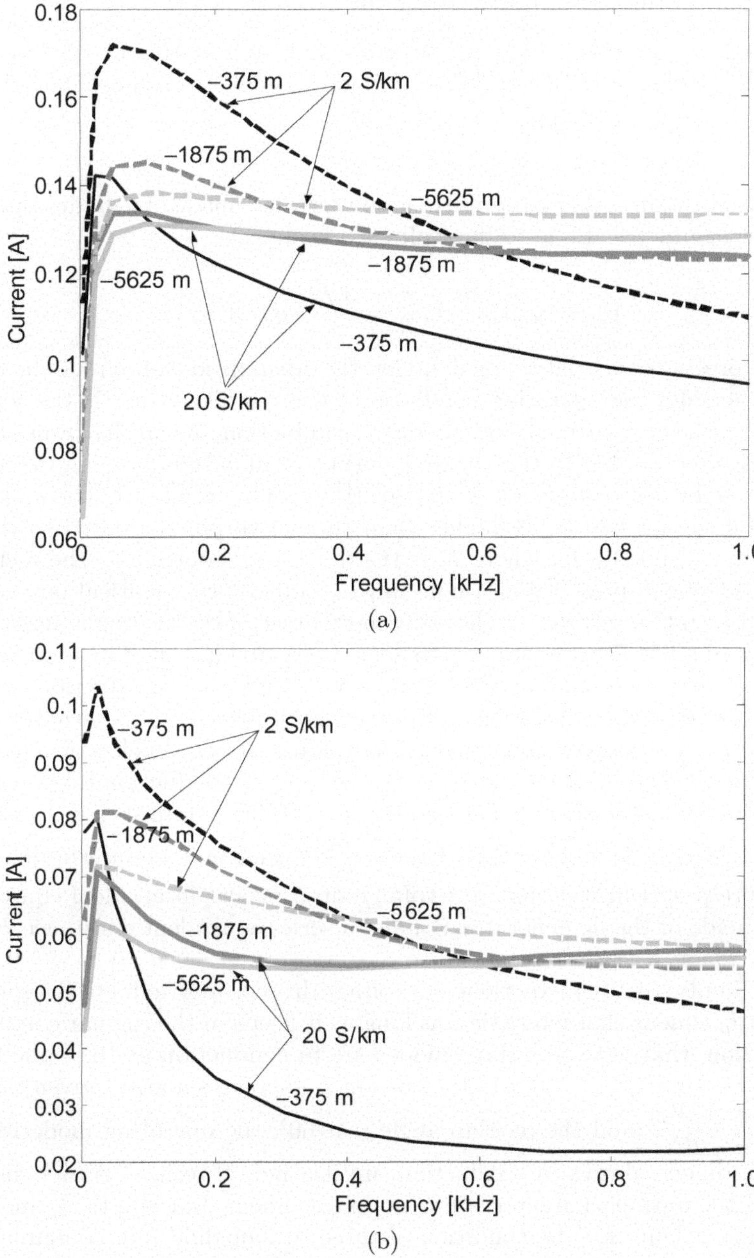

Fig. 2.13 Sensitivity analysis of return circuit currents on track 1 of an AT 2x25 kV system versus rail-to-earth conductance g_{Re} (nominal value 2 S/km): (a) through-earth conductor, (b) overhead conductor

2.1.10.1 Principles of Rectification

The ideal output voltage V_{d0} (in no load conditions at the dc side) for a p-pulse rectifier supplied by a p-phase symmetrical system of sinusoidal voltages with line-to-line voltage V_{ll} is $V_{d0} = \sqrt{2}V_{ll}\,\frac{6}{p}\,\sin\frac{p}{6}$.

6-Pulse Rectifier. For a 6-pulse reaction rectifier (made of 6 diodes supplied by three symmetrical line-to-line voltages) we have

$$V_{d0} = \sqrt{2}V_{ll}\frac{6}{\pi}\sin\frac{\pi}{6} = \frac{3\sqrt{2}}{\pi}V_{ll} \approxeq 1.35\,V_{ll} \tag{2.5}$$

This expression was determined under the assumption of zero inductance on the ac side; the operating condition of the rectifier group is the *normal condition*, the one with only two diodes in conduction, one in the lower part of the bridge (connected to the negative output terminal) and one in the upper part of the bridge (connected to the positive output terminal). The apparent power on the ac side is 5% higher than the active power related to the dc side power $P_{d0} = V_{d0}I_d$, where I_d is the dc current sourced by the rectifier. The effect of the presence of a non-null L_c inductance, identical on each ac side phase and largely due to the stray inductance of the feeding transformer, called *commutation inductance*, is to delay the switching off of a diode while a new one starts is conduction period, thus overlapping the conduction state of two diodes in parallel, insisting on the same output terminal, alternatively on the positive and negative terminals. The amount of time in electrical degrees at the fundamental during which the contemporary conduction takes place is called *overlap angle* μ or u. The conduction overlap produces a reduction of the average dc side voltage $V_d = V_{d0}\frac{1+\cos\mu}{2}$, that may be interpreted also, in a more practical way, like if a voltage drop occurs in an ideal equivalent circuit made of the dc generator V_{d0} and a series equivalent resistance $R_{eq} = 3X_c/\pi$, where $X_c = 2\pi L_c$.

So a 6-pulse rectifier experiences a voltage drop proportional to the sourced current I_d (linear characteristic), as long as it keeps in the regular commutation region, that is two or three diodes are in conduction at the same time, and this occurs for $\mu \leq 60°$. As the current I_d increases above a given threshold $I_d^* = \frac{\pi}{12}\frac{V_{d0}}{X_c}$ and the overlap angle $\mu > 60°$, the operating mode is that of three diodes always in conduction, until a new threshold is met: in this interval the output characteristic is no longer linear, but elliptical, and gets to be linear again for the condition of three to four diodes in contemporary conduction, and so on. The normal rating of a rectifier group is well within the linear region, so that the output current $I_d < I_d^*$ for any normal condition and even for overloading conditions. The prescribed overloading conditions for a rectifier (either 6- or 12-pulse, as well as for other static converters) are established in the EN 60146-1-1 [152]: for the most severe duty class (VI) the

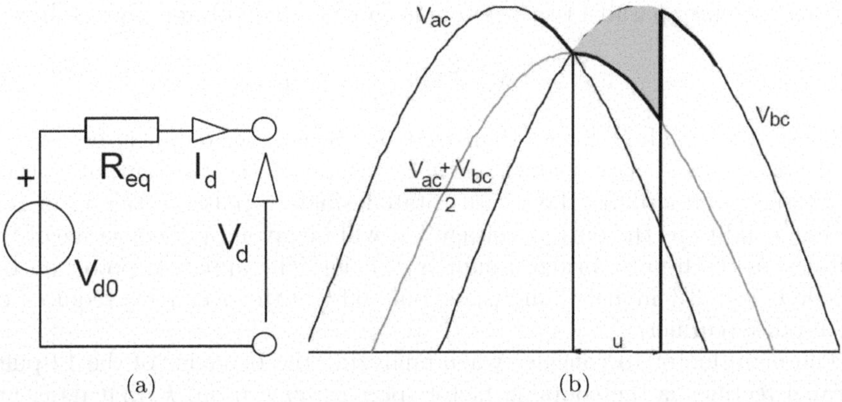

Fig. 2.14 6-pulse rectifier: (a) equivalent output circuit for the linear region and (b) overlap description

output current may reach 1.5 the rated current for 2 hours of service and 3.0 time the rated current for 1 minute of service.

12-Pulse Rectifier. From the point of view of the inter-connection on the dc side, a 12-pulse reaction rectifier may be obtained by either a series or a parallel connection of two separate 6-pulse rectifiers, fed by two three-phase voltage systems with a phase displacement of 30° at the fundamental; the two feeding voltages are obtained by means of a star and a delta connected secondary derived in principle by two separate transformers, but almost always belonging to the same double secondary winding transformer. The series connection makes the same current flow in the two 6-pulse bridges and the output voltage is the sum of the two instantaneous voltages; the series 12-pulse rectifier operates on the same principle of the 6-pulse rectifier and its behavior may be described as done before for the 6-pulse rectifier. The parallel connection on the contrary needs a dedicated development. While the mean output voltages of the two 6-pulse bridges are the same, the instantaneous voltages are different and cause the circulation of an internal current limited to the secondary windings of the feeding transformer and the two bridges themselves. This current has a zero mean value and pulses at six times the fundamental; the only limiting factor is the secondary reactance of the transformers, so that in case of a nearly unitary reactance ratio (see below (2.7)) with the transformer reactance concentrated on the primary side the current may reach a particularly high value. To control this phenomenon, not only a minimum secondary reactance is normally specified (limiting in turn the short circuit current amplitude), but also an inter-phase reactor connected between the negative terminals of the bridges and the load is adopted.

For a 12-pulse parallel rectifier the output no-load voltage is given by

$$V_{d0} = \sqrt{2} V_{ll} \frac{12}{\pi} \sin \frac{\pi}{12} \approxeq 1.398\, V_{ll} \tag{2.6}$$

and it is thus slightly larger than that of the 6-pulse rectifier, because in ideal conditions it corresponds to the envelope of the two output voltage waveforms produced by the phase rotated single 6-pulse rectifiers; in real loading conditions the output voltage V_{d0} will be given by the average of the voltages of the 6-pulse bridges, namely $1.35\, V_{ll}$. The apparent power on the ac side is only 3% higher than P_{d0}, so only 60% of the oversizing required by the 6-pulse rectifier.

The main factor to consider when analyzing the behavior of the 12-pulse parallel rectifier is the coupling factor (or *reactance ratio*) k, that describes how the commutation reactance is distributed between the primary and the secondary sides:

$$k = \frac{X_p''}{X_s + X_p''} \tag{2.7}$$

where X_p and X_s are the primary and secondary reactances, and the double apex stands for "seen on the secondary side"; X_p and X_s are to be determined from the transformer datasheet that reports the results of the binary short circuit tests, between the primary and either of the two secondary windings, the other left open (indicated as 1-2 and 1-3), and between the two secondary, windings the primary left open (indicated as 2-3). Taking $k = 0$ brings all the reactance to the primary side, implying that the two secondary windings are decoupled and behave like two separate transformers.

The output characteristic, as determined in [605], is shown in Figure 2.15.

The selection of a particular value for the reactance ratio during the design and specification of the rectifier groups is usually chosen keeping in mind that a large value close to unity helps in limiting the amplitude of the short circuit current, and this is particularly important in metro applications, where the substations are quite close, of relevant nominal power and operating at a reduced nominal voltage. On the contrary, for railway applications, substations are separated by several kilometers and short circuits occurring far from the substation terminals have a reduced current amplitude, the limiting factor being the impedance and resistance of the interposed traction line; the most important factor is that of avoiding that such far short circuits get undetected by protective devices, being "confused" with abrupt current absorption produced by intense tractive efforts.

The parallel connection of two 6-pulse bridges seems possible and plausible because they have the same output voltage. Yet, this is only true for the average output voltage V_d, because the instantaneous voltages $v_d(t)$ are different being the two bridges displaced by 30° electrical degrees, and this creates a transient short circuit condition between the two with the circulation of a current internal to the two bridges, that recloses through the transformer

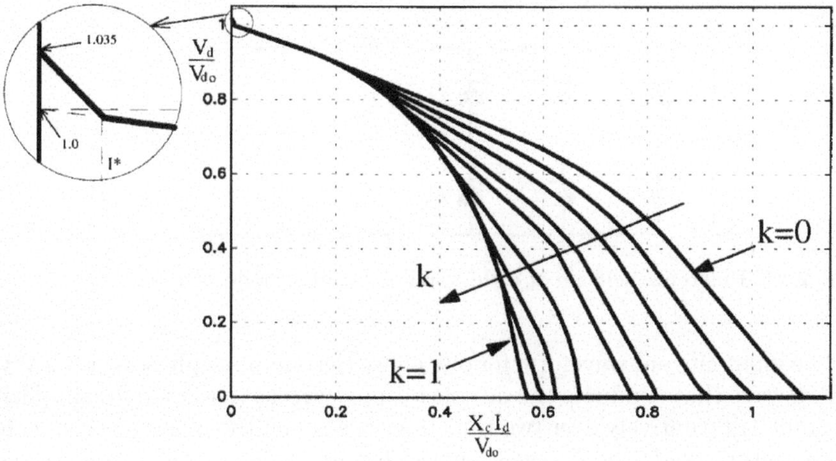

Fig. 2.15 Output characteristic of the 12-pulse rectifier as a function of the coupling factor k

secondary windings. This current is responsible for extra losses and a worse utilization of the transformer; the transformer reactance limits the amplitude of the circulating current, but for more control an inter-phase reactor is often added. The effect of the presence/absence of the inter-phase reactor (also called "inter-phase transformer") was treated extensively in [661], where the bridge waveforms are analytically derived.

The complete expressions for steady state and short circuit conditions may be found in [605], that is to authors' knowledge one of the most accurate and complete treatments available.

2.1.10.2 Harmonic Spectra

It is known that rectifiers are the main sources of harmonics in a dc railway and thus the input and output spectra of 6-pulse and 12-pulse rectifiers are analyzed to some extent in the following paragraphs. The attention is focused on diode rectifiers, so the assumption of a firing delay angle $\alpha = 0$ is applied, even if references to complete formulations are presented in addition to the given simplified expressions. The dc output voltage, for its impact on the traction supply line and on trackside equipment, is treated more extensively than the ac side current harmonics; also uncharacteristic harmonics are considered because they are not adequately filtered by tuned filters, if present.

Basic Waveforms and Operating Conditions. In this paragraph, the main elements concerning the operating modes and the quantification of the impact on the ac side at the fundamental are discussed. The reference scheme is shown in Figure 2.16.

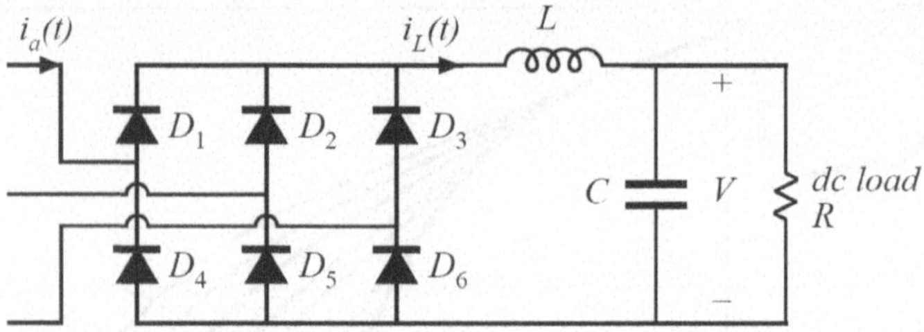

Fig. 2.16 Reference scheme of a three-phase 6-pulse diode rectifier

The input current waveform for one phase (let's assume phase a) is ideally a square wave that conducts the dc side output current (assumed for simplicity constant) alternatively over two time intervals, when the phase a is connected to the output by D_1 or D_4, as shown in Figure 2.17.

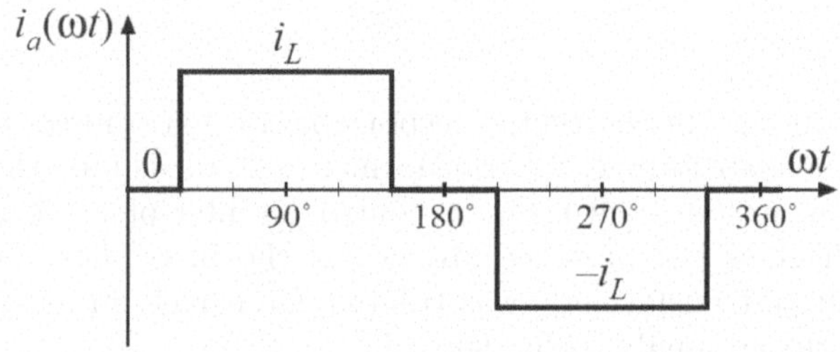

Fig. 2.17 Ideal input phase current waveform and related quantities

The input phase current resembles a square wave but the 60° wide missing conduction intervals around 180° and 360° are responsible for the elimination of triple harmonics from the waveform spectrum, thus improving the THD and the Distortion Factor with respect to a single-phase full wave rectifier: approximately 32% against 48% for the former, 95.5% against 90% for the latter.

In reality, the dc output current profile is not constant and contains the natural output ripple of the rectifying process (that will be quantified in the paragraphs below) and is shown in Figure 2.18(a), always under the general assumption of an approximately constant dc current (the superimposed ripple is magnified for visibility, but the output current is always above the zero line). If the output inductor is too small then the bridge operates in the so called discontinuous mode with impulsive charging of the output capacitor (as shown in Figure 2.18(b)) and the waveform distortion increases dramatically.

Power Factor and Displacement Factor. The displacement factor $\cos\varphi$ for a diode rectifier is defined starting from the active power P_1 at the fundamental with $P_1 = V_1 I_1 \cos\varphi$, where V_1 and I_1 are the rms voltage and

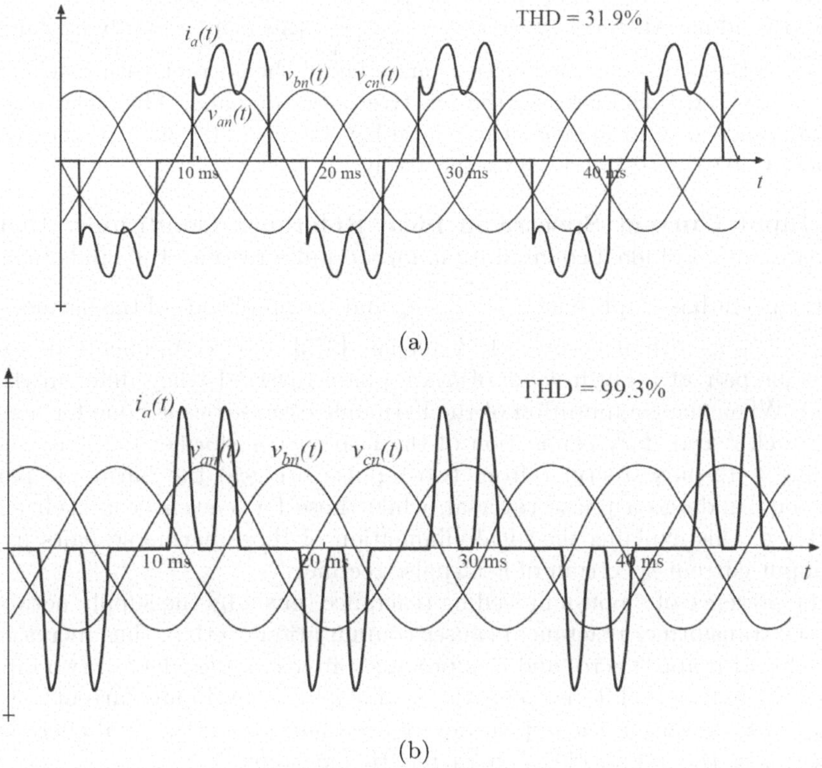

Fig. 2.18 Waveforms for (a) continuous and (b) discontinuous conduction modes; the example considers a 60 Hz power supply as it is evident from the shown time axis

current phasors at the fundamental frequency. The total active power P, that includes all the relevant harmonics, is defined as $P = \lambda S$, where S is the total apparent power and λ is the power factor; the power factor includes harmonic distortion and the ratio with the displacement factor is called the deformation factor $\nu = \lambda / \cos\varphi$.

For a 3-phase rectifier with zero commutation reactance, the output dc power $P_d = P_{d0}$ corresponds to the active power at the fundamental. With the approximation that the supply ac network is stiff with a short circuit power much larger than the power absorbed by the rectifier, only the input current is harmonically distorted and we may write that the total apparent power $S = \sqrt{3} V\, I_1$, while the apparent power at the fundamental is $S_1 = \sqrt{3} V\, I_1 = 3/\pi\, S$ and the deformation factor is $\nu = 3/\pi$.

If the phase commutation reactance is non-null, then $P_d = P_{d0} - \dfrac{3X_c}{\pi} I_d^2$ and the added voltage drop across the commutation reactance modifies the displacement factor as $\cos\varphi = 1 - \dfrac{3X_c}{\pi}\dfrac{I_d}{V_{d0}}$. If the effect of the firing delay angle α is added, then $\cos\varphi = \cos\alpha - \frac{3X_c}{\pi}\frac{I_d}{V_{d0}}$; this can be briefly explained observing that the delay produces a phase lag of the current with respect to the voltage and thus increases the reactive power demand. The effect of the overlap angle μ is to increase this phase lag: for $\mu = 20°, 30°, 40°$ the $\cos\varphi$ reduces to 0.97, 0.935 and 0.89 respectively.

AC Input Current Spectra in Ideal Reference Conditions. Under the assumption of ideal commutation, for a 6-pulse rectifier the fundamental input current has amplitude $I_1 = \frac{2\sqrt{3}}{\pi} I_d$ and the amplitude of the harmonics is $I_n = I_1/n$ with $n = 6k \pm 1$, k integer [375]. The components of each harmonic pair at a given value of k are phase reversed (they differ in their signs). When the computation of the harmonic expressions is done for a star and a delta secondary connection of the feeding transformer (as it is when forming a 12-pulse rectifier from two 6-pulse bridges) the harmonic pairs for k odd undergo a phase reversal, while those for k even keep their sign unaltered: this explains the ideal elimination of those harmonic pairs from the input current spectrum of a 12-pulse rectifier.

The presence of a non-null feeding reactance (given by the supply network and the transformer reactance) causes commutation overlap, thus increasing the value of μ above zero, and this produces an averaging effect of two input phases at a time, while keeping the symmetry of the input current waveform, thus preventing the appearance of even harmonics; the final effect is a reduction of the amplitude of characteristic harmonics.

On the contrary the diode rectifier (operating implicitly with a firing delay angle $\alpha = 0$) produces harmonics larger than any controlled rectifier, that has a positive and larger firing delay angle α. The complete set of curves as a function of α and μ is reported extensively by [469, 20] and [375]. The relationship between α, μ and the commutation reactance X_c (considered above in sec. 2.1.10.1), is recalled for the reader's convenience:

$$\mu = \arccos\left(\cos\alpha - \frac{2X_c I_d}{\sqrt{2}V_{ll}}\right) \tag{2.8}$$

where V_{ll} is the rms value of the line-to-line voltage, I_d is the mean rectified dc output current.

Curves for a non-null firing delay angle $\alpha \neq 0$ are presented as well, since they represent the mode of operation of on-board thyristor converters used for old ac vehicles. In the following figures, for each characteristic harmonic component, the relative amplitude with respect to the fundamental component I_1, versus the firing delay angle α between 0° and 90° and for different values of the commutation overlap angle μ is shown.

DC Output Voltage Spectra in Ideal Reference Conditions. As already discussed, the dc output voltage spectra contains the so called characteristic harmonics, that for a 6-pulse rectifier are given by the $n = 6k$

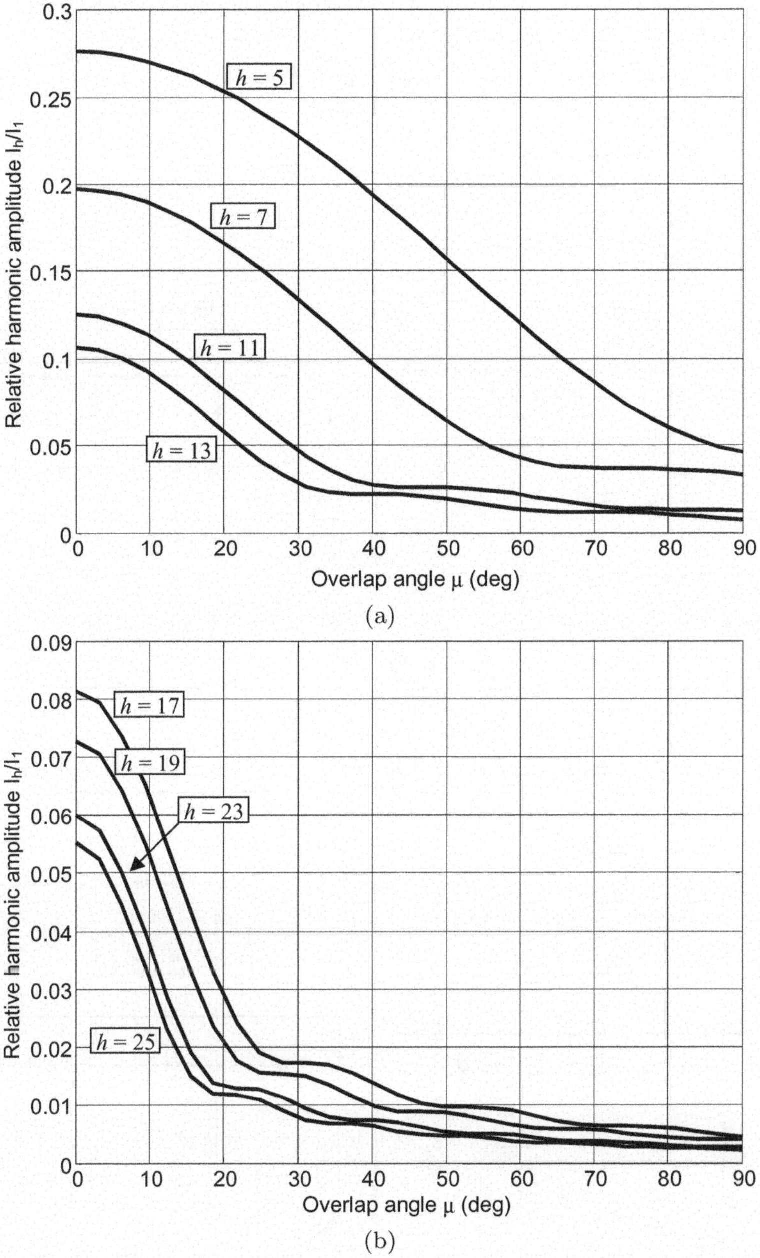

Fig. 2.19 AC input current harmonics for $\alpha = 0$ vs. overlap angle μ: (a) 5th to 11th characteristic harmonics, (b) 17th to 25th characteristic harmonics (ideal conditions of constant and flat DC current)

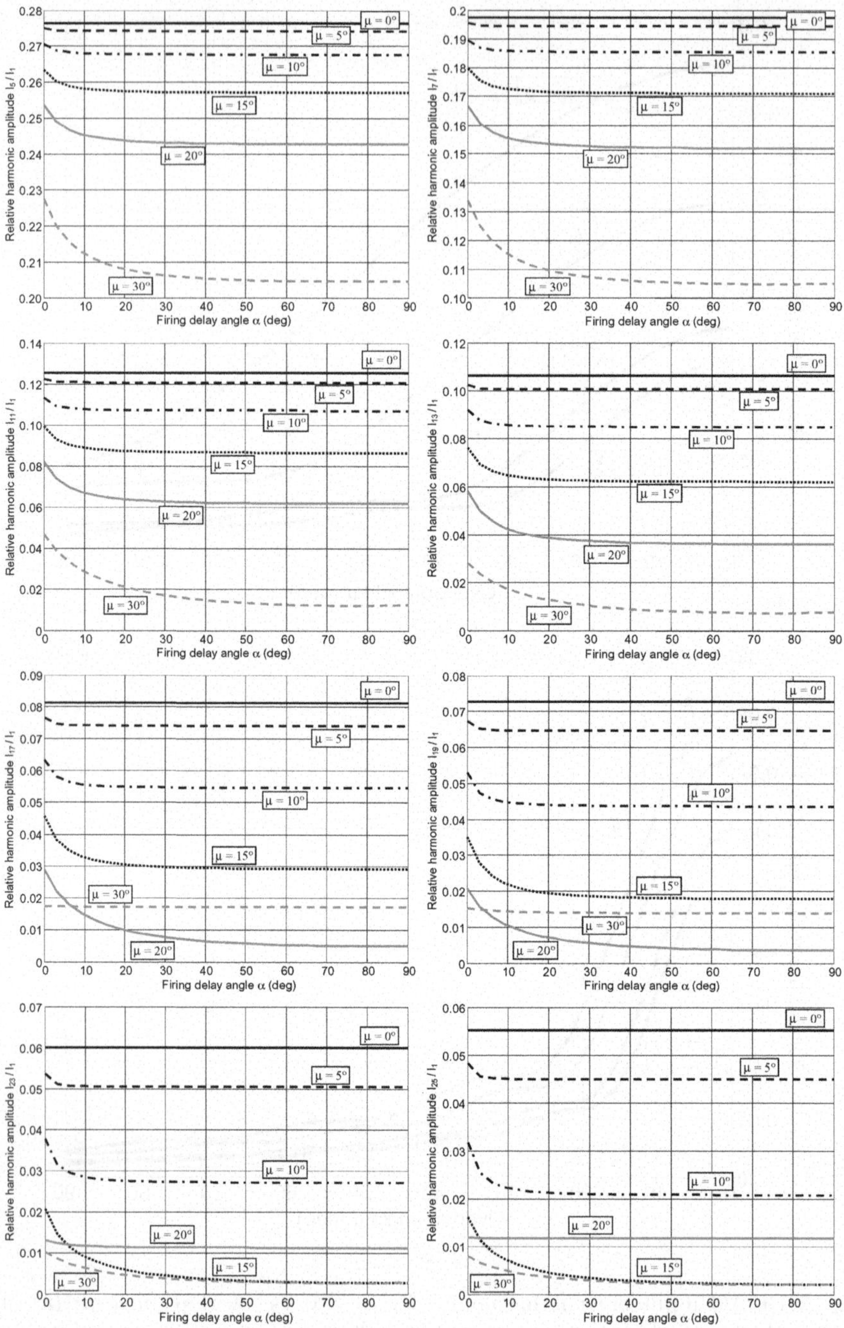

Fig. 2.20 AC input current harmonics for $\alpha \neq 0$ vs. overlap angle μ (ideal conditions of constant flat DC output current); the different used values of $\mu = 0°, 5°, 10°, 15°, 20°, 30°$ are indicated in each figure for different line types).

multiples of the fundamental frequency, k integer. More precisely, the formula for the prediction of the harmonics amplitude under the most general situation of an arbitrary overlap angle μ and firing delay angle α is

$$V_n = \frac{V_{d0}}{\sqrt{2}(n^2-1)} \left[(n-1)^2 \cos^2((n+1)\mu/2) + (n+1)^2 \cos^2((n-1)\mu/2) + \right. \\ \left. -2(n+1)(n-1)\cos((n+1)\mu/2)\cos((n-1)\mu/2)\cos(2\alpha+\mu)\right]^{1/2} \tag{2.9}$$

Ideally, considering a diode rectifier with $\alpha = 0$, for $\mu = 0$, the square bracket vanishes and the characteristic harmonics amplitude is proportional approximately to $1/n^2$. In [20] it is shown that the harmonic content increases rapidly for thyristor rectifiers with increasing α, as it is shown in the Figure 2.21. Equation (2.9) is plotted first, for a diode rectifier with $\alpha = 0$ vs. overlap angle μ and shown in Figure 2.21 for the first four characteristic harmonics of a 6-pulse configuration, namely 6th, 12th, 18th and 24th ones.

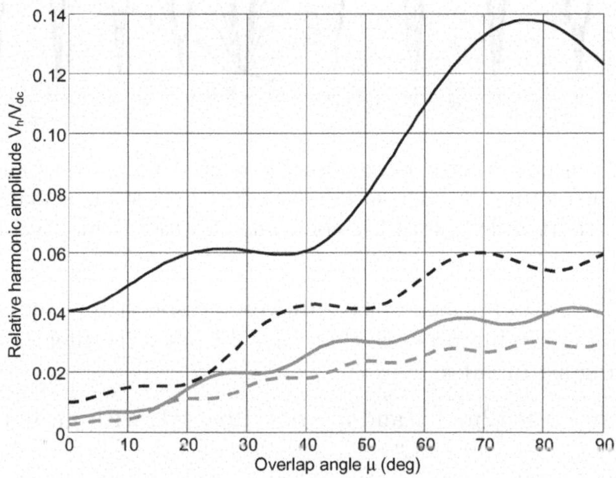

Fig. 2.21 DC output voltage harmonics for $\alpha = 0$ vs. overlap angle μ (ideal conditions of constant flat DC output current): 6th harmonic (black solid), 12th harmonic (black dashed), 18th harmonic (gray solid), 24th harmonic (gray dashed).

The same eq. (2.9) is plotted also for a thyristor rectifier with $\alpha \neq 0$ vs. overlap angle μ in Figure 2.22 with a separate graph for each of the same four characteristic harmonics.

Negative Sequence and Harmonic Components in the Supply Voltage. Let's assume that the three-phase voltage system on the ac side of the 6-pulse rectifier is perturbed by a harmonic disturbance of order h, with amplitude V_h in per unit of the fundamental voltage, and either positive or negative sequence. This disturbance gives rise to non-characteristic

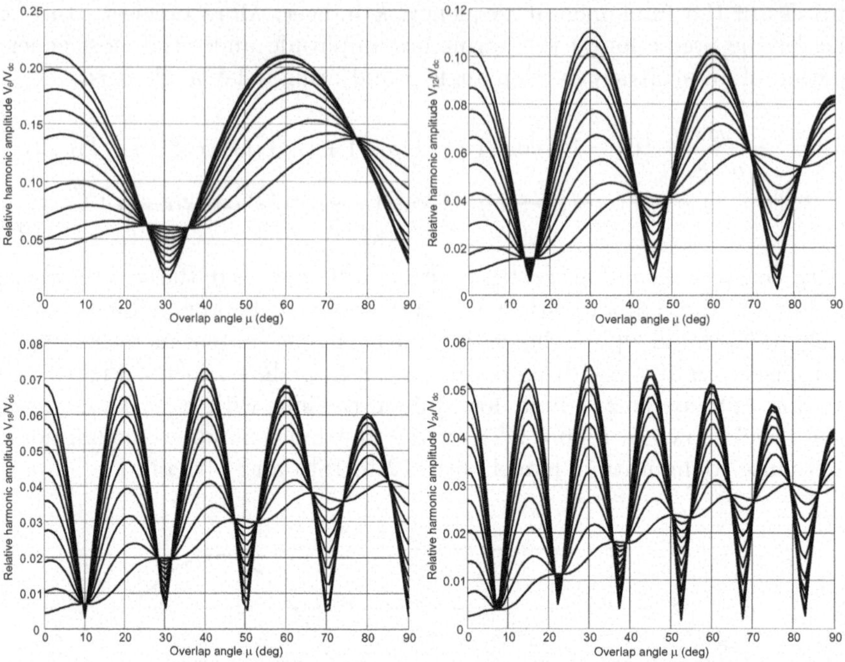

Fig. 2.22 DC output voltage harmonics for $\alpha \neq 0$ vs. overlap angle μ (ideal conditions of constant flat DC output current): (top left) 6th harmonic, (top right) 12th harmonic, (bottom left) 18th harmonic and (bottom right) 24th harmonic

harmonics of order n appearing in the dc voltage spectrum, due to the interaction with the rectifying process, that for a 12-pulse rectifier is summarized by the following set of rules:

1) if $h + n = 12p_1 + 1$ and $h - n = 12p_2 + 1$, with p_1 and p_2 integers

$$V_n = \begin{cases} V_h \dfrac{h\sqrt{2}}{h^2 - n^2} & h > n \\ V_h \dfrac{n\sqrt{2}}{n^2 - h^2} & n > h \end{cases}$$

2) if $h + n = 12p_1 + 1$ and $h - n \neq 12p_2 + 1$, with p_1 and p_2 integers

$$V_n = V_h \frac{1}{\sqrt{2}(h+n)}$$

3) if $h + n \neq 12p_1 + 1$ and $h - n = 12p_2 + 1$, with p_1 and p_2 integers

$$V_n = V_h \frac{1}{\sqrt{2}(h-n)}$$

4) if $h + n \neq 12p_1 + 1$ and $h - n \neq 12p_2 + 1$, with p_1 and p_2 integers

$$V_n = 0$$

Unequal Phase Impedances. Let's assume that the phase reactances on the ac side of the 6-pulse rectifier are not perfectly matched, but that they differ from a reference value X by a small deviation $X_R = X(1+\delta_R)$, $X_S = X(1+\delta_S)$ and $X_T = X(1+\delta_T)$, where the three deviations may range between 0 and a maximum δ_0.

On the dc output voltage the produced harmonics of order n are always even and of equal amplitude and they occur for three extreme values, one at $-\delta_0$, one at 0 and the other one at δ_0:

$$V_{d,n} = V_{d0} \frac{I_d X}{2\sqrt{6}} \delta_0 \tag{2.10}$$

On the ac input current the effect is more complex and the complete solution may be found in [20].

Inter-modulation. Given the set of characteristic and non-characteristic harmonics that populate the output voltage spectrum of dc substations, the new EN 50238-2 technical document [115] draws the attention also to modulation effects that may occur in the substation diode rectifiers of dc systems when the load is weak. This means that any dc vehicle conducted emission causes inter-modulation products to appear in light-load conditions, at different extent depending on the switching mode of the diode bridges and the intensity of the involved components. There is thus the possibility of unsteady components located almost evenly in the low frequency range up to one or two kHz, changing with the load mix and vehicle position.

2.1.10.3 Output Filters

It was said that resonant output filters are connected at the output of rectifier substations in order to attenuate the most significant harmonics of the dc output voltage. These output filters are mainly of two different types: low pass resonant filters and band pass resonant filters, as shown in Figure 2.23.

These filters may be tuned in principle either by selecting a cutoff frequency above which lie the characteristic harmonics that must be attenuated (low pass type) or by positioning the resonance frequency onto one of the characteristic harmonics (band pass type). In both cases, the inductance and capacitance values are a-priori set to the designed values, and they do not take into account two important facts: first, the on-board filters of rolling stocks may combine, depending on the distance between the rolling stocks, changing dynamically the characteristics of the filters; second, the traction line itself contributes and modifies the filter response with its distributed inductance and capacitance.

In addition to the above, the traction line behavior and the pantograph impedance seen by the rolling stock are influenced by the effect of the substation filter, as pointed out in [507], where the architectures and parameters of the filters in use in several countries (Italy, the Netherlands, Poland, Russia,

South Africa, United States) are collected and analyzed (see Figure 2.23 and 2.24).

As it is well known for industrial supply networks, resonant filters may create spurious resonant frequencies with the existing network that may in turn amplify existing harmonic components of moderate amplitude. This is what happens with the Italian substation filter, designed for a no load resonance frequency of 108 Hz (see Figure 2.23), and brought to about 100 Hz by traction line and rolling stock effect; in order to keep power losses in the filter inductor low, the factor of merit of the filter is considerably large and this unduly amplifies an otherwise not relevant harmonic component, the 2nd harmonic, produced by transformer asymmetry. This behavior is known and is shown by analyzing the measurement results collected for several days over some conventional dc railway lines [519].

2.1.10.4 Short Circuit Current and Short Circuit Protection

The short circuit current waveform of a 6-pulse or 12-pulse rectifier is a complex waveform mainly composed of a transient DC component, a 6- or 12-pulse ripple and a slower oscillation at the supply fundamental frequency, as shown in Figure 2.26. The steady state short circuit dc component $I_{sc,0}$ for zero load impedance is a function of the commutation reactance of the feeding transformer $X_c = R_c + j\omega L_c$; for the 6-pulse rectifier $I_{sc,0}$ is uniquely determined as

$$I_{sc,0} = \frac{V_{d0}}{X_c} \frac{\pi\sqrt{3}}{9} \tag{2.11}$$

while for the 12-pulse rectifier it depends also on the coupling factor k

$$I_{sc,0} = \frac{V_{d0}}{X_c} \frac{\pi\sqrt{3}\,(4-3k)}{9(1-k)\sqrt{3(1-k^2)}+1} \quad \text{for } 0 < k < \frac{\sqrt{3}}{\sqrt{3}-1} \tag{2.12}$$

$$I_{sc,0} = \frac{V_{d0}}{X_c} \frac{\pi\sqrt{3}\,(8-\sqrt{3}k^2)}{9(1+k)(2+\sqrt{3}k)\sqrt{3(1-k^2)}+1} \quad \text{for } \frac{\sqrt{3}}{\sqrt{3}-1} < k < \frac{2}{3} \tag{2.13}$$

$$I_{sc,0} = \frac{V_{d0}}{X_c} \frac{2\pi}{3(2+\sqrt{3}k)} \quad \text{for } \frac{2}{3} < k < 1 \tag{2.14}$$

An approximate formulation valid for $0 < k < 1$ is

$$I_{sc,0} = \frac{V_{d0}}{X_c}(0.47k^2 - 1.1k + 1.2) \tag{2.15}$$

The peak value $I_{sc,p}$ depends on the resistive-inductive characteristics of the load circuit: if the load is purely resistive then there is no over-shooting and

Filter	Par.	Value	Diagram
Italy (IT)	L	6 mH	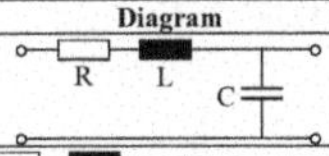
	C	360 µF	
South Africa (SA)	L	0.9 mH	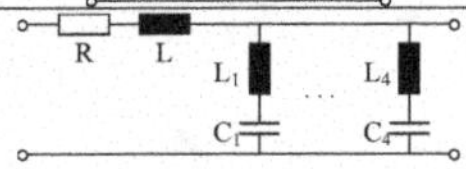
	L_1	2.814 mH	
	C_1	100 µF	
	L_2	1.173 mH	
	C_2	60 µF	
	L_3	1.564 mH	
	C_3	20 µF	
	L_4	1.759 mH	
	C_4	10 µF	
Poland (PL1)	L	6.15 mH	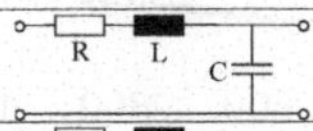
	C	900 µF	
	R	0.2 Ω	
Poland (PL2)	L	6.0 mH	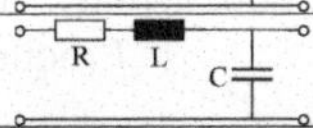
	C	800 µF	
	R	0.5 Ω	
Poland (PL6)	L	4.0 mH	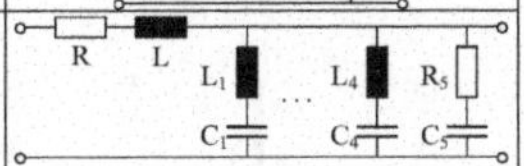
	R	0.2 Ω	
	L_1	2.34 mH	
	C_1	120 µF	
	L_2	1.76 mH	
	C_2	40 µF	
	L_3	1.04 mH	
	C_3	30 µF	
	L_4	0.88 mH	
	C_4	20 µF	
	R_5	50 kΩ	
	C_5	50 µF	
Poland (PL12)	L	1.8 mH	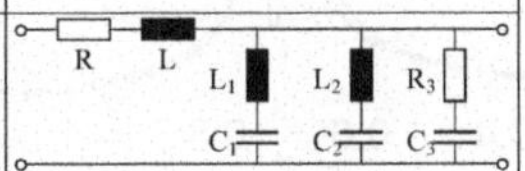
	R	0.2 Ω	
	L_1	1.173 mH	
	C_1	60 µF	
	L_2	1.759 mH	
	C_2	10 µF	
	R_3	50 kΩ	
	C_3	100 µF	
Russia (RU3)	L_a	5.0 mH	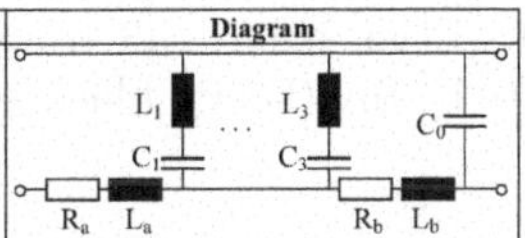
	R_a	0.2 Ω (est.)	
	L_b	3.0 mH	
	R_b	0.2 Ω (est.)	
	L_1	6.33 mH	
	C_1	400 µF	
	L_2	1.55 mH	
	C_2	400 µF	
	L_3	0.7 mH	
	C_3	400 µF	
	C_0	200 µF	
Russia (RU6)	L_a	5.0 mH	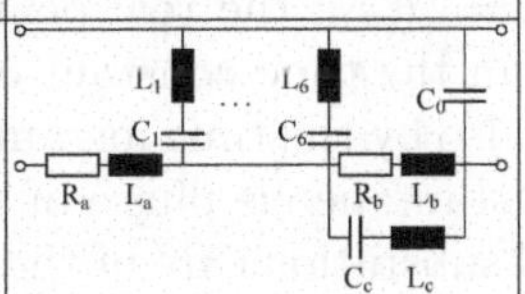
	R_a	0.2 Ω (est.)	
	L_b	5.0 mH	
	R_b	0.2 Ω (est.)	
	L_c	18.4 mH	
	C_c	18 µF	
	L_1	17.59 mH	
	C_1	144 µF	
	L_2	5.86 mH	
	C_2	108 µF	
	L_3	2.93 mH	
	C_3	96 µF	
	L_4	2.64 mH	
	C_4	60 µF	
	L_5	2.11 mH	
	C_5	48 µF	
	L_6	1.95 mH	
	C_6	36 µF	
	C_0	20 µF	
USA BART (US)	L	0.05 mH	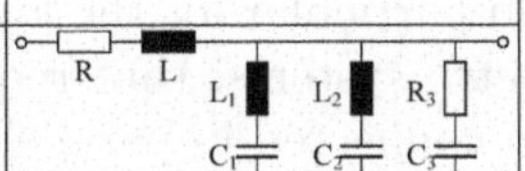
	L_1	0.3 mH	
	C_1	162 µF	
	L_2	0.5 mH	
	C_2	24 µF	
	R_3	1 Ω	
	C_3	10 µF	

Fig. 2.23 Schemes and parameters of substation filters

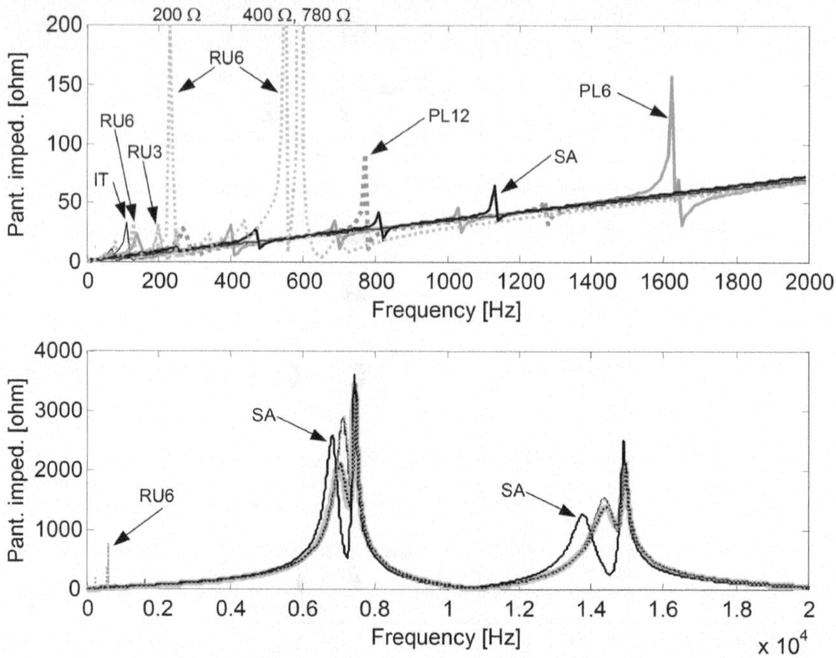

Fig. 2.24 Pantograph impedance vs. frequency at 6.666 km (1/3 of the substations separation); upper: PL6 (dark gray solid), PL12 (dark gray dotted), RU3 and RU6 (light gray dotted), IT (black solid), all others (black solid); lower: RU6 (gray dotted), US (black dotted), SA (heavy black solid), all others (black solid), short circuit (gray band)

$I_{sc,p} = I_{sc,0}$; if the load is purely inductive then the peak value is twice the steady dc value $I_{sc,p} = 2I_{sc,0}$; the real peak value for a given load depends in reality not only from the time constant of the load itself (inductance over resistance ratio), but also by the time instant at which the short circuit occurs with respect to the instantaneous phase of the ac supply voltage.

Protections for DC substations are of the high speed type (HSCB) with or without rate of rise detection and distance to fault measurement. Triggering switches on the rate of rise (also called "di/dt trigger"), before the current reaches the maximum is particularly important, not to say vital, for the protection of the line. Based on expressions above and some extreme, but not absurd, values of the parameters the peak value $I_{sc,p}$ may reach values of 100 kA and more, in particular for the lower supply voltage levels (such as 600 and 750 V) in metro systems, that require more than one rectifier (for example two or three) connected in parallel.

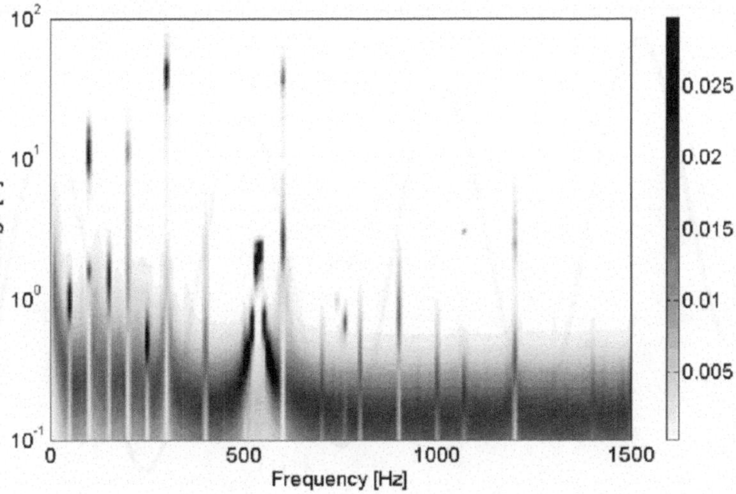

(a)

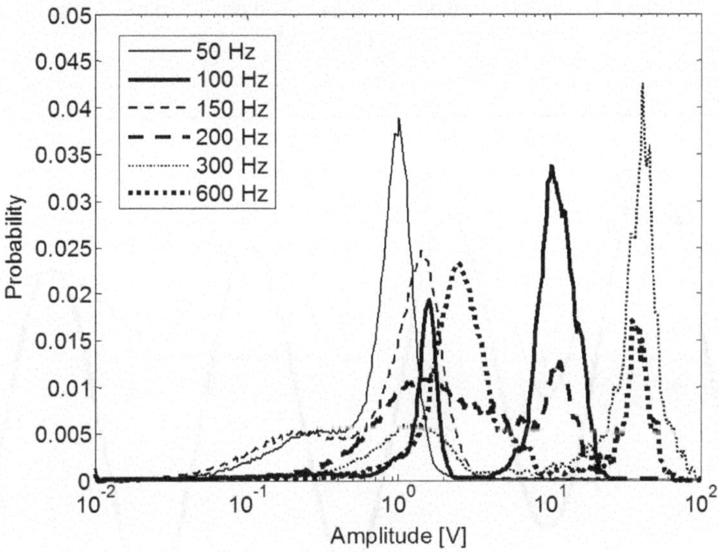

(b)

Fig. 2.25 Pantograph voltage spectrum distribution

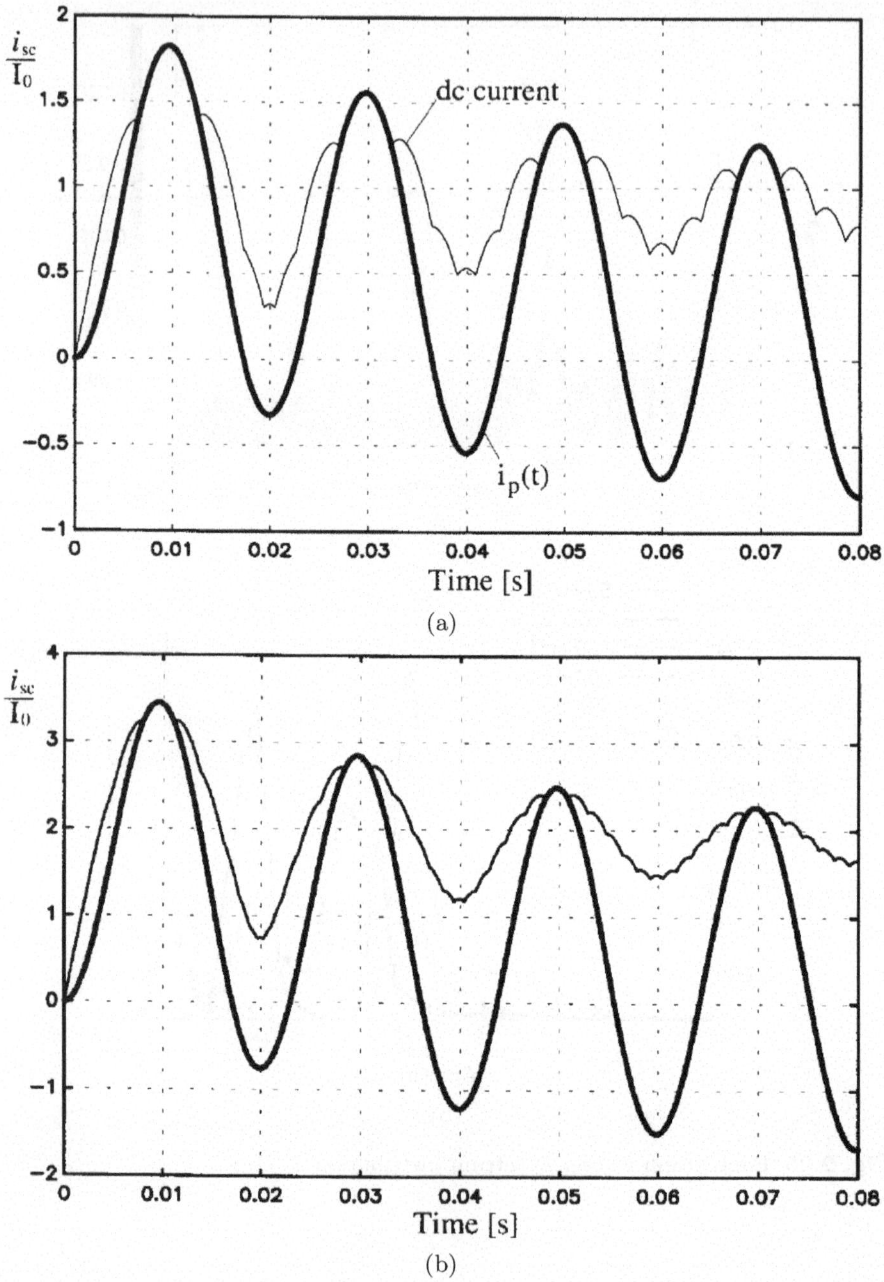

Fig. 2.26 Theoretical short circuit current waveforms for (a) 6-pulse and (b) 12-pulse rectifiers

2.2 Trains and Vehicles

The main characteristics of electrified and diesel vehicles and locos are reported in the following, as well as some peculiar behaviors of train consists with respect to conducted emissions and the interactions with the traction line impedance.

2.2.1 Train Propulsion Systems

Electricity and diesel powered trains are the most common modern trains. The main difference between these two train types is how the train is powered. Whether the primary energy source is grid electricity or diesel fuel powered generators, the power conversion process from energy source to propulsion power for the traction motor and to auxiliary electric power generation for on-board electrical systems is very similar. The process of converting grid and diesel energy sources to useable power is shown in Figures 2.27and 2.28. The power generated from the diesel generator or electrical grid is usually not directly suitable for the traction motor and the onboard electrical systems. It has to be converted into the correct voltage and frequency.

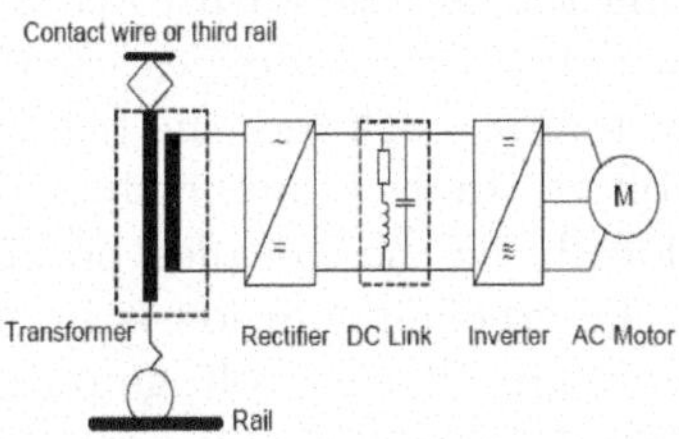

Fig. 2.27 Propulsion Systems Block Diagram.

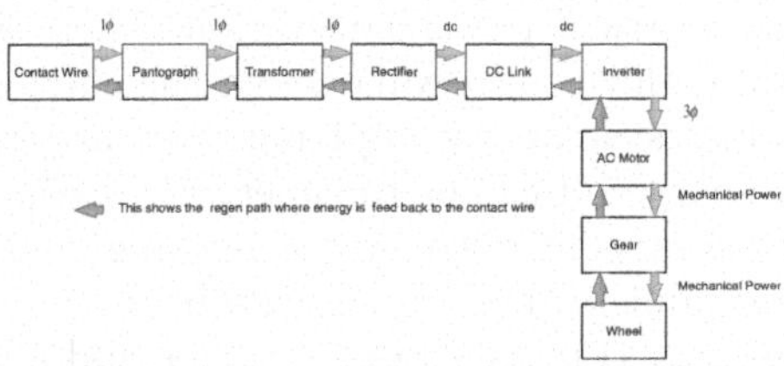

Fig. 2.28 Propulsion system energy flowchart

As shown in Figures 2.27 and 2.28, the power conversion process involves devices like dc-links and converters. In electricity powered trains, the power grid voltage has to be transferred down to a lower voltage by a transformer. Circuit breakers protect the grid and train in case of any malfunction of the electrical system. Since the power subsystems and power converters operate at high voltages and currents, their control and communication lines must be galvanically isolated and immune to the electromagnetic fields generated by the converter. Isolation and electromagnetic interference (EMI) immunity are important characteristics that ensure reliable, safe and stable train operation. In this section, emphasis will be given to Electrical Multiple Units (EMU), although the EMI issues discussed herein are common and independent of the primary source of energy.

The term EMU refers to a self driving train unit capable of coupling with other similar units. An EMU can be controlled from one cabin and is characterised by its speed, safety, comfort and high efficiency as features. An EMU implies an ac propulsion systems, which may either be 15 kV/16.67 Hz or 25 kV/50 Hz or a derivative of the these. The ac propulsion systems is composed of pantograph, main transformers, rectifiers, dc-link, inverters, ac motors, gear boxes and wheels. In simple terms, ac electrical power, supplied via Overhead Transmissions Lines, is collected by a pantograph and then transformed via elements previously mentioned before the eventual resultant mechanical output force of the train wheels propels the train.

In the case of an ac traction supply the pantograph is connected to a propulsion transformer that distributes the electricity to several secondary coils; the main transformer regulates the voltage to the desired level following which the ac to dc conversion is performed by the rectifier controlled by a Pulse Width Modulating (PWM) signal. A dc-link filter – which is composed of a DC-link capacitor, second harmonics filter, and midpoint earthing unit – then smooths the dc-link voltage and keeps the system voltage at the desired level attenuating harmonics and resulting ripple; following this the dc power is inverted to a 3-phase power. It is common to apply PWM control at the inversion although the PWM control is provided by a separate set of auxillary control circuits.

When the train is operational, electric power is converted to mechanical power as described above; however during braking, the mechanical power is converted to electrical power typically following the reverse path to that of the electrical power generation. There are currently two strategies used to deal with this reserval energy: the energy is let to flow back into the traction supply system, by applying a pantograph voltage that is larger than the catenary voltage, or the energy is dissipated locally on the so-called braking resistors, controlled by the braking choppers. The second solution is in any case implemented and used whenever the maximum allowed voltage on the catenary is likely to be overcome and network instability may appear..

The term EMU refers to a self driving train unit capable of coupling with other similar units. An EMU can be controlled from one cabin and is characterised by its speed, safety, comfort and high efficiency as features. An EMU implies an ac propulsion system, which may either be 15 kV/16.67 Hz or 25 kV/50 Hz or a derivative of the these. The ac propulsion system is composed of pantograph, main transformers, rectifiers, dc-link, inverters, ac motors, gear boxes and wheels. In simple terms, ac electrical power, supplied via overhead transmissions lines, is collected by a pantograph and then transformed via elements previously mentioned before and the eventual resultant mechanical output force of the train wheels propels the train.

The pantograph is connected to a propulsion transformer that distributes the electricity to several secondary coils; the main transformer regulates the voltage to the desired level following which the ac to dc conversion is performed by the rectifier controlled by a Pulse Width Modulating (PWM) signal. A dc-link filter - which is composed of a dc-link capacitor, second harmonics filter, and midpoint earthing unit - then smooths the DC-link voltage and keeps the system voltage at the desired level with respect to ground;

following this the dc power is inverted to a 3-phase power. It is common to apply PWM control at the inversion although the PWM control is provided a by separate set of auxiliary control circuit.

When the train is operational, electric power is converted to mechanical power as described above, however during braking, the mechanical power is converted to electrical power typically following the reserve path to that of the electrical power generation. There are currently two strategies used to deal with this reversal energy:

- Brake Resistors: When speed of an inverter-controlled ac motor is reduced, the motor acts as a generator, feeding back energy to the frequency converter. As a result, voltage in the intermediate circuit of the inverter increases. When a specific threshold is exceeded, the energy must flow to an external braking system in order to avoid drive failures. Braking resistors are designed to absorb such energy and to dissipate it into heating. The use of brake resistors allows drives to fulfill the requirements of particularly severe duty cycles, for example those featured by frequent braking, long lasting braking or impulsive braking.
- Regenerative Braking: A braking system used by locomotives and trains fitted with electric traction motors where the motors become generators and the current developed is fed into the overhead line or third rail for use by other trains or for return to the supplier. This strategy is most common with modern with new rolling stocks as its energy savings benefits fulfills the green technology philosophy and provide economic benefit.

To take account of regenerative braking, the propulsion system is designed to be a bi-directional system such that during braking the mechanical energy is transmitted back to the substations. In regenerative braking, the ac motors works in generator mode, propulsion inverter works in rectifier mode and the rectifier works as a single phase inverter.

Electromagnetic disturbance can be generated at any of the stages, described above, between the propulsion supply and electric motor; such electromagnetic disturbance can then be propagated via cables as well as radiated via loop circuits. Typically, the electromagnetic disturbance so generated is mitigated through the application of filters at the inverter output - however this solution is not always effective or economical feasible. Another solution is the replacement of copper cables with fiber optics cables; this is implemented in order to achieve isolation and electromagnetic immunity, since fiber optics technology offers both galvanic isolation and near perfect immunity to electromagnetic interference.

2.2.2 Vehicle Characteristics

Various kinds of characteristics are considered, that encompass mechanical and electrical elements and that help in better defining vehicle performance and operation. A vehicle or a locomotive, if supplied electrically by the traction supply line, transforms the electrical power into mechanical power and the transformation process is characterized then by the combination of the electrical and mechanical variables and parameters, by their nominal values, by their ranges of variation, and by any restriction and limit imposed by standards, by the need of safe, reliably and efficient operations and by practical considerations.

The electrical variables and parameters may be synthesized as:

- pantograph (or sliding shoe) voltage and current at dc or at the fundamental supply frequency for ac systems, for which also the displacement factor is of concern; these two variables define precisely the absorbed active (and reactive) power at dc or the fundamental frequency;
- additional distortion components that may be classified as harmonics (of a reference fundamental frequency, also for dc systems) or other components related to the on-board converters;
- the line voltage and the absorbed current are subject to an articulated set of nominal values and limits, that characterize the electrical interface, and that ensure on one side the prescribed vehicle performance and on the other side a limited impact on the supply network and on the other vehicles.

The mechanical variables and parameters that are relevant for the definition of the train operating conditions and in relationship to the electrical variables above are:

- train speed defines the operating conditions of the on-board traction drives, and thus characterizes the expected conducted emissions;
- the required tractive effort depends on the train aerodynamic resistance at the travelled speed and the applied acceleration and determines the absorbed power;
- in case of braking the reduction of mechanical energy produces an equivalent electrical energy through the regenerative operation of the traction drives that is either injected back into the supply network or dissipated on-board via braking resistors or a combination of both.

2.2.2.1 Pantograph Voltage, Current and Absorbed Power

The pantograph voltage and current waveforms define completely the instantaneous power that transits at the pantograph interface. For simplicity, only one pantograph is considered, while two (or even more, in some rare cases) may be active, and, in the case of sliding shoes used in third rail systems, there are usually one or two active shoes for tractioning vehicle. The instantaneous power $p(t)$ is given by the product of the instantaneous voltage $v(t)$ and current $i(t)$. The integral of the instantaneous power over a given time interval gives the total energy absorbed or released by the train. Correspondingly the energy may be computed from the pantograph power $P(f)$ defined in the frequency domain and again given by the product of voltage $V(f)$

and current $I(f)$ spectra, by the Parseval's theorem[5]; the product should be extended to the entire frequency range, but practically only the fundamental is considered by most standards and a few additional harmonics are required to make accurate energy and efficiency estimations. The pantograph active power P_1 at the fundamental is thus defined as $P_1 = V_1 I_1 \cos\varphi$, where V_1 and I_1 are the rms voltage and current phasors at pantograph and $\cos\varphi$ is the displacement factor, all at the fundamental frequency. The total active power P, that includes all the relevant harmonics, is defined as $P = \lambda S$, where S is the total apparent power and λ is the power factor; the power factor includes harmonic distortion and the ratio with the displacement factor is called the deformation factor $\nu = \lambda/\cos\varphi$. All these elements are derived more or less directly from the theory and foundations of ac supply systems for transmission and distribution, while for dc traction systems the identification of the harmonics and harmonic distortion turns into a more general investigation of the spectrum components produced by the rectifiers of the supply substations and the front-end converters of rolling stock. The voltage ripple includes all the relevant ac components [523, 516] and together with the current ripple may be used to define the total input power, so that something similar to the power factor can be defined.

To ensure stability especially for weak supply networks or in abnormal operating conditions, trains needs to adapt automatically the absorbed power depending on the line voltage: an excessive power absorption at low voltage increases the amplitude of the pantograph current above the maximum allowed value for the current capability, a displacement factor lower than unity either inductive or capacitive interacts with the same traction supply and may increase the reactive power flow. The standard EN 50388 [124] establishes limits and rules for the power factor as a function of the absorbed power and the line voltage:

- the capacitive power factor is not limited for line voltage between U_{min1} and U_{max1} as per EN 50163 [99]; the power factor shall not be capacitive for line voltage between U_{max1} and U_{max2};
- the lower limit of the inductive power factor is specified as a function of the absorbed active power, equal to 0.85, 0.93 and 0.95 for $P \leq 2$ MW, $2\text{ MW} < P \leq 6$ MW and $P > 6$ MW respectively; it is underlined that at the lowest active power levels the auxiliary loads represent a not negligible term and must be accounted for.

The absorbed pantograph current is also subject to limitations depending on the country and on the type of line (high speed or conventional), related to the current capability of the overhead distribution system.

For other electric transportation systems, such as light railways, metros, tram lines and trolley buses, there are no similar standardization level and

[5] Parseval's theorem refers to the result that the Fourier transform is unitary; loosely, that the sum (or integral) of the square of a function is equal to the sum (or integral) of the square of its transform.

limitations: the main reason is that the rolling stock characteristics are evaluated by the owner, most often during the design phase of the entire system, and compatibility is thus included in the design or procurement activities; moreover, the line is a closed line with an extension that may span over a portion of a city, the whole city or also its surroundings, but with no extraneous rolling stock coming from other parts of network, that for railways is potentially infinite.

2.2.3 Traction Converters

Traction converters are briefly reviewed, as they represent the most relevant source of emissions for the analysis of the interference onto track devices, such as signalling circuits. The world of traction converters is articulated and complex: only the principles are considered, in order to derive general rules for the evaluation of converter emissions, while a complete and comprehensive treatment is almost impossible for the continuous evolution of technical solutions, and we rely on specialized references, such as [644].

2.2.3.1 AC Supply Converters (Rectifiers, 4 Quadrant, Input Transformer)

Many different traction drives have equipped the locomotives and later the electric vehicles running on ac supply systems.

Commutator ac motors, in particular for 16.7 Hz systems, controlled first by a tap-changer transformer and later by a thyristor ac line converter, are no longer developed nor purchased; the Class 112 locomotive ordered by DB on 1991 represents probably the last example of this type of traction drive, even if some items may be still in use.

Another example of old drive architecture is that based on dc motors fed by phase controlled rectifiers; the general operation and the ac harmonic pattern of a thyristor controlled rectifier have been already treated while considering dc substation rectifiers in sec. 2.1.10.2.

Both drives are based on commutator machines and the reduction of manufacturing and maintenance costs, together with the need for improved reliability, led to their abandon in favor of synchronous and induction machines. A famous example of implementation based on the former is represented by the combination of a current source inverter commutated by the same synchronous machine and its counter-emf, and equips several versions of the French TGV-A, using a “Sybic” converter, with a single-motor bogie (the motor has a double star-connected stator winding) fed by a 12-pulse converter, with an auxiliary starting circuit; the line voltage at 25 kV 50 Hz is applied to the on-board transformer, but the dc link voltage after the rectifiers on the secondary windings is rated at 1500 V dc, to be compatible with the dc supply lines. This type of drive produces a large current distortion, even more critical for a lower supply frequency, such as for 16.7 Hz systems, and the necessary dc link reactors represent a critical element in terms of cost, weight and power losses. Years of service showed that this drive is however robust and reliable, thanks also to the possibility of using medium voltage thyristors, much more reliable than the GTOs at the time of the design and production. Types of synchronous machines other than the classical rotating field one, such as permanent magnet motors, haven't found satisfactory application until now, mainly because of the cost of permanent magnets and related machining and mounting.

On the contrary, the induction machine, and in particular its squirrel cage version, is the most diffused even if of relatively new conception and introduction: the first realizations appeared at the beginning of '70s; for about ten years the static converter was mainly of the force commutated thyristor type, such as for the DB Class 120, the first ICE exemplars, between late '70s and early '80s. Due to limitations of GTOs (in voltage, current and switching capabilities) the first realizations of voltage source inverters (VSI) and auto-sequentially commutated current source inverter (ASCI) appearing in the early '80s went to equip light railway and metro vehicles. Some other realizations adopting GTO technology, and thus with a limited switching frequency and relevant low order harmonics, are the Swedish Rc5, the Italian "Pendolino" ETR 450, the French Class 15000 (for 25 kV 50 Hz), Class 7400 (for 1500 V DC) and Class 22500 (for both types of supply). To overcome the limited voltage capability with respect to the supply values of available semiconductors, several attempts have been made to apportion the supply voltage over many switching devices arranged in series connection: in the early '70s it was proposed to add a voltage level to the usual two-level inverter by a connection to the virtual neutral point at the center of the dc-link capacitor bank [315]; a further development opened the series of Neutral Point Clamped topologies in the early '80s, that offer higher voltage levels, but at the cost of a more complex architecture and control, that for railway applications have never found ground for development and widespread application [558].

Four-Quadrant Converter (4QC). In 1974 the 4QC is proposed in the literature [465] as an effective way of absorbing power from a single-phase line with a nearly sinusoidal current waveform, thus ensuring a high power factor and a controllable displacement factor at the fundamental. The 4QC is built around a PWM bridge on the secondary winding of the loco transformer connected as a floating load, between two twin converter phase arms, and equipped with a trap resonant filter connected in parallel to the output dc-link capacitor, to block the 2nd harmonic ripple. Its harmonic pattern is that of a PWM single-phase inverter around the switching frequency, whose value depends on the used semiconductors; from the slower GTOs used at the beginning, the more modern IGBTs can ensure a 500-1500 kHz switching frequency for a nominal power of the single converter of several hundreds kW. Even with a relatively slow switching frequency of GTO based 4QCs, designed and manufactured in the '80s, the overall switching frequency seen on the primary side of the on-board transformer is quite high; the Re 460 first released in 1987 uses 200 Hz modules in number of eight, connected to four transformer windings and resulting - if correctly synchronized - in a primary switching frequency of 1600 Hz [288].

To cope with higher supply voltages also 4QC may be implemented as "three point" type (the terminals on the dc link side are connected also to the capacitor bank intermediate point, if available), while improving also the shape and spectrum of the input waveform fed back to the input transformer and the network. The ac output of the 4QCs are connected on either side of dedicated transformer windings, so supplying them with a positive polarity, a negative polarity or a null voltage (the short circuit is possible if the "three point" technique is implemented). On the primary side a harmonic filter may be necessary for specific needs concerning frequency intervals in use by signalling circuits and related limit masks, and two solutions were proposed in the past and are actually used [617]: the filter composed of a series inductor rated for the full primary current and a shunt RC network is located on the high voltage side before the on-board transformer; the filter is connected on a purposely built filter winding on the on-board transformer at a convenient low voltage.

Control Techniques. The basic technique suitable for slow switching devices is that of the square wave modulation (simply referred to also as "square wave inverter"), with the semiconductor devices switched on and off once per period of the output waveform; this implies the slowest switching speed, compatible with the characteristics of large power GTOs in the early '70s. Shaping of the output waveform is not possible and its amplitude is directly related to the dc-link voltage, so that this technique needs a controllable dc voltage by means of a thyristor rectifier. Thus the emission pattern on both the ac input and ac output sides is rich of low order harmonics of considerable amplitude.

An improvement was achieved by a more complex control scheme, called "harmonic elimination" control, where there could be selected a slightly larger switching frequency, pushing it to the maximum value attainable and determining as a consequence the optimum values of the switching instants over an output fundamental period (called "switching angles"), for the minimization of a given performance index: total harmonic distortion, a weighted version of it, the amplitude of specific harmonics, etc. The optimal values were determined by means of optimization algorithms, such as least mean square methods; it is possible to show that different optimization schemes have produced different harmonic patterns, with related pros and cons, in terms for example of achieved harmonic amplitudes and behavior in case of load unbalances [506][598].

With the appearance of faster switching devices PWM and vector modulators have been widely adopted, with adequately high switching frequencies and a correspondingly short cycle time of the modulator digital hardware, for all the related calculations. The first form of PWM is the analog implementation as described in [554]: a modulation signal is compared with a triangular carrier at higher frequency, where the ratio of the two frequency values is named "frequency ratio", and determines the number of switching events determined by the crossing values detected during comparison; guard times are imposed to avoid a sequence of too close switching events, beyond the allowed switching capabilities of the used devices, in terms of both pure switching times and power losses during commutations. In general, inverter control methods and architecture always implement rules to minimize the number of commutations, to avoid instantaneous short circuits during switching of more than one device at a time, and to avoid floating sub-circuits, that may reach dangerous voltages.

Early in the '80s there appeared the first attempts of digital implementations of converters control schemes. The PWM digital implementation may be considered as the first relevant example of a digital modulation scheme, facing and solving all the problems related to: first, the speed of execution of the main modulation task and any other auxiliary task (depending on the hardware speed, on the code optimization, and on the required switching frequency); second, errors of the analog-to-digital and digital-to-analog conversions, in terms of amplitude errors and related accuracy; third, and maybe more important even nowadays, synchronization and round-off timing errors, producing non-stationary time shifts, that are responsible for troublesome and hard-to-tackle beating components, located at low frequency, that produce torque ripple, unwanted interference and are hard to be filtered out. In industrial applications the attention is more focused on the first problem, since low frequency torque pulsation may stress long shafts and supporting bearings, in terms of torsional torque and vibrations. In the railway field, the attention is also directed towards conducted interference to low frequency signalling devices, such as power frequency and DC track circuits and general converter instability. When more complex control schemes are included, extending the digital control from the static converter to the whole drive, possibly including the synchronization of the front-end converter on the input transformer side and the drive converter, the problem of spurious artifacts due to the digital limited precision (on the amplitude and time axes) is even more articulated.

In rotor-field oriented control schemes, the stator current is decomposed into two components, one determining a stator flux parallel to the rotor flux and another one orthogonal to it, that produces and controls the developed torque. Both components are nowadays software controlled by the converter modulator and ensure accurate operation and torque control in static and dynamic cases. The rotor flux is not a stator variable and is in principle unknown: it could be measured, but the best practice is to determine it indirectly, based upon the measured terminal voltages and currents and a machine model, that can be of the adaptive type (to reduce any initial estimation error and to follow any parametric change, due for instance to aging, temperature, operating conditions, etc.).

Converter Emissions. It is intuitive that the least emissions spectrum is expected from a pure induction motor fed directly by the (almost sinusoidal) line voltage; in this type of locomotives the regulation was obtained by simple methods (such as the control of the external resistor in case of a wound rotor motor or by means of a tap-changer transformer) and the results are reported for comparison in [420], Chap. 2: the residual current spectrum components are originated partly by the only sensible non-linearity of the circuit, i.e. the saturation of the motor magnetic circuit, and partly by a pre-existing distortion in the same supply voltage.

Always in [420], Chap. 2, there are the results for locomotives equipped with a thyristor rectifier driving a dc motor: the ac current spectrum components in this case are slightly attenuated with respect to the theoretical values reported in Fig. 2.20, for the presence of the on-board transformer. Different cases are reported for three different sources of information: Swedish, French and Hungarian Railways. The results all refer to a locomotive without on-board filter and the equivalent impedance of the harmonic ac current generator is represented by the motor and transformer impedance; the envelope of the harmonic components for the various cases is comprised between I_1/h^2 and $3I_1/h^2$, where I_1 is the fundamental and h is the harmonic order.

In case an on-board filter is used, its attenuation is to be taken into account as well as its contribution to the equivalent input impedance [617]; the filter in parallel to the motor through two thyristors in conduction reduces the impedance on the secondary side of the on-board transformer and thus the overall locomotive input impedance.

When an ac induction motor driven by a static converter is considered, the thyristor rectifier may be maintained as the front end converter, or a more modern totally controlled bridge can be used operating as a four-quadrant converter, supplying the dc link, which static inverters are connected to. This is the scheme of a modern locomotive, that represents quite a common solution starting back from the '70s and in use in many countries with different types of semiconductors, modulations and switching frequencies. For this reason a omni-comprehensive picture of the harmonic emissions patterns and of the expected amplitudes is almost impossible; the common factor is that the emissions are characterized by odd harmonics of the supply frequency and they are particularly intense around the the twice the switching frequency, normally in the range of 1.5 - 2 kHz. The DB 101 locomotive, featuring a rated power of 7.6 MVA and dating back to the '90s (when the switching speed of semiconductor devices was in any case still limited), can push the first characteristic harmonic group at 1467 Hz. The switching frequency of each of the four 4QC converters is slightly larger than 183 Hz and thus 4x183x2 = 1464 Hz. It is possible that when some of the 4QC are out of service (degraded configuration) the individual switching frequency must be changed, possibly increased, to keep the overall harmonic distortion (or psophometric current, as it is sometimes indicated) under the limit.

2.2.3.2 On-Board Input Transformer

The input transformer is a key element in defining the behavior of converter emissions since, besides the turns ratio applied to each component vector in passing from the secondary side to the primary side and then through the pantograph to the line, it determines heavily the impedance change with frequency and characterizes the equivalent generator representation. An inductive behavior may be expected up to one or few kHz, thus keeping the original converter impedance almost unchanged; above it the transformer starts resonating with its own parasitic capacitance terms, distributed along the windings, between the primary and secondary one and towards the enclosure, thus determining differential and common mode stray capacitance terms. Moreover, since there are many 4QCs for traction and auxiliaries connected to the various secondary windings, reciprocal influence and coupling further complicate the equivalent circuit. The auxiliary windings are many, used to supply both locomotive loads (fans, pumps, air conditioning, heating, etc.) and train coaches (low voltage line, air conditioning, heating). Standard values of auxiliary windings are 220 or 380 Vac (for both 15 kV 16.7 Hz and 25 kV 50 Hz machines), 800-1000 Vac for 16.7 Hz and 950-1500 Vac for 50 Hz supply respectively; the number and type of traction secondary windings depends on the architecture of the 4QC converter, usually with an even number of modules (two or four) and a rated voltage around 1000-1200 Vac, depending on design choices and modularity. A classic example of a very symmetric architecture is the Re 460 locomotive by ABB where the eight 4QC converters are supplied two at a time by four windings 1783 V 630 A each [288]. On some locomotive models a secondary winding is also purposely added for connection of harmonic filter, rated thus at a lower voltage than the primary high voltage side. In the audiofrequency range the 4QCs feed significant emissions around the PWM frequency and its multiples [617], supported by transformer resonances, that may increase the equivalent impedance, increasing the amount of emissions coupled to and propagating along the traction supply system. The measurement and identification of the transformer frequency response is thus very important to predict the behavior of the locomotive in terms of harmonic conducted emissions and to optimize the filter design.

2.2.3.3 DC Supply Converters (Chopper, Step-Up/Down Front End Converter)

The natural propulsion drive for vehicles supplied at dc was a dc motor connected directly to the full line voltage, or in a series combination to reduce the terminal voltage for each motor. It is known that the dc motor has a large starting torque and its regulation is quite straightforward, being the result of the product of the field and armature windings currents. The regulation at origin however, without the use of static converters, has always been a tradeoff with power dissipation, and to this aim series/parallel combinations are preferable to pure rheostatic regulation of the armature winding; since the change of the series and parallel combinations of the available motor units was to be done while tractioning, a certain amount of series rheostatic burden was always present to limit the flowing current in the instantaneous short circuit during commutation and reconfiguration. Field winding regulation has lower dynamic performances and in some cases requires a rearrangement of the field winding itself for partialization of winding turns. However, in several old models the field weakening option was available for "economic" operation in combination with the armature winding regulation; moreover, the reversal of the movement direction was always made by changing the polarity applied to the field winding using few electromechanical contactors and for a lower dc voltage. All these motor configurations were characterized by almost negligible emissions, the only relevant, but anyway low, components produced by the commutator arcing being located at a frequency proportional to the rotational speed.

Chopper as Dc Motor Converter. DC motors had great benefit from the appearance of dc/dc converters, commonly named "choppers". For the nominal values of the motor terminal voltage and the line voltage, the mostly widespread converter was the step-down converter, with a GTO in series with the line source, chopping the line current absorbed by the motor load, and thus reducing after the output filter across motor terminals the mean value of the supplied dc voltage. With this class of regulating choppers the problem arose of the input line harmonics, appearing as a function of the load current I_2, of the switching frequency f_{sw} and of the duty cycle s. The typical input current waveform is a square wave with an on-interval T_{on} that is given by $T_{on} = sT$, where $T = 1/f_{sw}$ is the switching period; the harmonics are thus both even and odd, with a proportion that depends on the value of the duty cycle, but the amplitude in general decrease with frequency approximately linearly with the harmonic order.

For practical reasons regarding the minimum turn-on and turn-off time of the series GTO (or thyristor, with an auxiliary circuit to switch off and additional constraints on minimum times) the duty cycle was limited between two values s_{min} and s_{max}, and cannot reach the extreme values of the interval 0-1.

While driving a dc motor in all its operating conditions, beginning with startup, the duty cycle and the switching frequency must meet the requirement of voltage regulation on one side and filter sizing on the other side. This led to a variety of solutions, including the partialization at startup of a unique switching frequency (such as with the French BB7200 operating at 300 Hz, the Italian E632 and E633 operating at 390 Hz and the Polish EN57T operating at interlaced 600 Hz [683]), different values of the switching frequency for different operating conditions and traveling speeds (the so-called "gear change", adopted for example in the Italian E656 model, evolution of the E632/E633), and the inclusion/exclusion of a series resistor.

The use of the so-called "line filter" at the locomotive input was in any case compulsory to reduce the harmonic emissions to an acceptable level; the line filter has been normally made of a single LC stage (occasionally, in some loco models two smaller cascaded stages) with a very low self-resonance frequency and a larger capacitor bank to reduce the bulky and dissipative input reactor. For the analysis of the chopper operation the used assumption is that of continuous conduction, thus assuming a value of L above a minimum threshold. The line current ripple is of course maximum for the duty cycle located amid the regulation interval at $s = 0.5$; the portion of chopper current affecting the traction line I_l is the one in series with the inductor L and may be computed with reference to the current divider represented by the parallel connection of the inductive reactance $jX_L = j\omega L$ and capacitive reactance $jX_C = 1/j\omega C$, both supplied by the unfiltered current spectrum I_1 distorted by the chopper operation. For an accurate calculation the line impedance Z_l must be included, and for the considered frequency range it is prevalently inductive, so that $L_t = L + L_l$.

$$I_l = \frac{1/j\omega L_t}{1/j\omega L_t + j\omega C} I_1 = \frac{1}{1 - \omega^2 L_t C} I_1 \tag{2.16}$$

The value of the on-board inductor cannot be too small since the line inductance L_l is subject to large variations proportional to the distance of the loco position from the nearest substations and this would cause an unacceptable variation of the filter attenuation and filter resonance frequency. Usual values of L and C range from a few to a tenth of mH and mF, with a self resonance frequency around ten to forty Hz. In [420], Chap. 2, several examples of line filters are reported from Belgium ($L = 18\,\text{mH}$ $C = 1\,\text{mF}$, $L = 26\,\text{mH}$ $C = 1.135\,\text{mF}$, $L = 9.6\,\text{mH}$ $C = 2.8\,\text{mF}$, $L = 12.5\,\text{mH}$ $C = 2.95\,\text{mF}$), Norway ($L = 11\,\text{mH}$ $C = 2.4\,\text{mF}$) and France ($L = 2\,\text{mH}$ $C = 2\,\text{mF}$,$L = 2\,\text{mH}$ $C = 8.23\,\text{mF}$). The self-resonance frequency of the first two groups ranges from 26 to 37 Hz, while for French filters it is 79 and 39 Hz respectively. In the first two countries an additional trap filter is required and integrated into the line filter to block the 50 Hz component of the line current spectrum to meet the limit related to the interference to 50 Hz track circuits; the trap filter is often of the parallel resonant type in series with the inductor, but also a series resonant shunt filter has been used. In modern locomotives and EMUs (Electrical Multiple Units) it is often required that a 50 Hz trap filter is included in each static converter, even auxiliary ones, thus increasing its weight and size, but distributing the required stopping of the 50 Hz component. It is remembered that dc vehicles are subject to a specification of minimum input impedance as per UIC leaflet 550-3 [666].

Chopper as a Front-End Converter. For modern dc locomotives, where the ac induction traction motors are inverter driven, the common DC link is supplied by one or more choppers often PWM controlled. If two or four choppers are stacked and the PWM modulation pattern is interlaced by a proper phase shift, then the ripple is reduced on both sides (the dc input side and the dc link side) and the first relevant harmonic is moved to a multiple of the fundamental switching frequency.

In the case of the more recent E404P by Ansaldo [486] the four interlaced GTO choppers are operated at 260 Hz switching frequency (for an equivalent overall switching frequency of 1040 Hz) and supply the 2.4 kV dc link. Details of the chopper model and some simulated waveforms can be found in [486].

Induction Motor and Driving Inverter. The 3-ph. ac induction motor proved to be more robust, cheaper and lighter than the dc motor, and the mandatory driving converter for speed and torque control outperforms the dc chopper and all the auxiliary circuits. The induction motor has been used in many different configurations, either directly connected to the line voltage or supplied by a front-end chopper; this means that in the latter case the chopper emissions are overlapped to inverter emissions, thus complicating the emissions spectrum. The driving inverter may be of the voltage source or the current source type depending on the type of semiconductor device: GTOs or IGBTs in the first case and thyristors in the second case; the type of switching device has impacted also on the type of modulation and on the switching frequency, ranging from a low frequency square wave modulation up to a Pulse Width Modulation strategy with a switching frequency of one or few kHz; the overall switching frequency has been increased also by adopting interlaced modulations applied to modular architectures, where often the single converters divide the full dc link voltage [554].

Different solutions have been implemented by the manufacturers for the optimization of the VSI inverter driving the traction motors, in terms of control techniques, sizing of most critical components (such as filters and dc link capacitors), stabilization by reduction of tractive effort demand and wheel slip control, traction motor architecture. A tradeoff is necessary between the maximum line voltage (the nominal value is 3000 Vdc, but the non-permanent maximum value may be as high as 3900 Vdc, below 1 s for EN 50163 and 2 s for UIC 550 [664], and the prescribed maximum voltage starts to increase to values that may be around 4 and 5 kV, depending on standards and duration) and some characteristics of the static converters and their components (maximum continuous and transient voltage with respect to the same rating for current and switching frequency). For this reason it is quite common to see two series connected inverters on the locomotive dc link sharing the voltage, with the dc link supplied by a single stage (or again a double stage) front-end chopper. Alternatively, a three-level NPC configuration may be adopted approximately doubling the maximum dc link voltage for a given power switch voltage rating, thus coping with widespread 3.3 kV IGBT modules in place of higher rating older GTOs.

Bombardier (at that time AdTranz) pushed the modularity even further, designing a modular traction motor (double-star induction motor) and claiming benefits from the availability point of view [251]. Of course the design is complex, since in normal operating conditions the two applied waveforms and the resulting magneto-motive forces (MMF) add in the motor air-gap, with a perfect balance and compensation, but when the failure of either of the two motor windings or connected converter occurs, symmetry and correct motor operation may be compromised: particular shrewdness is needed for a correct distribution of winding turns in the machine slots along the air-gap.

The inverter operates in three different modulation regions (Indirect Self Control, ISC, Direct Self Control, DSC[6], and Field Weakening, FW), that are characteristics of many inverter drives and of the almost whole Bombardier production of induction drives for the dc market. In the ISC mode the inverter operates such as a PWM inverter with a large enough carrier-to-modulating frequency ratio with a circular trajectory of the flux in the direct and quadrature space. When the inverter moves to DSC mode, the trajectory follows an hexagon, to maximize the rms of the fundamental applied to stator windings; yet, the corners of the hexagon are folded towards the center, in order to reduce flux ripple (and torque ripple) located at the 6th harmonic. In FW mode the inverter operates as a square wave inverter and the harmonic pattern on the dc side is dominated by 6th order components [506]. When DSC is operated with an asynchronous switching frequency additional components appear on the dc side as a by-product of the beating between various multiples of the switching frequency and stator frequency. The ISC characteristic harmonics are the even multiples of the switching frequency together with sidebands of the stator frequency, similarly to a PWM modulation [554].

Converter Emissions. In almost all locomotives the chopper (be it the main dc motor driving converter or the front-end converter) is configured with multiple n units run in interlaced mode, with a phase shift of $2\pi/n$ each; the resulting harmonic emissions are thus located at frequency values that are multiples of the equivalent switching frequency $n\,f_{sw}$, while at the other multiples of the basic switching frequency f_{sw} there are residual components due to imperfect compensation of the various chopper units. The usual configuration is with a chopper group of $n = 2$ and it is replicated for each motor bogie; of course, the compensation between two different chopper groups is less accurate because of slightly different wheel speed and possible slip.

2.2.3.4 Diesel Locomotives

There have been a progressive change-over to electrification and electric vehicles in the last fifty years, fostered by the commonly agreed savings in cutting

[6] The DSC (also known as Direct Self-Regulation) [242] can apply 9, 6 or 3 synchronized pulses for waveform half cycle, featuring lower losses, better torque characteristics and reducing (or eliminating) the motor choke.

operating costs: with a progressive mortgaging of initial costs of the electric installation, the two solutions yield the same operating costs after 6-10 years of service and the electric one gets to about one-third of the diesel one after 25-30 years. This notwithstanding, diesel finds application in marshalling yards, low traffic country-side lines, long lines with very low headway, such as in United States and Canada [298], half-developed lines in third world countries, for which the electric solution is still not practical and most of all is contrasted by the relevant initial capital investment: if speaking of a whole line and not of a single yard, electrification requires additional clearance in horizontal and vertical directions on existing structures and installations to find the place for poles and overhead system, the electric equipment and devices may be quite expensive depending on periods for the cost of metals, in particular copper, and a series of new issues must be adequately considered: immunization of existing signalling, grounding, stray currents, etc..

The traditional solution for a diesel-electric locomotive is a diode or thyristor rectifier fed by the on-board alternator and driving a dc motor; in this case the drive architecture and its control are of the same type of old dc locomotives. A modern scheme with a Voltage Source Inverter fed induction motor is more flexible, features higher performance and may be also competitive with diesel-hydraulic machines from the point of view of energy efficiency and overall costs. One aspect that was remarked in the past during the transition from dc to ac motors was the high adhesion capability of the latter, due to the fact that if an axle suddenly speeds up from poor adhesion, it will immediately reduce tractive effort since the ac motor slip will be reduced; however, this characteristic can be obtained equally well with a dc motor if separate excitation is utilized. The increase in the number of motors led sometimes to more economic solutions for the distribution of electric power: not only a common dc bus coming from the main rectifiers fed by the on-board generators, but also a common ac bus is possible at the inverters output, which the ac motors of a bogie or all the ac motors of the locomotive are connected to. This of course reduces the ability to control the tractive effort on the loco axles independently and requires very close wheel diameter tolerances between axles; increasing the motor nominal slip to counteract this phenomenon normally reduces the power efficiency as well [298]. Examples of this kind of architecture are the F 202 loco for Amtrak (USA), ME 07 for TCDD (Turkey), ME 10 for Steel Works (Iran) and DE 1003 for Ruhr Coal Ltd. (Germany).

A peculiar aspect of diesel vehicles is that they represent an isolate electric network from the on-board generator to the traction motors and other loads: field control of the synchronous alternator makes it possible to adjust the dc link voltage supplied to the traction inverters more conveniently than in a catenary or third-rail system, where the supply comes from far substations. This important degree of freedom can be used to control the flux in the traction motors as a function of many variables (first of all speed and load), in combination with inverter control, providing opportunities for enhanced operating characteristics and efficiency [458].

It is underlined, however, that modern diesel locomotives must be equipped with auxiliaries and converters to supply the trailed coaches, in particular if they are passenger coaches. For example, purely diesel locos have no limitations in Germany, but in the presence of the heating line (for passenger trains), operated at 22 Hz to avoid interference with its harmonics, a limit mask of 2 A is anyway applied with respect to the interference to track circuits, as it will be considered in sec. 3.1.1.1.

2.2.4 Auxiliary Converters and Other On-Board Loads

Many loads and auxiliaries are supplied by separate converters that are supplied directly by the line voltage for dc vehicles and by one or more dedicated secondary windings for ac vehicles. Their rated power depends on the type of vehicle (a locomotive, that supplies also the trailed coaches, or an independent electric unit, used for a variety of applications, such as high speed trains, metro train consists, etc.) and the supplied loads and auxiliaries. Heating, air conditioning, ventilation, lighting are the normal loads required for passengers and they are normally supplied by a common train bus at LV level and industrial frequency, namely 3-ph. 380 V and 50 or 60 Hz; in some cases the train bus is at a higher dc voltage (even the line voltage for applications at 750 V) and the 3-phase power is derived by inverters locally to each coach. Auxiliary services are also needed for the locomotive (or electric unit) itself, for heating and ventilation of traction converters, for supply of electrical and electronic equipment for control and signalling functions, and again they are usually supplied by static converters for various kinds of supply voltages, either dc (for example 48 and 110 V) and ac (for example 3-phase 380 V).

2.2.5 Train Consist

In this section there are reviewed EMC relevant phenomena pertaining to a whole train consist or to more than one vehicle or coach. The reported information and considerations are of course a synthesis and schematization of a vast range of cases, that are often solved on a case by case basis, being characterized by specific elements that cannot be easily generalized. For this reason in the following the conclusions have general, but generic, value and need to be supported by practical considerations and specific information to be used.

2.2.5.1 Current Circulation through Axles and between Coaches

Until now the sources emissions have been considered punctual, so represented by an equivalent generator with a precise location in space and connection to the rest of the network. This in general doesn't correspond to the real situation and there are some cases where the real connection and current circulation represents a specific issue, as it is briefly seen in the following paragraphs.

The vehicle chassis and enclosure are far from being an ideal equipotential surface and they experience the circulation of several return and equipotentialization current components that follow parallel paths through the various metallic elements, bars, joints, to the point of lowest potential closer to the running rails. The leaking of current components to the enclosure occurs whenever a common mode current appears, and it's the effect of the combination of:

- poor isolation of the source from the vehicle frame/enclosure, because of a very large mating surface, various mistakes such as touching screws, damaged insulation, etc.;
- use of the wrong grounding scheme, such as multi-point grounding in place of single point grounding, as it happens when a grounding connection is supplied through a direct connection of the equipment enclosure and vehicle frame and also from the shield/screen of the entering supply cable or any control cable or by the characteristics of the load itself (let's consider for example a motor that, even if without explicit ground paths of the phase terminals, is terminated to ground through its windings stray capacitance);
- unbalancing of supply currents entering/exiting the equipment from the purposed input terminals, caused for instance by improper routing of the same cables, differences in the loading impedance and possible resonances due to the internal circuits, etc.

Such a current distribution not only represents a threat to vehicle equipment and thus a problem of internal EMC, but also influences the distribution of the return current through the vehicle wheels and similarly through the wheels of the whole train consist. It is easy to see that the equipment considered above may be located in one of the trailed coaches, or one of the cables carrying current on its shield/screen may be running to or from one of the trailed coaches. Moreover, since the resulting current path depends on the mutual coupling between various elements belonging to the vast ground circuit and the emissions from the source are influenced by impedance symmetry and resonances, the distribution of the return current through the wheels is expected to change with frequency.

Besides these stray effects, wheels are subject to the most relevant circulation of the traction return current and its harmonics produced by the traction converters and supplied traction motors. On-board traction drives are distributed over the various vehicle or locomotive bogies following the popular schemes of motored/trailed axles; this means that all the wheels are interested by a portion of the traction return current, but, again, because of unwanted coupling (even by on-board filters inductors) and current leakage through converter and motor enclosures, current may distribute unevenly between the wheels.

Last, auxiliary loads on trailed coaches or on the units or a fixed composition train divert a significant percentage of the absorbed current for ventilation, air conditioning and heating. Even if modern schemes use a three-phase train bus for the isolated supply of such distributed loads, so that the circuit recloses at the source of supply, there are many examples of static converters located on coaches and supplied directly with a dc voltage bus along the train in the 1-3 kV range (depending on the traction supply scheme and chosen intermediate voltage level) or directly at the line voltage at 750 V for many metro applications. In this case the return current is subject to the same considerations done for the enclosure of a single vehicle. In old applications and for many trains trailed by diesel locomotives the return current is purposely driven on the running rails instead of using a dedicated negative conductor, producing a 100% leaking current. Moreover, since for diesel locomotives the electric supply is represented by an on-board generator driven by the prime motors and thus subject to widely variable rotational speed conditions, the dc voltage obtained after rectification may have very poor characteristics and large distortion at standstill and low speeds, even if the power absorbed by loads on coaches is not reduced accordingly, because independent on the tractive conditions and speed profile (heating and air-conditioning).

The uneven distribution of the return current and the different geometries of the wheel circulating current at different frequencies represents a very elusive issue for the evaluation of the interference to signalling devices, and in particular track circuits while the train is entering the section: the shunting is partly in front of the track circuit unit (be it the transmitter or the receiver) within the "correct" track circuit section and several axles are still shunting the previous track circuit, while the current distribution between the various axles may be changing, producing several problems of intermittent detection. The same happens for inductive loops that may be seriously disturbed by the local wheel current circulation, not to mention the stray magnetic field produced by on-board filter inductors.

2.2.5.2 Summation of Harmonics

The problem of extending to multiple units the results related to a single unit has been already addressed by the standard EN 50121-3-1 [77, 78] specifically for the measured psophometric current (see sec. 3.2.2.1). There the attention is focused on how to extend the results within the same train consist, composed of multiple units (e.g. EMUs): the proposed relationships indicates a proportionality of the total current with n or square root of n, where n is the number of units; the former is suitable for synchronized units, so that the phases of the components at the same frequency from several units are almost equal and the sum is performed "in phase", while the latter is suitable for random phase distribution, so that the amplitudes, assuming uniform phase distribution, are summed by a root-mean-square approach.

The standard states that for dc systems, where the most significant term of the psophometric disturbance is produced by the dc ripple of substation rectifiers, the sum is done in phase; when the substation ripple is not prevalent, the root-mean-square method is indicated. For ac systems, where modern on-board converters are often synchronized by the supply voltage waveform, the sum is done in phase for four quadrant converters (4QC), while using the root-mean-square method if the converters are of different types and in particular of the phase controlled type. In all cases, when the converters operate in interlaced mode, the standard generically states that the total resulting psophometric current is lower than that of a single unit, because of the mutual compensation and cancellation of various components.

These rules need a more articulated treatment, since the starting point of the rationale reported in the standard was that the units belong to the same train consist and for modern ac converters synchronization is assumed. Yet, vehicle emissions are a combination of one or few front-end converters that collect the conducted emissions of traction inverters, auxiliaries and other loads; if it likely that different front-end converters on multiple units of the same train consist are synchronized by seeing the same ac line voltage, it is not for other converters fed separately by a local dc voltage supplied by the 4QC themselves, and for this reason some spectral components may result in a root-mean-square combination even if high level synchronization may be assumed. On the contrary, different trains in the same supply section might be different from the point of view of the conducted disturbance pattern and on-board sources characteristics; moreover, they are subject to a variable phase displacement of the supply voltage depending on the relative distance. Moreover, if the problem of interference to signaling circuits from multiple trains is considered, the subject is extended to the identification of summation rules for each individual spectrum component, rather than for the psophometric current alone. For industrial supply networks the problem was considered in [24], by accounting for the statistical distributions of current harmonics in the presence of several industrial loads; a similar analysis is done for traction supply systems by [682, 686], but the attention is focused on the characterization of the substation loading rather than on the distribution of the disturbance components along the traction line and possible interference to signalling systems. The problem is considered in [300] where, first, the various sources are classified for their degree of synchronization and frequency components distribution, and, second, the characterization of the traction line connecting the various sources is considered, in terms of attenuation and phase rotation.

Classification of Sources. In [300] the sources are classified as:

- *Unsynchronised Emission Sources with Fixed Frequency.* The interference currents falling in this category have a fixed frequency and either fixed or variable amplitude. A typical example is the interference pattern produced by fixed frequency chopper vehicles on DC supplied lines. Each vehicle has its own, autonomous chopper controller with an internal reference oscillator. The choppers of all vehicles are nominally operating at the same frequency, but they are not synchronised. So, emissions from different vehicles may have similar or even identical frequency, but random phase shift.
- *Synchronised Emission Sources with Fixed Frequency.* The interference currents of this category have the same frequency for all synchronised sources. The converters can be synchronised through a common controller, via dedicated synchronisation lines between different controllers, or via the common AC supply voltage. The phase shift between the disturbance signals from the different sources depends on the quality of the synchronisation and on the interlacing, i.e. the intentional phase shift introduced to eliminate some interference components. Examples of sources for this category of emissions are: i) multi-phase choppers on the same vehicle; ii) multiple units with synchronised and interlaced choppers; iii) synchronised and interlaced four-quadrant converters of the same traction vehicle; iv) four-quadrant converters of technically identical vehicles operating in the same AC feed section. An approximate estimation of the error for different methods of converter synchronisation is: i) converters controlled by the same controller: 100 ns typ.; ii) converters driven by different controllers, but synchronised by a dedicated synchronisation line: max. 0.2 deg. at 16.7 Hz $= 33.3\,\mu s$; iii) converters on different vehicles, synchronised through the supply frequency: max. 1 deg. at 16.7 Hz $=$ 0.166 ms.
- *Uncoupled Emission Sources with Variable Frequency.* The most common sources of variable frequency disturbance are motor inverters driving a three phase induction motor with variable speed. If the converters operate independently, they will normally produce emissions at different frequencies and will not add up at all. Examples of sources for this category of emissions are: i) traction inverters of two independent vehicles (not in the same train consist), ii) auxiliary converters driving induction motors at variable speed, e.g. cooling blowers, compressors.
- *Coupled Emission Sources with Variable Frequency.* Variable frequency inverters can be coupled even if they are not directly synchronised or electrically connected. The speeds of the different wheel-sets of the same vehicle and hence the inverter frequencies are almost identical. It is unlikely that the frequencies will be identical over a longer period of time, because already small differences in wheel diameter, load or adhesion conditions will cause differences in the motor supply frequency and hence, in the frequency components injected by the inverter back to the dc link [287]. Even if two inverters operate at the same output frequency, their interference currents

would not necessarily be in phase. Examples of sources for this category of emissions are: i) traction inverters of two bogies on the same vehicle; ii) traction inverters of two vehicles in multiple unit operation; iii) auxiliary inverters on the same vehicle, driving different induction motors, but with the same set value for the motor speed (e.g. oil pumps or blowers).

- *Modulation Products between Fixed and Variable Frequency Sources.* The disturbance components in the line current of a vehicle may be the modulation product between a fixed frequency chopper or four-quadrant converter, and a variable frequency load on the DC link side of the chopper or converter. The most significant effect will be the modulation side-bands on both sides of the characteristic harmonic components of the fixed frequency converter; these side-bands are of reduced amplitude but with variable frequency and they contribute to the pollution of that part of the frequency range between fixed harmonic components. The fixed frequency components act as carriers for the modulation side-bands. The modulation side-bands from two different vehicles would be at the same frequency and in phase, only if both the fixed frequency sources (i.e. choppers or four-quadrant converters) and the variable frequency sources (i.e. the load variation or the motor inverter frequency) are at the same frequency and are in phase. The synchronisation of only one of the sources (e.g. the fixed frequency choppers) is not enough to produce synchronous modulation side-bands that would add up linearly. Examples of sources for this category of emissions are: i) low frequency load variation caused by bad adhesion conditions produces side-bands on a chopper or a four-quadrant converter vehicle; ii) motor inverter harmonics of a four-quadrant converter vehicle: the motor inverter produces variable frequency components, the four-quadrant converters generate the fixed frequency pattern acting as carrier frequencies at odd multiples of the supply frequency.

Propagation through the Traction Line. In [300] an ac traction line was considered and attenuation and phase rotation curves are derived per unit length as a function of frequency, in order to give reference values for any reasoning on superposition and composition of harmonic patterns from distant sources.

As an example, the curve shows that a 10 kHz signal will be reduced by approximately 4% at a distance of 10 km from the vehicle and by 40% at a distance of 100 km.

The phase rotation is approximately 0.03 degrees per meter at a frequency of 10 kHz; this means that a 10 kHz signal in the return current will have a phase shift of 300 degrees at a distance of 10 km from the source. The value of the phase rotation per-unit-length is $6.04\,10^{-5}$ deg/m at 16.7 Hz and $1.57\,10^{-4}$ deg/m at 50 Hz.

Composition of Multiple Sources. When multiple sources and the composition of their conducted emissions at a given point along the track are to

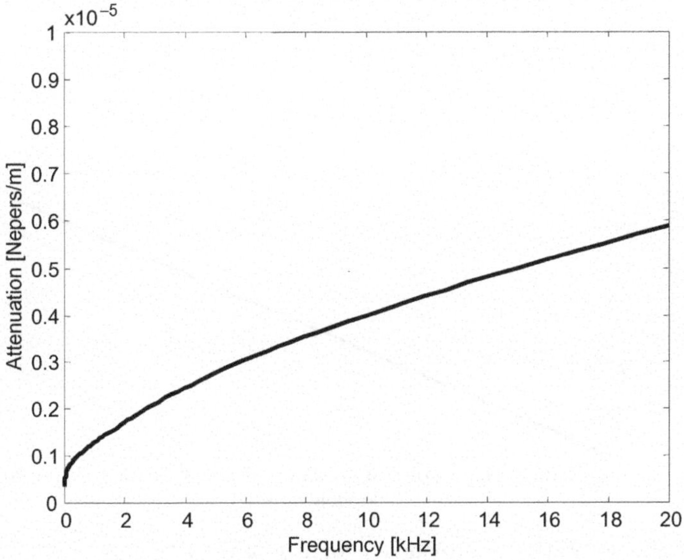

Fig. 2.29 Attenuation vs. frequency for 1 m of track

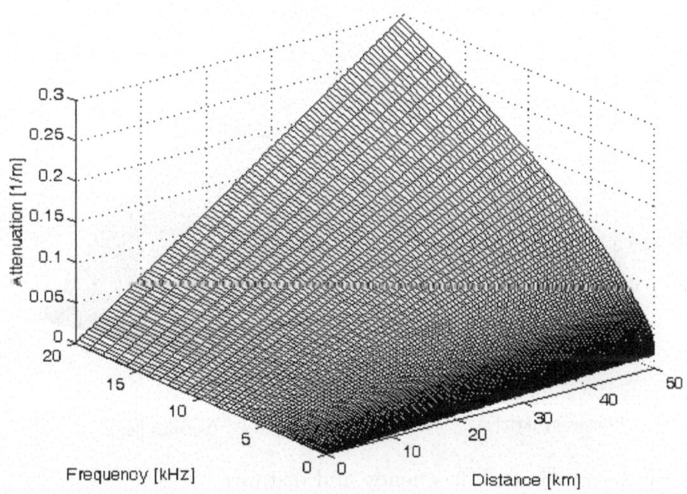

Fig. 2.30 Attenuation vs. frequency and distance

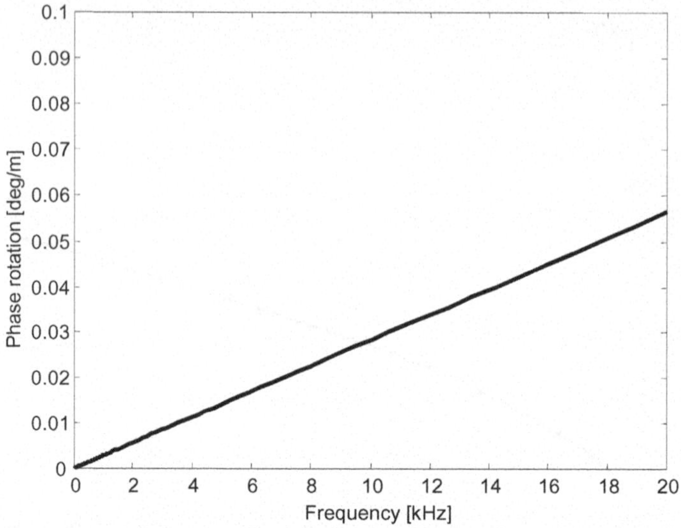

Fig. 2.31 Phase rotation vs. frequency for 1 m of track

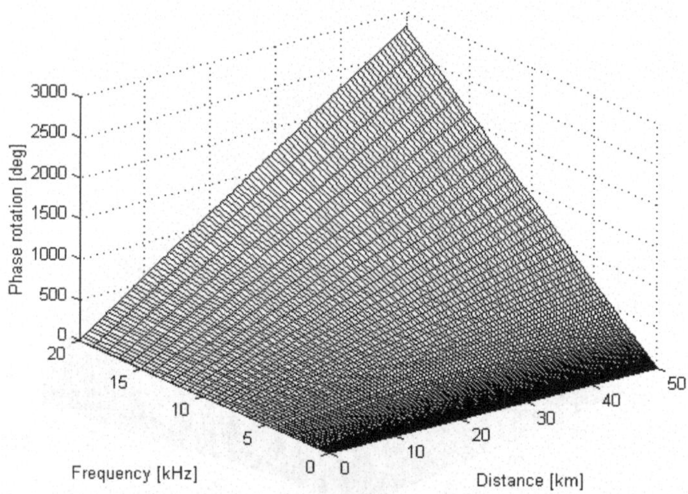

Fig. 2.32 Phase rotation vs. frequency and distance

be evaluated, the reciprocal synchronization, phase displacement and attenuation must be considered, in order to correctly determine the phase relationship and the vector composition. It is possible to distinguish different degrees of synchronization for sources belonging to the same vehicle or to different units in the same train consist or to different and distant trains; from the maximum time error between the reference time frames of the two sources, the equivalent phase distribution is estimated, considering it as a random variable with uniform distribution.

- *Interference from Non-Synchronised Sources.* Interference from non-synchronised sources may be generated at exactly the same frequency, but there is usually a random phase shift between the signals at the source. However, influencing events common to all sources may have a synchronisation effect, even if the sources are not normally synchronised. Typical influencing events are supply voltage fluctuations, mechanical oscillations on a train, modulation of the tractive effort set value for all traction units of a train. These events may only cause a quasi synchronous reaction of the vehicles at relatively low frequency, typically up to the bandwidth of the control system. For modern vehicles with static converters and microprocessor control this is approximately 300 Hz. The following tendencies may be outlined:
  - for frequency less than 300 Hz, linear summation of interference from different sources may be pessimistic, but no justification for other methods was found; further investigations are necessary for a final recommendation;
  - for non-synchronised sources, root mean square summation is indicated for interference above 300 Hz.
- *Interference from Synchronised Sources.* Interference from different synchronised sources (e.g. choppers with a synchronisation line, four-quadrant converter vehicles supplied by the same AC voltage line) has a fixed frequency pattern at multiples of the supply frequency and/or of the converter switching frequency. In between the odd and even harmonics, the interference is caused by modulation side-bands and by non-synchronised variable frequency sources, such as the motor inverter. The side-bands may be subject to the same influencing events and therefore the same addition rules apply as explained above. For the choice of the summation rule, the behaviour of the carrier, i.e. the fixed frequency harmonic, needs to be considered.
  - Effect of Synchronisation Error at the Source: The first effect to be considered is the synchronisation error at the source, i.e. at the converters of two different vehicles. For vehicles with a dedicated synchronisation line, this error is typically 16.6 μs (half of the maximum of 33.3 μs). This error causes a phase shift of 90° at 15.1 kHz. At 90° phase shift, the geometric summation (root mean square) is correct. So, it is justified to recommend root mean square summation for interference above 15 kHz even for converters with a dedicated synchronisation line. For vehicles synchronised only via the AC supply voltage, the typical synchronisation error is 83.3 μs (half of the maximum error of 166 μs). This error causes a phase shift of 90° for an interference signal of 3 kHz. For frequencies above 3 kHz, it is therefore justified to recommend root mean

square summation for interference from different vehicles synchronised only via the AC supply voltage. It is underlined that the values of the synchronisation error used in this section are typical values; slightly different values apply for different manufacturers with different vehicles and traction drives.

- Effect of Supply Voltage Phase Rotation: The second effect to be considered is the phase displacement of the supply voltage at the fundamental frequency. Fig. 9 shows the phase rotation of the 16.7 Hz supply voltage as a function of distance. As a consequence two identical vehicles will produce phase shifted interference because of the phase shift in the voltage supplying the two vehicles. Fig. 2.34 shows the frequency vs. distance curves for a phase rotation of 90° and 180° caused by the phase shift in the supply voltage. These curves do not take into account the synchronisation error at the source defined above. As an example, the line voltage at 16.7 Hz seen by two vehicles standing at a distance of 5.51 km from each other will have a phase shift of 0.33°. This phase shift at the fundamental frequency will cause a phase shift of 90° at the higher harmonics near 4509 Hz. It is therefore justified to use root mean square summation for frequencies above 4.5 kHz for vehicles standing at a distance of 5.51 km. For vehicles standing at a larger distance, root mean square summation is justified already for lower frequencies, e.g. 1002 Hz for vehicles at a distance of 24.78 km. The curve for 180° phase shift shows the distance for each frequency at which the two signals will have opposite phase and therefore compensate each other. It would be reasonable to consider vehicles in the same feed section only up to this distance for the consideration of interference at a given frequency. As an example, for interference at 3 kHz, only vehicles in the range of 16.5 km should be considered.
- Effect of Attenuation and Phase Rotation on Return Current: The third effect to be considered is the attenuation and phase rotation of higher frequency components in the return current. This means that the return currents from two identical and fully synchronised sources would not add up linearly if the sources (vehicles) are distant from each other. The superposition is calculated for two identical interference sources as a function of the distance between the two sources. For each frequency, the distance $x*$ at which the root mean square summation of the amplitudes of both sources is equal to the correctly calculated amplitude corresponds to the distance producing a phase rotation of 90° between the two signals, which is the curve for $\varphi_j = 90°$ in Fig. 2.33 and the black curve of Fig. 2.34. For phase rotation larger than 90° the root mean square summation method is justified, since it gives pessimistic results for the total current compared with the correct addition of the vectors.

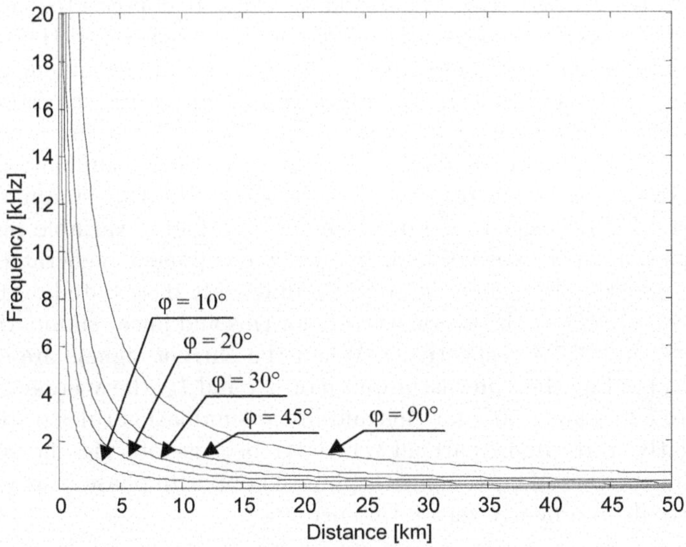

Fig. 2.33 Constant phase rotation loci as a function of frequency and distance

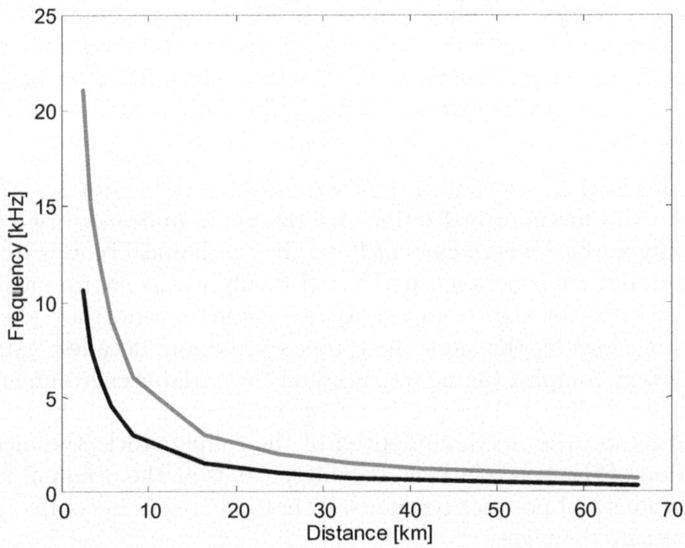

Fig. 2.34 Frequency vs. distance for 90° (black) and 180° (gray) phase rotation

- *Error related to the choice of summation rules.* It is presented the framework for the theoretical evaluation of the error related to the choice of either summation rules. Let i_n, $n = 1...N$ be a set of currents at the same frequency with amplitude I_n and phase displacement φ_n; the error is then evaluated for all termsi_n with equal moduli $I_n = I$. For small phase displacement (all φ_n are smaller than ±45°) the linear summation rule is almost exact (less than 6% error for $N = 2$). The rms summation rule is preferable for larger phase displacement (or highly variable phase displacement, as in the case of high frequency components generated by two moving sources). For example, with φ_n uniformly distributed over $[0; 2\pi]$, $I_n = 1.0$ and $N = 6$, the average errors for rms and linear summation rules are 100% and 357% respectively. When the current signals are only two ($N = 2$), holding the same assumption on φ_n and I_n, the average errors increase to 520% and 750% for rms and linear summation rules respectively. The relative error figures are all very large because of a significant contribution of the case of currents with (almost) opposite phase displacements, which result in a nearly zero total current.

2.2.6 Current Collection System

Current collection is performed by means of various types of sliding contacts, supported by pantographs, sliding shoes, trolleys, depending on the traction system and its voltage and current levels, on the method for supply distribution and on the speed of circulating rolling stock. The interaction between the fixed part (i.e. contact wire or third rail) and the sliding part (i.e. pantograph, trolley or sliding shoe) is a very complex phenomenon:

- from a mechanical viewpoint, wearing and materials duration are determined by the mechanical friction, by the spots and arcs originating at the mating surface where current flows, by mechanical bumps due to the dynamic interaction between parts, and finally to the environmental conditions (widely variable from country to country), and they are subject to investigations by the same operators since many decades; pantograph design is very complex taking into account the variable environmental conditions, the mechanical properties of the contact wire and the suspension mechanism and the maximum speed of the rolling stock; the mechanical interaction of the two relatively moving parts is at the origin of mechanical vibrations and possible resonances, that may result in components and infrastructure damages;
- from an electrical viewpoint, the phenomena extend from low frequency to high frequency both conducted and radiated: the electric arc behavior is much different in ac systems during polarity inversion at each supply voltage cycle, and also rectification may occur, with the appearance of a d component; the current intensity and pantograph speed determines the time duration and dynamic behavior of the arcs, that in turn determine

the high frequency emissions, conducted along the contact wire or radiated by the same contact wire and by the pantograph circuit.

2.2.6.1 Overview of the Mechanical and Electrical Characteristics

Several static and dynamic tests are prescribed for a new vehicle before the entry into service (see EN 50215 [111], sec. 9.13) aimed at verifying the current collection system. The tests must cover all the possible conditions and configurations in terms of the number of raised pantographs (or sliding shoes in contact), speed, train length and composition, distance between current collectors, direction of motion, and a check shall be made that "satisfactory current collection takes place, without damage, abnormal wear or vibration of either the collector or the supply in accordance with EN 50206-1 or EN 50206-2." During test runs particular attention shall be made to the electrical and mechanical operation of the current collector and associated electrical circuits over neutral sections and gaps in the third-rail. Also the aerodynamic effect is to be checked for adverse effects on raising and lowering of the pantograph and the current collection operation.

The dynamic interaction between the pantograph and the overhead line is a complex issue, that requires both theoretical calculations and simulations and experimental verification, as prescribed by the EN 50317 [122].

The maximum train speed (also called "line speed") is classified into categories by the EN 50367 [123]: AC1 and DC1 (conventional and connecting lines) up to 160 km/h; AC2 and DC2 (upgraded lines around 200 km/h) with speed between 160 and 220 km/h; AC3 (an added category for upgraded AC lines) and DC3 (DC operated high-speed lines) with speed between 220 and 250 km/h; AC4 (AC operated high-speed lines) with speed above 250 km/h. For each of these line categories the EN 50367 specifies minimum and maximum contact wire height and the maximum deviation of contact wire from the centre of the track because of lateral wind (it is remembered that the contact wire is normally staggered with a zig-zag fashion between successive suspension poles to distribute the wearing on the sliding contact). Additional and more refined characteristics are the height and skew of the pantograph head, the pantograph width and its shape.

The properties of the contact strip in terms of accepted materials, contact force and current collection capability are then closer to the following treatment on the electrical behavior; these properties are detailed in standards EN 50206-1 [108] and EN 50206-2 [109], regarding the materials, the construction and the characteristics of the pantograph, and EN 50317 [122], presenting the methods for the measurement of the said characteristics and the main metrological properties. The load presented to the contact wire by the whole train is subject to limitations in particular at standstill where the creation of a hot spot can break the contact wire (the current collection point is fixed so that the overheating is concentrated in the same point without any cooling by the air circulation): for AC1 to AC4 systems the current limitation is 80 A; for DC1 to DC3 systems the current limitation is 200 A for 3 kV and 300 A for 1500 V systems respectively.

The EN 50388 [124] specifies on the other side the current limitation of the European lines, conventional and high-speed, to which the train must comply during movement, that is reported in the Table 2 of the standard: for a DC 3 kV conventional line the maximum current ranges from 2.5 to 4 kA, and is increased to 5 and 6.8 kA for 1500 and 750 V lines; for AC conventional lines the limitation is always below 1 kA; for high speed lines both DC and AC the current limitation is raised to the respective maxima, with AC lines ranging up to 1500 A for both 16.7 and 50 Hz systems.

The limitations on the collected current amplitude are accompanied by specifications on the required contact force value and its deviation with respect to the assigned contact force - speed curve of recommended values; these values is a compromise between lowering the contact resistance at the collection point, the mechanical robustness of the pantograph and of the contact wire with its suspension system, and the possibility of creating adverse dynamic effects during train movement. These limitations are given in sec. 7.2 of the EN 50367 together with quite a long and articulated series of special national conditions in its Annex B.

In the following subsections the attention is focused on the peculiar behavior of the electric arc at the current collection point and its influence on the dynamic electromagnetic behavior of the ensemble of the rolling stock and traction line. Due to the high electrical load the contact surface features a high current density, even higher in dc traction systems that are operated at lower nominal voltages. As a result, current collection is accompanied by a bright visible flashing arc moving along the pantograph, especially when the train speed is high and draws heavy current. Weather largely influences pantograph arcing: a thin film of moisture is present between the contact surfaces and under fair weather conditions it acts as a lubricant, enhancing the smoothness of the sliding contact; on the contrary, during cold winters this layer becomes frozen, reduces the surface smoothness and produces also mechanical stress to the parts of the current collection system.

2.2.6.2 Electrical Behavior at Low Frequency

The low frequency characterization of the current collection system and the effect of the electric arc ignition and extinction has been reported extensively by the authors working at University of Uppsala and the ABB and Bombardier Research Centers, where a test setup for laboratory measurement of electric arc behavior was designed and implemented [544, 545, 546].

Electric Arc in Dc Systems. The test setup used to generate electric arc at a simulated contact wire - pantograph interface with variable nominal current, supply voltage and relative speed is described extensively in [545].

A rotating wheel simulates the sliding contact of the contact wire and the staggering movement is simulated with an electric motor moving alternatively in the two lateral directions. A large rotating generator supplies the ac supply voltage, that will be used directly applied to the test setup for the ac tests and that here is fed to a diode rectifier. The rated ac voltage can be as high as 6.5 kV with 10-100 A of supply current, while the d supply is limited to 800 V and 10-20 A (the authors do not explain this difference, but it is believed that this fact was due only to contingency, maybe because of a limited rating of the rectifier diodes or limitations in any dc current switch or relay). The DC circuit is characterized by a very large inductance (2.7, 4 H) that the authors justify "to sustain the arc for longer duration", but also probably to filter out the rectifier ripple. The mechanical structure is conceived to provide the required variation in the air gap between the contact wire and the wheel.

Preliminary tests of the breakdown voltage of the air gap were made with the following results: 3-4 kV/mm, lower with rotating wheel and with a difference between opposite polarities that increases with the air gap thickness.

The main reason to study this class of transients in dc systems is the possible unwanted tripping of protective breakers, observed in the following cases:

- the Stockholms Localtrafik (Sweden) line operated at 750 Vdc with the overhead line in iced conditions; two recommendations were given: adding a series inductor, which will cause faster breakdown of the dielectric ice layer during arcing, and stopping the regenerative breaking was recommended during the iced condition;
- the Appenzeller Bahnen (Switzerland) line operated at 1000 Vdc when operating on an upward slope on the mountain terrain, again during winter; the adopted countermeasures were the circulation early in the morning of old locomotives without sophisticated circuit breakers, and to use two pantographs, the first one to clean the catenary and the second one to collect the traction current.

The results reported by the authors are less complete and detailed than those for the ac case. The conclusion on the influence of supply voltage polarity is that when the pantograph is cathode (thus with a positive supply voltage, as it is in real systems) the arc root is forced to move quickly, following the zig-zag motion, and there are more transients and interruptions of the power supply; the phenomenon is worsen since the sliding contact is convex and at the extremes the air gap thickness increases, causing the arc extinction with higher probability. So, the conclusion is that for the arcing phenomenon the best solution is when the supply voltage is negative!

Electric Arc in Ac Systems. The same setup was used for tests with the ac supply voltage, provided that the diode rectifier and the feeding inductor are removed. The observed phenomena in the recorded waveforms are:

- zero-current intervals after a zero-crossing in the current waveform $I(t)$, marked as $\tau_{p \to n}$ and $\tau_{n \to p}$ for the intervals during a positive-to-negative and a negative-to-positive transition of the current respectively ($\tau_{n \to p} > \tau_{p \to n}$); transients in the electric arc voltage drop $U(t)$ occur during the said intervals and at the end of the time interval the voltage drop returns to nearly zero;
- changed slope after a zero-crossing of the current, where the change is higher in the positive half period compared to the negative one; the two time intervals are barely noticeable and no overshoots or transient changes are remarkable; the voltage drop $U(t)$ keeps at a higher value compared to the first case.

For a purely resistive circuit (and, in particular, a resistive load), regarding the dependence on v_{line}, for low values there are reduced overshoots and a symmetric behavior in both polarities of $U(t)$ and lower values of $\tau_{p \to n}$ and $\tau_{n \to p}$, with little differences between them; regarding the dependence on I_{rms}, for larger values there are reduced overshoots and a larger value of $U(t)$, especially in the negative polarity, and lower values of $\tau_{p \to n}$ and $\tau_{n \to p}$, but with some differences between them. If an inductive component is added to the load, the effect is similar to that observed for a large I_{rms} current. Going to the influence of the material used for the sliding contact of the pantograph, carbon, and also copper and aluminum, were investigated: carbon as a larger cathode surface area, a higher temperature and more efficiency to emit thermal electrons; copper and aluminum exhibited also larger $\tau_{p \to n}$ and higher overshoots in $U(t)$, so that carbon is still the preferred material from the purely electric arc behavior, besides the self-lubricating property and a conducting oxide.

DC Rectification in Ac Systems. In [544] the appearance of a dc component on ac systems during current collection, due to the peculiar behavior of the electric arc, was considered: DC components due to pantograph arcing can saturate the traction feeding and vehicle transformers, and may interfere with DC-fed track circuit. Because of the sliding motion between the two electrodes (located at the contact wire and at the pantograph), both in the lateral direction (slow, due to stagger) and in the longitudinal direction (fast, due to train movement), arc roots at both electrodes move along their respective surfaces at different speeds, having different electrothermal conditions; this leads to an asymmetry between both polarities of the voltage drop $U(t)$ and current $I(t)$ waveforms. Both the pantograph arcing, its polarity-dependent nature, the $U(t)$ and $I(t)$ waveforms, and arc root movements are influenced by several parameters like the speed with respect to the line v_{line}, the collected rms current I_{rms}, the supply voltage $U_0(t)$, and the inductive part of the load (the supply line voltage is assumed variable because of the interaction with the arc current and the supply impedance). The two variables $U(t)$ and $I(t)$ were recorded with a 100 kS/s sampling rate, and put in relationship with the physics of the electric arc phenomenon and the setup variables and external conditions:

- $I(t)$ dependence on arc reignition: two phenomena are recognized, dielectric reignition and thermal reignition:
  - dielectric reignition has a typical signature pattern of zero current intervals of duration $\tau_{p\to n}$ and $\tau_{n\to p}$, for positive-to-negative and negative-to-positive polarity change, after the current zero crossing; to reignite the arc, a minimum reignition voltage is required, which depends only on the contact gap parameters like gap length, contact material, etc., and independent of the circuit parameters;
  - as the nature of reignition becomes more and more thermal, the values of the time intervals and the corresponding line voltage values tend to decrease; however, the arc current shows a change in current slope and current peak value, as well as in the voltage drop waveform;

  the authors conclude that "the mean duration of zero current from negative to positive crossover (where copper conductor becomes cathode and pantograph becomes anode) is clearly larger than the reverse", that is $\tau_{p\to n} < \tau_{n\to p}$.$U(t)$ dependence on arc reignition: the voltage drop between the two electrodes depends mainly on arc length (cathode and anode voltage drops are normally less than 20 V), and hence, on the movement of the arc roots along the contact wire; even if $U(t)$ may show wide variations because of air-gap changes, the convex surface of the pantograph, its zigzag motion, surface irregularities, and the stochastic nature of the arc, $I(t)$ was observed not to change correspondingly; the voltage drops for $p \to n$ and $n \to p$ transitions are much different for dielectric reignition, getting closer as reignition becomes more and more thermal.
- DC current component due to asymmetry: since $\tau_{p\to n} < \tau_{n\to p}$, the net calculated dc current over one supply period will be negative, indicating a net current flow always in the direction from the contact wire to the pantograph; practical and worst case value may be estimated as follows: if a full wave rectification occurs in the worst case the dc component is 32% of the peak value of the ac supply current; observed values for heavy loaded freight trains are around or larger than 50 Adc, that is about 10 to 20% of the peak ac current.
- DC voltage drop component due to asymmetry: the arc voltage drops are different in the two polarities and the final net dc value depends also on the displacement factor of the load.

2.2.6.3 Electromagnetic Emissions at High Frequency

The electric arc at the sliding contact is responsible for high frequency intermittent modulation of the absorbed current, causing first, conducted emissions along the supply distribution system itself, and, second, overall radiated electromagnetic field emissions, directly from the arcs themselves and indirectly from the conductors of the current collection system and of the supply system.

There exists a series of electric arcs to transfer the current from the third rail or overhead contact wire to the sliding contact of the vehicle shoe or pantograph. There are always several arcs at any instant of time, each carrying ordinarily a few tens of amps, depending on the same supply current value; the formation and extinction of the electric arcs generates conducted and radiated radiofrequency noise. In general, from a phenomenological viewpoint the arc behaviour depends on several factors:

- arc length L, which is related to the inter-electrode gap length and the relative speed of the electrodes, that is the linear speed of the vehicle;
- total arc current I_{lf} at low frequency, which is defined by the traction current and how it is apportioned onto the active current collecting elements; it is preferred not to identify any relationship with higher frequency current components, more difficult to predict and influenced by the impedance of the supply line and of the rolling stock equipment;
- electrode construction and geometry: flat cathodes reduce arc stability; two dimensions arc stabilisation is achieved when the distribution of arc current on the cathode surface is symmetrical around the arc root, and this is obtained with a hollow cathode; the shape of the anode surface, on the contrary, has little effect on arc stability.

Radiofrequency emissions are related to the average arc lifetime (AAL), which is a direct measure of arc stability, defined by the lifetime of an individual arc as the time interval from arc initiation until arc self-extinction. Some previous measurements in the literature [473, 530] indicate that dc arcs are quite stable up to 150 km/h for 300 A arc current and 3 mm inter-electrode gap, that are operating conditions overestimated with respect to those of metro and light railway systems, with nominal speed values around 90-100 km/h, and an equal or larger current of several hundreds Amps, that produces a more stable arc.

A simplified model of the arc radiofrequency emissions is a short wire antenna (or Hertzian dipole) of length L, carrying a current I, with the length L defined as the inter-electrode gap length. The radiofrequency current intensity I can be at this point estimated to be nearly constant (equal to I_0) for a broad frequency range [DC, 1 MHz], as confirmed by direct comparison of the values of electric field at 3 and 10 m distance predicted by this model with laboratory measurements performed on free-burning arcs in air [473, 530]. Above $f_0 = 1\,\text{MHz}$ the current I must fall with a more than linear law of the kind $1/f^n$, with n approximately between 1.5 and 1.8 and normally about 1.6, to make the electric field strength E be in accordance with measurements and data presented in the literature. Hence the expression for current I is

$$I = \begin{cases} I_0 & f < f_0 \\ I_0 (f_0/f)^{1.6} & f > f_0 \end{cases} \tag{2.17}$$

If the far field assumption holds, the model of radiation is that of an elementary Hertzian dipole (see Chap. 7) and the electric field strength E can be expressed simply as

$$E = \frac{\beta I L}{4\pi\varepsilon_0 r\, c} \tag{2.18}$$

where

$\beta = 2\pi/\lambda$ is the phase constant;
ε_0 is the dielectric constant of free space;
r is the distance between the electric arc and the measuring point;
c is the speed of light in free space.

A sample electric field profile from a free burning arc is shown in Fig. 2.35.

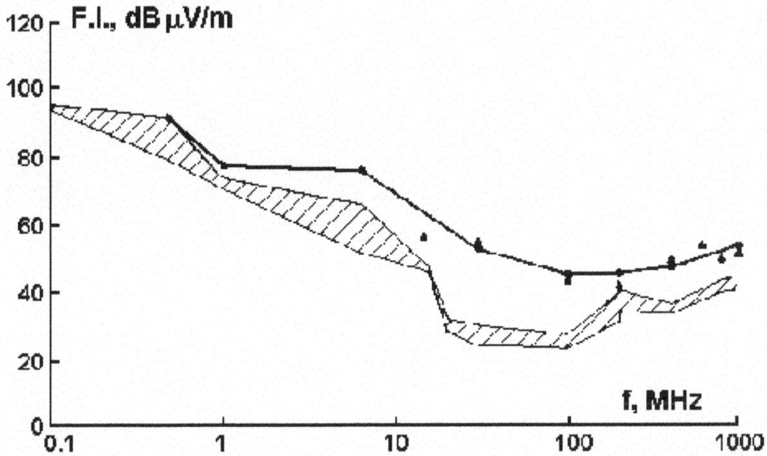

Fig. 2.35 Electric arc emissions from a free-burning arc (10 mm long at 100 A arc current) versus frequency (background noise shaded), measured at 3 m distance

There are no literature references for the characterisation of electric arc emissions above 1 GHz; it is believed that a satisfactory estimate is to prolong the value at 1000 MHz in the GHz range and this should represent a worst case estimation, since it is sensible that all the phenomena have an intrinsic low-pass behavior.

To conclude, in general, the total noise level at the receivers of various radio communication devices provided by the direct radiofrequency emissions of the electric arc depends at a first approximation on the gain and directivity of the victim receiver antenna, the bandwidth B of the receiver. To make a more precise characterisation of the radiation source, a dynamic model of arc behavior, a dynamic model of the mechanical system for current collection, and an accurate electromagnetic model of the environment, taking account of absorbing and reflecting surfaces and of scattering phenomena, must be adopted. It must be remembered also that the electric arc emissions are a highly intermittent phenomenon, that would result in a much lower E-field amplitude when measured with an average, rms or quasi-peak detector.

If the analysis is extended to the traction circuit in order to calculate the e.m. radiation, the arc itself is modeled as an equivalent generator and several models may be selected for the rest of the circuit: closed form expressions (such as those for a long wire antenna in sec. 7.3.3), a wire frame approach with a model based on the method of moments, or a finite element method model may be adopted including also the infrastructure and various dielectric materials (sleepers, ballast, platform, etc.).

Chapter 3
Signalling and Communication Systems

The signalling system is an essential part of a railway. The essential purpose of the railway signalling system is:

- to maintain a safe distance between following trains on the same track;
- to safeguard the movement of trains at junctions, and when crossing a path which could be taken by another;
- to regulate the passage of trains according to the service density and speed required

However, the signalling system also affects the sort of service that can be run on a section of railway. The signalling system thus has to perform in a safe manner, but perhaps more importantly, if something goes wrong it has to fail in a safe manner. When railways were first introduced as a means of transportation, there were no fixed signals and Policemen wandered around the stations changing the points at junctions and giving instructions to train drivers by coloured flags by day and oil lamps by night. There was no means of communication with adjacent stations, so there was no way of knowing whether it was safe to let a train set off for the next station. All that could be done was to give the Policeman an egg-timer and a train was allowed to leave a station a fixed time interval after the previous one. There were amazingly few accidents, probably because train speeds were not very high at that time.

Very soon fixed signals on post began to appear, which consisted of a board, with different railway companies using different shapes. These boards were provided so they could be turned end-on to the driver. If a driver could see the board he had to stop at it. If it is turned end-on, so that it was nearly invisible, then he could proceed. The lack of a positive proceed display was not very satisfactory so that the next development was to have a positive indication for both the stop and proceed indications, or "aspect" as they become known. The most famous version of this improved signal was the "Disc and Crossbar", where a disc being visible gave the proceed signal or a rectangular crossbar being visible the stop aspect. In 1841, the semaphore

A. Ogunsola and A. Mariscotti: Electromagn. Compat. in Railways, LNEE 168, pp. 95–177.
springerlink.com

was introduced and this soon became the standard mechanical signal (at least in the UK); soon after that, the signal box was invented with the objective of controlling points and signals from a central point at the station. Levers were connected to the signals by wires and to the points by rodding. This development meant that Policemen no longer had to wander about, as all the controls were concentrated in one place, thus inventing both the Signal Box and the Signalman. Connecting the levers together in such a way that they were physically locked, unless it was safe to pull them led to the development of interlocking - a major element of signalling - by the 1880's interlocking became a legal requirements for all stations in the UK.

About the 1880's, simple electrical communications between stations had become possible, these were implemented by bells using a Morse code like system. This required an electrical connection along the trackside - the so called telegraph wire and poles. This in turn led to the development of the block system, which was a system of communication and needle indicators between stations which when used under a very strict procedure, ensured that only one train can be on the track between one signal box and the next at any one time. This section of block was known as the "the block" and the system known as the "Absolute Block". With the implementation of the absolute block system and interlocking, a high level of safety was achieved, however accidents still happened because signalmen forgot where trains mights be standing for long periods; this, in the UK, led to the introduction of the infamous Rule 55 which required a guard or fireman to walk to the signal box to formally remind the signaller of the presence of train, if a train was stopped at a signal for more than three minutes. Another signalling invention that followed was the development of track circuits - electrical circuits in the rails that could detect, in a safe manner, that a section of track was not occupied by any rail vehicle. Development of signalling systems has continued over the years, with Communication Based Train Control system replacing track circuit train control. In the preceding sections, we provide a general overview of signalling systems in so far as they contribute the description of the railway electromagnetic environment.

Different examples indicate that electro-mechanical signal-interlocking has about 130 years (St. Paul station in London, UK, 1876; Victoria terminal, UK, 1883; several places in the United States, between 1876 and 1889; color light signals, USA, 1904), while purely mechanical interlocking is even older and the first reported example dates back to 1843 (Bricklayers Arms Junction in England, by C.H. Georgy).

3.1 General Functions: ATP, ATC and ATO

The three acronyms stand for Automatic Train Protection, Automatic Train Control and Automatic Train Operation respectively. A simple definition is not an easy task if all the different functions and relationships are to be included. The old definitions given at the beginning of '70s by UIC of these terms were: the ATO is the system covering the acceleration, running and braking of trains automatically; the ATC effects an emergency brake operation if the permissible speed is exceeded; no ATP definition was given. Later on in the early '90s within the ETCS project, UIC gave the following definitions, which represent an update and improvement of the previous ones: ATP is the continuous or discontinuous speed monitoring with or without cab signalling; the ATC is made of the ATP plus continuous display in the cab of variables aiding the driver to drive as closely as possible to the monitored speed limit and/or supervision curve when braking. It is evident that while the first set of definitions could be inadequate for modern systems, the second set of definitions is probably too closely related to the ETCS project and put too much emphasis on the cab signalling, that is the train-to-wayside communication. In general ATP is that system and that set of techniques that ensure that the train is operating within its speed limit, fulfilling requisites of safe distance, safe clearance and position; concerning the elements on-board, the ATP is also responsible for the safe opening of doors. The ATC performs its functions at a higher level, supporting driver's actions, keeping speed and distance within the prescribed limits, that may be communicated by trackside infrastructure. The ATO is a complex system, depending on the implemented features: traveling comfort, energy efficiency and optimization of power consumption, timetable optimization, etc.

Cab signalling may be defined at various levels of performance and complexity, from simple signal repetition (implemented with the help of coded messages picked up by on-board detectors to show to the driver the status of track signals in the cab), possibly integrated by a supervision of driver's behaviour and vigilance and his/her ability to react to warning signals, to a more complex information exchange, including speed limits, track status, track gradient, etc., allowing the calculation of the theoretical train speed and safe braking distance. Going further the same parameters may be used to optimize the train travelling times, energy efficiency and line traffic capacity, thus supporting ATC and ATO functions.

3.1.1 Train Detection Systems

It is important to detect the presence of trains for several reasons: maintaining trains headway, locking points (also called "switches") and selecting a route, announcing the arrival of trains at stations, coordinating the information appearing on displays and given by loudspeakers, ensuring the use of free tracks, thus avoiding situations of conflict and those that might be potentially unsafe. Some of them require that the operations are fail-safe at all times, and signalling devices dedicated to such operations are often referred to as "vital". A more detailed definition of train detection leads to a classification of the devices depending on the required function: detecting the presence of a train on a specified track or detecting the passing by of a train at a specified point; track circuits and loops implement the first function and axle counters and treadles the second one.

3.1.1.1 Track Circuits

Track circuits were invented more than one century ago in the United States and then used through the world. The track circuit is the basis of modern signalling; its first purpose is to prove that a section of track is clear of all vehicles, while its second purpose is to detect the presence of a train and to lock the route ahead of it and so ensure its safe transit. The track circuit also provides the indications on the signalling control panel, which advises the operator of the whereabouts of all vehicles on a given section of the line and enables the train describers to keep in step with train movements.

The characteristics common to all types of track circuits are: the detection principle is that of a drop in the received voltage due to the shunting effect of the axle of the rolling stock entering the protected track section; the general architecture is that of a signal generator (the transmitter, TX), an interposed circuit that includes the running rails and a receiver (RX). Track circuits operation and performance is analysed under several infrastructure conditions, such as: current leakage to earth due to distributed rail-to-earth conductance for dry and wet conditions, variable grounding conditions again due to dry and wet conditions, accidental disconnection of bonding connections, and even rail break. The assessment of the safe operation of a track circuit is a complex process that under the normal and assumed exceptional infrastructure conditions applies a series of steady and transient disturbances both in simulated and real conditions for various rail-wheel contact conditions: rolling stock weight per axle, regularity of the wheel surface and rail-wheel mechanical mating, rusty surface on the rail web. In summary, the two main parameters representing the infrastructure and testing normal conditions for a first benchmarking and parametrization of the TC are the ballast conductance (or resistance) and the shunt resistance (simulating axle resistance).

In the track circuit calculations and the subsequent analysis of the results, it is essential that not only the main signal path be taken into account, but also possible shunting paths and the mutual coupling of the circuits, including those both on the same and on adjacent tracks. The latter is due to the asymmetry in the rail line or cables impedance, caused by external connections to the rails (such as grounding straps, inter-track jumpers, etc.), accidental shorting to adjacent track rails, damage to joint isolation or absence of rail joints, cable and rail bonds breaks, etc.

There are many types of track circuits and they may be classified with respect to various characteristics: the signal path from the transmitter to the receiver, distinguishing between end fed and center fed; the use of one or both running rails as part of the interposed circuit; the type of used signal (steady or transient) and the signal frequency (dc, low frequency, audio frequency);

modulation or coding applied to the transmitted signal. In the following, we describe the main track circuit types, cross-referencing other types and their properties, where required.

Single-Rail Track Circuits. Single-rail track circuits requires an additional return path to close the circuit between TX and RX, but it leaves the other running rail free for use for traction return current without requiring special coupling circuits. These types of track circuit were used quite often in the past, due to their simplicity and because the rolling stock, at the time, did not represent a significant threat in terms of distortion of the traction return current and conducted disturbance. Single rail track circuits are not floating and half of the circuit is in common with the traction current return circuit, suffering significant conductive coupling and leakage to earth. Manufacturers often specify for the installation of their products a minimum grounding resistance to limit the phenomenon of leakage to earth through the normal grounding of the return rail to suspension poles and earthed equipotential conductors.

There exist single-rail TCs both at power frequency and audio frequency, even if the former are the most common and there exist a few examples of the latter (UM 71 and FTGS 46). At audiofrequency the coupling to the running rails is accomplished also by means of S-bonds and O-bonds, the directivity and efficiency increased by the presence of the insulating rail joints on the signal rail. Single rail TCs are still in use on old low traffic lines and at stations.

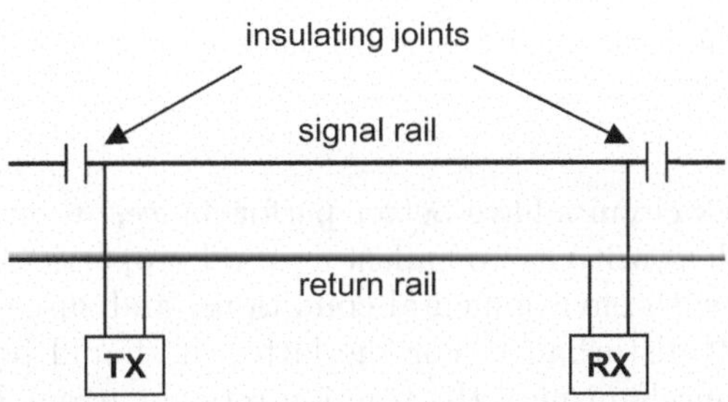

Fig. 3.1 Single rail track circuits

Double-Rail Track Circuits. Double-rail track circuits are the most modern and widespread track circuits circuits, operating on a wide frequency range, from power frequency to audiofrequency, and also using impulsive waveforms, for the so-called "high voltage" or "high voltage impulse" track circuits. They represent a broad category of balanced or symmetric track circuits, using various solutions for isolation of adjacent track circuit sections (mechanical insulating rail joints, tuned resonating circuits for the so-called "electrical insulating rail joints", composition of S-bonds and tuned circuits) and to couple transmitter and receiver to the track (direct connection to the running rails, transformer coupling, an additional winding of the impedance bonds, S-bonds and O-bonds either directly or inductively connected to the transmitter and receiver).

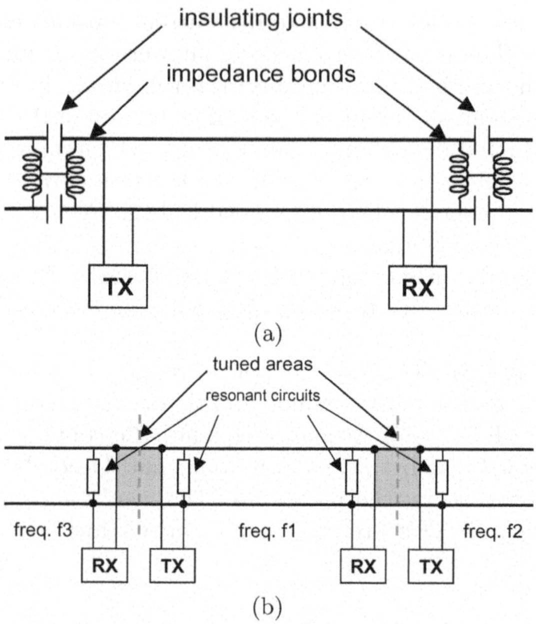

Fig. 3.2 Double rail track circuit, using (a) mechanical insulating joints, (b) electrical insulating joints

Double rail track circuits have better performances in terms of maximum track circuit length, sensitivity to ballast and soil properties and with respect to their variation with environmental conditions, and susceptibility to traction return current disturbance. For the latter, it should be noted that the primary coupling mechanism of the traction return current I_t onto the track circuit may be described as a common mode (c.m.) to differential mode (d.m.) transformation: the traction current is a common mode current with respect to the running rails representing the analysed circuit, since it flows in the same direction, reclosing to the substation. The reason for such a transformation may be found in the normal asymmetry of a double track line, where the external running rail is further from the other track than the internal one, and has a different location with respect to the other conductors of the return circuit, from which a different geometrical inductance – the different inductive coupling with respect to the opposite track – is also a source of c.m. to d.m. transformation for the return current in the opposite track, whilst a d.m. to d.m. transformation normally exists as the natural inductive coupling of two rectangular turns lying on the same horizontal plane; then, differences in the longitudinal resistance due to defects (loose connections, oxidation, etc.) in the joints may add a small percentage of unbalance relevant at the lowest frequencies; at higher frequencies the transversal capacitive coupling between adjacent tracks and in particular the internal rails produces a further unbalance and a signal transfer between adjacent tracks, that also depends on the terminal impedances.[1]

3.1.1.2 Pre-shunt and Islanding Tracks

When using insulating rail joints each track section pertaining to a track circuit (also called a "block") is defined with high accuracy and train axles standing within that track circuit section are not sensed by the adjacent sections. This feature allows a precise alignment of the block section to route and track elements (switches, signals, etc.), and for this reason the track circuit blocks protecting switches and stop signals are almost always equipped with insulating rail joints.

If a jointless track circuit is considered, where the electrical separation between track circuit sections is obtained by means of resonant circuits, there is always some deal of approximation in defining the position of the train axles with respect to the status change of the TCs on the track, because depending on the TC and track characteristics there is a "gray zone" outside the block section, where axles can however shunt to occupy the block section. We can view this zone, called "pre-shunt zone", as a track stub prolonging from the track circuit transmitter or receiver location, in electrical parallel to the track circuit section and tuning units. When the first or the last train axle is within this zone and shunts the two running rails, there is a short circuited track stub in parallel, that lowers the total equivalent impedance and thus reduces the voltage drop at the receiver, possibly below the threshold that turns the block section to occupied. The zone length depends upon the axle shunting resistance and the electric contact with the rails, on the ballast resistance (that is the rail-to-earth conductance), on the track circuit length and the resulting track impedance (also in case compensating capacitors are used), on the impedance of transmitter and receiver units and of the tuning circuits, to cite the main factors. The impedance of the stub is mainly inductive (given by the loop impedance of the two running rails) with minor or significant capacitive compensation in case only the rail-to-rail capacitance is accounted for, or compensating capacitors are added to the track. Since for long track circuits, the length of the pre-shunt zone may be as long as a hundred metres,

[1] The terms of near-end crosstalk (NEXT) and far-end crosstalk (FEXT) for two coupled differential circuits depend on the combination of the characteristic impedance of the two circuits, of the mutual and self terms of the inductive and capacitive coupling and of the terminal impedance values, so that in brief small terminal values enhance inductive coupling terms and large terminal values enhance capacitive coupling terms [599].

it includes at least one compensating capacitor (if any), and this results in a slightly resonating behaviour modulating the intensity of the pre-shunt while the axle is approaching or leaving, in a way different from that of a purely resistive-inductive circuit, that would be monotone (the shorter the stub, the higher the shunting effect). Lastly, even if obvious, for such a pretty short stub length the axle resistance can influence the overall stub impedance: axle resistance may be highly variable since real resistances may be as low as a tenth of mΩ, but any safety demonstration must be carried on with the limit shunt value, that depends on the administration and local procedures and standards may be as high as 0.5 Ω; the inductive reactance of the stub section over the lower audiofrequency interval ranges between 1 and 4.5 jΩ approximately.

C.m. to d.m. and d.m. to d.m. Transformations in a Track or between Adjacent Tracks. Two tracks are considered (see Fig. 1): the odd track (left) is the inducing track, where the source current is circulating. Two rail currents may be defined, namely $I_{o,r1}$ and $I_{o,r2}$ and the c.m. and d.m. components may be defined as $I_{o,cm} = (I_{o,r1} + I_{o,r2})/2)$ and $I_{o,dm} = (I_{o,r1} - I_{o,r2})/2)$ respectively. The rail-to-rail distances may be easily expressed as $r_{11} = s + d$, $r_{22} = s + d$, $r_{12} = 2s + d$, $r_{21} = d$. Now we proceed with the calculation of the rail-to-rail inductive coupling term, taking one rail at a time of the odd track with respect to either rail of the other (even) track. The single source rail x ($x = 1, 2$) produce a non-null magnetic field on the rail circuit determined by rail 1 and 2 of the other track: mutual coupling M_x may be computed starting from the net flux across the rail circuit

$$M_x = \frac{\mu}{2\pi} \ln \frac{r_{x2}}{r_{x1}} \tag{3.1}$$

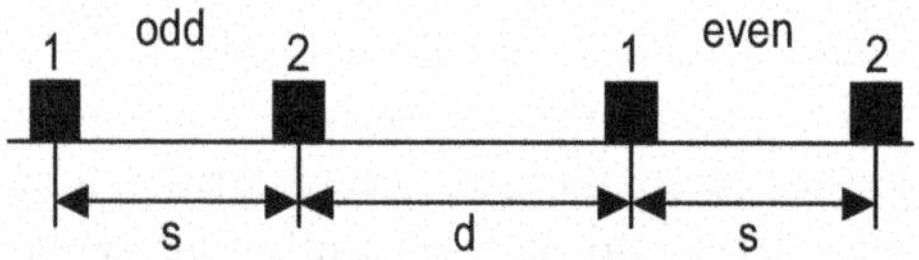

Fig. 3.3 Double track line section: odd track (left), even track (right) and distances

Depending on the desired transfer function (c.m.-to-d.m. or d.m.-to-d.m.) the two coupling terms M_1 and M_2 for the odd track rail currents must be summed or subtracted. More generally, given two arbitrary rail current I_{o1} and I_{o2} on the odd track, they may be decomposed into an equivalent c.m. and d.m. track current terms $I_{o,cm}$ and $I_{o,dm}$, that produce a d.m. induced voltage on the even track given by

$$
\begin{aligned}
V_{e,dm} &= j2\pi f\ [(M_1 + M_2)I_{o,cm} + (M_1 - M_2)I_{o,dm}] \\
&= jf\mu\left\{\left[\ln(\frac{r_{12}}{r_{11}}) + \ln(\frac{r_{22}}{r_{21}})\right] I_{o,cm} + \left[\ln(\frac{r_{12}}{r_{11}}) - \ln(\frac{r_{22}}{r_{21}})\right] I_{o,dm}\right\} \\
&= jf\mu\left\{[\ln(2s+d) - \ln(d)]\, I_{o,cm} + [\ln(2s+d) - \ln(s+d) + \ln(d)]\, I_{o,dm}\right\} \\
&= j2\pi f\ [M_{cm}I_{o,cm} + M_{dm}I_{o,dm}]
\end{aligned}
\tag{3.2}
$$

The two terms M_{cm} and M_{dm} correspond thus to

$$
M_{cm} = \frac{\mu}{2\pi}\ln(1 + 2s/d); \quad M_{dm} = \frac{\mu}{2\pi}\ln\left(\frac{(2s+d)d}{(s+d)^2}\right) \tag{3.3}
$$

For a standard High Speed Railway Line of the Italian network these terms are: $M_{cm} = 0.126\,\mu\text{H/m}$, $M_{cm} = 0.019\,\mu\text{H/m}$.

By manipulating (3.2) so that the odd track rail currents $I_{o,r1}$ and $I_{o,r2}$ are used, rather than c.m. and d.m. components, we obtain

$$
\begin{aligned}
V_{e,dm} &= \frac{\mu}{2\pi}[\ln(2s+d) - \ln(d)]\left[\frac{I_{o,r1} + I_{o,r2}}{2}\right] \\
&+ \frac{\mu}{2\pi}[\ln(2s+d) - \ln(s+d) + \ln(d)]\left[\frac{I_{o,r1} - I_{o,r2}}{2}\right] \\
&= \frac{\mu}{2\pi}\ln\left[1 + \frac{s}{s+d}\right] I_{o,r1} + \frac{\mu}{2\pi}\ln\left[1 + \frac{s}{d}\right] I_{o,r2} \\
&= M_{r1}I_{o,r1} + M_{r2}I_{o,r2}
\end{aligned}
\tag{3.4}
$$

Again, for a standard Italian HSRL the two terms are: $M_{r1} = 0.0531\,\mu\text{H/m}$, $M_{r2} = 0.0725\,\mu\text{H/m}$; it is easy to verify that the sum and difference of these two terms give the two M_{cm} and M_{dm} terms calculated above.

To determine the odd to even tracks coupling, in terms of induced current, the rail circuit impedance is needed. The impedance of the track circuit is given by the following terms: rail circuit impedance Z_{rc}, transmitter and tuning devices impedance Z_{tx} and receiver and tuning devices impedance Z_{rx}. For a general approach Z_{tx} and Z_{rx} may be neglected and focus is on the Z_{rc} term, which is proportional to circuit length; it is clear that Z_{tx} and Z_{rx} represent a minimum impedance value including for track circuits of negligible length. The general expression for Z_{rc} is computed starting from the self impedance of the single rail and by subtraction of mutual coupling terms between the two rails of the same track

$$
Z_{rc} = 2(R_{11} + j2\pi f\, L_{11} - R_{12} - j2\pi f\, L_{12}) \tag{3.5}
$$

where R_{11} and L_{11} are the p.u.l. values of the self resistance and impedance of a single rail; R_{12} and L_{12} are the p.u.l. values of the mutual resistance and inductance of the two rails.

Given the frequency range of interest, these values may be calculated quite accurately by Carson's expressions[2] and adjusted with results obtained

[2] Carson's equation is treated in detail in Chapter 7.

through measurements , that take into account the influence of frequency and rail magnetic permeability on the rail self resistance and internal inductance terms, the latter to be added to the purely geometrical self inductance term. These values have been determined for a broad frequency range (covering all the known track circuit applications) and they are shown in Table 3.1.

Table 3.1 Rail circuit p.u.l. electrical parameters

Freq. [kHz]	R_{11} [mΩ/m]	L_{11} [μH/m]	R_{12} [mΩ/m]	L_{12} [μH/m]	Z_{rc} [mΩ/m]
0.1	0.22	2.10	0.09	1.21	0.25 + j 1.11
0.3	0.51	1.90	0.29	1.10	0.44 + j 3.02
1.0	1.38	1.73	0.98	0.98	0.80 + j 9.41
3.0	3.63	1.59	2.93	0.87	1.40 + j 27.22
6.0	6.83	1.51	5.84	0.80	1.98 + j 53.62
10.0	10.96	1.46	9.69	0.75	2.57 + j 88.63
15.0	16.07	1.41	14.49	0.71	3.16 + j 132.25
20.0	21.09	1.38	19.25	0.68	3.68 + j 175.77

Detailed calculations are now developed for the 1-20 kHz frequency interval, where simulations have been performed and comparison with experimental data is possible. The p.u.l. Z_{rc} expression by interpolation of the values in Table 3.1 is

$$Z_{rc} \simeq 0.80 f^{0.5} + j9.41 f^{1.08} \tag{3.6}$$

where f is the frequency in kHz.

If experimental data obtained during audiofrequency track circuits tuning and field verification are considered, a conservative estimate of the Z_{tx} and Z_{rx} terms is about 1 Ω or larger at the operating frequency; this term becomes negligible for track circuit length larger than approximately 50-100 m at minimum depending on frequency.

Under the following simplifying (and over-estimating) assumptions (Z_{tx} and Z_{rx} are neglected; the real part of the rail circuit impedance is neglected, with an error between 32% down to 4.2% over the considered frequency range), the induced d.m. current on the even track may be considered constant with frequency (both the induced voltage and the rail circuit impedance are proportional to frequency) and given by:

$$\begin{aligned} I_{e,dm} &= \frac{V_{e,dm}}{Z_{rc}} = \frac{j2\pi f\,[M_{cm} I_{o,cm} + M_{dm} I_{o,dm}]}{2(R_{11} + j2\pi f\, L_{11} - R_{12} - j2\pi f\, L_{12})} \\ &< \begin{cases} [M_{cm} I_{o,cm} + M_{dm} I_{o,dm}]\,/2(L_{11} - L_{12}) \\ [M_{r1} I_{o,r1} + M_{r2} I_{o,r2}]\,/2(L_{11} - L_{12}) \end{cases} \end{aligned} \tag{3.7}$$

If the values shown in Table 3.1 are used, then (3.7) simplifies to

$$I_{e,dm} < \begin{cases} \left[M_{cm}I_{o,cm} + M_{dm}I_{o,dm}\right]/2(L_{11} - L_{12}) & = 0.09I_{o,cm} - 0.014I_{o,dm} \\ \left[M_{r1}I_{o,r1} + M_{r2}I_{o,r2}\right]/2(L_{11} - L_{12}) & = 0.038I_{o,r1} + 0.052I_{o,r2} \end{cases} \tag{3.8}$$

This expression indicates a maximum c.m.-to-d.m. transformation of 9%. The relationships above have been computed for a straight line without transversal connections between conductors and discontinuities; on a real line equipotential bonding and impedance bonds are present at predefined distances (normally about 0.5 to 1 km) and these elements modify the theoretical current distribution, by moving a significant portion of the rail current to the other conductors of the return circuit. Finally, it is underlined that a few % of c.m.-to-d.m. transformation gain applied to a specific frequency component of the traction return current (in general limited to a small percentage of the component at the fundamental frequency) produces a significant differential in-band signal on the rail circuit, applied to the track circuit receiver. An approximate estimate, based on a 500 Arms amplitude of the fundamental component at 50 Hz, gives:

1. approximately 1/4 of the traction return current flows through the rails and represents the c.m. current on the odd track;
2. the traction return current components in the audiofrequency range may be approximately hundreds of mA or a few A (however, it is about 1% of the fundamental in all cases);
3. this c.m. current may produce a maximum d.m. disturbance through the track circuit of a few % up to 9%, that is tens of mA up to a few hundred of mA;
4. this d.m. disturbance represents a significant portion of the usual operating current of this class of signalling circuits, often tuned around 500, or even 1000, mArms of shunted (i.e. the track circuit is artificially occupied) receiver current.

The results of some simulations performed with a distributed parameters simulator are shown in the following figures (Figure 3.4 to 3.9) for a 2x25 kV 50 Hz system, equipped with impedance bonds and earth wire and through earth collector, thus quickly taking away the return current from the running rails; the return current flowing from train axles is normalized to 1 A. Three transformation gains are defined as:

$$\begin{aligned} k_e &= I_{e,dm}/I_{e,cm} \\ k_o &= I_{o,dm}/I_{o,cm} \\ k_{oe} &= I_{e,dm}/I_{o,cm} \end{aligned} \tag{3.9}$$

The k_{oe} term indicates the left term in (3.8(a)), while k_o and k_e indicate the % unbalance between differential (not symmetrical) and common mode (symmetrical) rail current components, as shown above in (3.9).

The increase in circuit unbalance at line ends is due to the fact that the reference test line is not a real line and all the conductors are cut at test section boundaries: these points should be neglected. The drop in the gain values in Figure 3.6 occurs at train position; again in Figure 3.9 the same drop occurs at 11000 m plus an additional drop due to line resonance (the injection point is no longer in the middle of the test section, but close to one of the line ends and the length of the resonating line is approximately doubled [507, 314]). It is easy to see that the k_{oe} gain is always smaller than the theoretical maximum value 0.09 computed above and it attains the theoretical value only in small part of the considered test section.

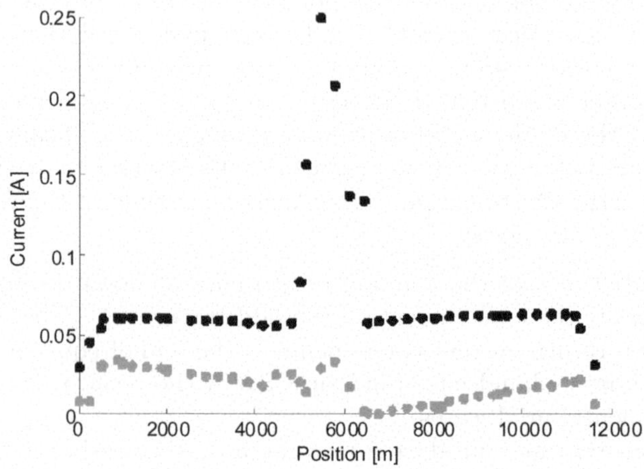

Fig. 3.4 Odd track equivalent currents: c.m. (black) and 10 x d.m. (gray), train pos. 6 km

The numerical and theoretical results shown above have been validated through experimentation: the largest values of a set of sample readings of the d.m. current circulating in the track circuit on the opposite track along the HSRL between Turin and Novara were between 8 and 10% (a slight influence of the rail-to-rail conductance terms may be expected in practical cases).

It is possible to conclude that double rail circuits are the preferred track circuit type and IBs represent a valid means of achieving common mode and differential mode component separation; they are however not immune from symmetry problems, since a double track traction line is not symmetric by itself with respect to each track. Moreover, the two tracks behave like parallel rectangular turns on the same horizontal plane and a great deal of magnetic coupling is expected: both differential mode and common mode components on the culprit track, where the traction current is flowing, give rise to a disturbing differential mode components on the adjacent victim track; while the former represents the unbalanced traction return current and the operating current of the TCs on the culprit track, the latter corresponds

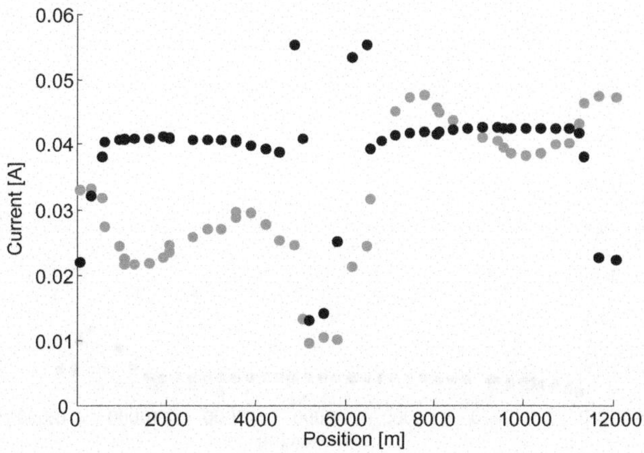

Fig. 3.5 Even track equivalent currents: c.m. (black) and 10 x d.m. (gray), train pos. 6 km

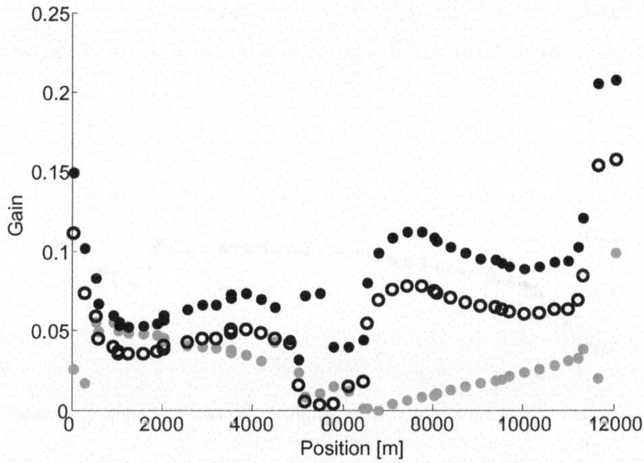

Fig. 3.6 Transformation gains: k_e (gray dot), k_o (black dot) and k_{oe} (black circle), train pos. 6 km

to almost the full traction return current. Theoretical and practical values as high as 9% have been identified in the audiofrequency range, so that this transformation/coupling mechanism represents a relevant factor in estimating the real operating conditions of TCs on track and in deciding the limits for interference.

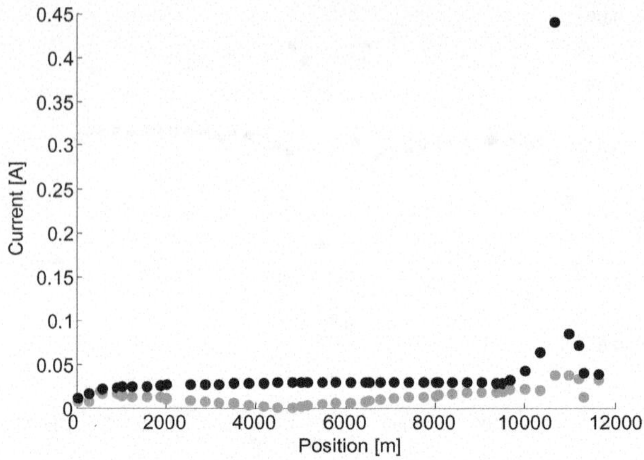

Fig. 3.7 Odd track equivalent currents: c.m. (black) and 10 x d.m. (gray), train pos. 11 km

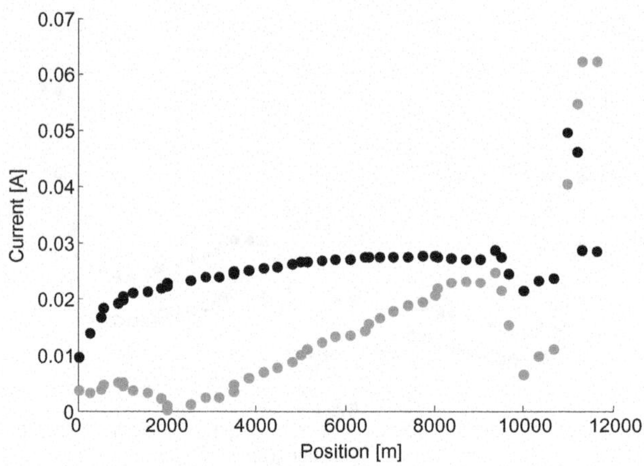

Fig. 3.8 Even track equivalent currents: c.m. (black) and 10 x d.m. (gray), train pos. 11 km

3.1.1.3 Types and Classification of Track Circuits

In this section, a classification of track circuits is presented, dividing them into categories determined by the operating frequency: DC, power frequency up to some hundreds Hz, audiofrequency (further divided only for practical reasons into low and high frequency intervals), very high frequency, impulsive. Track circuits are treated with the help of manufacturers' documentation and with reference also to technical reports and annexes of applicable standards (in particular EN 50238). With respect to EMC, in addition to basic

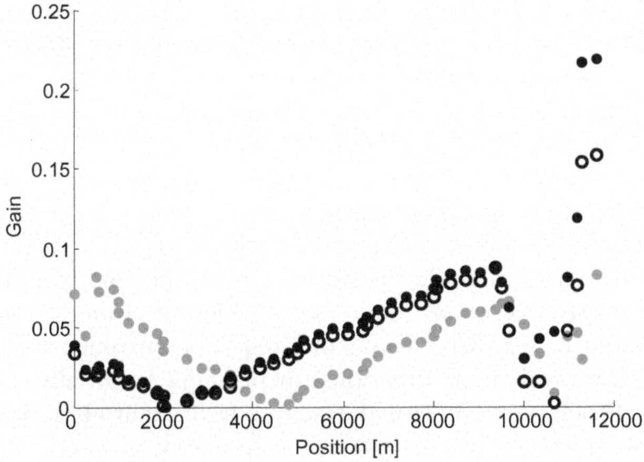

Fig. 3.9 Transformation gains: k_e (gray dot), k_o (black dot) and k_{oe} (black circle), train pos. 11 km

reference information on operating frequency, signal amplitude, length, type of circuit, minimum shunting resistance and ballast resistance, track circuit susceptibility to direct conducted interference is the most relevant element, both within and outside the operating frequency interval. In order to reduce in-band direct interference on the TC signal, a modulation is normally applied to audiofrequency TCs (FSK modulation being the most robust modulation scheme in use) and the coding rate is low enough to result in a compact required bandwidth and to mask out most of the transients (that by definition have a broad frequency spectrum and can potentially affect a wide frequency interval). With respect to incoming disturbance, the receiver bandwidth is defined and limited by a selective bandpass filter with a free response time constant compatible with the adopted modulation and coding scheme. As an additional mitigation, modern track circuits react to excessive noise over the operating bandwidth at the receiver by turning the track to occupied and ensuring a fail safe behavior even in cases of receiver jamming; the channel noise is measured continuously between received messages, thus ensuring a continuous control of the track circuit channel. Furthermore, there are track circuits with a redundant digital link between transmitter and receiver and the correct received message is checked against a "clean" copy of it, sent over a safe channel (a wayside cable) rather than through the track. Before the use of digital transmission, this function was implemented by the two coils of a vane relay. It is underlined however that for analog track circuits (that represent still the largest portion of installed track circuits) another relevant element of the overall EMC picture is the wayside cable connecting the transmitter and receiver to the interlocking cabin, where relays are located; we will see that the relays may be a susceptible component of the whole chain, and a wise choice of those connected to the field is always necessary.

Since track circuits represent a relevant part of the EMC and interference analysis and there are several types and categories, manufactured and installed over the last 40 years, they are addressed by dedicated sections in the following. When the immunity level of a track circuit is declared by indicating a limit curve (the so called "limit mask") for the current flowing in the track over a frequency range containing the operating bandwidth of the said track circuit, several distinctions and considerations need to be put forward: the interfering current for direct conducted coupling onto the receiver is the net differential track current, while the return current flowing from rolling stock axles includes both a common mode and a differential mode component, as already discussed in sec. 3.1.1.1 and in [508]. The current leaving the axles divides in the left and right directions, in principle originally as a common mode type and then partially transforms into a differential mode current due to various asymmetries (intrinsic track asymmetry, ballast properties, rail joints differences, specific track arrangements, impedance bonds saturation, and so on), that may reach outstanding values for worst case configurations (such as a broken rail, a rusty rail, etc.). The relationship between the return current and the impinging receiver current (or resulting voltage) is called "transfer function" [112, 114, 508]. It is evident that a wide dispersion of its values may be expected, especially if the influence of parasitic elements at higher frequency is also taken into account; railway operators and infrastructure managers have different approaches to determine the reference worst case and the safety margins applicable to their assets, and this may result in even different limits of the return current spectrum for the same track circuit!

Another very practical, but relevant, issue is the complex behavior of loose contacts, oxidized surfaces, and in particular bad wheel-rail contact. Most bonding connections are required to be doubled and redundant, so that they ensure a longer duration and higher availability, but this doesn't prevent the occurrence of loose contacts, noise and over-heating if inspections and maintenance fail to do their own duty. More troublesome and commonplace is the bad wheel-rail contact and its consequences: the contact surface is very tiny and the mechanical movement (especially at low speed close to switches) doesn't ensure that the surface is cleaned from rust and oxides, so that there is the possibility of a high resistance electric contact in series to the train axles offering a poor shunt resistance to the track circuit; since the behavior is intermittent and high resistance shunt doesn't occur steadily for a long distance, the problem may be solved by introducing delays in the interlocking logic to avoid premature freeing of the block section, in particular, for level crossings, as it is briefly addressed in sec. 3.1.4. In some countries the surface of the rail head is modified by adding a soldered/brazed profile in the proximity of the points to protect (e.g. switches and level crossings), to increase the vertical pressure and improve the electrical contact with the wheel.

3.1.1.4 DC Track Circuits

They are probably the first to be used from an historical point of view and they are still in use.

These track circuits are cheap and simple, and have the advantage of a negligible longitudinal voltage drop with a zero inductive reactance of the running rails. A redundant transmitter may be easily implemented with a rectifier fed by an ac supply and a battery may supply the transmitter dc voltage, even in case of ac supply shortage. A low pass filter may equip the receiver in order to reject any ac traction fundamental and harmonic components; of course there are transient ac signals (such as the inrush current of a vehicle transformer) with a remarkable dc component that may produce interference. External stray currents originating from nearby dc traction systems or even heavy geomagnetic effects (on long track circuits) may couple a significant interfering dc current.

These TCs exist both in end-fed and center-fed configurations. The maximum TC length set by the infrastructure operators adopting this type of TC depends on the assumed ballast resistance and shunt resistance: several railways (SBB, Network Rail, Banverket) specify a ballast resistance ranging between 1.0 and 2.5 Ωkm and a shunt resistance ranging between 0.3 and 1.0 Ω, establishing from 1000 to 2500 m of maximum length.

For the installation of dc TCs however mechanical insulating joints are mandatory to isolate the TC section from the rest of the system or from adjacent dc TC sections. For the traction return current to flow back through the rails, impedance bonds for a dc component cannot be used, so that a single-rail arrangement is needed.

To improve TC immunity to transient dc components, additional delays are included in the receiver, besides the time constant of the low pass filter, using a slow to pick up repeat relay, such as a motor relay. The dc TC in use in Denmark on the 25 kV 50 Hz lines (Regional railways) has immunity levels specified as 4 A continuous and up to 15 A for a time interval shorter than 1.5 s. In Finland the return current limitation is 3.8 A for a time interval of 10 s or longer and it is extended from dc to 2.4 Hz. In Sweden a 10 A current limit is specified over a similar bandwidth, from dc to 2 Hz.

As mentioned previously, dc track circuits are used in ac electrified railways and their use is limited to areas where there is no risk of a wrong side failure due to false operation of a track relay by extraneous direct current, such as could be encountered if the railway is in close proximity to a dc electrified railway line. For ac immune dc track circuits, the iron circuit is modified (to make it ac immune) by the fitment of two copper slugs over the cores near the pole pieces and a magnetic shunt is fitted between the cores above the copper slugs. When direct current is applied to the winding the flux produced attracts the armature as normal, but some flux will be diverted through the magnetic shunt. As a result, the operating power required will be increased. When an alternating current flows in the coil, circulating currents in each slug oppose the alternating flux. The alternating flux consequently finds it difficult in establishing itself across the air gap and is largely diverted through the magnetic shunt. The alternating current flux therefore has very little effect on the armature.

3.1.1.5 AC Power Frequency Track Circuits

This type of track circuit is quite widespread, being its straightforward implementation on dc railway lines relied upon the availability of 50 (or 60 Hz) power supply. Their use have been extended to AC railways, but with the track circuit frequency changed to values other than the power frequency fundamental in other that the track circuits do not suffer from interference originated externally or within the same railway system: it is evident that a 50 Hz track circuit frequency is not the best choice, because of the many sources of interference, even in a dc system: the on-board auxiliaries operate at 50 Hz and any leakage or unbalance current component represents a potential threat, that requires expensive and bulky trap line filters for all types of converters installed on-board, not to mention the presence of stray current from external electrical installations supplied from the nation wide 50 Hz supply grid.

The TX unit is normally coupled to the rails by means of a feeding transformer with a series limiting resistor, in case of short circuit. The RX unit is again coupled by means of a pick-up transformer and a series protective resistor. This technique was extended also to old audiofrequency TC operating in the lower portion of the frequency range. The following types of AC track circuits are generally in use:

- conventional ac track circuits using double-element vane relays, operating for example at 50 Hz in dc traction areas or $83\frac{1}{3}$ Hz or 105 Hz in ac traction areas, especially replacing dc TCs when there is a possibility of interference from a dc source in the same area; their use is also extended to non-electrified lines with much less constraints on the selection of the operating frequency; the most relevant characteristic is the simplicity and the possibility of reaching extremely large TC length values, such as in the case of Microtrax TC [675] ensuring from 6 to 9.5 km length depending on the type of rail and ballast leakage resistance;
- tuned Reed track circuits which are used in similar circumstance to conventional ac track circuits and also when long track circuits are required in non-electrified areas; their operating frequency is in the range of some hundreds Hz;
- jointless track circuits of either the voltage or current-operated type which are used on plain line in continuously welded rail areas;
- special purpose track circuits, for example pulsing types (designed to break down a rust film on the rail surfaces) and overlay track circuits, which are designed to detect the arrival of a train at a given point within a longer track circuit.

The above is a simple classification of ac track circuits, however there is a wide range of operating frequency values, to adapt to the specificity of each country in terms of used traction supply frequency, its harmonics and other components characteristic of power applications; as in the large variety of TC, different types of modulations are applied, either phase or frequency modulations [113].

The power frequency values (in a broad sense) vary widely, from as low as some tens of Hz to about 300 Hz, the largest values located at the boundary between power frequency and audiofrequency:

- UGSK3 (208.75, 225.45 and 242.15 Hz) and UGSK95 (137.5, 175 and 225 Hz) in use in Austria and Switzerland, the latter with a 4 A current limit with 4 Hz bandwidth and 0.5 s integration time;
- GRS (ATBEG) at 75 Hz in Belgium;
- KOA at 50, 75, 125, 175 and 275 Hz in use in Czech Republic all with a 100 mA current limit defined over a 12-14 Hz bandwidth around the center frequency;
- 77 Hz in use in Denmark with a 4 A limit for 4 s time duration or longer over a 15 Hz bandwidth around the center frequency;
- 25 Hz, 75 Hz, 83 Hz and 125 Hz all in use in Finland, with a 1 A rms current limit for 1 s time duration, all over a 6 Hz bandwidth around the center frequency;
- 50 and 83.3 Hz in France have absolute limits of 1.2 and 0.7 A respectively over the 1500 V dc network, tolerating special transients (such as pantograph operation, switching on of supply, etc.) if of less than 0.5 s duration; for 50 and 83.3 Hz in Italy the absolute limit is lower (400 mA) with a similar consideration of transients;
- 42 and 100 Hz in Germany all with a 2 A current limit defined over a 4 Hz bandwidth around the center frequency for a 0.5 s integration time;
- GRS 75 Hz is used on 3 kV lines to Belgium and the current limit is specified as 0.5 A between 70 and 80 Hz and increasing to 1 A at 65 and 85 Hz, and up to 3-4 A around 55 and 95 Hz, where measurements are made with a bandpass filter similar to the victim track circuit and a 0.2 s integration time;
- the 95/105 Hz TC on Norway lines is characterized by a 2 A current limit defined with a 1 Hz resolution bandwidth over a 6 Hz frequency interval around the center frequency for a 1 s integration time;
- the OTS, OTL, OTZ at 50 Hz equip the PKP Polish lines and the current limit is specified as 2.4 A between 48 and 52 Hz, increasing up to 6.2 A at 40 and 60 Hz with a 0.2 s integration time; the 50 Hz type DrL56 (Alcatel/Marconi), GRS (Alstom), JRV 10207 (Bombardier), UT.1 (Dimetronic), Ms-46653 (Enyse).

There are two principle methods of immunisation:

- *Double element relay.* Double element relays of the vane type require two supplies, called "local" and "control", which must produce fluxes having

components at right angles in order to produce a positive torque to operate the relay vane. If the supplies are not at the same frequency the net torque will alternate to beat frequency and the relay will not operate. The relay will therefore only respond correctly to "control" current of the same frequency as the "local" current. It is usual to observe caution in 50 Hz traction areas to prevent the supply to the local coil of the relay becoming contaminated with 50 Hz current. The "local" supply for relays can be confined to specified relay rooms or may be screened by use of earth leakage detection equipment or 50 Hz detection equipment can be used.
- *Filters.* This technique involves the use of an oscillator and amplifier at the feed end producing a given frequency which is passed through a relay end filter before being amplified and used to operate the track relay, which is usually of the DC tractive armature type. Enhanced security is afforded if adjacent and parallel tracks use different frequencies.

To increase their immunity, these TCs are normally associated to motor or vane relays, in order to include either a delay (to mask out transient disturbing components) or a matching control between transmitted and received, plus a delay.

Reed track circuits are typically used in areas where interference from both dc and ac traction is present and in non-electrical areas where long track circuits (such as in tunnels) are required. It is a single frequency track circuit that is usually installed in single rail mode on a ac railway or double rail mode on a dc railway. The salient feature of a reed track circuit is a pair of mechanically coupled reed filters at either end; the pair at the feed end ensures that energy only at the desired frequency is fed to the rail, while the relay and filters respond only to the same frequency. The reeds can be tuned with considerable precision and possess a high Q, such that frequencies in common use are pitched only 3 Hz apart. The frequencies of operation for reed track circuit manufactured by Alstom are 363, 366, 369, 372, 375, 378, 381, 384, 408, 417, 423 Hz. The use of different frequencies for adjacent and parallel track gives freedom from faulty operation, caused by insulated joints failure.

3.1.1.6 AC Audiofrequency Track Circuits

The higher operating frequency of audiofrequency track circuits enables the adoption of various solutions to couple the signal to the rails and to separate adjacent sections, because of the more handy size of inductors and capacitors:

- coupling to the rails is accomplished by means of:
  - resonating transformers, often integrated into the impedance bonds, with the transmitter and receiver connected to an auxiliary secondary winding, made resonate with a series/parallel circuit arrangement;
  - cable bonds (of the "S" or "O" type), with the transmitter and receiver conductively or inductively coupled to either half of the bond;

- section isolation is accomplished by means of:
  - mechanical insulation rail joints, installed with an impedance bond for traction current circulation;
  - the same cable bonds, that give directivity to the signal and in conjunction with a resonating lumped circuit produce the required separation, in terms of different track impedance at the two operating frequency of the left and right section;
  - resonating lumped circuits in direct connection to the rails, tuned on the two operating frequencies and forming the so-called "tuned areas";

  the last two isolation methods are normally grouped under the term "jointless TCs" or "electrical rail joint TCs".

Some practical problems arise when the operating frequency is as high as an audiofrequency and the applied modulation has a limited frequency deviation, so that the frequency channel spacing is conveniently tight:

- when the track circuit signal is generated by an analog circuit, the frequency value may be inaccurate and most of all may drift with aging and following environmental changes; if the modulation is applied with an analog circuit its frequency and the modulation ratio may drift in the same way;
- in case of track non-idealities (for example modification to the infrastructure elements or increased rail or ballast conductance, as in the case of contaminated ballast), or in case of the impedance bond saturation, the tuning units may be slightly de-tuned.

A benefit of the use of a higher frequency is that the jointless separation of TC sections is possible because of the more favorable sizing and efficiency of resonant units [467, 518, 583], but the problem of "early shunt" (occurring when the train axle is approaching the TC section, but is still outside it) is even more troublesome: early shunt occurs when vehicle axle is sensed by a TC before the axle enters physically the TC section and this represents a problem for the accuracy of the location of the train, in particular in shunting areas and close to switches; the track length at the TC section ends exposed to "early shunt" depends on many factors and parameters, in particular the TC sensitivity in terms of maximum shunting resistance and the track leakage conductance [674].

Another peculiar problem of audiofrequency TC is the longitudinal inductive voltage drop along the running rails, that lowers the available receiver voltage and thus limits the maximum TC length; we could say that without any of the solutions that are considered hereafter the length between TX and RX would be limited to a few or some hundreds meters, as it occurs for level crossings TCs or switch and shunt area TCs. Two solutions have been proposed and are normally adopted:

- the TC is driven by positioning the TX in the middle (center-fed configuration) and two RX are located at the two ends, thus doubling the standard length up to nearly 1 km [585]; the benefit of such a configuration is that the two TX-RX sections can be monitored independently and train location is more accurate with more degrees of freedom in route formation with benefits for interlocking and circulation;
- the track is compensated by means of capacitors (called *compensating capacitors*) positioned evenly at a prescribed separation distance, with the aim of flattening the frequency response of the track, capacitively compensating the inductive longitudinal voltage drop.

Track Capacitive Compensation. The insertion of a periodic pattern of capacitors produces a series of distributed resonance frequencies along the frequency axis and at each of them the voltage frequency response between TX and RX gets rising thus compensating the monotonic low pass profile of the purely inductive track (in reality the track itself possesses a own rail-to-rail capacitance in the order of one to some nF/m (see sec. 7.5.7) and thus a self resonance frequency located at a much higher frequency, normally above 20 kHz for 1 km length). The effect of an increasing number of compensating capacitors over a fixed 1 km TC length (thus reducing the length of each track cell between two adjacent capacitors) is shown in Figure 3.10, taken and adapted from [518]. The determination of the transfer function and the effect of the variations of some track parameters were also considered in great detail in [583].

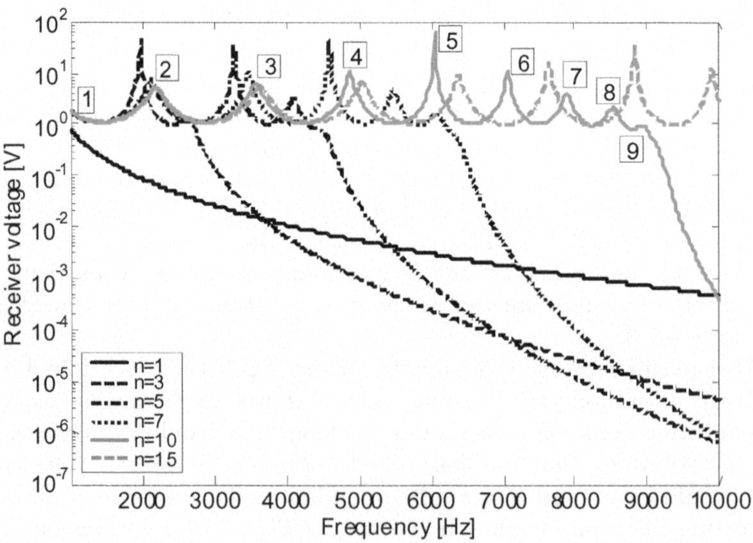

Fig. 3.10 RX voltage for different numbers n of compensating capacitors over 1 km TC length

The determination of the right value of the compensating capacitors was done by the manufacturers by analyzing the behavior of the track cell formed by two adjacent compensating capacitors and the track in between, leading to a range of capacitance values of about a few μF to a ten of μF for cells of about a hundred meters and different types of TCs and desired compensation degrees. The mutual effect between the track inductance and the two added capacitors determines the main self resonance, but the loading effect of adjacent cells sharing one of the two capacitors in turn modifies this first estimate, so that the overall behavior is obtained by analysing all the mutual effects of the entire track section [466, 583, 518]. It is to be noted that the first and the last track cell experience a highly asymmetry, being loaded on one side by the TX and RX impedance respectively; to this aim several solutions have been investigated with many site experiments on real tracks; to sum up the first and last cells as seen from the TX and RX are usually half-length cells, or put in another way the cells have constant length and TX and RX are connected to the track only at their center.

Last consideration is that the use of compensating capacitors requires that the latter are included in the safety analysis of the TC, that is there is the need to show that the TC apparatus including any cabin logic circuit is able to detect any disconnection or degradation of such capacitors that is affecting the TC frequency response, as it is required for broken rail detection in many countries. This analysis is even more complex because the modifications to the frequency response and thus to the RX voltage on the assigned specific TC frequency channel are the result of the interaction of all the considered elements: track, the variability of its parameters, compensating capacitors, adjacent TCs and TX and RX impedances and tuning units. The method and the results for this analysis are seldom published and for further reading reference can be made to [224, 518]. In synthesis the analysis and the experimentation must ensure that whatever the faulty capacitor the RX voltage is always on the safe side of a procured occupation. The problem is not trivial if the mutual influence between cells is considered and the fact that the accidental disconnection of a capacitor creates a cell with twice the length; in [224], in particular, some considerations are reported on the optimal impedance termination at TX and RX to maximize the detectability of such events.

Train-to-Wayside Communication. Even if the train-to-wayside communication may be implemented in almost all TCs, depending on the required amount of information to be exchanged, AF TCs lend themselves to an efficient message exchange, with baud rates as high as 200/300 baud in some cases. The FSK modulation is quite robust, has a favorable spectrum occupation (if Continuous Phase Modulations, and in particular Minimum Shift Keying, is used the occupation is minimal) and may be decoded either by an optimal correlation receiver or a sub-optimal one, made of a bandpass filter bank. Optimal reception relies upon orthogonal symbols that may be decoded with a correlation receiver. It can be shown that FSK among digital modulation is the one with the best performance when detected incoherently with respect to optimal reception [606]. When an entire message is considered, with the FSK symbols encoding the message bits (for a binary FSK one message bit corresponds to one symbol), the code words must be conceived for a series of properties that are particularly relevant when sending commands and moving authorities to a train entering a TC section:

- correct operation: a specific code word w_i corresponds to a command; there are n different commands, and one is repeated continuously when the train enters the TC section;
- unsafe failure - wrong sync: synchronization may be an issue since the train enters the section asynchronously with respect to the message frame; a synchronization error implies that a word w_r is recognized that does not correspond to the transmitted word w_t and the recognized word w_r thus overlaps to the sequence of alike words w_t, one after the other; codes where no overlap between any two code words is itself (in the absence of bit errors) a valid code word is said a "comma-free" code;
- unsafe failure - right sync, wrong decoding: a code word affected by one or more bit errors may be decoded wrongly and another valid command may be understood, with a correct synchronization, so the time origin that is kept unaltered from the transmitter to the receiver;
- safe failure: the corrupted word may be recognized as an invalid command.

From a railway safety point of view, the safe failure is only relevant for system availability, while unsafe failures are potentially dangerous. The probability of one of the considered failures depends on the single bit error probability P_e (that in turn depends on the signal-to-noise ratio, on the transmission rate and on the decoding method), on the word length in terms of number of bits and on the number of words in vocabulary, and their Hamming distance (that is the number of corrupted bits needed to pass from one correct word w_i to another correct word w_j). Thus word sets (of size M) that form a vocabulary of commands are designed for the maximum distance, that among others is ensured by increasing the word length n, that has the side effect of reducing the number of transmitted commands in the unity of time, or "command rate" r_c. On the other hand, as mentioned above, the probability of unsafe failures is reduced if the same code word is repeated many times, let's say Q times, but at the cost of further reducing the commands rate. Finally the minimum command rate is related to the maximum train speed and TC length. Ref. [302] is a good reference for detailed calculations of error probabilities for some vocabularies and coding schemes of example; here, we simply say that for a small enough P_e value, $(1 - P_e)$ may be approximated to unity, and the probability of an unsafe failure is given simply by $P_{uf} \cong (P_e)^d$, where d is the Hamming distance between the two words and for an optimized vocabulary corresponds to the average Hamming distance, constant for all words. The probability of a safe failure P_{sf} is larger, since at least one corrupted bit is missing, so that the new corrupted word does not belong to the vocabulary;

in reality all the d combinations of $d-1$ errors must be summed together, and so for the $d\,(d-1)$ combinations of $d-1$ errors, and so on; it is thus evident that P_{sf} is - as expected - much larger, except for very large P_e (approaching unity), as it is shown in [302][309]. Since both right and wrong side failures, related to unsafe failure[3], are a matter of probability, they both are acceptable if their probabilities are below an agreed threshold, that through the years has been set in the order of ten years and ten thousands years respectively.

Examples of Specifications of Susceptibility and Limits of Interference. AF TCs installed in European railways cover a wide range of frequencies from about a thousand Hz and up to about 20 kHz; the largest variety of operating frequency, circuit topologies and modulations can be observed in the lower portion of this frequency interval, with different products spanning over more than thirty years of design and production [130]:

- Specifications SNCF IN2724: the French railways specify a 30 mA over a ± 50 Hz bandwidth around the center frequency, excluding a small frequency interval of 10 Hz around the center frequency where a 70 mA profile applies (for UM71 C, UM71 CB, UM71 CTVM PSE, UM71 CTVM 300 track circuits); analogously the 30 mA limit is replaced by 100 mA and the 70 mA by 500 mA for the UM71 CTVM 430, UM71 SEI track circuits; the test procedure is very complex (a moving window Discrete Fourier Transform is computed, with at least 80% overlap and frequency resolution of 1 Hz, to cite the main elements) and there is no clear indication of the integration time that is normally used by the other administrations;
- the Jade AF by Alstom operates globally on several frequency channels, selected and configured for the specific country and application: 2206, 2500, 2841 and 3157 Hz in the Netherlands, with a current limit of 0.8, 0.7, 0.7 and 0.6 A at the center frequency with an integration time of 5 ms, so without nearly any compromise for fast time-varying harmonics occurring during transients and variable rolling stock operating conditions; a second version also in the Netherlands on the Betuwe-Route operates at 1600, 1900, 2200 and 2500 Hz has a slightly lower current limit (0.5 A) at the center frequency with the same integration time; in Switzerland the selected frequencies are 1595, 2206, 2500, 2841, 3157, 3472 and 3750 Hz with a 0.5 A current limit defined over a 90 Hz bandwidth around the center frequency for a 0.5 s integration time;
- FTGS 46 by Siemens, operating at 4.75, 5.25, 5.75 and 6.25 kHz has a 1 A current limit over a bandwidth of 200, 206, 214 and 220 Hz for the four channel frequencies respectively with an integration time of only 40 ms, so quite demanding also for transient time-varying harmonics produced by rolling stock; the FTGS 917 version operating at 9.5, 10.5, 11.5, 12.5, 13.5, 14.5, 15.5 and 16.5 kHz has a lower current limit of 0.33 A over a bandwidth

[3] More precisely unsafe failure indicates a wrongly received command, but not all commands lead to either a right or a wrong side failure.

ranging from 360 to 510 Hz over the considered frequency channels and same integration time, but higher than other AFTCs operating in the same frequency interval, such as GLS9/15 and EON7 in Germany;

- ME 3091 is an AFTC in use in Spain operating with the channels located every kHz between 11 and 19 kHz, but with a much larger bandwidth of 1200 to 3000 Hz, so that one channel overlaps to one or two adjacent channels, that cannot thus be freely used; yet, the declared limit is of 1.5 to 1 A for the various channels;
- TI-21 is an old product by AdTranz, now Bombardier, in use in particular in Spain and UK, that adopted the same specification for the limit of interference: the TC operates at 1699, 2296, 1996, 2593, 1549, 2146, 1848, 2445 Hz of nominal channel frequency values, where two channels between TX and RX are open at ± 17 Hz from the nominal value, both with a ± 6 Hz bandwidth specified at -3 dB; the arrangement of the nominal frequency values is such that the distance in frequency between two channels used in adjacent sections is maximized to avoid channel crosstalk; as almost all AFTCs, the additional countermeasure against interference is the use of a low frequency modulation of 5 Hz applied to the transmitted signal, to filter out any amplitude modulation due to time-varying harmonics, that imposes a long integration time set to 2 s for the evaluation of the interference, so that the corresponding low limit of only 91.4 mA is not so critical, as compared to other limits.

AF TCs operating in the upper part of the frequency range are more modern systems, with a common approach to frequency channel division: CBDAC and FTGS917, for example operate between 9 and 17 kHz, with frequency channels centered every kHz starting from 9.5 kHz, a channel bandwidth around 200 to 600 Hz (that depends on the code rate and on the possible use as a train-to-wayside communication link) and a guard band for channel separation.

Examples of TCs operating on higher frequency intervals are:

- the Polish EOC operating at three frequency values of 26150, 28700 and 31500 Hz with a 78 mA current limit over a bandwidth of about ± 600 Hz and an integration time of 0.2 s; in the remaining part of the frequency interval in between and around the occupied frequency channels the limit is 250 mA;
- Alstom high frequency TCs of the Jade family are used in the Netherlands on 25 kV 50 Hz lines at the Europort, from which the name "Havenspoorlijn"; they operate at two frequencies, 49082 Hz and 67232 Hz, with a current limit of about 300/400 mA from the central frequency down to an interval of ± 500 Hz around it; their frequency suggests that they could be used to monitor switches and on very short track sections.

TCs used in overlay for level crossing control are somehow different and may be very different, since the objective is the cost reduction and the flexibility of installation, in terms of both minimum distance from the other elements of the track (impedance bonds, other TCs, and so on) and ease of interfacing to the existing ATC/interlocking:

- OTC45 operating at 10 kHz in use in the Netherlands;
- SJ-24S operating at 800 Hz manufactured by GEC-Alstom in use in Spain;
- AFO by Ansaldo [674] operating at various frequencies between 870 and 2365 Hz with a modulation frequency ranging from 18 to 49 Hz for increasing carrier frequency, with a classification into two groups that allows to mix circuits of much different carrier and modulation frequencies; TC length is above 600 m for the frequencies below 1900 Hz and 250 m for the two larger frequencies of 1945 and 2365 Hz;
- ATT 20 by Ansaldo [674] operating at three frequencies, 12.28, 15.0 and 20.0 kHz, with a modulation frequency of 390 Hz and a maximum length of about 100 m;
- ME 3015 operating at 15 kHz with a 1 A current limit over a 1200 Hz bandwidth around the central frequency, ME 3047 operating at 14.5 kHz with a 0.5 A current limit over a 1000 Hz bandwidth around the central frequency, and ME 3091 with channel frequency every kHz from 11 to 19 kHz and a current limit of 1.5 down to 1 A over different bandwidths.

In general any audiofrequency TC can be used as overlay, provided that it has some features, such as the correct logic interface and the electrical compatibility with the existing TCs on the main line, in terms of selection of frequencies, mutual loading and other adverse effects.

3.1.1.7 Impulse Track Circuits

This type of track circuit was developed for bad shunting conditions where the rails have been sparely used or where vehicles have been standing for a considerable time. Basically, these track circuits apply high voltage pulses to the rails to break down the rust film. In [597], it would appear that HVI track circuits are also used to deliver traction current on a 25 kV electrified railway through the rails. There are a few types of HVI track circuits used in railway applications:

- *Jeumont Type.* The jeumont track circuit is designed for use on non-electrified and track electrified by ac or dc. At the feed end, the transmitter consists of an oscillator which outputs 80 V 3 ms impulse at the rate of three per second. These are detected at the receiver end by a special relay having a ring iron circuit, which responds only to energy of the waveform developed by the transmitter. The waveform consists of a well damped positive pulse of brief duration followed by a negative pulse of lower amplitude; the integrals of both pulse being equal, so removing any dc component and possible saturation of iron cores. The result is that a symmetrical waveform cannot operate the relay and thus wrong-side failures due to the fundamental or harmonics of the traction supply can not occur. Jeumont track circuits is capable of operating track circuits of up to 2 km long.

- *ITE Type by Alstom.* The Alstom track circuit is designed for use on both non-electrified and AC and DC electrified tracks. The track circuits are short in electrified areas, being up to 200 m in DC areas, and up to 300 m in AC areas. They are therefore suitable in electrified areas where space to accommodate the tuned zone for jointless track circuit is limited, such as points and crossings. They are also suitable for areas where there is poor wheel-rail contact. The track circuit operates by applying a short high voltage impulse to the rails, typically 100 V, three times per second. The operating pulse waveform is asymmetric with the ratio of the positive to negative voltage being 7:1; this waveform is generated by charging a capacitor to a high voltage and then discharging it through a thyristor and a track transformer. The inductance of the feed and relay end track transformers, together with that of the rails, causes the positive pulses to be followed by a negative undershoot;
- *Pulsat type by Ansaldo.* The use of an even higher peak voltage (the nominal value is 180 V) ensures not only a higher immunity (that depends also on the wave-shape and other factors), but also a longer TC length (to be used also as line TC, for example around stations) and the capacity of perforating oxidized layers and ensuring a good low contact resistance between the wheel and the rail. The transmitter is a pulse generator emitting a positive pulse followed by a negative one, coupled to the track by a step-up transformer, with a circuit to regulate the pulse repetition frequency, nominally at 4 Hz, and a receiver using an amplitude detection threshold and an additional check on signal polarity.

3.1.1.8 Axle Counters

An axle counter track circuit is a track circuit that works on the principles of absolute block - a section of line is deemed to be clear when any train which has entered the section has also been shown to have left it. A typical system consists of a counting head (or "detection point") installed at each end of the section, and as each axle passes the head at the start of the section, a counter increments. A detection point comprises two independent sensors, therefore the device can detect the direction of a train by the order in which the sensors are passed. As the train passes a similar counting head at the end of the section, the counter decrements. If the net count is evaluated as zero, the section is presumed to be clear for a second train.

This is carried out by safety critical computers called "evaluators", which are centrally located, with the detection points located at the required sites in the field. The detection points are either connected to the evaluator via dedicated copper cable or via a telecommunications transmission system. This allows the detection points to be located significant distances from the evaluator. This is useful when using centralised interlocking equipment but less so when signalling equipment is distributed at the line-side in equipment cabinets.

Traditional axle counters are of the mechanical type, with a pedal moved by the edge of the passing wheel. It is well known that these devices suffer extreme environmental conditions, in particular snow and ice, and may be subject to various kinds of mechanical breaks. They may still be found in very old transportation systems and in some low performance applications, such as in a depot, loading or marshalling yards.

The modern axle counters are all of the magnetic type, with the passing wheel that modifies the magnetic reluctance of a tuned magnetic circuit, through which a transmitting and a receiving coil are coupled: the transmitter sends a sinusoidal signal in the form of a magnetic field that is coupled to the receiver; the signal toggles normally between two frequencies so that the detection of reluctance change due to the passing wheel is accomplished as the detection of a relative change of amplitude between the two.

3.1.1.9 Inductive Loops

The inductive loops are designed for installation between railway running rails and they provide communications between trains and trackside equipments. The form of the loop is an eight-shaped circuit with a width equal to the rails gauge or half of it (one of the two longitudinal sides of the loop may be on the track axis) and the length of each turn that is limited normally to some meters or some tens of meters; not all the turns of the loop are of the same length, because they are also related to a special additional coding. Crossovers in the same loop are sensed by the on-board antenna since the polarity of the picked-up signal reverses: if the crossovers are placed at a given fixed distance, the train is able to know its speed and to reset any accumulated error in the odometry; moreover crossovers can be used for accurate train stop in front of some signal, at platform or at platform screen doors in automatic drive metros and light railways.

An on-board antenna detects the carrier flowing in the loop and even the coded messages in more modern versions. The coded messages may bear moving authorities, other control commands, train status, etc.; a high carrier frequency (most often between some and several tens of kHz) is possible since the p.u.l. parameters and the length of the loop cable are more favorable than those of the running rails, so that the operating frequency can be pushed higher than for TCs; with a high carrier frequency also the symbol rate on the channel is higher and more complex messages can be exchanged.

The susceptibility of these systems is complex since the on-board antenna may pick up any return current component in the rails as well as stray magnetic field emissions from power apparatus on-board, and in particular motor cables and filter inductors; on the contrary, the loop on the track plane is susceptible to the stray magnetic field appearing beneath the vehicle caused by any power apparatus or component along the whole length while it is travelling over the loop itself, with a certain deal of modulation caused by the train speed with respect to the fixed loop.

The TWT by Ansaldo operates on two different adjacent frequencies for wayside-to-train (up-link) and train-to-wayside (down-link) communication in the ranges 30-36 kHz and 50-56 kHz respectively; a FSK modulation is used with a data rate of 1200 baud [8].

The LZB (Linienzugbeeinflussung) is probably the older loop-based system (that however integrates also cab signalling and train protection capabilities, so going beyond the original inductive loop system) and dates back to the '60s. Mainly used in Germany and Austria, and also Spain, it is characterized by a standard crossing distance of 100 m, 50 m at the boundaries of a LZB controlled section; the maximum length may be as long as 12.7 km, but to enhance robustness and system availability modern LZB is divided into 300 m sections, supplied separately by repeaters. The trains are equipped with receiving antennas and the telegrams are sent to the train using a FSK modulation with 1200 bps baud rate, operating around 36 kHz with a frequency deviation of 0.4 kHz; the train replies with telegrams transmitted back at 600 bps baud rate using a FSK modulation around 56 kHz with a frequency deviation of 0.2 kHz. Call telegrams from trackside to train are 83.5 bits long, response telegrams sent back by the train are 41 bits long.

3.1.1.10 Balises

Historically, there were many products based on point information and transmitting data to (and possibly from) the train by means of an electromagnetic field at different frequencies. The first attempts were using magnetic field at one or few kHz, as in the case of Indusi (Inductive Signal Protection, from German), based on the transmission of magnetic field by on-board coils at selected frequencies (500, 1000 and 2000 Hz), coupling to tuned resonant circuits trackside. Coupling between the on-board and trackside circuits is detected by the on-board apparatus and coordination between a speed decrease and a delayed braking takes place: the correct sequence down to a complete stop is made by a first 500 Hz point for an initial speed reduction, a further speed reduction is indicated by the 1000 Hz point and the complete stop at the signal is indicated by the 2000 Hz point.

Modern balises are devices that represent points of information transmitting to on-board receiving apparatus over a radiofrequency channel. The transmitted information is in the form of a message, called a telegram, that transports the information identified at the beginning of this section for cab signalling in coded bit fields, together with the balise identification code and additional bits for error correcting capability. The effort of the manufacturer and of the standardization bodies has been to keep the telegram as short as possible, and correspondingly the time interval needed for transmission, copying with increasing train speeds; the track length available for transmission depends on the radiating characteristics of the balise and on-board antennas (gain and radiation pattern) and on the sensitivity and transmitted power of on-board receiver and balise transmitter respectively. A passive trackside balise is energized by the approaching train, by the so-called telepowering, occurring over a different frequency channel; the time required to have the balise in full operation, so that a message can be transferred to the train, must be compatible with the maximum train speed and the length of track where a reliable transmission can occur.

EBICAB, originally derived from Ericsson system, is a trademark of Bombardier. It is a precursor of the balise-based signalling and its introduction dates back to the '80s. It was designed to operate in the ISM band at 27 MHz with transmission of data to trains occurring at 4.5 MHz at about 50 kbps, with 32 bit packets for EBICAB 700 (encoding 12 bits of information) and 255 bit packets for EBICAB 900 (encoding 180 bits of information); the latter is evidently one of the reference designs considered for the Eurobalise.

A more modern implementation of some of the Eurobalise features is represented by the Italian SCMT ("Sistema Controllo Marcia Treno"[4]), coexisting with the much older RS ("Ripetizione Segnali"[5], based on coded currents sent along power frequency track circuits of the 50 Hz type, see sec. 3.1.1.5) and integrating functions on the on-board subsystem (SSB). The physical layer operates on the same frequency bands of the Eurobalise; the system is equipped with two on-board antennas for redundancy and some degree of repetition of lost messages.

3.1.2 Signals

The aim of all signal systems is to prevent derailment or collisions while the rolling stock travels at the prescribed or assigned speed. The safety function of signals is implemented traditionally by visual means (coloured lights or semaphore arms) or by a combination of visual and audible commands directly in the driving cab. A further level of enforcement in the same driving cab may be foreseen, should the driver fail to respond in the due time to cautionary or danger signal aspects. Like any other safety related component signals are required to be fail-safe, in that the failure of any component cannot result in a potentially hazardous situation, but must lead to what is considered a safe default configuration[6]. When also human failure is to be taken into account, for example while passing a danger signal, various forms of Automatic Train Protection (ATP) take place.

---

[4] Italian, system for the control of train movement.

[5] Italian, with the meaning of signal repetition on-board.

[6] It is observed that the "safe default configuration" is in a railway philosophy the condition of the complete stop, but there are situations in which this could not correspond to the highest level of safety, like, for instance, during an accident in a tunnel with the risk of fire: in this case the trains shall be recovered as fast as possible.

The type of signals implemented on a route depends on the circulation/traffic rules of the infrastructure owner and of the specific country: speed signals indicate by their aspects what is the speed the driver must follow (speed reduction may occur for several reasons, such as reduced braking capability on the route section ahead, or the simple fact that the next signal ahead is at danger); the indication of the target speed may be given by a colour code on the signal aspects or by numerals that indicate quantitatively the prescribed speed or a combination of both. Finally, it is worth noting that alternative to speed signalling there is route signalling, where the indication is on the line to take, which permanent speed restrictions are related to. In practice, there are many situations in which the two approaches are mixed and, for example, speed indication is accompanied also by a form of indication of the line on which the train is being routed, and vice-versa.

There are many type of signals depending on their function. Speed signals have been already considered above; other signal profiles are running signals (that control train movement from block to block), stop signals (that require trains to stop at or before the signal post), warning signals (indicating a successive stop signal and for this reason named advance or warning signals), repeating signals (in case a signal cannot be sighted at a sufficient distance, for example, because of a curve), contraflow signals (to warn the circulation on a track is occurring in the a direction opposite to the normal one, and this may happen both in normal use and in case of track works or emergency), shunting signals (used at low speeds to bring train together, for example, for attaching them to form a longer train consist or to take them to a depot, but also simply to allow access and admit a train to an occupied track at a platform), level crossings track signals (protection may be ensured by normal running signals, or in case of automatically operated level crossings there is a variety of choices depending on operators).

The coding of aspects is achieved by means of color, relative position and flashing. Colors are many, from the main and universally interpreted ones (such as green, yellow and red, usually associated to free line, warning and stop concepts, yet with several differences among operators and countries) to other more specific colors (such as white, blue, violet, etc.). Flashing has a varying meaning, usually associated to caution or to a temporary situation, but the examples span from contraflow signals and speed reduction even if the track is free, for flashing green, to caution, reduced braking distance, or even drive on sight, for flashing yellow, to even further speed reduction, extreme caution, level crossing not operating, for flashing red.

The case of lamp failure is being considered and treated in various ways through the years by the railway operators: single or double filament bulbs, separate redundant bulbs, control of filament health status (by either a dedicated procedure repeated at prescribed times or continuously, by monitoring the power absorption, or using a light detector, such as a photo-diode), recently improved by the introduction of LED (Light Emitting Diode) lamps, with increased redundancy and improved reliability. Concerning reliability the figures given as prescriptions for suppliers or measured on-site are from about 2000 to even 10000 hours of operation, with the largest numbers for halogen lamps, freshly introduced at the time these figures refer to.

The supply and the control signals are as well important for the analysis of safety factors. In the past voltage levels and frequencies for supply were conveniently selected similar to those used for other parts and circuits. The signal operation was that of a series of resistive loads (the lamps), driven by multi-pair cables with each pair carrying the voltage level for lighting up the aspect; because the voltage level at the source is fixed at the nominal value and the cable impedance may be highly varying depending on the signal distance and thus cable length, fine regulation is always provided locally by means of a rheostat or variable tap transformer or a combination of the two to accommodate for luminosity equalization. More recently, with the introduction of modern interlocking systems and following more demanding safety and interference analysis, control signals and even power supply are represented by "coded" signals, that mainly consist of overlapped frequency channels each used to drive a separate aspect or to convey a more complex code that is locally decoded, or are possibly digital signals sent over a copper or fiber cable. Frequency channels are selected not to overlap with the known disturbance associated with the traction circuit (supply fundamental and harmonics) and rolling stock (spectrum of conducted emissions); frequency separation is obtained by means of band-pass filters.

The analysis of interference starts from the calculation of the induced voltage on the signal cable over its whole length (possibly adding conductive coupling in case insulation fault and leakage are included in the analysis). Then, the results in terms of voltages at the supply frequency and relevant harmonics are compared to the operating modes and susceptibility of the signal. The determination of the latter is often performed identifying the minimum current to make a signal visible during worst case conditions, e.g. during night.

3.1.3 Switches or Point Machines

A pair of tongue rails with their stock rails is termed as *point*. A set of point and crossing leading from one line directly onto the other line is a *turnout*. Two sets of points and crossings (that is two turnouts), leading from one track to the other and vice-versa is a *crossover*. A turnout is composed of the following elements (see Figure 3.11):

- tongue rail, also called switch rail, a tapered movable rail made of resistant steel to withstand the wear;
- stretcher bar, provided to maintain the desired opening at the toe of the tongue rails;
- stock rail, the fixed rail against which the tongue rail operates and houses with its toe;

- outer straight lead rail, the outer running rail representing the continuation of the stock rail to the running rail of the straight track outside the point area;
- inner straight lead rail, the inner running rail representing the continuation of the stock rail to the running rail of the straight track outside the point area;
- outer curve lead rail, a curve rail connecting the tongue rail on the outer side to the crossing leaning on a heel block;
- inner curve lead rail, a curve rail connecting the tongue rail on the inner side to the crossing leaning on a heel block;
- check rail, to guard the wheel flange while leaving the point;
- gauge tie plate, to keep the gauge maintained in particular for sleepers that are not made of steel.

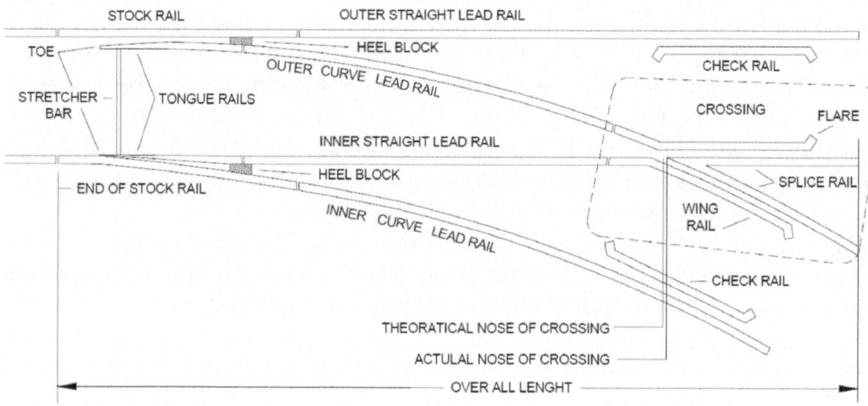

Fig. 3.11 Mechanical structure of a turnout

A turnout is designed as right end or left end depending on the direction of diversion of the traffic. The direction of the point is called *facing* when the vehicle approaches to face first the thin end of the tongue rail. Conversely, the direction of the points is said *trailing* if the vehicle approaches first the crossing before entering the tongue rail. Depending on the tongue rail and lead rail curvature the turnout is characterized by a limit speed for the vehicles passing over it.

When a point is operated, it is freed from a locked rest position, it is moved by a oleopneumatic or electric transducer or drive and it lands on the other rest position where it is locked; if for any reason (dirt, a stone, a) the rest position is not reached there is the need of operating the point in the reverse direction to bring it again to the original safe rest position; an unlocked position is a dangerous unallowed position that can lead to a derailment.

With a closer look to the electrical elements and to the control of the point, the following elements can be identified:

- one or many electric motors, that directly through a mechanical apparel or indirectly through an oleodynamic or pneumatic circuit apply force to the point moving parts;
- for a mechanical point machine, other components are represented by a clutch, gears and various mechanical parts to apply the force with the required direction and intensity, and at a row of points that depends on the mechanical characteristics of the point, such as length and weight of the moving parts;
- for an oleopneumatic point machine, a complete hydraulic circuit is composed of pistons, electric valves, pipes and other mechanical parts, again to apply the force to the point rail elements; these types of point machines are normally used for high speed points that require smaller curvatures and thus are longer and heavier;
- electromechanical devices to lock the tongue rails in either positions;
- various contacts to remotely monitor the occupation of the positions, to set the direction of movement of the tongue rails, to monitor the status of the same electric motors, like overcurrent or end-of-run sensing; relays are normally used to implement all these functions, and also to monitor the relays themselves, in order to achieve fail-safe operation; the interface with the interlocking and station apparatus is implemented with the so-called two-wire, three-wire, five-wire control interfaces, depending on manufacturer's and administration philosophy;
- all points have a disengagement option and movement is applied locally by a crank handle.

Supply Characteristics. The electric motors may be of the dc or ac type (either single or three phase), supplied at various voltage levels such as 110, 144, 150, 220 and 380 V. Lower voltages, such as 24 and 48 V dc, and the industrial 380 V three-phase are also available. For example 24 V has been used in Australia to interface with solar power supplies, mostly useful since the electric loads on such railways are quite dispersed and a dedicated distribution network is challenging and anti-economic [535]. The supply voltage is applied to the motor by means of a contactor and a second contactor is used to reverse the polarity (exchanging two phases in the three-phase case) to reverse the direction of motion. The rated power of the electric motor typically ranges from 1 to 2 hp, so from 0.7 to 1.5 kW. When operated, a point machine is characterized by a current absorption that is proportional to the required torque (depending on the rail weight, on the lubricating conditions, on the external temperature and possible friction, the presence of snow, etc.) and lasts for a time interval that is determined by the required movement and the gear ratio, set also by factors that oppose the tongue rail movement. The time duration of the operation (called "switch over time") is

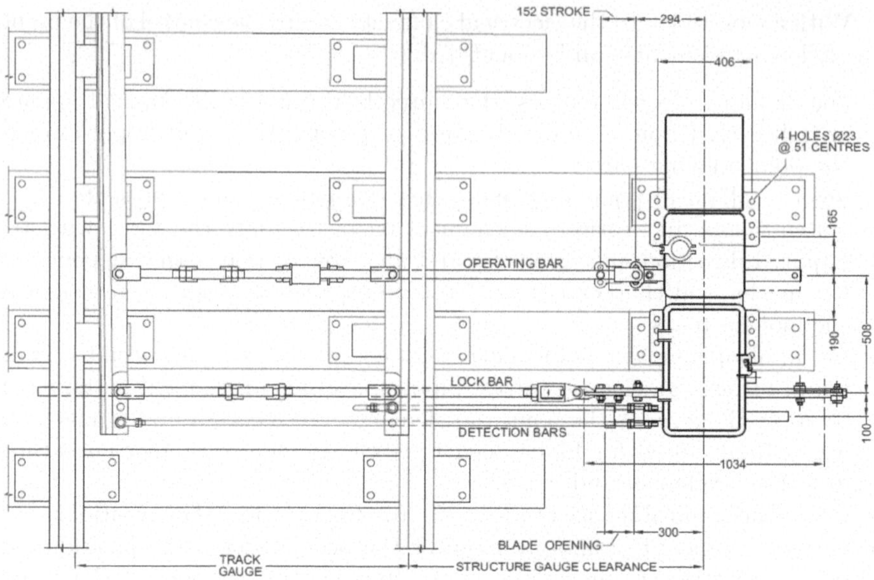

Fig. 3.12 Electrical point machine and tongue rails

Fig. 3.13 Pictures of point machines

widely variable from a few seconds to almost half a minute [673]. It is evident that current absorption and wire sizing is a relevant aspect since about 5 to 15 Arms are likely to flow during operations for various supply voltage and rated power combinations. Usually current absorption is maximum at the beginning, when the motor starts from standstill with maximum load due to static friction, and at the end of the maneuver, when the tongue rail has reached its final position and the rotor is almost blocked (this situation is quickly removed by the relays for overcurrent protection, end of run detection and jammed point detection). A point machine may be located far from the cabin with the supply source, thus requiring a long run of cable, possibly affected by a significant voltage drop. The motor connection is almost fixed for ac type, while series, series-multiple or series-split configurations are possible with dc motors, and they must be decided and implemented during installation.

Immunity to External Disturbance. Concerning the immunity of point machines to external disturbance and the possibility of undue movement of the tongue rail, it is first noted a change of statue from the rest position by the tongue rail is considered potentially unsafe, this might occur even if the required energy, to instigate such a change, is much less than that normally required to for a complete maneuver. It is sufficient and quite possible that the external disturbance lasts only for a fraction of a second, so again much less than the duration of the switch over time. This "opens the door" to possible interference from short circuit events, that inductively couple onto the point machine supply line. The components that characterize the short circuit current on dc and ac traction systems are the supply fundamental (either dc, 16.7, 50 or 60 Hz), a transient low frequency component (due to the unidirectional component) and some harmonic ripple (for dc traction systems at the characteristic harmonics of the substation rectifiers). The induced voltage on the victim cable is proportional to frequency, so that any dc component may be immediately ruled out.

A review of point machine datasheets indicate that point machines have been designed with immunisation in mind with dc type point machine being classified as "ac immune", implying that a balanced steady ac waveform of high enough frequency is not able to produce any net torque and movement. However, if the time duration of a first positive or negative peak is long enough – as for the unidirectional component –, then the time-amplitude product is relevant even if the waveform is globally "ac like". This aspect is particularly critical for several reasons: the lack of data concerning the immunity of these machines to such transient waveforms, the intrinsic difficulty in modeling these transient phenomena with the most common frequency domain approach, and the variability of several parameters that all influence the amplitude and time duration of the transient unidirectional component. On-site tests on a 144 V dc machine confirmed a sort of hyperbolic susceptibility curve, with higher pulsed voltage levels allowed for shorter times and vice-versa, with a constant of about 10 Vs at short intervals (below 50 ms) up to 15 Vs for longer intervals (around 100 - 150 ms).

On the contrary, ac type point machines are a more modern device and we can state that ac machines, and in particular three-phase ones, are immune to the vast range of induced disturbance considered above. This statement is supported both by intuitive considerations on the operating principles of the electric motor and its behavior with respect to common mode induced voltages rather, and by the evidence gained by several on-site tests, where very large voltages were successfully applied with limits only imposed by the voltage withstanding capability of the machine and wiring (up to 1200 V rectangular pulses of variable time duration on a 220 V 3-ph. machine).

Some point machines manufacturers are BBRE (UK), Cogifer (France), Voest-Alpine (Austria), Siemens (Germany), Union Switch & Signals (now Ansaldo STS), Westinghouse (now Invensys).

3.1.4 Level Crossings

Train presence at level crossings is normally monitored by the combination of two systems: a safe train detection system, such as a track circuit of adequate length, and a mechanical or electric treadle, for redundancy and to detect train direction during train passage. The combination of devices outputs with the required delays and redundancy checks is performed by relay logic inside the interlocking system.

The track circuit may be simpler and shorter than those of the line type and operate in overlay with possibly existing full line track circuits (see sec. 3.1.1.6). For this reason, the track circuits for level crossings have been designed with high or low operating frequency (above or below the first interval of audiofrequency track circuits, but above the power frequency track circuits, i.e. 10 to 20 kHz and 250 to 1000 Hz respectively). Level Crossing TCs have been in use for many years and predates the use of audio frequency track circuit, they are therefore simply and not sophisticated in design: they have a very simple structure, the coupling to the rails being through a bandpass filter and a transformer, with a very simple modulation or no modulation at all, since there was no need to transmit codes, and immunity to conducted disturbance was not of concern. For this reason there are many cases of renewals of old lines that begin with the most safety critical devices: single-rail track circuits at stations and level crossing track circuits. General requirements for modern level crossing track circuit are:

- operation in overlay with the full line track circuits, that in the '80s and '90s spread for type, operating frequencies and modulation characteristics, thus requiring more and more flexibility in the selection of their own operating frequency;
- compatibility with various track grounding methods;
- availability of short to moderate lengths in the range of 20 to 200 m;
- no requirement for Insulated Rail Joints, to ease overlay and to minimize the impact on the infrastructure and on existing signalling circuits.

Treadles are devices able to sense train passage and are normally placed at the end of the level crossing area for the normal traffic direction, or doubled and placed in the symmetric opposite position for counter-flow traffic. They may be either mechanic (the most common at the time of installation and with benefit for cost reduction) or electrical. They are quite robust devices with respect to interference and to environmental agents; it is funny to think that a relevant problem in urban areas are the kids standing on the mechanical treadles!

The task of a LCTC or a treadle is to free the level crossing when the train has passed, leaving it to road traffic and pedestrians, so that there is the exigency to release the level crossing as quickly as possible. On the contrary the level crossing barriers are lowered when the train is approaching and this is detected by the full line signalling, so adequately anticipating the train passage.

Level crossings are a particularly sensitive part of a transportation system, where minimum delays or malfunctions are immediately perceived and where safety is particularly important, since the behavior of cars, car drivers, pedestrians are all exogenous variables, that tend to escape a "pure railway logic". For these reasons their correct operation is continuously monitored in a number of ways: sensing absorbed current, for amplitude and duration of barrier maneuvering, as an indication of their correct operation; measuring time intervals between various events, such as time from train entering the controlled area to lights switched on, or time from lights switched on and train passage in front of the barriers, total road closed time, etc.; checking the coordination of the various signals on the railway line and on the road. Moreover, the reliability of the information supplied by the level crossing signalling devices may be increased in favor of safety by overlapping treadles and LCTCs and by applying coordination at the interlocking level with the line signalling, to ensure the redundancy of the sensitive information and to prevent various procured free conditions, at the expense of retarding the release of the same level crossing.

Examples of LCTCs are all overlay TCs that were considered in sec. 3.1.1.5, and in particular in sec. 3.1.1.6.

3.1.5 Relays

Even with the recent widespread use of digital techniques, relays continue to find wide use in the railway. As such, an overview of relays is provided in this section, this is particularly so as they represent one of the EMC relevant elements that are to be taken into consideration when performing interference analysis.

Basic mechanical elements and naming conventions are briefly stated to support the following discussion:

- contacts states are usually named "high" and "low" or "front" and "back", depending on the movement with respect to the observer; when the relay is de-energized it rests on the low (or back) position and these contacts are thus said to be normally closed (NC);
- contacts may be independent (if they are free of closing/opening with their own dynamics) or dependent (if a group of contacts is solidly bonded and their movement is synchronous, and they reach their final position

all together); this is done to ensure that even in the case of mechanical damage or melting of a contact surface the contacts are not brought to an unsafe configuration that does not correspond to the issued command signals; this feature is very often used in logic functions when an event is propagated onto two or many logic chains that cannot be brought in an ambiguous state;

- moreover, for a contact or dependent contact group the manufacturer specifies if the final state is reached (making or establishing the contact) before the old state is left (breaking the contact) or not, thus identifying a break-before-make (more common) or a make-before-break property;
- contacts are mostly regular contacts for normal service, that as per operator's specifications correspond to some Amps of nominal current even if the relay is used inside a cabin for the implementation of logic functions only; this increases in general the robustness, the reliability and the expected service life of the unit; additional or service contacts may be present, often in dependent connection with regular contacts, for other logic functions; the surface finishing is normally silver because silver oxide is a good conductor as well and its formation with use and aging doesn't increase the electrical contact resistance and doesn't produce overheating; heavy duty contacts may be supplied as well with specific shaping and other techniques for the extinguishing of electric arc formed during opening of large dc currents;
- coils put the contacts in movement by applying magnetic field to the magnetic apparel; different coils are normally available that may correspond to different number of turns so allowing to control the force that is applied to the contacts and the moving speed by selecting one of the coils or a series combination; we will see that separate elementary coils allow to apply a different force while picking up and releasing with different moving speeds thanks to the use of current direction selective devices, such as diodes;
- fast/slow pickup and release operations, where all combinations are allowed to accommodate for specific functions that are beyond a mere algebraic logic, such as slow release while a train is leaving a level crossing area to avoid races in the various logic chains triggered by the event;

Relays may be classified in several ways, depending on their operating voltage levels, the coil resistance, the accepted input waveforms, the energy required to move the anchor, the mechanical apparel. There are major categories based on the type of supply that are further subdivided for additional characteristics obtained by either mechanical or electrical means; vane relay, despite being an ac relay, has been kept separate because of the peculiarity of the operating principle.

3.1.5.1 DC Relay

The coil is a dc supplied coil, where the magnetic field created by the dc current is able to move the armature to either position depending on the polarity of the applied signal; a rest safe position is however ensured by the elastic force of an opposing spring in case of supply shortage or cable interruption. They are normally used to sense the output of a receiver indicating free and occupied conditions; one of their peculiarities is that there exist dc relays that are strongly ac immune and thus to any induced interference picked up by signaling cables.

3.1.5.2 AC Relay

The coil is an ac supplied coil, where the magnetic field created by the ac current is able to move the armature to a position of minimum reluctance, but without the selection of the polarity, so that the return to the rest position is ensured by the elastic force of an opposing spring. DC relays are usually preferred because of the control of the polarity driving the armature towards both positions and the possibility of backup supply by means of batteries.

Sometimes ac relays are equipped with an internal rectifier located at the terminal entrance inside the case, so that they of the same kind of a neutral dc relay. The main use of ac relays is that of sensing the presence of the ac supply and react in case of supply black-out switching to a backup battery system.

3.1.5.3 Vane Relay

An aluminum vane (low mass density, high electrical conductivity) free to turn around a shaft is positioned between two polar expansions, one for the coil supplied by the track signal the other one supplied by the local signal (this technique is used in track circuits with code duplicated on a safe separate cable laid down in parallel to the track from the transmitter to the receiver). The coils may be split into two separate windings, not only to accommodate for different driving voltage levels, but also to insert a phase-shifting or filter device; filters on local and track inputs may be used to prevent interference from power supply lines and traction current distortion by rolling stock emissions.

The operating principle of this electro-mechanical system is based on the eddy current effect produced in the aluminum material by a transversal magnetic field; if the magnetic field were moving along the aluminum plate, the circulation of the eddy currents would produce a force by reacting with the magnetic field. In a vane relay the two coils are supplied out of phase and the resulting magnetic field mimics the moving magnet and produces rotation of the aluminum vane; for the maximum torque the track and the local signals must be out of phase by a quarter of a cycle (90°). A phase-shifter may be needed to optimize the phase relationship between the two, depending on the real ballast condition, the electrical parameters of the track and so on; in some cases a compensating impedance is added to accommodate a variable length of the track circuit.

Typical working voltages are between 50 and 110 V ac for the line input and a fraction of volt to some volts for the track input [3]. The coil resistance values are between a hundred and a thousand ohm for the line coils and a few ohm for the track input (the track coil results in series to the receiver input through the connecting cable with its own impedance, and for this reason its resistance shall be compatible with the usually specified values of the receiver impedance). Vane relays were often used for power frequency track circuits and the operating frequency of old electromechanical units is normally between 50 and 150 Hz, equal to 50, 60, 83.3, 100 and 125 Hz; the most modern electronic units implement the same characteristics, but in principle they could be readily adapted to any desired value.

3.1.5.4 Motor Relay or Timer Relay

Motor relays (and now more modern electronic timer relays) were conceived in order to provide long accurate delays following a signal transition to a predetermined level.

Electronic versions are microprocessor controlled and are able to operate for a vast range of input signals, both dc and ac, at the most common frequency values (the difference of frequency value usually being only relevant for the input filter setting); moreover, they are able to implement much longer delay times up to several minutes, with accurate time trimming to a fraction of a second.

3.1.5.5 Coil Arrangement

There exist one-coil or two-coil relays, where the connections between the coils are made on the back of the plug-board for series, parallel, or single coil operation. The possibility of reconfiguring the coil arrangement gives additional degrees of freedom with respect to the following aspects:

- force exerted on the armature, that combined with the mechanical stiffness and the elastic force of the compression and compensating springs give rise to slow pickup/release operation; for example slow pickup or release are normally obtained using only one coil, even if it is underlined that the control of the time response is not so accurate, being mutual influence between the driving line impedance, the absorbed current and the voltage drop, and the instantaneous force exerted on the armature;
- change of overall coil resistance, for ease of interfacing to incoming cables or other relays;
- possibility of introducing additional electrical elements to modify the electrical time response or the circuit behaviour, such as adding series RC circuits in parallel to the coils, trimming the switching front of the incoming signal (as shown in Figure 3.14(f)), or using diodes to discriminate between pickup and release, shunting one of the coils for a given polarity

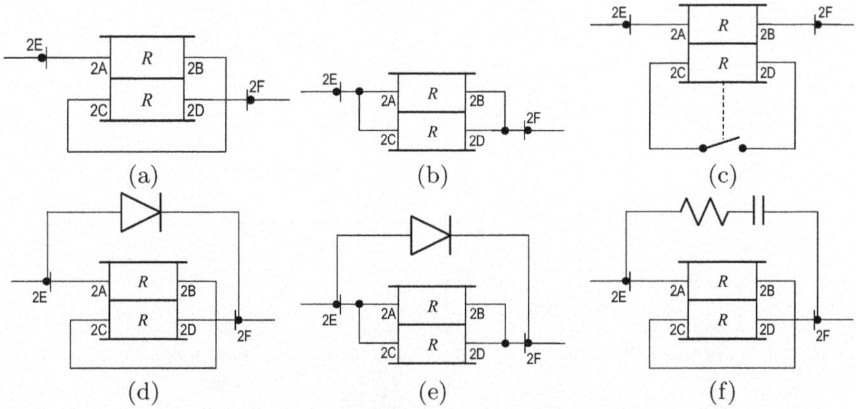

Fig. 3.14 Electrical schemes for various dc relay coil configurations

of the incoming dc current and of the applied dc voltage (as shown in Figure 3.14(d) and (e)).

- neutral and polarized (or biased) relays (dc relays), that reacts to a signal of either polarity or to a signal of the right polarity respectively; the latter is implemented by including a magnetic in the magnetic circuit of the armature, that opposes to the ampere-turns produced by the voltage applied in "wrong" polarity;
- latched or impulse operated relays (dc relays), that change status after a voltage of correct polarity is applied for a long enough interval (voltage impulse) and then they keep the last position of the armature, usually because of an internal magnet; an impulse of opposite polarity brings the armature in the other position where it rests again because of the internal magnet attraction; the latch function may be implemented by supplying the coil in parallel through one of the normally open contacts, that once closed supplies the coil and is thus retained in that position until the interruption of the continuity on the parallel supplying circuit by another relay [2].

Maybe the most natural classification for interlocking applications is with reference to their safety properties, that is if they can be employed in safety relevant circuits or not, or in other words if they can implement a prescribed safety function with or without the help of other auxiliary relays, that realize a logic function:

- vital track relays, that implement a combination of characteristics against external disturbances, thus improving the immunity of the relay itself: a) input voltage level above or below a given threshold, b) input voltage frequency close to the nominal one within a prescribed bandwidth, c) input

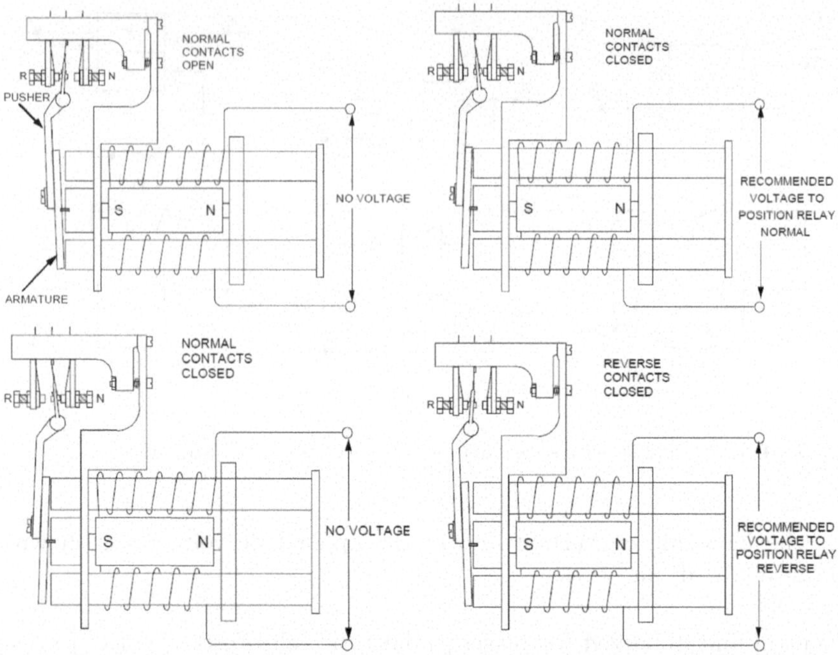

Fig. 3.15 Operation of a DC impulse operated relay (Alstom, P1457 [2]): the relay is in the rest position and no voltage applied (upper left); the armature is moved to the right attracted by the applied signal (upper right); with no voltage applied the armature stays in the new rest position (lower left); a signal of reverse polarity brings the armature in the previous position (lower right).

voltage phase matching a control signal phase within a prescribed tolerance, d) pickup/release delay, acting like a low pass filter against even intense transient disturbance [3];

- non vital relays, not adequately implementing the characteristics above a prescribed level of immunity and thus prevented from being connected to devices and cables from the field; they are used for other logic functions in a protected environment such as the interlocking cabin.

To this aim it is remembered that in the past relays were designed, manufactured and tested to meet the specifications and needs of one operator, and thus they have inherited the classification and the naming conventions of the operator itself. For this reason manufacturers have given their products, now sold all over the world, characteristic names, such as "Q" relays.

While fail-safe operations and an overall safety level may be ensured by accurate design of the *relay logic* at the interlocking level, EMC and evaluation of possible interference focus on the relay as an electrical device, connected to other relays within a cabin or to cables connected on the other end to field devices. To this aim relays may be synthesized as an impedance (with the resistive part given approximately by the coil resistance and the inductive part determined by the magnetic circuit, subject to movement by the way) with associated voltage levels in steady conditions for leaving the safe positions and corresponding time-voltage curves in dynamic conditions for the same event to occur. From a general point of view both positions (high and low) are to be considered, accompanied, more for interpretation purpose than for evaluation, by the associated level of relevance of the criticality: for example, in a very straightforward interpretation, the high position may be in relationship with a free track condition, while the other in the low position may be in relationship with occupied track condition[7].

Distinction has been made between steady and transient conditions:

- a disturbance is considered steady if lasting for more than any mechanical transient response of the relay itself and any associated delay, so that increasing further the duration of the disturbance doesn't influence the resulting behavior; the relay susceptibility is described as a voltage threshold above which the relay leaves its rest position, possibly moving the anchor to the other rest position, or better, below which there is no unsafe movement and the rest position is firm;
- a disturbance is considered a transient if its duration is comparable, and its impact on the relay behavior is to be evaluated by an equivalent area or energy concept; it must be remembered that a relay is an electromechanical system with a qualitative low pass characteristic given by the combined action of the coil time constant, the anchor inertia, the spring constant, any initial and viscous friction, additional delays (like in motor relays) and non-linearity due to possible saturation of magnetic materials; so the expected behavior of the thresholding principle is an approximate hyperbolic law where longer durations must be accompanied by lower amplitudes and vice-versa.

The relevance of transient events is related to the fact that they may convey higher energies, even if for shorter time intervals, such as during short circuits on the traction line and associated induction effects. A frequency domain analysis indicates that transient low frequency harmonics appear and they may couple onto dc and low frequency track circuits and relays. Considering the electrical coupling on the track circuit and the electromechanical dynamics of the relay, the analysis of all the possible coupling modes to obtain a comprehensive safety case is rather complex.

---

[7] The low level that is normally reached by gravity is always associated with the safe condition, that usually corresponds to assuming a track occupied, leading only in case of mistake to a delay and traffic inconvenience (right side failure).

3.1.6 Communication Based Train Control (CBTC)

CBTC is defined as a continuous automatic train control system utilizing high-resolution train location, independent of track circuits; continuous high capacity, bi-directional train-to-wayside data communications; and train-borne and wayside processors capable of implementing vital functions. CBTC is generally used to cover a wide range of communication based train control technologies, such as:

- ATCS: Advanced Train Control Systems
- ARES: Advanced Railway Electronics Systems
- AATC: Advanced Automatics Train Control
- ATACS: Advanced Train Administration and Communication System
- ATS: Automatic Train Stop (ATS-S first, and then ATS-P) [529]
- CBS: Communication Based System
- EJTC: East Japan Train Control system [529]
- ERTMS: European Rail Traffic Management System
- ETCS: European Train Control System
- PTC: Positive Train Control
- PTS: Precision Train Control
- TBS: Transmission Based Signalling
- TBTC: Transmission Based Train Control System

The basic standards for the definition of Communications-Based Train Control (CBTC) system was issued by the IEEE at the turn of the millennium. These definition are mostly dedicated to functional performance, such as the on-board bus types for local (L) and between coaches (T) communication, among the installed signalling devices (IEEE Std. 1473 [390], Std. 1474-1 [391], 1474-2 [392] and 1474-3 [393]), with the definition of the rules and requirements for the implementation of ATP, ATC, ATO and ATS[8]. The existing CBTC systems in use are very different and the differences derive from the method used for the train-to-wayside communication in terms of used frequencies, type of modulation, adopted protocols: Siemens with its Uragan system is using a proprietary physical layer at 900 MHz, Thales is working on a wireless band at 2.4 GHz using a Frequency Hopping Spread Spectrum (FHSS) modulation, to name a few.

Communication Based Train Control technology has been around since the late 1960s, with the development, in Europe, of a loop based system (i.e. LZB and SELTRAC), employing a continuous or semi-continuous circuit placed parallel with the track with one wire against a running rail and the other wire centred between the running rails. This asymmetric loop arrangement is common for most loop based CBTC system, although there are a few locations where the loops are arranged in a symmetrical position with both

[8] Going from a local low level to an overall high level perspective, ATP is Automatic Train Protection, ATC is Automatic Train Control, ATO is Automatic Train Operation and ATS is Automatic Train Supervision.

loops set at a distance from the running rails. More recently, CBTC has been developed using wireless communication technologies, most utilising GSM-R or a proprietary cellular WAN concept. The technology spectrum is divided between those implementing a loop system (otherwise known as Inductive Loop) and wireless systems. A detailed description on ERTMS (including ETCS) is given in the following subsections.

3.1.6.1 European Rail Traffic Management System

This section is dedicated to the architecture and functions of the system, but the technical details of the various elements are reported in other sections. The European Community has mandated ERTMS (European Rail Traffic Management System) as the system that must be adopted by all railways for all new high speed lines and will be progressively adopted also for conventional lines. It represents one of the most advanced Automatic Train Protection and Cab Signalling systems ever designed for use on main line railways. Many motivations are behind ERTMS such as: enable the unrestricted movement of high-speed rail traffic across borders, standardise the signalling interfaces and make them less country specific and be commercially attractive by encouraging an open market in equipment. ERTMS was identified as the new signalling system for inclusion in the Interoperability EU Directive 96/48 issued in July 1996. At that time the results of several EU funded projects (Morane, Eirene) were available. but no harmonised standard was available. The signalling-related part of the systems is called ETCS (European Train Control System) and is designed to be a highly advanced ATP and Cab Signalling system. The driver is provided with a comprehensive display interface independent on a specific manufacturer, minimizing the need for specific route knowledge and providing a protective safety envelope: this is called *Operational Interoperability*. From a low level technical point of view, the ERTMS motivations and expectations condense into the so called *Technical Interoperability*, that is on-board and trackside key subsystems can communicate with each other regardless the manufacturer. The ETCS system is composed of the following subsystems:

- EUROCAB on board computer plus Man Machine Interface (MMI) and Odometry;
- Specific Transmission Module (STM) – An interface module to provide compatibility with legacy signalling technologies, e.g. an ASFA STM to allow the train to operate over existing ASFA ATP equipped lines in Spain;
- EUROBALISE is the pan European spot transmission sub-system for transmission between the wayside and on-board ATC/ATP equipment; the Eurobalise can be a passive or active transponder system;
- Balise Transmission Module (BTM) – The train carried interface to the EUROBALISE;
- EUROLOOP – Is the semi-continuous transmission system that uses leaky coaxial cables to transmit safety related information between the wayside

infrastructure and the train; and vice-versa. Its main function is to transmit and/or receive signal through the air gap;
- Radio infill – Performs the same function as the Euroloop, but utilising a GSM-R radio link;
- EURORADIO – The data and voice radio system using a special variant of GSM modified for railway use (i.e. GSM-R);
- Radio Block Centre (RBC) – The interlocking interface and limit of movement authority management sub-system. The RBC also uses EURORADIO to communicate with the trains in its control area.

To address different operational contexts, referring to different line speeds, traffic patterns and route complexity, three *application levels* have been defined (Level 1, Level 2 and Level 3), with different sets of features and functions; backward compatibility with existing signalling systems in the absence of an ERTMS infrastructure was foreseen as level 0. Level 1 exists in reality in two versions, one without and one with infill by means of a loop before a balise.

3.1.6.2 ETCS Level 0

Level 0 covers operation of ETCS equipped trains running on lines or sections of a line not equipped with ETCS. For example an unfitted line being upgraded in stages will have already upgraded sections of Level 1 track with the yet to be upgraded sections as Level 0. Only a few basic functions are implemented, e.g. level transition and supervision of maximum train speed. No ERTMS/ETCS trackside equipment is used except for Eurobalises to announce level transitions and other specific commands.

Level STM is used to run ETCS equipped trains on lines equipped with other national train control and speed supervision systems. Train control information generated trackside by the national train control system is transmitted to the train via the communication channels of the underlying national system and transformed on-board into information interpretable by ETCS.

3.1.6.3 ETCS Level 1

ETCS Level 1 is a fixed block intermittent (or spot) supervision system. The trackside equipment consists of encoders (Line-side Electronic Unit, LEU) and Eurobalises. The LEUs are programmed with fixed infrastructure information and combine this with the variable information taken from the aspects displayed by the signals (e.g. by sensing the current flowing in the signal lamps or the position of the control relay contacts) and then pass a serial message to the Eurobalises for transmission to the train. The balise transmission takes the form of a telegram containing the ETCS Moving Authority (distance to go, speed restriction, gradient, static speed profile, temporary speed restrictions, driving mode, etc.) and other applicable control data; there are thus two types of balises: controlled balise, that needs a communication cable, and fixed balise. An on-board antenna receives the balise transmissions.

The ATP functions within the trainborne ETCS subsystems calculate a safe speed profile on the basis of the received ETCS data and known train braking parameters. An on-board odometry system senses the movement of the train to provide the ATP functions with the means to check that the speed and distance limits defined by the Movement Authority are not exceeded. The odometry function uses tacho generators, radar sensors and accelerometers to determine speed and location. The outputs of these sensors are combined to form most advanced and most retarded locations to ensure that factors such as wheel slide/slip cannot lead to unsafe estimates of actual location. The odometer error band is reset to zero every time an Eurobalise is passed. The Level 1 trackside subsystem does not know which train it is sending the information to. ETCS Level 1 can be readily overlaid onto an existing conventional signalling system without the need to renew existing signalling and modify the interlocking system. Being an intermittent transmission system, a train is unable to take advantage of a signal aspect stepping up until it reaches the balise group at the signal; in other words, the limits of the braking curve received by the previous Eurobalise shall be maintained until the train passes the next Eurobalise where the train computer receives updated information. To mitigate against this, *infill* can be provided between the distant and main signal, that is between the two Eurobalises. The basic idea of all infill systems is to close the gap in information provision between the distant and the main signal to mitigate the traffic performance degradation problem caused by the Eurobalise spot transmission system. Both intermittent and semi-continuous infill options are available.

Eurobalise. The Eurobalise is one set of technical solutions for Balises used in an ERTMS/ETCS installation. An Eurobalise is a Balise that fulfills the mandatory requirements of the European Train Control System Specification. A Balise is a wayside transmission unit that uses the magnetic coupling in the air gap between a transmitter and receiver for conveying data and energy. In the Eurobalise transmission system, the Eurobalise uses 27 MHz band for tele-powering and 4.5 MHz band for the Up link transmission. The magnetic field is mainly vertical, and the transponder is located in the centre of the track. Its main functions is to transmit and/or receive signals through the air gap. The Balise is a single device mounted on the track, which communicates with a train passing over it.

Information can be transmitted both ways, i.e., Up-link and Down-link data transmission. Information transmitted from an Up-link Balise to the On-board Transmission Equipment is fixed or variable depending upon the application (Up-link data transmission). Information can be received by a Down-link Balise from the train (Down-link data transmission). Spot transmission is when a transmission path exists between the wayside equipment and the On-board Transmission Equipment at discrete locations. The information is provided to/from the train only as the Antenna Unit passes or stands over the corresponding Balise. The length of track on which the information is passed and received is limited to approximately one meter per Balise.

Euroloop. The infill is implemented by means of a loop (Euroloop) that receives data from the LEUs through a loop modem. The infill information is transmitted by the Trackside Equipment only in the presence of a train. A leaky coaxial cable, the Euroloop Leaky Cable (ELC), of up to 1000 m length is fastened to the inner or outer side of rail's web (the vertical height from the foot of the rail to its running surface) and is used as trackside magnetic coupling device. Trackside signals are still required as the driver has to observe the status of the signal when approaching a balise group to know whether to stop the train in rear of the balise group (signal at "stop") or to pass over the balise group (signal at "proceed").

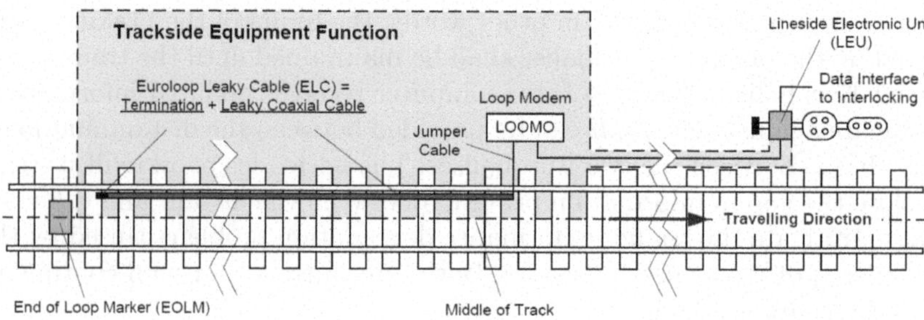

Fig. 3.16 Euroloop installation

The On-board Equipment function consists of an Antenna Unit (AU) function and a Loop Transmission Module (LTM) function with the Loop Receiver (LR) function and the Loop Decoder (LD) function. The Trackside Equipment function is made up of the Loop Modem (LOOMO) function and the Euroloop Leaky Cable (ELC) function.

The signal modulation of the ELS employs Direct Sequence Spread Spectrum (DSSS) technology:

- to allow for Code Division Multiple Access (CDMA) among adjacent ELS installations in the air gap;
- to increase the ELS' immunity with respect to narrow-band interference generated by infrastructure noise sources e.g. the train;
- to mitigate multi-path propagation of the ELS signal between the transmitting and receiving antenna.

A set of 16 different spread spectrum codes of length $Nc = 472$ are used, enabling 16 nearly orthogonal transmission channels within the same frequency

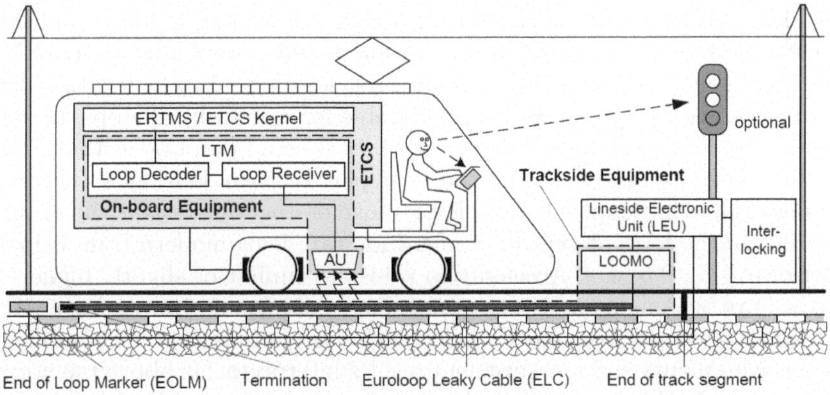

Fig. 3.17 Trackside & On-board Equipment of Euroloop Subsystem

range. The proposed centre frequency is 13.5475 MHz ±30 ppm. This choice prevents interference with the Eurobalise system covering the frequency band from roughly 1 MHz to 7.4 MHz and the tele-powering signal at 27.095 MHz ± 5 kHz. The carrier is modulated in a non-coherent way by the DSSS modulated data signal using Binary Phase Shift Keying (BPSK) modulation. The details of the selection of the Euroloop band and center frequency may be found in [252].

There are four factors that affect the time for which the loop transmits: the length of the loop, the speed of the train, headway (time between trains), if the train stops on the loop then the length of time for which the train stops and the associated acceleration/deceleration rate.

Typical loop lengths are 300 m up to 500 m and exceptionally up to 1000 m. The minimum length of the leaky coaxial cable to transmit a long a telegram is shown in Table 3.2.

Table 3.2 Minimum of length of ELC for telegram transmission

Speed	Length of ELC
180 km/h	15 m
300 km/h	25 m
500 km/h	42 m

Train speeds for a suburban route are generally up to 120 km/h and for intercity routes up to 200 km/h. Speeds of 200-300 km/h are normally only applicable to newly built high speed routes. The headway (i.e. the time between trains) depends on a number of factors. In the peak hour this will generally be higher than an average calculated over 24 hours. Very busy lines in the centre of cities, where a number of routes come together, might have a headway of 3 minutes or shorter, although this timing is more applicable to metro systems. A busy suburban station would expect a headway of 5-10 minutes in the peak and a 15-20 minutes service off-peak. The length of time for which a train may stop is not predictable for unscheduled stops, however station stop times for most trains will be between 45 and 60 s. Here trains are shut down before beginning the next part of their journey so it can be assumed that the loop will be off for most of the time when the train is standing on the loop. A typical deceleration rate for a modern train is in the order of 5% i.e. 0.5 m/s. Acceleration rates are similar or slightly higher.

Test have been made with a pure sinusoidal tone at 6.9, 10.9 and 14.9 MHz, with an emitted electromagnetic field much more intense than the normal to ease measurements and measurement results interpretation above the normal ambient noise. The results of the measurements performed both in the near field (10 m distance in front of the leaky cable) and in the far field (about 1 km in various directions) were extrapolated to the real Euroloop signal intensity and spectral distribution and are summarized in Table 3.3.

Table 3.3 Extrapolated e.m. field intensity from Euroloop in near (10 m) and far field (Loc 1, 2, 3 at approx. 1 km distance); in far field conditions the equivalent magnetic field intensity is given by $H = E/377\Omega$ or -51.5 dB.

Frequency [MHz]	10 m [dBμA/m]	Loc. 1 [dBμV/m]	Loc. 2 [dBμV/m]	Loc. 3 [dBμV/m]
6.9	-9.3	-12.1	-7.3	-17.1
10.9	-2.6	-17.8	-9.4	-10.3
14.9	-9.2	-14.6	-14.0	$<$-23.7

As far field measurements demonstrated, the Euroloop behaves like a travelling wave antenna with a gain of about -40 dBi over the operating frequency range; its large dimensions and the travelling wave behavior make it impossible to translate accurately a far field value from the near field values, thus justifying the far field measurements.

3.1.6.4 ETCS Level 2

ETCS Level 2 is a fixed block continuous speed supervision system. The trackside equipment consists of centralized of Radio Block Centres (RBCs) and distributed Eurobalises. GSM-R Radio replaces the Eurobalise to transmit the information to the train. Balises are still used, but only as a means to initialize and periodically recalibrate the on-board odometer (fixed balises). The RBC takes information directly from the interlocking on the aspects displayed (and therefore on the status of the line ahead) in place of the LEU. The RBC is programmed with fixed infrastructure information and combines this with the variable taken from the interlocking to select and then pass a serial message to the GSM-R radio for transmission to the train. As GSM-R provides bi-directional (duplex) data transmission, real time data regarding train locations and speeds can be made available to the signalling control and train describer function. Though trackside signals are no longer required, some could be kept at key locations as a fall-back and also for use by non-ETCS trains. The RBC that provides the information to the trains knows each ETCS controlled train individually. It provides continuous speed supervision and also protects against overrun of movement authority. Train detection and train integrity supervision are performed by the trackside equipment of the underlying signalling system (interlocking, track circuits, axle counters, etc.). The system does not require trackside signals but they are often retained as a backup in case of system failure or in case not all trains operating on the line are ETCS equipped.

3.1.6.5 FIS and FFFIS

Functional Interface Specifications (FIS) interfaces are only defined on a functional level, meaning that the logical data-flow is specified, but not the physical interface. A FIS does not therefore ensure exchangeability, as the physical interface might be different from application to application.

Form-Fit Functional Interface Specifications (FFFIS) interfaces are defined on a functional and technical level, meaning that the detailed physical characteristics of the interface are also specified. A FFFIS ensures exchangeability on that interface (but not necessarily of a single component).

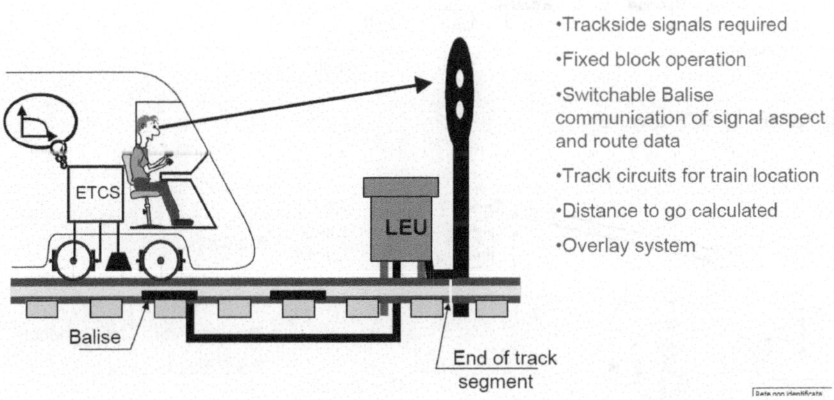

Fig. 3.18 ETCS level 1 without infill

Table 3.4 represents a synthesis of ERTMS/ETCS features described above and gives the whole picture of the implemented functions and of the most relevant elements. The three ETCS levels are compared, and Level 1 is further subdivided distinguishing between with and without infill option.

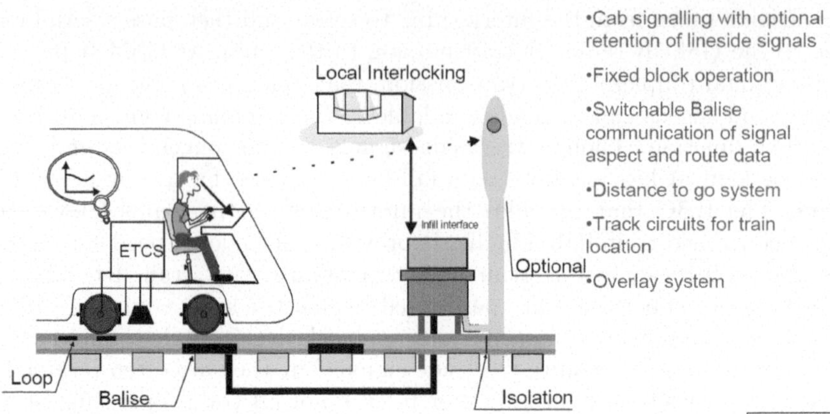

Fig. 3.19 ETCS level 1 with infill

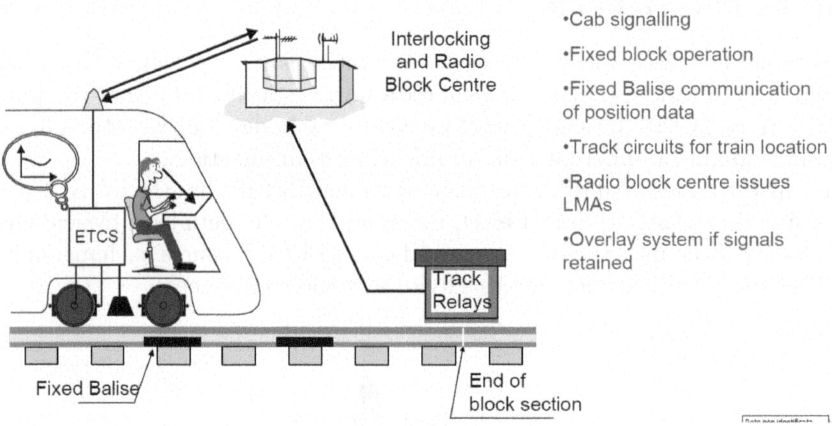

Fig. 3.20 ETCS level 2

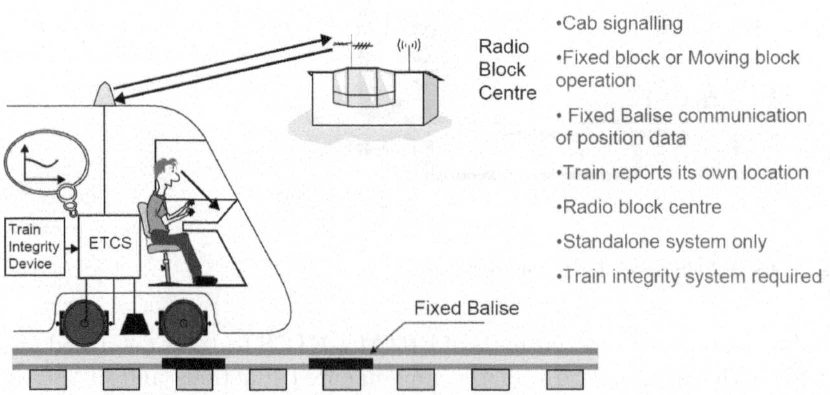

Fig. 3.21 ETCS level 3

Table 3.4 Synthesis of ERTMS/ETCS features

Feature	Level 1 w/o infill	Level 1 with infill	Level 2	Level 3
Implementation	Overlay on system that controls train separation	Overlay on system that controls train separation	Overlay or standalone (if separation controlled by RBC)	Standalone only
Interoperability	Technical	Operational	Operational	Operational
Trackside Signals	Yes	Optional	No (optional for Level 1 compatibility)	No
RBC	No	No	Yes	Yes
ERTMS MMI	No	Yes (control only)	Yes (control only)	Yes (control and advice)
Block Sections	Fixed	Fixed	Fixed	Fixed or Moving
Radio Communication	None	A Radio infill is Specified	Bi-directional to RBC	Bi-directional to RBC
Movement Authority	EuroBalise (switchable)	EuroBalise (switchable)	Radio based	Radio based
Overspeed Protection	Yes	Yes	Yes	Yes
Trainborne Location	Not required	Not required	Determined by train using fixed EuroBalise (Note 3) and by track circuits	Determined by train using fixed EuroBalise and on board odometry
Track Occupancy	Track circuits or axle counters	Track circuits or axle counters	Track circuits or axle counters	Reported by the Train (Note 5)
Train Integrity	Track circuits	Track circuits	Track circuits	Trainborne
Odometry for Train Position	Yes	Yes	Yes	Yes
Functions required from conventional Interlocking	Train separation; route set and lock; points and signal control	Train separation; route set and lock; points and signal control	Train separation; route set and lock; points control	Route set and lock; points control (Train separation made by RBC)

3.1.7 Cables

Many different types of cables are used in a modern railway transportation system: Medium Voltage cables, either single core or 3-phase, Low Voltage supply cables, for various kinds of supply and current ratings, signal cables for generic applications, network cables for data transmission, telephone cables, coaxial cables for radio, mobile and video applications. These cables may be purchased in different versions such as armored, shielded, double shielded, etc. Cables may be sources of disturbance, depending on the characteristic of emissions of the connected equipment, in terms of spectrum of emissions, source impedance, differential and common mode transfer functions, where the cable itself is part of the propagation and coupling path. On the other side, based on the same parameters, cables may represent a relevant coupling path for external disturbance. In this section the focus is on the general characteristics of cables that are used for EMC analysis and for the definition of cable segregation criteria.

It is underlined that most of the cable classification codes and criteria are still related to the past practice of a specific railway authority and are thus not harmonized and fully interchangeable with those of another country.

3.1.7.1 General Properties of Cables

This section reports an attempt of classification of cables, dividing them in power and signal cables, and focusing on the electrical and electromagnetic properties as they are normally referred to by the users, manufacturers and standards.

Power Cables. Cables for power applications may be of the single- to five-core type, with the single and three-core types the most common, even for very large current and voltage ratings, while the other types are used only for low voltage applications. The impedance/admittance matrix representation (that represents a common approach for cable modelization and simulation) is based on the self and mutual per-unit-length parameters, the former (longitudinal resistance and inductance, core-to-screen capacitance) given in the cable datasheet, the latter (mutual inductance and resistance, mutual capacitance between cores or screen termination in case of screened or armored cables), depending also on the cable arrangement and layout and normally given by the manufacturer in the datasheet as well for standard geometries.

Signal Cables. When considering signal cables of the multi-core or multi-pair type for use over a wide frequency range, the attention is focused also on derived parameters, such as:

- characteristic impedance, attenuation, differential capacitance:
  - the characteristic impedance, as defined in sec. 7.1.2, is relevant for the propagation of signals of relatively high frequency compared to

the length of the cable, as it happens for long cables of some km between track devices and the interlocking cabin; cabin apparatus and track devices (such as tuning and coupling units for track circuits) have impedance values matched to the characteristic impedance of cables, either directly or by means of a coupling transformer;
- the attenuation is the real part of the propagation constant (again defined in sec. 7.1.2) and for long trackside cables it represents an additional factor to consider; it is often expressed in dB/m and dB/km and at low frequency corresponds exactly to the longitudinal voltage drop due to the loop impedance of a single pair, given by the conductors resistance and the "self less mutual" inductance;
- the differential capacitance is a direct measure not only of the loading caused by the cable on the connected signal generator, but also of asymmetries in multi-pair cables, such as those used for telephone networks and to connect signalling systems; the uniformity over the different pairs of the differential capacitance (as well as of the wire-to-shield capacitance considered below) indicates a good deal of symmetry and a reduced exposure to common mode disturbance, and it is a feature that is often required for many signalling cables;

• transfer impedance, shielding effectiveness and screening factor:
  - the transfer impedance is a common way to quantify the amount of external common mode disturbance that transforms into a differential mode disturbance on the internal conductors of the cable;
  - the shielding effectiveness is a property of a cable shield that is directly related to radiated phenomena and thus limited to the high frequency range; the shielding effectiveness of a conductive screen is defined in terms of ratio of the incident field and the field appearing at the other side of the screen, that is for a cable inside the dielectric carrying the internal conductors; it is thus evident that its application to cables and coupled interference is limited and may be used for the high frequency properties of the cable screen, especially concerning optical coverage;
  - the screening factor, as normally referred to in the CCITT standards [417, 418, 419, 420] and in some other national standards, is the ratio between the external applied longitudinal voltage and the net differential voltage appearing at the internal signal conductor(s); the screening factor represents an equivalent parameter that defines globally the shielding effectiveness of the cable shield, the symmetry of the cable, the amount of unbalanced stray capacitance, as it is done for the transfer impedance; on the contrary, the screening factor is normally applied at low frequency for the calculation of induction effects, both in normal and short circuit conditions, and for this reason the specification is normally done at the fundamental traction supply frequency (and maybe at a higher psophometric frequency) and also for large current intensity (or applied voltage) to rate possible saturation of the shield material;

- crosstalk between pairs: crosstalk between different pairs in the same cable is normally caused by deviations from manufacturer's design (wire twisting and distance between pairs) and by worsening of the capacitance and conductance terms; for cables in good conditions crosstalk is generally very low, in the order of 60-80 dB, as confirmed by the values measured on telephone cables of several km length at Florence station [684]: the crosstalk at 200 Hz and 3400 Hz was about 81 and 71 dB/km respectively, and approximately linearly dependent on frequency between the two.

3.1.7.2 Signalling Cables

In this section a classification of signalling cables is proposed; it is underlined that there is no unique classification depending on the specific country and administration, the adopted standards, the characteristics of the managed railway systems and the origin of the installed products.

- **Signalling Cables Type A1, A2 and A3:** These types of cables are used for free wiring in Relay Rooms, Signal Boxes, REBs (Equipment Boxes) and Location Cases. All Type A Cables have Class 2 stranded tinned copper conductors and are insulated with a flame retardant LSZH (Low Smoke Zero Halogen) compound. A1 and A2 Cables are single core cables and A3 Cables are multi-core. A2 and A3 Cables also have an LSZH sheath. These light-duty Signalling cables are generally for internal use – a smooth sheath makes them suitable for installation in ducting and between other cables.
- **Signalling Cables Type B1, B2, D1 and D2:** Type B and D Signalling Cables are used between Equipment Boxes, Signal Boxes, Relay Rooms and Location Cases to connect up signal equipment. They are found in external cable troughs or attached to cable posts and provide control and signal transmission – although 2-core B2 Cables (10-95mm^2) provide power for the signalling network in trackside situations. B1 and D1 Cables are used for maintenance within networks that comprise of B2 and D2 Cables. Both B and D Cables have Class 2 stranded tinned copper conductors – B1 and D1 Cables are single core cables and B2 and D2 Cables are multi-core cables. Whereas B1 and B2 Cables have EPR (Ethylene Polypropylene Rubber) insulation and HDPCP (Heavy Duty Polychloroprene) sheathing, D1 and D2 Cables have LSZH or EPR insulation and LSZH sheathing. Type D Cables are used in place of Type B Cables in tunnels greater than 200m in length – and locations designated as underground such as Birmingham New Street Station.
- **Signalling Cables Type C1, C2, E1 and E2:** Type C and E Signalling Cables are designed to carry control and signal information between the Location Cases and the Signalling Head. They often sit on the Ballast (stones within and around the sleepers – there's a train specifically designed to clean it) or can be clipped to cable posts. These single and multi-core heavy-duty cables do have the same insulation and sheathing

as Type B and D Cables – Type E being the LSZH version –, but they have flexible Class 5 stranded tinned copper conductors. C2 and E2 Cables are commonly referred to as Tail Cables. C1 and E1 Cables are used for the maintenance of these installations.
- **Signalling Cables Type C3 and E3:** Designed for use in the rail network's TPWS – or train protection warning system – these cables have two flexible cores with either EPR (C3 Cable) or LSZH (E3 Cable) insulation, a drain wire, and an aluminum screen.

3.1.7.3 Power Cables

As already said power cables may be distinguished by:

- the number of conductors (each assigned to a phase of the power system, to the neutral or protection earth),
- their section (that contribute to the determination of the ampacity),
- the used material for conductors (normally copper or aluminum), the dielectric material surrounding the conductors and ensuring the voltage insulation between themselves and with respect to external parts (paper and oil, polymers, such as XLPE, Polyethylene, etc., polymers loaded with metallic particles and other semi-conducting materials, such as graphite, etc.),
- the presence of a screen and/or an armor (with electrical and mechanical purposes),
- the type of the external protection (called sheath or cover, made of various types of plastic materials and also rubber),
- the overall mechanical behavior (consisting for example in the resistance to traction and strain and to compression, the bending characteristics, etc.),
- the resistance to chemical agents and to environmental conditions,
- the behavior in case of fire,
- the maximum operating temperature (that is used to determine the maximum tolerable steady and transient current amplitude),
- the voltage level as a steady voltage and the voltage withstanding capability in case of transient impulse voltages,
- the reliability characteristics.

Power cables, depending on their use and application, may be classified as:

- Medium Voltage power cables, operating normally between 10 and 40 kV up to a considerable current level;
- Low Voltage power cables, operating up to in principle 1000 Vac or 1500 Vdc, featuring a vast range of voltage values (depending on countries and applications) and current levels;
- Track-feeder cables, normally operated at the traction supply voltage levels (see sec. 2.1, and in particular sec. 2.1.6, 2.1.7 and 2.1.8) and connected between substation rectifiers, substation additional elements (such as filters, reactors and switches) and the traction line conductors; the current

rating is particularly high and the required cross section is normally beyond the capability of a single cable, so that cable bundle are used with many cables of 300 to 1000 mm^2 are used in parallel.

3.1.7.4 Telecommunication Cables

Communication along the railway and the connection of far apparatus between stations happen nowadays by means of a wide range of systems, and the necessary cabling system is very complex. For older systems dating back up to the '90s this occurred via a traditional fixed telecom network, connecting everything from stations and relay rooms to telephone points (both normal service points and emergency boxes). Copper pairs have been used extensively for this function, with cables of up to a hundred pairs and even more running trackside and dividing at junction points in smaller cables down the single equipment. For reasons concerning available bandwidth, cost, lack of maintenance and immunity to electromagnetic disturbance fibre optic have replaced large gauge copper cables for backbone large traffic connections. The multi-pair copper cables for telephone applications usually were and are still specified with solid plain copper conductors and PE insulation, together with a complex arrangement of insulating layers as a moisture/water barrier (petroleum jelly water-blocking compound, separators, sealing rubber, external rubber sheath, etc.) and a particularly robust armouring against mechanical stress (the same cable screen is robust enough to be relevant also for mechanical purposes, and then a corrugated steel outer enclosure or wound-up steel tape or polymer laminate tape are added). Armouring can have an impact also on the shielding characteristics, in particular in avoiding saturation and/or damage by self-heating of the internal screen under heavy current at supply frequency, thus extending the operating conditions of the internal screen at its nominal (or almost nominal) performance. There is a huge variety of such cables, but there is a general agreement in the specification of the screening factor, in order mainly to reduce audible noise in telephone applications: values between 35 and 45 dB are often required, even if a slightly different behavior is known for internal pairs in the cable center and external pairs close to the screen of large cables.

Other types of cables are in use in the most modern systems, where communication and connection of remote equipment occur by means of different networks and protocols:

- the high speed backbone replacing heavy gauge telephone cables is most recently a fiber optic redundant connection supporting SDH architecture, passing though an intermediate solution based for example on high speed Ethernet, again mostly on fiber optic media for reasons of attenuation and immunity to electromagnetic disturbance;
- TCP/IP protocol is widely used for a variety of things such as data network for ATO computers, SCADA computers, generic office operations, surveillance and even IP cameras, data collection for monitoring and control of

remote units (VHF radios, Passenger Information System and Passenger Announcement System, etc.), and it is often routed on four-pair Ethernet cables even for distances of a hundred metres, up to the first hub or repeater; this type of cable is quite diffused in the versions Cat. 5 and Cat. 6, supporting slightly different bandwidths and data rates, featuring no shield (Unshielded Twisted Pair, UTP), low quality shield (Foil-shielded Twisted Pair, FTP), high quality shield (Shielded Twisted Pair, STP) and including separate screens for each pair (Shielded-Shielded Twisted Pair, SSTP); the STP is usually taken as the reference for connections that are more than a few meters long and in particular in locations that cannot be considered absolutely safe from electromagnetic disturbance, and it features a screening factor of about 50 dB, and up to 60 dB for SFTP version (in some cases an external braided shield is added onto the internal aluminum foil screen);

- for lower data rate connection through a serial protocol (such as RS 232, RS 422, RS 485, CAN, etc.) the cable specified by the related standard or by the manufacturer of the communication devices is often used, that is a shielded cable with the right amount of pairs and possibly earth/reference conductors; the screening factor for such cables is not so different from the one above even if specifications may be highly variable, so that a conservative estimate could be between 40 and 50 dB;
- depending on the type of devices used for radio communications systems (antennas, leaky cable, power splitters or combiners, amplifiers, couplers, and so on) different type of coaxial cables and connectors are used, ranging normally between RG 58 for very short cable lengths and at the lower frequencies, when flexibility and light weight are required, RG 213 and RG 214 or equivalent, for moderate length and low/medium power connections over a wide frequency range, heavy gauge (half inch, 3/4", 7/8", etc.) low loss rigid cables for the final delivering of the transmitted power.

3.1.7.5 Bare and Screening Conductor Cables

In overhead line networks, electrical interference caused by the potential of a high voltage cable can interrupt communications in telecom cables. To soak up the magnetic field generated by the overhead line cables and minimise the effects of this interference, Bare and covered Screening Conductor Cables are installed. Screening Conductor Cable has a single core stranded aluminum conductor and a black PVC or LSZH sheath, and a Bare Conductor consists of a single core stranded aluminum conductor. Screening Conductor Cables are often referred to as FTN Screening Conductor Cables

3.2 Telecommunication Systems

The term "telecommunication system" encompasses a very complex and articulated system, that may be composed of different subsystems dedicated to different services, in some cases under different scopes of supply, and that embed different technologies in rapid evolution, operating with variable power levels and signal characteristics, over a vast frequency range, either in air or in cable (either copper or fiber optic). This is particularly true for modern metros, that operate in an urban context and are required to interface with several public, safety and emergency services. A first evolution may be seen to be occurred in the '90s when the radio system was required to support emergency services, such as police and fire department calls, in otherwise uncovered areas, like in tunnels or underground stations and technological areas; a viable solution was the implementation of a Tetra network, ensuring connectivity for personal mobile terminals, as a digital evolution of various types of older communication media, such as VHF and 430-460 MHz mobile terminals. At the turn of the century the increasing number of modern telecommunication services, almost all digitally implemented, has required the broadband interconnection by means of a high performance backbone, leaving aside custom solutions based for example on Ethernet lines (still used for low cost implementations of moderate performance systems); both for bandwidth needs and for electromagnetic immunity exigencies, fiber optic backbone lines, possibly in fault-tolerant double ring architecture and supporting the so called Switched Digital Hierarchy (SDH) has begun to appear. In the last years we have seen the widespread use of wireless technologies, not only as a customer support for the newest mobile terminals, but also for safety and security functions (e.g. signalling and trackside-to-train communications, and on board CCTV cameras interconnection, respectively).

3.2.1 Radio Systems

Radio systems began to appear as a support for emergency communication, ensuring the service in tunnels and covered areas to police, fire brigade, etc.. For this reason the first supported systems were of the VHF type in the 140-170 MHz frequency range and other licensed or unlicensed bands, such as 430 MHz band. More recently, the traditional walkie-talkie communication was embedded into intelligent terminals connected to a TETRA system (operating over the 380-420 MHz frequency interval and additionally close to the GSM band above 870 MHz), together with the possibility of external calls, data and short messages exchange, etc. In the last twenty years the support for civil communication in the GSM band has often been required as an additional specification, not only as a "fringe benefit" for passengers, but also as an additional mean to increase safety (passengers are usually the first to call in many occasions in case of accidents or aggressions and assaults). Finally, the so widely used wireless devices and communication systems have been brought to the modern transportation systems, first, as a support for train-to-wayside communication and for advanced signalling and, marginally, as a support to civil services brought along by passengers.

3.2.1.1 Public Mobile Support

Cellular phones intended as mobile terminals, principally for voice communication, but also for data transmission (short messages and e-mails exchange and multimedia data transfer), operate over several frequency bands, depending on the adopted technology and on national regulations and policies. In the following the most known protocols are considered and briefly reviewed on the basis of available standards and technical references; the attention is focused on the main architectural elements and the specific aspects that are EMC relevant, such as transmitters and receivers levels, accepted signal-to-noise ratios, minimum signal strength for satisfactory service implementation (the so-called "coverage"), past experience on interference and problems of operation.

GSM 900 MHz. Several countries have developed the voice communication protocols starting from GSM over the 900 MHz band: the frequency interval assigned to public services is located between 880 and 915 MHz for the receiving (or down-link) band, and between 925 and 960 MHz for the transmitting (or up-link) band. The characteristics that are relevant for the analysis of both electromagnetic emissions from GSM apparatus and electromagnetic susceptibility to electromagnetic emissions produced by nearby equipment are related to the physical interface: transmission power level, antennas characteristics, receiver susceptibility, minimum detectable field intensity. The nominal power of a mobile terminal is 1 W at this frequency range (and 2 W for the DCS service); the antenna gain is about that of a small dipole, 2.2 dBi, whatever the real implementation, as an external antenna (e.g. a short whip antenna) or an internal, integrated antenna (e.g. a microstrip antenna), because the correct use of the terminal requires an omni-directional radiation pattern. The receiver sensitivity is around –85 to –95 dBm, depending on the type of service, for static or slow moving applications [275]; the most critical situation is during hand-over, when the power decay profile versus distance from the signal transmitter becomes relevant. Depending on the type of implemented service and assigned priority, so on the criticality and safety relevance of the exchanged information and the availability of error correcting procedures (such as interpolation of speech samples or error correcting codes in the message frames), the minimum signal-to-noise ratio (SNR) may range from a few dB to about ten dB [607].

The GSM protocol, extended also to all other similar services operating at slightly a different frequency band or representing a successive improvement at the other frequency band of 1800-1900 MHz, is based on a subdivision of the available frequency interval into separate physical channels, distinguished for up-link and down-link; this part of the modulation and access to the transmitting medium is a Frequency Division Multiple Access (FDMA). While the fixed station (called Base Terminal Station, BTS, or Radio Base Station, RBS) can operate on several physical channels at once, a single mobile terminal selects a physical frequency channel and negotiate the access requesting the assignment of a slot number X within the physical channel, with a Time Division Multiple Access (TDMA) approach: the access and the transmission and reception is made cyclically by the single terminal during a time interval that corresponds to the slot duration at an instant of time where its slot begins. Faster communication, if allowed to the subscriber, can occur on multiple slots within the same frame. The time duration of the single bit (or symbol) is $T_{sym} = 3.692\,\mu s$ and the time duration of a slot is $T_{slot} = 576.92\,\mu s$ with about 156 bits in one slot (i.e. the length of the GSM message or speech chunk, including headers, synchronization and error control and correcting bits); a frame is made of 8 slots and thus lasts for $T_{frame} = 4.615\,\text{ms}$. The complete structure of the GSM TDMA is shown in 3.22.

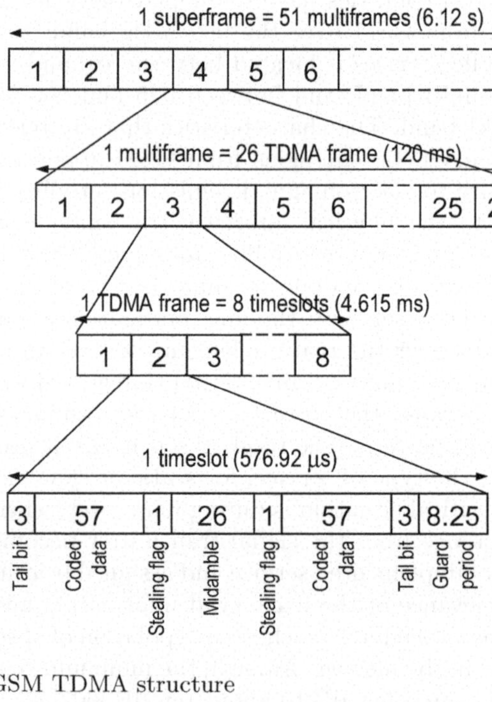

Fig. 3.22 GSM TDMA structure

The bandwidth B_{ch} of a physical channel is 200 kHz and the transmission of the channel symbols within a frame must fit the bandwidth restriction, with respect to the maximum symbol rate R and the frequency spectrum of the adopted modulation. The adopted channel coding was thus optimized with respect to bandwidth occupation, starting from the most robust and simple modulation, the Frequency Shift Keying (FSK); the GSM implements a Minimum Shift Keying (MSK) modulation, that consists of a FSK with phase continuity between adjacent symbols and is part of the greater family of Continuous Phase FSK and Continuous Phase Modulations (CPMs). Phase continuity is ensured by keeping memory of the last symbol and avoids phase jumps, that produce a significant spectrum spread, in the order of twice the symbol rate. The Gaussian property that forms the name of the adopted modulation (GMSK) is added in order to increase the time duration of the symbols that span over more than one symbol time interval T_{sym}, thus overlapping each symbol, with the tails of the tapered copies of the adjacent symbols; the shape of the tails is that of the sides of a Gaussian distribution. By increasing the symbol time T_{sym} without reducing the symbol rate R, the spectrum is even more compact around the center frequency; this particular setting for any CPM is in general called "partial response CPM" [606]. The possible consequent Inter-Symbol Interference (ISI) is avoided if a perfect synchronization between the symbols is maintained and decoding at the receiver is achieved with an optimal receiver, in the sense of the communication theory [606]. The raw Bit Error Rate, BER, associated to the symbols sent over the physical channel may be estimated on the basis of the amount of noise N_0 with respect to the energy conveyed into one symbol (or bit in this case) E_b, both identifying the equivalent of the Signal-to_noise Ratio, SNR: under the assumption of an optimal receiver (the so called "coherent receiver") for orthogonal symbols of a generic two-level modulation, the BER is given by the "error function" $erf(\cdot)$, also indicated by $Q(\cdot)$, applied to the square root of the ratio E_b/N_0 as $BER_{orth.} = Q(\sqrt{E_b/N_0})$, where the use of the error function - that is the integral of the Gaussian function from the argument to infinity - was allowed under a further assumption of a Gaussian noise distribution. For the MSK the argument is to be multiplied by 2, where for a GMSK a slightly lower increase is expected around 1.8: $BER_{GMSK} = Q(1.8\sqrt{E_b/N_0})$. From the BER it is conceptually possible to go back to estimate the error rate of the entire message or speech chunk or of the entire frame, even if the presence of Forward Error Correcting (FEC) codes complicates the treatment.

Other types of cellular systems are based on a similar scheme, even if operating on different frequency bands, and are thus not treated specifically in this book.

3.2.1.2 Global System for Mobile Communications - Railway (GSM-R)

The global deployment of GSM-R, a new standard for wireless rail telecommunications (endorsed and sponsored by the International Union of Railways), represents an immense opportunity for wireless telecommunications industry on a global scale. GSM-R is not a commercial telecommunications system and thus it is relatively immune to market variations and downturns. Since the system focuses on safety, reliability, prevention of accidents and the general well being of the traveling public, the GSM-R deployments are supported and funded by governments, multilateral finance institutions and public. GSM-R is a sub-system of ERTMS, it is used for communication between train and operational control centres. The system is based on GSM and EIRENE[9] - MORANE[10] specifications which guarantee performance at speeds up to 500 km/h, without communication loss [618].

GSM-R is essentially the same as GSM, with railway specific functionality. It operates in the 900 MHz band and provides a wide range of functions such as logistics, passenger information, on board signalling, rail communications etc. GSM-R uses standard GSM digital radio technology, enhanced with additional functions specifically for railway operation. GSM-R operates in exclusive bandwidths of 876-880 MHz and 921-925 MHz allocated for up- and down-links respectively. In certain countries these bandwidths may be wider. The GSM-R radio network is built up of track-side radio cell sites. The Base Transceiver Stations (BTS) comprise the radio equipment (transceivers and antennas) needed to service each cell in the network. It handles the radio interface to the mobile units, which can be handsets operated by personnel or train-mounted "terminal equipment". A group of BTS's are connected to a Base Station Controller (BSC) which controls a defined number of cells which are linked to it. The BSC's provide all functions such as handover, cell configuration and control of RF power levels in the BTS's. The BSC's are connected to Mobile Switching Centres (MSC) which provide overall control of the network, links to the Public Switched Telephone Network and the other railway intelligent networks for train location monitoring etc. The expected distance between BTS masts is typically around 4 km for high speed routes and 7 km for the rest. For flat terrains, it may be possible to extend this spacing further. A spacing of around 2 km may be required where the terrain is hilly or the track is not straight. Leaky feeders or micro-cellular BTS will be used to provide coverage in tunnels and at stations. Communications from the MSC's to the BSC's can be provided using dedicated rail fixed SDH networks. A mix of fibre optic cable and copper cable utilising DSL technology will provide the final links to BTS's and local elements along the line, including signalling and line-side phones [463, 663].

3.2.1.3 Terrestrial Trunked Radio (TETRA)

TETRA is a professional mobile radio and two-way transceiver (colloquially known as a "walkie-talkie") specification. TETRA was specifically designed for use by government agencies, emergency services (police forces, fire departments, ambulance), maintenance and security staff for public safety networks,

[9] European Integrated Radio Enhanced Network.

[10] Mobile Radio for Railway Networks in Europe.

rail transportation staff for train radios, transport services and the military. The TETRA standards are conceived for an operating band up to 1 GHz. TETRA is a digital radio standard developed and used in Europe with increasing international usage. In Europe, the frequency interval between 380 and 420 MHz, following the EU decision 96/04 [282], that assigns the following bands: 380 – 390 MHz paired with 390 – 399.9 MHz, the first 5 MHz of each band dedicated to emergency systems; 410 – 420 MHz paired with 420 – 430 MHz; 450 – 460 MHz paired with 460 – 470 MHz; 870 – 876 MHz paired with 915 – 921 MHz. The latter is overlapped with the GSM services used for GSM-R; moreover, in order to expand the number of providers of TETRA services, the band 872 – 876 MHz paired with 917 – 921 MHz is under consideration, since it was released by the recent closure of analogue telephony networks. TETRA bands are all very close to ISM unlicensed bands: wideband telemetry (car alarms and similar) in UK and rest of Europe uses 418 and 433 MHz with 0.25 and 10 mW EIRP limitations; European harmonized analog voice (PMR 446) uses 446 MHz with 500 mW EIRP limitation, and UK industrial/commercial telemetry uses 458 MHz again with 500 mW EIRP limitation; European harmonized telemetry uses the 868 – 870 MHz band again with 500 mW EIRP limitation. While TETRA has strict band limiting (considered at the end of this section), telemetry devices are not and TETRA may represent a relevant source of interference.

The transmitted power of Base Stations (BSs) and Mobile Stations (MSs) are determined according to the assigned power class; for BSs power ranges by steps of 2 dB from class 10 with 28 dBm (equal to 0.6 W) up to class 1 with 46 dBm (equal to 40 W), for MSs the step is 2.5 dBm starting from class 4L with 27.5 dBm (equal to 0.56 W) up to class 1 with 45 dBm (equal to 30 W); for QAM modulation two additional levels, 5L and 5, are added below 4L for 22.5 dBm and 25 dBm respectively.

TETRA also, like GSM, uses a TDMA access scheme. When using phase modulation the carrier bandwidth is 25 kHz, for QAM instead the carrier bandwidth may be 25 kHz, 50 kHz, 100 kHz or 150 kHz. The structure of the TETRA TDMA is shown in Figure 3.23, where the time duration of each level entity is reported (the two subslots dividing in two parts the timeslot are used only in the up-link).

The symbols in the time slot are 255 of 55.56 μs each if phase modulation is used or 34 of 416.7 μs each if amplitude modulation is used:

- phase modulation is of the type $\pi/4$-DQPSK or $\pi/8$-DQPSK for 36 and 54 kbps respectively, where the leading D stands for "differential";
- amplitude modulation is of the type 4-QAM, 16-QAM and 64-QAM; a multi sub-carrier approach is followed with 8, 16, 32 and 48 sub-carriers in 25 kHz, 50 kHz, 100 kHz and 150 kHz carriers respectively and the modulation symbol rate on each sub-carrier is 2400 bps.

The receiver sensitivity is reported in [268] for various operating conditions: with moderate fading (multi-path delays no greater than 5 μs) the receiver

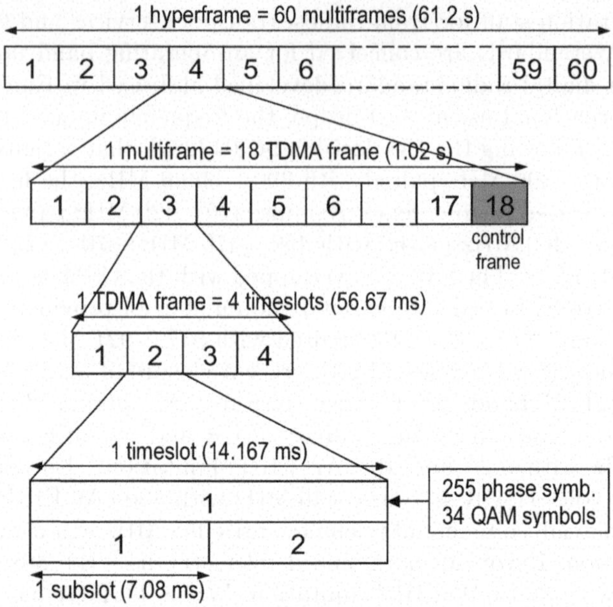

Fig. 3.23 TETRA TDMA structure used for voice and data

sensitivity of a mobile terminal is between -90 and -110 dBm depending on the modulation: for phase modulations the sensitivity is the worst between -90 and -95 dBm, the other more favorable values refer to QAM at different carrier frequencies.

Transmitter emissions outside the operating bandwidth due to spurious emissions and other types of noise, and not relevant for inter-channel interference but to interference to external victims, are subject to the following limits:

- for discrete spurious emissions -36 dBm measured over a 100 kHz bandwidth between 9 kHz and 1 GHz and -30 dBm measured over a 1 MHz bandwidth between 1 GHz and 4/12.75 GHz;
- for white band noise emissions and other type of emissions at different frequency offsets from the carrier specification of limit is given as "dB below carrier" or dBc and thus relative to the transmitted power, but a maximum absolute level ranging from -36 dBm to -70 dBm is given as well for increasing frequency offset from the carrier frequency (see [ref] sec. 6.4).

3.2.1.4 VHF and UHF Support for Emergency, Police, Fire Brigade Other Than Tetra

Many transportation systems might not adopt modern digital voice and data radio systems for various reasons: cost containment, old design and construction and no need for upgrades, compatibility with existing important analog radio networks. In this case the radio system is required to support analog modulations such as FM, AM, SSB, over the most widely adopted bands, such as the classical "VHF" (between about 140 and 165 MHz, depending on national regulations), 430 MHz and 860 MHz located in the UHF band. The most important function is the support of the communication network of police and fire brigade or maintenance teams, especially in emergency cases of various kinds; usually beacon antennas, at least for station areas, and/or a leaky cable for tunnels are purposely installed to ensure coverage and satisfactory communication quality. These analog FM modulations are normally characterized by a favorable signal-to-noise ratio and not so restrictive requirements on the communication quality, so that they do not represent a relevant victim. On the contrary the level of emitted electromagnetic field may be quite large close to each mobile unit (being the maximum transmitted power up to 5 W the E-field intensity at 1 metre distance can be as large as 15 V/m), to beacon antennas (same type of electromagnetic field emissions, with a slightly larger expected maximum E-field intensity accounting for a large transmitted power, even if a more favorable geometry of exposure) and to leaky cable (the E-field emissions repeat similarly along the whole tunnel and their value should be computed by numeric simulations, expecting a much lower, but more diffused, E-field intensity, around some V/m maximum).

3.2.1.5 Wireless Data Networks

Wireless services for customers are normally intended as radio LAN in the 2.4 and 5 GHz bands meeting the specifications in IEEE 802.11 publications; specifications span over several services, but presently the attention is focused on the basic wireless LAN services 802.11a, b and g [380, 381]. The same networks are used not only for support to customer services, but also for non-vital functions in particular at modern metro lines, such as interconnection of sensors for environmental control and in particular interconnection of "IP cameras" for surveillance. The three LAN services, described by the respective standards, operate at 5 GHz the former, and 2.4 GHz the other two. The adopted modulation schemes not only depend on the service, but also within the same service by the required/affordable transmission rate and the response of the environment, in terms of perceived quality of service. The basic principles of the said modulations are only reviewed here in order to define their channel occupation, spectral characteristics and thus their susceptibility/immunity. Relevant susceptibility is in-band (or slightly out-of-band) susceptibility and at this frequency range threats are represented on one side by similar radio LAN services located and operating nearby, and on the other side by the broadband intermittent noise originating in the same traction system, mainly by fast commutations and electric arcs.

The 802.11 networks use different types of modulations depending also on the data rate: Orthogonal Frequency Division Multiplexing (OFDM),

Frequency Hopping or Direct Sequence with Spread Spectrum implementation (FHSS and DSSS respectively). Receiver sensitivity is highly variable depending on several factors, such as the amount of fading, the used modulation, the data rate, the assumed inter-channel interference, the target packet error rate; moreover, its specification is not in the standard, but it is derived by simulations performed as internal IEEE working group evaluations and by manufacturers' data. As a general reference for 802.11a OFDM at the highest and lowest data rates of 54 and 6 Mbps the required receiver sensitivity is -64/66 dBm and -84/86 dBm, for different channel code rates, under conservative assumptions on the noise figure (10 dB) and the implementation margin (5 dB); for 802.11g the receiver sensitivity seems slightly better by a few dB for comparable data rates; for 802.11b, featuring lower data rates (the maximum value is 11 Mbps) the receiver sensitivity is better by another 3-4 dB.

Public wireless devices as sources on the contrary are less relevant, being limited to hand-held terminals with a transmission power limited to 100 mW (in some cases 200 mW). Based on a far field formulation (really justified for the considered wavelength), being the antenna omni-directional, the estimated maximum electric field at 1 m distance is 2.2 V/m; it is underlined that, unless the mobile terminal is not purposely locked at maximum transmitted power, it is difficult that this field value is reached.

3.2.2 Analog and Digital Phones

3.2.2.1 Analog Telephones

A specific mention is made to analog telephone, as they represent a very susceptible equipment and the justification for the widespread use of psophometric measurements. Analog telephones are now not implemented any longer in new systems, but they were requested until the late '90s for various reasons (redundant service, backward compatibility, economic convenience, etc.) and are still in use in many old systems. Their original function was the coverage of all the communications between the various places and officers, employers, operators and technicians; then, they have been limited to a backup system for more modern digital telephones, to connect inexpensively remote or ill-serviced places, and as an emergency communication system, in particular for emergency niches trackside.

Analog telephones are connected by means of copper pairs running inside multi-pair cables: the cables are normally twisted-pair cables, with a first aluminum or copper shield and an external armour, made of steel or lead; different technologies and materials are in use because of the wide temporal span of construction and installation. The standard termination impedance is 600 ohm, also indicated by the standards for the measurements; of course various equipment may differ slightly from the prescribed value. Moreover, long cables are usually patched many times and compensating networks may be introduced, so that the frequency response of cables and terminating loads may vary substantially and on-site determination of frequency response and overall per-unit-length line parameters are advisable before any measurement of coupled interference.

The weak side of analog telephones is their susceptibility in the audio band: audible whistles and rattles are in the range of mV and several disturbances produced by contactors and by time-varying harmonics manifest themselves as annoying audible clicks and modulated audio noise. Given the variable irritability of the human ear as a function of frequency, it was proposed to weight the perceived disturbance for its distribution versus frequency (i.e. frequency spectrum) and the so called psophometric weight tables were established through the years, following several measurement campaigns and evaluation of the subjective reaction to various kinds of coupled noise.

Psophometric and C-weighting Methods. Several standards have set down limits for audible interference and related procedures, starting with the first attempts in 1941; limits have been always associated to a weighting curve for the various frequency components of the interfering signal, based on the sensitivity of the ear and the response of telephone equipment: one weighting system is based on Bell Telephone and Edison Electric Institute (EEI) tables, the other one on the International Consultative Commission on Telephone and Telegraph systems (CCITT), now ITU-T. The weighted disturbance is determined with a general expression of the form

$$F = \sqrt{\frac{\sum[k(f)\, p(f)\, V(f)]^2}{\sum[V(f)]^2}} \tag{3.10}$$

where $k(f)$ is a multiplying factor to scale the weighting factor, $p(f)$ is the weighting factor and $V(f)$ is the spectrum of the voltage disturbance, all defined for the same frequency values. The value F is called Telephone Influence Factor (TIF) or Telephone Harmonic Form Factor (THFF), if computed with the C-message or psophometric weighting, proposed by Bell/EEI and CCITT respectively. The values of $k(f)$ are defined as $k(f) = 5000\,(f/1000)$ and $k(f) = f/800$ for the C-message and the psophometric weighting; the weighting curves $p(f)$ are shown in Figure 3.24, where for the exigency of overlapping the curves on the same scale the $k(f)$ factor for the CCITT curve is already included into $p(f)$.

The numeric value F is to be compared with a reference limit: in various publications CCITT proposed 1 mV, with an increase to 5 mV in special circumstances indicated in [422]; Bell Telephone proposed 0.173 V [490].

The Network Rail standard 50016 [578] also indicates a psophometric voltage limit of 1 mV over two frequency intervals, [16.66 - 6000] Hz and [50 - 60000] Hz for "audio interference" and "equipment susceptibility" respectively. The interpretation of the latter is based on the observation that being mostly outside the audible frequency range, the interference is interpreted as

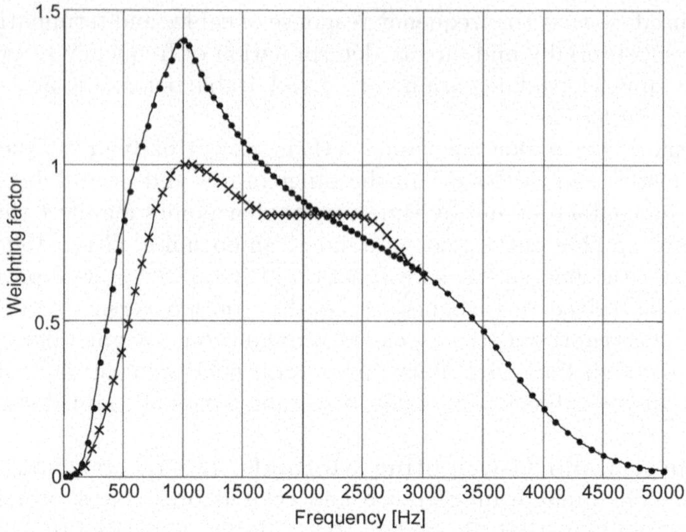

Fig. 3.24 C-message (cross) and psophometric (dot) weighting curves

a susceptibility aspect; however, since this susceptibility limit is not expressly assigned to analog telephones, it is not excluded that it could be applied also to digital telephone systems at a first approximation.

3.2.2.2 Digital Telephones

The term digital telephone covers any telephone equipment operating on a digital protocol, most often a PCM code over a copper pair (such as for ISDN connections) or over an Ethernet network (such as for IP telephones). The used cable is a twisted pair type possibly shielded, like for Cat. 5 and Cat. 6 network cables. Their susceptibility is the same as an Ethernet network, depending on the transmitted levels, the expected signal attenuation and dispersion over the network cables and the decision thresholds of the receiver.

3.2.3 Data Networks

3.2.3.1 Ethernet and 802.3 Physical Layer

Ethernet represents a short form recalling a brand name to address the set of specifications of the IEEE Std. 802.3: we may say that for the present treatment limited to the main characteristics of the physical layer the two definitions coincide. Ethernet uses a Manchester coded signal (thus ensuring a transition at each symbol time whatever the transmitted sequence) with a data rate and a voltage level that evolved from the old 10 Mbps format running on coaxial cables to the 100 Mbps format applied to differential pairs on multi-pair network cables. The format uses base-band signalling, so its spectrum extends in principle from dc to a maximum frequency that is about twice the adopted signalling rate. Each Ethernet frame begins with a preamble: 7 bytes of 10101010 plus 1 byte of 10101011; the 10101010 pattern, when Manchester encoded, yields a square wave with a transition in the center of each bit interval and the receiver synchronizes its clock to these transitions. The spectrum at dc and low frequency is negligible since first, the adopted format ensures a zero mean value of the sent message down to the single symbol, and second, in the 100 Mbps Ethernet format a differential Manchester encoding is adopted, where symbols are alternated for successive identical digits. The amplitude of the square wave is $\pm 0.85\,\mathrm{V}$, so that the energy per symbol (or bit) is $7.22\,\mathrm{V}^2/\mathrm{MHz}$ for a 10 Mbps Ethernet (in Figure 3.25 the vertical scale of the power spectral density is converted to dBm/MHz with a unitary reference resistance). The waveform shown in Figure 3.25 left is a zoom of the first 50 symbols sent out at the rate of 10 Mbps and generated as a random sequence; the spectrum shows a bandwidth of about 10 MHz, with the second lobe down by 10/11 dB with respect to the main lobe and a negligible dc component, about 40 dB lower than the main lobe.

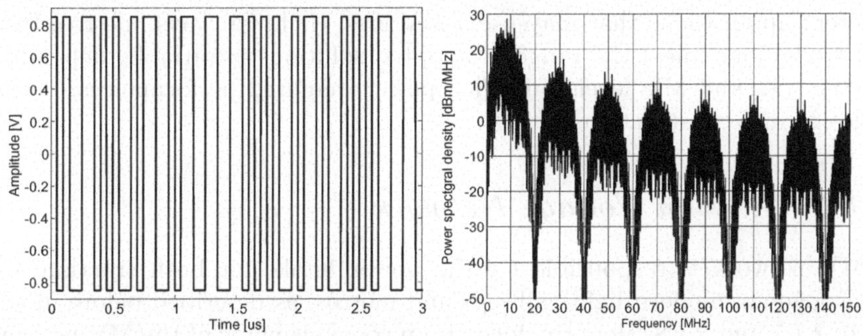

Fig. 3.25 Ethernet simulated waveform for a random sequence (without preamble) and related power density spectrum (the measurement unity is converted from $\mathrm{V}^2/\mathrm{MHz}$ to dBm/MHz, assuming a reference resistance of $1\,\Omega$.

3.2.3.2 SDH

Synchronous Digital Hierarchy, SDH, is a multiplexed digital network of serial data streams (data and calls) operating over an optical channel, or a standard electrical copper medium for low data rate channels. SDH, a transport oriented protocol, is able to host simultaneously within a single framing different circuits with different data rates and origin, such as for example ATM frames, Ethernet frames and PCM circuit-oriented calls. The SDH frame is complexly articulated and within it data and header are interleaved, so that the maximum latency for each data stream conveyed in the SDH frame is shorter than the time duration of the SDH frame itself. The SDH data stream is basically framed with STM-1 modules at 155.52 Mbit/s; an optional frame structure at lower rate (51.84 Mbit/s) is the STM-0. The time duration of the STM-1 frame is 125 μs, so that there are 8000 frames per second. Higher data rates are possible at upper frame levels, called STM-4 (four times faster than STM-1), STM-16 (sixteen times faster than STM-1), and so on.

SDH is nowadays the preferred solution for the communication backbone of modern transportation systems, in particular metros and light railways, where the spanning distance is however limited (up to however several km), there is a significant need to support several communication services and related protocols and the centralized approach to monitor and control is preferred. It's physical implementation can rely upon fiber optic, eliminating the necessity for emissions and susceptibility analysis and in particular giving more freedom in cable-way management.

3.3 Medium/Low Voltage Distribution Networks

The main elements that characterize the Medium Voltage (MV) and Low Voltage (LV) networks used in railway applications are briefly reviewed. This treatment does not want to be exhaustive nor complete, in the sense that for different applications that may use a MV level (light railways, metros, tram and trolley bus lines) the range of adopted solutions is variable, depending on countries and MV levels, installed power, required performances, period of design and construction.

3.3.1 Medium Voltage Network

A MV network is responsible for the power intake for both traction and other loads in modern light railways and metros, as described before in sec. 2.1.6, 2.1.7 and 2.1.8. Here the focus is on some elements of the MV network relevant to the propagation of low frequency interference and in particular the cables and the interface to the Point of Common Coupling.

3.3.1.1 Medium Voltage Cables and Cable-To-Cable Coupling

MV cables are a source of magnetic field emissions and disturbance for nearby equipment and cables, due to the flowing current in normal and exceptional service conditions.

Normal service conditions imply a sinusoidal current at the supply fundamental and superimposed harmonics, due mainly to the traction rectifiers and their ac input current spectrum (see sec. 2.1.10.2). Usually calculations are done at the maximum supply current in admissible overloading conditions and for worst case harmonic spectra, assuming for instance the values shown in Fig. 2.19 for a zero or minimum value of the commutation overlap angle. The results are used to ensure for example that the magnetic field level keeps smaller than the one adopted for the immunity test pertaining to the assigned electromagnetic environment, or to estimate the level of noise coupled to nearby cables, such as for example telephone cables.

Exceptional service conditions correspond to extreme overloading due to out-of-service of one or more rectifier groups or substations and possible deterioration of their harmonic patterns for lack of symmetry up to short circuit conditions, considered for a time interval that corresponds to the worst case tripping time of protection relays (circuit breakers). In the first case the analysis is carried on in the same way as for the normal service conditions. For the short circuit the attention is focused mainly on the dramatic increase of the flowing current in terms of fundamental current alone. The results are mainly used to ensure that the induced voltage on nearby cables keeps below the prescribed limits: the induced voltage is mainly in terms of common mode voltage and has impact on the electrical safety of personnel and equipment, and shows up also as a differential overvoltage when cable asymmetry and related screening factor are considered. The limits for the electrical safety are specified for example for railway applications (EN 50122-1) and for telecommunication installations (ITU-T K.68), with a voltage-time curve where the induced voltage and tripping time must be reported , while the limits for interference are normally derived by the CCITT Vol. VI [422] and psophometric voltage [578], that are treated in more detail later on in Chap. 6.

3.3.1.2 Conducted Interference to the Point of Common Coupling (PCC)

The harmonic emissions of all distorting loads (most of all the traction substations equipped with rectifier converters and some other power converters for auxiliary loads, such as ventilation and air conditioning) connected to the MV network produce a distortion of the voltage at the PCC, that depends on the ratio of the source and load rated powers (the so called short circuit power ratio) and on the intensity of the individual harmonic emissions, that in turn may depend also on the operating conditions and load level. The basic concept for the calculation is straightforward: the source of emissions is at first modeled as a current source, that is a current generator at each frequency of emission $f_h = h\, f_1$ with intensity I_h, where h is a positive integer that indicates the harmonic order with respect to the fundamental frequency f_1. This approach is reported in IEEE Std. 519 [375], where static converters, in particular rectifiers, are modeled by formulas for the prediction of harmonic amplitude depending on the operating conditions (firing delay angle, commutation overlap and DC load characteristics).

The methods and the models used for the analysis are not reported extensively here, since this topic pertains to the classical load flow and harmonic load flow studies on electric networks, and the cited standard itself, as well as other famous books [20], may be used as a valid reference. The propagation of harmonic disturbance and the caused voltage distortion along the distribution network (normally at medium voltage for several metros and light railways, at high voltage for heavy railways) and at the PCC can be evaluated and analysed by means of load flow programs, once the following information has been fixed: amplitude of the harmonic disturbance (see sec. 2.1.10.2 for substation rectifiers), short circuit reactance and equivalent resistance of the feeding transformers, per-unit-length parameters (including capacitance) of power cables, that may depend on the adopted layout and have margins for optimization, short circuit power and homopolar capacitance[11] of medium or high voltage intakes, and finally impedance or component values of harmonic filters, that may be installed or that must be designed and sized to reduce harmonic distortion.

There are many cases of resonances in the medium voltage network of metros and light railways, due to the interaction of the cable capacitance with the inductance of the cables themselves and of substation transformers and thus worsening the voltage harmonics and the voltage distortion in the high frequency range around 1.5 or 2 kHz. Moreover, harmonic filters of the resonant type may reduce the amplitude of voltage harmonics around their resonance, but causing further resonances on the network at higher frequencies, where the limits for harmonics are even lower. These phenomena can be analyzed as detailed in [20][375] with the help of custom code or commercial simulators, normally used for harmonic load flow analysis.

3.4 Other Services

In this section there are described the technical services that equip a modern transportation system and that may be prominently or marginally EMC relevant.

3.4.1 Passenger Information System and Public Announcement (or Address) System

These two systems are the so called PIS and PAS, delivering information to passengers on-board and waiting on the platforms, halls and concourse at stations, concerning train times and locations, trip/destination information, changes and updates, special announcements, announcements related to safety and emergencies, etc. Recently for modern transportation systems they are designed and implemented in an integrated way, as it should be considering that the main purpose is to transmit information to passengers with different means. The installed equipment consists of information technology equipment, mainly computers, displays and network facilities, accompanied by audio products for the diffusion of speech messages and for the

[11] Also called “zero sequence capacitance”.

control of the acoustic quality (sound processors, amplifiers, loudspeakers, microphones). The displays are mainly of the flat type, such as LCDs, for graphical information (such as maps, diagrams, timetables, and also advertisements and entertainment content) and simpler dot matrix displays are used, mainly on-board, for short service messages, such as the name of the next station, the time of arrival, short service messages. Information is sent through serial networks (such as for example RS232, RS 422, CAN bus for dot matrix displays and also IP addressing for graphical displays) or supplied by a standard VGA connector by a local computer. The audio equipment is responsible for the diffusion of announcements, regarding for example the destination of the approaching train, information on delays and changes of platforms, the next station for on-board passengers, but also safety messages, like the ones on unattended baggage or other police controls, or emergency messages, such as in case of fire or other mis-haps. It may be used also for entertainment, such as the diffusion of music, and for advertising, sending short commercial messages or broadcasting a dedicated radio channel.

The main issues from an EMC viewpoint are on one side formal (the adopted product standards are focused on information technology equipment and audio products, so with potential differences and normative gaps due to the different issue dates and the considered electromagnetic environments and coupling phenomena) and on the other side practical (both displays and audio products and the related network hardware may be very susceptible devices and caution must be adopted with respect to cable segregation, power supply filtering, and in particular for on-board installation).

Concerning the former, information technology equipment is normally disciplined by EN 55024 standard [149] for immunity, with a lack of full correspondence of tests and levels even with the residential environmental standards EN 61000-6-X; for audio products the standard (EN 55103-2 [151]) is even older with a similar lack of correspondence. This problem is treated extensively in Chapter 5 at sec. 6.2.3.1, 6.2.3.2 and 6.5.

Concerning the latter, the perceived and the minimum required quality of a display or audio service must be clearly stated and described, so that installation practices, countermeasures and on-site tests can be designed, prepared and planned in advance and harmonized with the rest of the system. These systems in fact represent usually one of the weak points from a quality of service point of view, even if they are not really severely affected by electromagnetic interference and they usually can carry on their purposed functions even with degraded performances. The only standard that addresses both functional performance and minimum immunity purposely for PIS on-board rail passenger vehicles is the IEEE Std. 1477 [396]; the standard encompasses all the forms of passenger information, so also non-electric items like exterior side and end signs, interior signs and displays, visibility of characters and their size, are included. The requirements specified for electrical interfaces are: power supply levels, at 28, 37.5 and 74 V with different current ratings for each display, announcement system, etc., and accompanied by the withstanding capability with respect to electrical short circuits, reverse polarity, ground faults, and overvoltages in terms of a 2.5 kV transient carrying an energy of 80 Ws. The latter specification on transient overvoltage is to be evaluated with respect to its CENELEC counterpart (EN 50121-3-2 for on-board equipment), specifying an Electrical Fast Transient level of 2 kV and a Surge level of 2 kV line-to-ground and 1 kV line-to-line for "battery referenced ports": while the amplitude and energy more or less corresponds to that of the surge waveform (the 2 kV peak waveform with 1.2/50 μs rise and half time if referred to the generator internal impedance of 2 Ω gives something more than 100 Ws, while for the used coupling resistance values of 10 Ω and 40 Ω the resulting energy is correspondingly lower at about 20 and 5 Ws), it is not clear how the IEEE standards requires that the overvoltage is applied (line-to-line, line-to-ground, and with which impedance).

3.4.2 Ticketing and Fare Collection Systems

Depending on the level of automation of the considered transportation system, ticketing and fare collection systems may be simply constituted by attended or self-service terminals at the ticket office, that are standard and safe products from an electromagnetic point of view, almost always far from significant sources of electromagnetic emissions. However, if additional access control is implemented (denial of access to unauthorized passengers, for safety reasons and to prevent scams, monitoring of passenger flows, access to platforms in correspondence with train arrival and coordination with platform screen doors, etc.), then further network support is required; even in this case significant electromagnetic interference may be excluded, being the used cableways and the installed equipment in quiet portions of stations, normally far from significant sources of emissions. Increased robustness to EMI could be required to prevent jamming and intentional interference by attackers, aimed at disrupting a control function and gaining access to otherwise forbidden areas.

3.4.3 Platform Screen Doors

This system represents an important safety element, that prevents the direct access of passengers to the track, and as a consequence any accidental fall on the track, contact with live conductors and throw of objects on the track itself. Platform screen doors (PSDs) are coordinated with the train arrival and coordinated with a series of elements of the Automatic Train Control (ATC) and Operation (ATO) systems: train stop is made correspond to PSDs by means of first a beacon or a tag that is serviced by a dedicated signalling function; then, train doors are open if the platform screen doors sensors detect a safe alligned position. When passengers have entered and exited the coach or vehicle, the doors are closed with a verification of the absence of any object or human obstructing the door operation, using different levels and sensitivities for the detection of the mechanical interference and potential pinching by the closing doors; door closing is coordinated with warnings on-board and at platform, so interacting again with ATC and ATO for the issue of warnings and for the subsequent detection of safely closed doors. The PSDs are driven by various types of electric drives, based often on permanent magnet motors for fine and smooth regulation and large initial torque to leave the rest position; these drives are in turn driven by a control system, that receives the status of all the deployed sensors and communicates at a higher level with the other subsystems (ATC and ATO). Since the location and connection of sensors, drive elements, control equipment and communication devices are widely different from manufacturer to manufacturer and several elements are located inevitably close to the track at a distance shorter than 3 m, it is evident that the EMC issues may be many, considering also the safety role of this system: all the control lines must be protected against the highest grade of interference as dictated by the standards (EN 50121-4) and door sensors may be subject to particularly intense emissions from not only the traction conductors, but also vehicle elements and equipment, due to the reduced distance during train passage. In particular, sensing of mechanical interference to door closing is performed (as in other applications, such as car windows) by sensing the absorbed current and most often this is accomplished by Hall effect sensors, that in principle are sensitive to stray magnetic fields. Another type of devices sensitive to magnetic field are proximity sensors of the inductive type used for example to sense the presence of people across the door or to align the vehicle and platform doors. The location of PSD elements clearly identifies the applicable standards, and if more benign immunity test levels and a quieter environment is aimed at for the control part of the PSD system, then its location must be in an equipment room, further away from the track zone; yet, in order to keep the correct environmental conditions for equipment ports, the connections between track devices and the equipment room need to be adequately shielded and protected, for instance using a metallic conduit for increased protection.

Allignment of train doors at platform (and recently with PSDs) is accomplished with a vast range of solutions. The first method ensuring safe opening of train doors at platform was the use of AF TCs in overlay, possibly supported by lumped devices such as antennas on track and on-board requiring an approximate alignment, with a sufficient accuracy for alignment of train with platform. The requisite on the accuracy has increased with the appearing of PSD, requiring first the use of additional loops of shorter length with "eight-like" shape in traditional inductive loop products and then the use of tags of various kinds. Modern tag devices operate on magnetic matching of permanent magnets and focused detecting coils, ensuring a target braking accuracy of a few cm or a tenth of cm. Alternatively, the use of radiofrequency electromagnetic fields on the principle of RFID devices ensures a similar target braking accuracy. These devices may be led back to short range devices as far as EMC standards are concerned (see sec. 6.2.3.7).

3.4.4 Ventilation and Air Conditioning

Ventilation and air conditioning, including heating, is a system that extends to all parts of stations, public areas, technical rooms, and also on-board. The power part of the system is made of induction drives driving fans and compressors, with normally a LV three-phase supply for station installations, while on-board various solutions are adopted, from dedicated static converters supplied directly from the dc train bus, to the use of a more standard three-phase LV train bus for smaller loads. The control part may be dedicated, with specific control panels, especially on-board, but more and more commonly the system is governed by the SCADA system at all fixed installations; this requires a standard interface for the control of power elements and the interconnection of several sensors for environmental monitoring. This part will be treated altogether in the next section dedicated to SCADA.

The power part of the system is often responsible for characteristic emissions on-board, especially if one or more of the usual good practices are ignored:

- disconnectors and relays are highly discouraged for control of power elements, and in particular inductive loads, such as air-conditioning compressors and other type of motors, such as fans, since they produce intense inrush currents and overvoltages and steep transients during commutation;
- inverter drives shall be used instead, even if more expensive, to avoid both inrush and transient phenomena, but probably causing a not negligible increase of harmonic emissions at low frequency and responsible for some of the vehicle frame/enclosure currents; in this case the interference to signalling devices may be of concern over a broad frequency range, in terms of both conducted and radiated coupling;
- cable segregation shall be enforced, in particular for nearby susceptible equipment, such as video and displays, audio products, and signalling equipment, since inductive coupling may be relevant especially at the high frequency of the disconnectors induced transients;
- when installed on-board, static converters in particular shall meet the requirements for grounding to vehicle frame: they usually need to be galvanically insulated to prevent uncontrolled leakage of current components produced by commutations and to force all the return current through the supply connector, where the EMI filter is installed and where any additional countermeasure in case of interference may be more easily applied; current leakage on the contrary may be subject to local loops and resonances, that modify its intensity and distribution for variable frequency. It is underlined that improper grounding can occur also though unduly oversized common mode capacitors. This policy should be extended to all low

power and medium power converters located on-board, that normally are considered as minor electromagnetic sources, but can represent the source of many uncontrollable enclosure currents.

3.4.5 Escalators and Elevators

Escalators and elevators are normally not relevant as disturbing loads and are really immune to external interference, in particular for the usual location at stations, between floors in the station building, from the concourse down to platforms, but at sufficient distance from the track. In general, a local control panel is responsible for fine speed regulation and reaction to malfunctions and accidents, such as emergency halt, but it is normally remotely controlled by the SCADA system. The most commonly used drives are induction drives driven by static converters, but of low power rating (a few to a tenth of kW), so responsible for minor harmonic emissions in absolute terms onto the LV 3-phase supply network. Escalators and elevators have their own EMC product standard for emissions (EN 12015 [61]) and immunity (EN 12016 [62]):

- for low frequency emissions the standard allows a significant harmonic distortion, specifying % limits that are much higher than those specified by the EN 61000-3-2 for low power low voltage equipment; the conducted radiofrequency emissions are the same for Class A equipment, as reported in the EN 55011, while radiated radiofrequency emissions are even higher, with a 10 dB margin on Class A equipment, with a 50/57 dBμV/m limit; from a worst case point of view the expected radiated emissions may raise some concern with respect to the demonstration of compliance with nearby equipment;
- immunity specifications are particularly severe, since the location of such a system may be so close to the track or to other relevant sources of disturbance: in particular the radiated radiofrequency immunity is verified at an electric field intensity ranging between 10 and 30 V/m and the 2 kV level is prescribed for EFT and Surges, so that a high degree of immunity is expected.

3.4.6 SCADA

In a modern highly (or fully) automated transportation system (such as a light railway, a metro or a tram line) the Supervisory, Control And Data Acquisition (SCADA) system is a very articulated and complex system, depending on the number of implemented functions and the extension of the supervision and control to several subsystems and phenomena:

- air ventilation at stations and tunnels, both in normal conditions and in case of emergency (fire); the controlled elements are for example the main fans, the ventilation circuit in the station halls and concourse, on the platforms, technical rooms, etc. with particular attention to the emergency procedures (fire rescue plan), not to favour fire and its propagation: in case of fire in the station air barriers on the platform and the coordination of air flow between inlets and outlets is able to limit the propagation of fire in particular in areas where people is present;

- the fire can be detected by several types of sensors and techniques, such as thermocouples and pirometers, infrared cameras and sensors to remotely monitor and detect fire initiation, smoke sensors;
- fire extinction is performed also by means of the dedicated water circuit (equipped with pumps and valves controlled by the SCADA itself) and additional dust or CO2 systems for areas where water cannot be used or is ineffective (at electrical installations, in the presence of specific flammable substances, etc.);
- electrical networks at all levels: traction supply and substations, including protection (by circuit breakers and switches trigger and coordination) and monitoring (by measurement of voltage and current levels for diagnostic and control); low voltage for power; low voltage for lighting of public and technical areas inside and outside, of displays and panels, of trains; emergency lighting and emergency supply;
- anti-intrusion and alarm systems, for stations and tunnels, by interaction with the specific equipment for recording of doors opening and closing and badge accesses, for video surveillance, for volumetric control (such as infrared, microwave or ultrasound sensors);
- escalators and elevators, to monitor the status and the general conditions and to control their operation (starts and stops, parameters such as speed and jerk, etc.), in particular to facilitate evacuation in case of fire in a coordinated way with the air ventilation and fire extinction subsystems;
- other services, such as sewage and drainage, equipped with electric pumps and valves;
- depot specific installations, besides lighting, electric power, air ventilation and so on, such as vehicle cleaning and washing, maintenance and repairing; in this case the installations are usually controlled locally and the SCADA action is limited to monitoring;
- telecommunication system, subdivided in all the subsystems already considered, with which the interaction may range from simple monitoring (in case of sophisticated apparatus such as radiofrequency transmitters and receivers for wireless and radio coverage, high speed digital networks, etc.) to control (in some cases, such as for PABX and telephone networks);
- passenger information system (PIS) and passenger announcement system (PAS) are strictly controlled and coordinated by the SCADA in particular in case of emergency and in any case for normal operation, synchronized with the ATO, that controls the status of the circulating trains, the timetable and delays;
- platform screen doors, that are controlled locally for the safe and correct synchronization with the arriving train and positioning of vehicle doors,

but are monitored and controlled at higher level for coordination with PIS, PAS and ATO (to let passengers access the platform at the right time) and with various alarms coming from the anti-intrusion system, the fire alarm system, etc.
- access control, ticketing and fare collection systems are monitored as well with diagnostic purposes and to enable/disable terminals and booths in a coordinated way.

The subsystems considered above are manufactured and supplied by different companies, so the first level of automation is often supported by different solutions that implement locally the basic control functions. This broad class of equipment is characterized by and subject to an equivalent broad range of EMC standards and phenomena (see Chap. 6), that need a harmonization process with the assigned electromagnetic environment and related environment/application standards. Monitoring, supervision and high level control are on the contrary implemented by the SCADA system, that from an EMC point of view consists of information technology equipment and network equipment, is subject to the related standards and is normally hosted in a benign environment of the residential or light industrial type.

Chapter 4
EMC Management

4.1 Introduction

A railway is a complex engineering system and thus the construction of a new railway infrastructure or the modernisation (refurbishment) of an existing railway infrastructure requires systems overview. Electrified railways can require very high output (up to several MVA) electrical systems and power electronic systems that are characterised by their non-linearity (producing harmonics). In an electric railway the locomotives, which move with the trains, have to be supplied via sliding contacts from a supply line, called the catenary or overhead, or a trackside conductor rail, which is installed along the track. The current generally returns to the substation via the rails, a separate return conductor and via the earth. A detailed description of the railway and its core components is covered in Chap. 2.

Problems of electromagnetic interference (EMI) arise not only within the locomotive and the power supply but also in associated systems, such as signalling and communications systems. Non-electrified railways such as diesel electric traction may also be a source of electromagnetic noise. The normal and abnormal operating modes of such systems may be a source of electromagnetic disturbance, which may impact the normal and safe operation of all other systems within its zone of influence. The railway environment contains many electromagnetic disturbance sources and receivers and the close proximity of these within the railway environment presents a unique environment, in which good EMC Engineering practice must be applied to achieve system wide compatibility.

Railway systems failures linked with the advent of electrified railways and computerised signalling systems are documented in open literature. Such system failures (e.g. EMI to safety signalling systems resulting in abnormal operation, safety hazards from induced voltages, etc) have the potential to cause railway accidents and are usually classified as a high risk to railway safety. Reducing the risk of incompatibility to an acceptable level is the remit of Railway EMC Engineers and this remit is executed, in part, through

A. Ogunsola and A. Mariscotti: Electromagn. Compat. in Railways, LNEE 168, pp. 179–190.
springerlink.com

the documentation of the process(es), control procedures, organisation and responsibility matrix, requirements identification, apportionment and management, identification of validation and verification processes to be applied as well configuration and change control elements in what is known as the EMC Management Plan. In this book, a distinction is made between an EMC Management Plan and an EMC Control Plan; for large complex engineering projects, such as those encountered in a railway (Design and Build, Re-signalling Project, etc.) the EMC Management Project serves as the top level document in which the process for managing EMC throughout such a complex project is documented. It usually defines the overall EMC organisation, responsibility matrix, defines the demarcation of responsibilities, defines the overall EMC objective of the project, states the overall project EMC requirements, the acceptable documentary evidence to be produced by all relevant stakeholder and the overall EMC schedule. The EMC Management Plan also specifies the requirements for sub-contractor deliveries and this includes the requirements and content of an EMC Control Plan; the EMC Management Plan is therefore typically prepared by the main project contractor for a design and build project, or by the overall system integrator. An EMC Control Plan is usually prepared, by a sub-contractor, in response to a EMC Management Plan, it may also be a fist tier contractual deliverable in response to an Invitation To Tender (ITT). This Chapter deals with the development and implementation of a EMC management process for railway applications.

4.2 The Objective of Railway EMC Management

The safety critical nature of a railway requires that the following objectives form part of the design brief; the railway systems must be designed to perform the following [589]:

- to operate without suffering wrong-side failures;
- to operate without suffering an unacceptable level of right-side failures;
- not suffer permanent damage, except in extreme circumstances;
- to operate without endangering life.

The objective of EMC is to reduce the spurious interactions within and between systems below the threshold of interference to the systems' functions and performance. This undertaking may take the form of a stand-alone project or of an adjunct project within the system-development project. The development of a generic EMC solution to a class of systems (e.g. design rules for integrated circuits) warrants a major stand-alone effort whose cost is justified by the savings thereafter. However, a recovery effort for an (operating) system that is already adversely impacted by electromagnetic disturbance and requires retrofit (design or layout) in order to achieve the desired level of compatibility and performance – is highly cost sensitive. The cost of such an effort is proportional to the system's complexity. It entails diagnosis, analysis, measurements, changes in architecture, signal design and routing, layout and packaging. The cost of ensuring the EM compatibility effort with the system development process, from the conception phase right through to the disposal phase, is proportional to the accrued cost of the system development. The rate of interference-risk reduction is proportional to the resources allocated and disruption of the project. These amount to the equation of the project burn-rate due to EMI mitigation:

$$BR_{EMC} \propto -C\mid_t \frac{dR_{EMI}}{dt} \tag{4.1}$$

Where BR_{EMC} is the burn rate of the project due to EMI mitigation, $C\mid_t$ is the accrued cost of the project and R_{EMI} is the risk of system failure due to electromagnetic disturbance. While the accrued cost derives from the project strategy, the second term depends on the EMC strategy: a *posteriori* fix approach penalizes the project in high cost and in delay. A reduced risk approach by over-designing the EM isolation between circuits may cost in system performance, weight and cost. An attempt to achieve compatibility by adaptively tailoring the design, layout and signals' structure of the system constituents, requires an interactive diligent effort across the project working groups and throughout the project development. The rate of reducing EMI risk, dR_{EMI}, is bounded not only by the level of effort but by the shear lack of information on potential spurious interactions. This is a particularly important point to note for distributed complex systems such as a railway, where possible spurious interactions in the system are not known at the onset and reliance on past experience is an important element. Only interactions anticipated from the conceptual architecture can be inferred during the conception phase. More information can be extracted during validation, when critical tests are carried out. The signal paths and system configuration, as developed through the project, also create further spurious interactions. Late unveiling and mitigation of these cost a higher already-committed effort (lost cost), as portrayed in Figure 4.1. The interactive EMC effort is fed by the project available design information and samples to test, and feeds back recommended changes in system block diagram, signal flow, layout and packaging. Early EMI unveiling and mitigation obviously reduce the cost of any changes needed and disruption of the project pace.

With respect to EMC therefore, the overall objective is that all systems procured and installed within the railway environment are electromagnetically compatible with each other (intra-system compatibility) and that the railway is electromagnetically compatible with it's neighbour (inter-system compatibility). Furthermore, as electromagnetic disturbance may result in an unacceptable level of right-side failures, there is the objective to achieve the apportioned system and sub-system Reliability, Availability and Safety targets. To achieve these, it is important to ensure that EMC engineering forms an integral component of the project management process and is

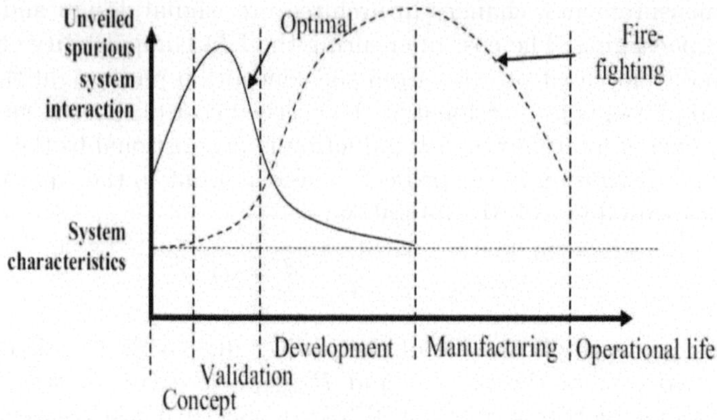

Fig. 4.1 Cost implication of Identification of Unintended EMD per Life Cycle Phases

considered from the outset of the project (and that the stakeholders involved are fully aware of their obligations) through to commissioning and end of life.

4.3 EMC Management Process

A project may be defined as “a temporary endeavor undertaking to create a unique product or service”. A certain value is perceived to be gained upon completion, and resources are planned toward this goal (cost of project). The process of project planning is analytic: sub-projects (phases) are defined and work packages organized and are inter-related by progress and by resource allocation. The progress is measured by the accomplishment – increasing the value toward the objective. However, if such an achievement depends on reducing the risks ahead it is not actually an achievement rather a risked expenditure (in cost and in time). A true measure of progress is thus the reduction of risk to achieve the project value. The optimal rate of risk reduction in the project from its onset to completion is obtained considering the concurrent rate of change of the project expenditure – its burn-rate. The cost lost during an incremental time (as in the dashed brackets in Figure 4.1) is the product of the risk and the project burn-rate

$$R\left(t\right)B\left(t\right)dt=\triangle\left[Lost\ Cost\right] \tag{4.2}$$

Minimizing the lost cost over the project entails a concave risk-reduction curve and an escalating burn-rate curve accordingly. This model advocates expediting the reduction of risks before launching high-cost investments. The direct consequence is that the project pace is bounded by the remaining risk, and is not an independent parameter. Time-squeezing techniques of project management (as in CPM – Critical Path Method) are then effective only for low risk projects or phases thereof. It is important to bear this in mind, when defining an appropriate EMC management strategy for railway projects. The impact of project risks in complexity, novelty (in system integration and application) and technology uncertainty on their management style have been shown by Shenhar and Bonen [630], EMI risk may be mapped by the same dimensions as that shown in [629]. The meanings of these axes would be different, however. The EMI risk lies in the system implementation: signals' generation, structure and paths, system layout and packaging. It may be high for otherwise low-risk project. The cost of its mitigation is typically a small fraction of the project cost, but its share in the project risk may be high. The synchronization of the EMC effort into the project management has to consider both maps and their respective management styles. Early identification and mitigation of the EMI risk, while deferring high-cost project activities until EMI mitigation is underway, appears to optimize cost and time of the project. A major part of the EMC management activity is identifying the failure/ disturbance modes and ranking the interaction map by probability of failure, then mitigating failure modes according to the project policy of reliability. The techniques for mitigating known interactions are within good engineering practice in most cases and do not constitute substantial risk. The changes required in the system may be costly, however, and delay the project. The logical sequence of the EMC management is:

- Information gathering including the identification of EMI coupling
- Information analysis including modeling and simulation of the coupling
- Creation of the EMI risk matrix and ranking of the hazards
- Measurements of system prototype
- Re-evaluation of the EMI Risk
- Application of EMI Mitigations
- System Validation test
- Evaluation and Qualification

The sequence may then re-iterate when further information is unveiled along the system development. EMC activity impacts the project in each of these phases. It interacts across the project's work groups and creates unpredicted dependencies between them. Frequent information update across the project is needed. System/ sub-systems and simulators have to be made available for observation tests, model verification tests and then validation tests. Changes in the system may have to be implemented along the development path. The management of synchronizing these with the project depends on the project characteristics and risk level. Low-risk projects are planned to optimize resource allocation. Risks are mainly logistical and their probability is calculated based on statistics of past experience. The EMI risk, however, is lack of knowledge of the nature of the potential problem, the knowledge gap is does not lend itself to statistical assessment and aa-priori synchronization of EMC into the project work packages and logical structure.

The objectives described above are best achieved through the implementation of an EMC Management process. Clause 5 of EN 50121 Part 1 [73], specifies a requirement for the EMC strategy to be documented in an EMC Management Plan, the referenced Clause further points to EN 50238 [112] for the management process, however the EN 50238 deals with inter-system compatibility issues with emphasis on the compatibility between rolling stock and train detections systems.

For a successful EMC strategy, the EMC management process needs to commence at the conceptual stage of the works and followed through the operations and maintenance phase. In this respect, Figure 4.2 reflects a typical EMC Management life-cycle [222] for a design and build railway project.

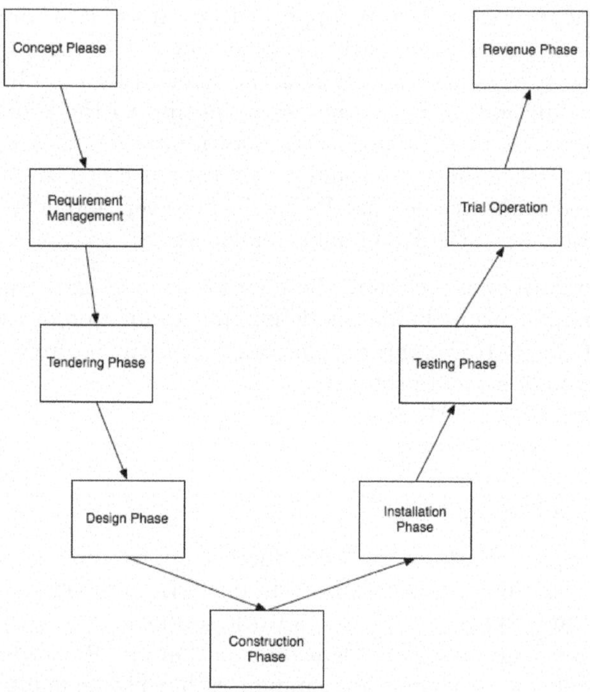

Fig. 4.2 Typical EMC Management Life Cycle

The EMC Management life cycle presented in Figure 4.2 has been intentionally presented in a V life cycle form, aligning it with the project life cycle typically applied in a railway project. The purpose in defining the project life cycle is to establish a framework for meeting the stakeholders' needs in an orderly and efficient manner. This is usually done by defining life cycle stages, and using decision gates to determine readiness to move from one stage to the next. Various life cycle models such as the waterfall, spiral, Vee, and agile development models are useful in defining the start, stop, and activities appropriate to life cycle stages. The Vee model is used to visualize the system engineering focus, particularly during the concept and development stages.

The Vee highlights the need to define verification plans during requirements development, the need for continuous validation with the stakeholders, and the importance of continuous risk and opportunity assessment. It details the sequence of steps in a project life cycle development. It describes the activities and results that have to be produced during product development. The left side of the "V" represents the decomposition of requirements, and creation of system specifications; the right side of the V represents the integration of parts and their verification; V stands for "Verification and Validation". Furthermore, EMC engineering falls under the scope of specialty engineering as part of System Engineering [588]. Specialty engineering integrates the specific system attributes and disciplines into the acquisition process and access (and conforms) various system attributes. This element includes Safety Engineering, Reliability Availability and Maintainability (RAM), Human Factors, Electromagnetic Compatibility (or Electromagnetic Environment Effects, E3), Quality Engineering and Information Security Engineering. Specialty engineering analysis describes technical details of the design from a particular perspective and often requires specialized skills. These analysis, which are performed through the system life cycle, help the program to define requirements and design features and/or describe characteristics of the design and related operations in support of validations and verification, requirements and functional analysis.

4.3.1 Concept Phase

The Concept Phase is executed to assess new business opportunities and to develop preliminary system requirements and a feasible design solution. The first step is to identify, clarify, and document stakeholders' requirements. During the Concept Phase an in-depth study, that evaluates multiple candidate concepts and eventually provides a substantiated justification for the system concept that is selected is undertaken. As part of this evaluation, mockups may be built (for hardware) or coded (for software), engineering models and simulations may be executed, and prototypes of critical components may be built and tested. Prototypes are helpful to verify the feasibility of concepts and to explore risks and opportunities. These studies expand the risk and opportunity evaluation to include affordability assessment, environmental impact, failure modes, and hazard analysis. Key objectives are to provide confidence that the business case is sound and the proposed solutions are achievable. The system EMC engineering effort undertaken at this stage will include an assessment of the electromagnetic environment along the proposed alignment (or an existing alignment) - this will include the identification of external electromagnetic source and potential victims neighboring the alignment. For a brown field project, the identification of electromagnetic sources and potential victim will include systems within the railway environment as well as those external to the railway. The identified sources and victims are categories with respect to their frequency of operation, separation distance to the alignment and field strength. A preliminary hazard analysis will be undertaken to identify and detail potential hazards and mitigations - this will feed into detailed EMC analysis to undertaken at later stages.

4.3.2 Requirements Phase

This phase identifies and develops all the requirements and ensures that they are met throughout the system life cycle. It is a series of iterative tasks performed by a multifunction team. Defined requirements are managed, through a requirement management tool, through verification to ensure compliance with customer needs and expectation, communication of allocations and management of change. Requirements management is an interactive process, which typically consists of the following steps:

- establishing a requirements management plan;
- requirements elicitation;
- developing the vision document;
- creating use cases;
- supplementary specification;
- creating test cases from use cases;
- creating test cases from the supplementary specification;
- system design.

As in system engineering, the first step in EMC management is the establishment of EMC requirements. The requirements are a combination of applicable regulatory requirements and operational environment specific requirements. Operational EMC requirements tend to be driven by contractual reliability and quality requirements and are also very important from business point of view, whilst regulatory requirements are driven by the protection requirements of the EMC regulation. It is usually beneficial to list all the electromagnetic disturbances that the system is likely to encounter (or be exposed to) in its intended operational environment. The railway environment is abundant with many sources of electromagnetic interference. The emission spectrum and the frequency content of some of these sources have been adequately addressed in Chap. 6. Some of these sources are transitory dependent, i.e. pantograph - contact wire and the magnitude of the interference may not be easily predictable. The electromagnetic environment list (or database) should include details of test methods, test specifications and a methodology for demonstrating compliance. At the beginning of a project, a complete list of all possible EM threats may not be easily identifiable or readily available.

The requirement must also take into account the existing installation conditions and the safety critical nature of the existing and new installations. For a brown field project, the railway environment may consist of equipment that predates the compatibility limits specified in international standards and thus may have a lower susceptibility profile and/or may exhibit a higher emission spectrum. Such equipment, referred to as legacy equipment, continues to provide valuable service to the railway and as such EMC requirements will need to be designed around such equipment. Translation of system level EMC requirements to sub-system (or component) level and installation level requirements is another important issue in the design and installation of a complex distributed system.

In summary the EMC engineering activities at this phase of the project involves the:

- development of the general EMC Management and specific technical requirements;
- development of EMC interface specification;
- documentation of the design criteria and standards to be used;
- documentation of the associated documents;
- preparation of relevant schematics that underpin key EMC strategies.

These aspects must include consideration of lightning protection, earthing and bonding, and environmental design.

4.3.3 Tendering Phase

In this phase, the EMC engineering activity generally involves the packaging of EMC requirements (developed during the requirements phase) into a overall project tender package. In addition, tender assessment criteria will be developed to guide the assessment of the EMC content of the Tenderer's bid. This usually involves an assessment of the Tenderer's understanding of the EMC requirements, an assessment of their approach to EMC Management, an assessment of their EMC experience on a related project, an assessment of their approach to system issues to name but a few.

4.3.4 Design Phase

As part of the synthesis element of system engineering, balanced EMC solutions must be developed to meet the set of defined requirements. The solutions may be a mixture of both electrical and mechanical design and/or complex design solutions, tailored to deal with a particular problem at a specific installation site. The safety critical nature or function of a system (or sub-system) should be reviewed in line with the design solution. All EMC design solutions should be recorded in an EMC Control Plan, along with the technical rationale supporting the design solution. The use of commercial off the shelf (COTS) products and legacy equipment should be analysed and assessed against the EMC requirements of the system. It maybe necessary to accept a slightly reduced severity level or failure criteria for a COTS depending on its functions, safety criticality and intended environment of use. With COTS and legacy equipment, there is usually little or no control over the internal design. As a distributed system, the railway relies on a massive network of metallic conductors and structures for interconnecting power, signalling and telecommunications. These metallic conductors and structures act as propagating paths for electromagnetic interference and there is a need to design and install them, in such a manner as to ensure that propagation of adverse electromagnetic disturbance levels are kept to a safe limit. Grounding, cable design and management issues, as well as all other interface issues should also be covered in the EMC Control Plan.

From a client view point, EMC activities during this phase of the project include:

- stakeholder management;
- EMC design reviews;
- gap analysis and mitigation;
- numerical and computational analysis;
- EMI Hazard close out.

4.3.5 Construction Phase

A key EMC activity during this phase is the EMC assurance activity. In this context, EMC assurance covers all the qualification activities required in response to the EMC requirements. This implies testing the system (or sub-system) to a set of EMC requirements specified at the beginning (and/or modified during the system's life cycle). For complex systems, verification may be a combination of laboratory test, in-situ test and numerical modeling. Testing regimes can be divided into two separate phases: the design verification phase and the product verification phase. The design verification test assesses the compliance of the sub-system to the EMC requirements specified; this phase of EMC test should be performed early and often to enable the refinement of the sub-system. Certification of subsystems may be necessary if national regulations require so. Product verification testing seeks to assess the performance of the complete system, for complex large distributed system a complete system test may be impossible. At product validation testing, the area of interest tends to be the system interfaces, such as cables, usually experiencing the most severe threats, and connected ports. At this phase of the test, targeted EMC test are conducted to assess performance against specific EMI which are unique to the environment of use and installation.

4.3.6 Installation Phase

Prior to this phase, it is advisable to detail any test and test procedure deemed necessary to ascertain the integrity of the proposed installation. These would normally be driven by the need to demonstrate the functional performance of the item being installed and that any associated risk is as low as possible. It is equally possible subject items of installations to specific EMC test, that perhaps was not possible to conduct in laboratory due to the size of the installation or the complexity of the system functionality under investigation. In most cases, EMC activity at this phase is limited to workmanship assessment of the installation ensuring that best practices being followed and that agreed EMI mitigations are being implemented.

4.3.7 Testing Phase

During this phase, a number of system EMC test are scheduled and performed; this typically includes in-situ system test, system and signalling compatibility. In-situ testing of the distributed system may be required to demonstrate compliance to operational EMC requirements; such tests are difficult to conduct to a high degree of confidence and repeatability due to the statistical behaviour of the interference, the varied system configuration and the possible modes of operations.

4.3.8 Revenue Phase

Before commencement of revenue, it is expected that all test and commissioning works have been completed and that a robust case for safety, including EMC aspects, has been prepared, independently assessed and accepted. Therefore, EMC activities during this phase involves configuration management of the railway electromagnetic environment, trend analysis, determination of permissible limits for specific safety critical assets as well as development of code of practices, design criteria and rules, and company standards.

4.4 EMC Testing

Testing occurs throughout the life cycle of the system: in diagnostics, validation, compliance/ verification and regulatory compliance. The nature or scope of testing varies with the application. Regulatory compliance is a well defined measurement procedure in terms of the test configuration, test equipment, measured parameters and threshold values. Diagnostic, validation and verification measurements are system-specific, tailored to the system complexity and state of knowledge. They sample parameters of the system state between the measurement probes, and serve to build models that describe the system and relate excitations to responses. The initial EMI Coupling identification, listing all potential EMI-related sources-to failure modes coupling, represents the lack of models to predict system response. The respective scope of measurements required for characterizing the spurious interactions/ failure modes - is substantial, and its reduction ("culling") is essential. It is obvious that Monte-Carlo sampling the measurement ports does not increase reliability of any element that was not tested, and is not a viable testing model. A model-based culling employing an iterative sequence of measurements and analysis has been applied successfully, reducing the matrix size by two orders of magnitude. Alternative models were tested computationally for their response to a structured excitation (e.g. swept frequency, shaped pulse), and compared to measurements. The measurements were used for choosing the models for each interaction, thus gaining a hierarchical topology based on dominant interaction models. This verified topology is then used to cull the weak interactions and schedule the measurements accordingly. The measurement program is a significant effort that requires preparation and coordination with the project activity. The system under development has to be made available, including the respective labor for its operation. The higher the innovation in the system, the more measurement phases may be required. High complexity necessitates the tests of more layers (subsystems) and their combination. Major test facilities for exposing systems to external radiation, serve to test global susceptibility and system qualification. They are not effective for diagnosis, where local interactions (e.g. skin penetration) are sought. This requires local probing [47].

EMC testing is made mainly during the construction phase, when procurement of equipment and apparatus occurs. In this case EMC testing consists of applying the tests for emission and immunity prescribed by the applicable standards on the products that are procured and installed: if they are certified by the manufacturer then they have been tested at origin for emission and immunity following the applicable product and/or environment standards and they are suited for specific installations and applications; analogously, when the equipment undergoes relevant modifications the manufacturer is responsible for the repetition or integration of tests to reissue the certification; in both cases the tests may have been executed years before the procurement; if the equipment is not a commercial product, but is rather a custom product or a modification of an existing product for the present application, then the tests are due at a laboratory before the equipment is shipped and installed.

Chapter 5
EMC Assurance

A railway is a complex engineering system and thus the construction of a new railway infrastructure or the modernisation (refurbishment) of an existing railway infrastructure require a system overview and a consideration of safety. In the context of electromagnetic compatibility, the complexity is due to the close proximity of safety critical systems (and systems performing safety related functions) to electromagnetic disturbances, the majority of which are generated by other railway systems themselves. Electromagnetic interference, whether intentional or not, can cause a wide spectrum of difficulties including momentary, minor inconveniences to system failures in electronic/electrical equipments. In considering EMI risk in railways, it may be necessary to take in account the structural factors in facilities (i.e. control and equipment rooms) with respect to architecture, materials and topography, as this plays a significant role in environmental protection plan. During design and maintenance, significant consideration is given to local weather phenomena, soil, environmental and structural factors based on historical data, as a quantifiable risk. In this book, safety critical systems are railway assets, whose improper function, as influenced by electromagnetic interference, would result in loss of life or injury, while systems performing safety related functions are those that perform a specific function or functions to reduce the risk level to that which is considered tolerable. Reducing the electromagnetic interference (EMI) risk is therefore a major concern in the railway industry.

5.1 EMC and Safety

Traditionally, the discipline of EMC and Safety has been treated separately, but the growing use of electrical, electronics or programmable electronic systems in railway system, performing safety and control functions, has led to a re-think of this relationship, particularly so with the publication of IEC 61508 series of standards [333] - [332]. IEC 61508 specifies the requirements for

A. Ogunsola and A. Mariscotti: Electromagn. Compat. in Railways, LNEE 168, pp. 191–215.
springerlink.com

ensuring that systems are designed, implemented, operated and maintained to provide the required safety integrity level (SIL). Four SILs are defined according to the risks involved in the system application, with SIL4 being used to protect against the highest risks. The standard specifies a process that can be followed by all links in the supply chain, so that information about the system can be communicated using common terminology and system parameters. Safety assurance requires a detailed analysis of all reasonably foreseeable risk including that posed by exposure to electromagnetic fields. Risk mitigation is required for all identified EMI risk that could have an adverse effect on safety. For safety assurance. the traditional EMC approval processes are enhanced to reflect the safety implications, EMC Engineers need to be familiar with tools and processes required for demonstrating safety from an EMC perceptive.

For any system, which could pose a hazard to people or property, if incorrectly controlled and/ or designed, safety questions must invariably be addressed, and these questions must include the issue of safety in the face of external electromagnetic disturbance. Safety in the railway can be subdivided into two areas: (1) functional safety, which is directly related to apparatus performing a safety related function and (2) personal safety issues, which may lead to degradation of health of railway users (and maintainers) and/or loss of life.

For safety systems, the basic principle of design are [589]:

1. to operate without suffering wrong side failures;
2. to operate without suffering an unacceptable level of right side failure;
3. not suffer permanent damage except in extreme circumstances;
4. to operate without endangering live.

In the railway these four principles are key to its operation. The operation of the railway is dependent on the railway administrator's ability to operate and maintain a safe railway. It is possible to have an apparatus that is safe, fails to operate correctly and complies with all legal requirements; however it would be deemed unacceptable to have an apparatus that is unsafe, operates as intended and complies with all legal requirements. Thus, from a safety assurance view point, all possible failure modes - their causes and consequences - must be assessed and mitigation provided. The railway administrator has overall responsibility for overseeing safety of the railway infrastructure. It discharges these responsibilities via a formal set of guidelines, which provide a systematic basis for the safe management of change. In the UK, for examples, these procedures are detailed in a series of documents entitled "Engineering Safety Management" [681], known in the railway industry as the "Yellow Book". The aim of these documents is to ensure that any change to be introduced on the railway is managed safely and that the safety arrangements are adequate for the proposed undertaking. It is required that the safety implications of any proposed change be shown to be acceptable before implementation of that change. Proposed changes are formally defined and submitted with safety justifications to a change management system. It is the responsibility of the change instigator to ensure that the potential hazards and risks has been identified, assessed and shown to be As Low As Reasonably Practicable (ALARP). This justification is provided in the form of a Safety Case, a document that provides the prime assurance that all hazards and risks have been assessed and shown to be ALARP. The Safety Case must contain considerations of EMC issues and its impact on the safe operation of the railway.

The principle of ALARP ensures that the risk of any system with serious consequences in terms of human loss and injuries, is kept to a level which is As Low As is Reasonably Practicable. ALARP defines three risk levels:

- Intolerable Risk, which cannot be justified or accepted, except in extraordinary circumstances;
- Tolerable Risk, which can be accepted only if risk reduction is impractical or if the cost or risk reduction greatly exceeds the benefit gained;
- Negligible Risk, which is broadly acceptable and does not require risk mitigating measures.

If risk is determined to be at the intolerable level, measures must be taken to reduce it immediately to a tolerable level. If risk is found to be at tolerable level, risk mitigating measures should still be applied, provided that a cost benefit analysis is in favor of it. The principle of ALARP describes the way in which risk is treated legally in some countries and in railway applications; ALARP is a fundamental component of the safety assurance process. In any case it is always necessary, whatever the cost benefit arguments, to demonstrate the application of good practice. Figure 5.1 shows the so-called ALARP triangle which illustrates these regions and Table 5.1 illustrates a risk classification matrix based on ALARP principle.

In some countries, the consideration of EMC issues are documented in what is referred to as the "Electrical, Electronics, Signalling and Communications

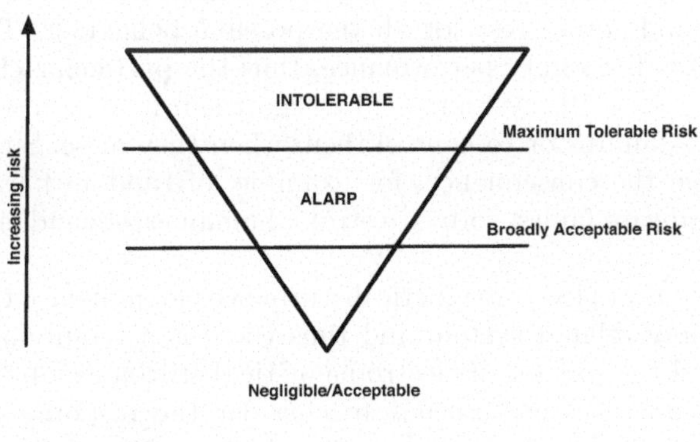

Fig. 5.1 ALARP Triangle

(EE&SC) Safety Case", the EMC Safety Case or the EMC Safety Justification. The intent of these documents, no matter what it is called, is to provide a link between the safety evidence and the safety claim(s) demonstrating that the safety evidence sufficiently supports the claim(s) being made. The kind of "safety" evidence that arises when developing or assessing a critical system is characterised both by its diversity and its uncertainty.

Table 5.1 Hazard Categorisation based on ALARP Principles

Frequency/Consequence	1 Very Unlikely	2 Remote	3 Occasional	4 Probable	5 Frequent
Catastrophic					
Critical					
Major					
Minor					
Negligible Risk		Tolerable Risk		Intolerable Risk	

5.1.1 *Safety Related System*

A safety related system is defined in [326, 319, 320] as a designated system that both implements the required safety functions, necessary to achieve or maintain a safe state for the equipment under control, and is intended to achieve, on its own or with other electrical, electronic or programmable electronic (E/E/PE) safety-related systems, other technology safety-related systems or external risk reduction facilities, the necessary safety integrity for the required safety functions. It includes all the hardware, software, human operators and supporting services necessary to carry out the specified safety function. The safety function(s) of the safety related system is specified in terms of functionality and safety integrity.

Safety functions are frequently performed by E/E/PE systems. These systems are usually complex, making it impossible in practice to fully determine every failure mode or to test for all the possible behaviour. This means it is difficult to predict safety performance, thus the particular challenge with respect to such E/E/PE systems is to design them in such a way as to prevent dangerous failures or to control them when they arise. Such dangerous failures may be the consequence, for example, of random hardware faults, systematic hardware faults, software errors, human errors and environmental influences.

It is accepted that electromagnetic disturbance can influence the functional safety of a safety related system and thus there is a requirement to assess and understand the impact of electromagnetic disturbance on safety related systems; such an assessment needs to consider the response of the safety system under normal and abnormal operating conditions, as well as under environment exposure and adverse electromagnetic disturbance.

The concept of IEC 61508 distinguishes between the consideration of the application and the design of safety related electrical and electronic control systems. The interface between both is the Safety Requirements Specification (SRS); it specifies all relevant requirements of the intended application:

1. definition of safety related functions, based on a risk assessment of the intended application (which function may cause a dangerous failure);
2. appropriate (SIL) for each safety-related function based on a risk assessment of the intended application;
3. definition in which environment the system shall work including the EM environment.

The safety related system intended to implement a specified function for a particular application has to fulfill the SRS for this application. Equipment intended for use in that application has also to fulfill this SRS.

5.1.2 System Fault

IEC 61508 [329] defines a fault as an "abnormal condition that may cause a reduction in, or loss of, the capability of a functional unit to perform a required function". Faults may be divided into systematic and random faults, and the nature of the faults will depend on the form of the elements concerned. The failure probability of single elements can be estimated from the knowledge of the strength (failure threshold) of the equipment and the stress which the equipment is exposed too.

5.1.2.1 Estimation of Immunity Threshold

The immunity threshold of a device can be measured as part of the EMC and Functional Safety qualification test, the aim here however is to establish the severity level above which fault may occurs. Such failure data is collected, stored and can be used in future EMI risk analysis. It is equally possible that the immunity threshold is estimated based on knowledge of and history of the device or a similar device. Such an estimation should be treated with caution as the response of the device to electromagnetic disturbance is random by nature. The use of estimation techniques is effectiveness when applied to the behaviour prediction of components, such as a micro-controller. Camp et al. [44] have shown that the strength (breakdown failure rate (BFR)) can be described by the Weibull distribution. The parameters of that distribution are however dependent on the shape of the radiated electromagnetic pulse. For different pulse shapes different breakdown behavior is expected. Using the window norms in frequency domain, the energy and amplitude efficiency factors [581] can be calculated. With the help of the energy and amplitude efficiency factors the breakdown behavior for one pulse shape can be estimated from the knowledge of the breakdown behavior for another pulse shape.

5.1.2.2 Estimation of the Stress

In addition to the immunity threshold, the knowledge of the electromagnetic disturbance that the device will be exposed is important and equally required. For that the principles of the electromagnetic (EM) topology can be applied. The main idea of EM topology is the same as in fault tree analysis: a complex system is decomposed into smaller subsystems. That way, the subsystems can be analyzed independently. In order to analyze the EM topology the whole system is divided into sub-volumes which are separated by surfaces. A surface can be represented by a transfer function from one volume to another. Having the absolute knowledge of the situation including information like the exact positions of the objects, material parameters, angle of incident of electromagnetic wave, there would be no need to analyze the system statistically. In reality, we don't posses the absolute knowledge of the situation but a certain degree of belief or expectation about the true situation and thus we have uncertainties.

The classical EM topology concept however, does not take into account uncertainties, thus it would need to be extended to do the same. There are two ways this can be achieved, the first method is to extend the topology concept in order to include the uncertainties and then to use the input events occurs the results as input data for the fault tree. The second method is to further extend the fault tree and to include the effects of shielding and propagation in this fault tree. By choosing the fault tree analysis, the whole system can be described by one model. There are however restrictions of the fault tree. The events in the tree are binary and thus continuous values, such as field strength or shielding effectiveness, can be included if the symbols of the fault tree are extended. However the more serious restriction is that a disturbance in the electromagnetic topology graph can take several paths to each system element. Trying to implement this parallel paths in the fault tree is difficult and for some cases even impossible. As a consequence, the classical topology should be used to describe the electromagnetic disturbance that the device is exposed too. The classical topology can be extended by by assigning the transfer function between two volumes uncertainties. This uncertainty can be described by single values such as the standard deviations or by whole probability density function. A classification of typical electromagnetic disturbances, disturbance and compatibility levels for various locations is given in IEC 61000-2-5. This should be used in the absence of actual data.

5.2 Hazard and Risk Analysis

The railway environment is abundant with many sources of electromagnetic interference. The emission spectrum and frequency content of some of these sources have been adequately addressed in literature [498, 17, 303, 10, 237, 628, 683]. Some of these sources are transitory dependent, i.e. pantograph - contact wire and the magnitude of the interference may not be easily predictable. A safety related system that is susceptible to external electromagnetic disturbance may be unpredictable and therefore potentially dangerous; unfortunately, being inherently random in nature, electromagnetic disturbance is difficult to characterize and its effects are very difficult to predict. Some of the special features of the railway environment are highlighted below:

- a variety of power supply configurations supplying high power to loads;
- close proximity and area of sources and victims circuits;
- methods of power transfer to moving trains;
- a fluctuating and imprecise system of current flow to and from the trains;
- high single-phase loads which may cause imbalance in the three phase system;
- the possibility of simultaneous generation of electromagnetic disturbance from several sources;
- moving sources within the same zone of influence;
- the variability of electromagnetic noise levels due to weather conditions.

The due consideration of the possibility that an adverse effect will be produced from exposure to electromagnetic disturbance *(EMI Risk)* and the process of quantifying and evaluating the risk (*EM1 Risk Assessment*) are crucial to the reduction and management of the risk itself. EMI risk only exists when EMI hazards exist and something of value is exposed to the EMI hazard. It is therefore important that a detailed assessment of the risk posed by the interaction of the railway's electromagnetic environment and the apparatus be undertaken and analyzed. Equally, one should assess the impact of the apparatus to the railway's neighbours, if by nature of its operation, it has the ability to influence the electromagnetic environment at large (i.e the radiated disturbance from a moving train impacting television set located outside the railway boundary). The classification and probability of malfunction is industry specific, however a generic classification applicable across all industries is defined in [330].

5.2.1 EMI Hazard Analysis

A failure has to be identified before it can be quantified and there are a number of formal procedures that can be implemented for this purpose. Hazard analysis involves the identification of hazards and the evaluation of possible scenarios leading to unwanted consequences. The hazard analysis stage is a very important part of a railway project risk management process, as no action can be made to avoid, or reduce, the effects of unidentified hazards. The hazard analysis stage also has the largest potential error with little or no feedback of those errors. Hazard analysis relies on a structured and systematic approach to identify potential; hazards. There are a large number of techniques that can be used to perform hazard analysis, however there are a handful of techniques, that are typically suited for identifying failures due to electromagnetic disturbances.

5.2.1.1 Hazard and Operability Study (HAZOP)

The HAZOP process is based on the principle that a team approach to hazard analysis will identify more problems than when individuals working separately combine results. The HAZOP team is made up of individuals with varying backgrounds and expertise. The expertise is brought together during HAZOP sessions and through a collective brainstorming effort, that stimulates creativity and new ideas and a thorough review of the process under consideration. A HAZOP can be used at varying times during the life cycle of a railway project, from concept right through de-commissioning, including hazard assessment of any modification proposed during the operational life of the railway. To produce a comprehensive evaluation of the process, a number of guidewords are combined with parameters and systematically applied to aspects of the system under consideration. A HAZOP can be time consuming and expensive and requires detailed design drawings in order to perform a full study.

5.2.1.2 Failure Mode and Effect Analysis

To understand how the components of a system contributes to the overall failure mode, a process known as Failure Mode and Effect Analysis (FMEA) is performed. The FMEA is an analysis method that was originally aimed at predicting reliability in systems and the purpose of the method is to systematically eliminate identified failures within a component. The technique behind FMEA is quantitative, since each identified failure mode is assigned a probability. The advantage with the method is that the analysis will be very detailed at a component level, which could be used to identify single faults or the need for changes in the design, such as implementing redundancy or fail safe technology.

A FMEA could be performed using a hardware approach or a functional approach. The hardware approach considers the failure modes of identifiable items of the hardware. The hardware approach focuses on the effects from a failure of each component since it is normally not considered as an appropriate approach when analysing the effects from electromagnetic interferences. The effects from electromagnetic disturbance are usually a result from disturbances in the operating conditions (currents and voltages) of components, rather than failures of components themselves. The functional approach of an FMEA is, on the other hand, a more appropriate method for investigating the effects of electromagnetic disturbances. With the functional approach, the method asks the question "In what ways can this function deviate from the specified requirement?". This approach identifies the more critical functions and will therefore require a higher level of immunity. When a FMEA is conducted, the work is emanated from a scheme showing how the component or function is connected to the rest of the system. Different failure modes are identified and a FMEA chart is filled in. In the FMEA chart, the causes and effects of each failure mode are analysed. An analysis of the critical behaviour is performed and the behaviour of the safety related system when the different faults are present is stipulated and actions are taken in order to compensate the various failure modes. This process is repeated after every change in the system. However, a bottom-up approach such a FMEA will overlook interacting contributions from different components. Since the bottom-up methodology considers all fault modes, including fault modes not relevant to electromagnetic disturbances, this will be an unnecessarily extensive and complicated method for complex systems.

FMEA has difficulty in identifying hazards, that require the failure of more than components, due to the complex interaction of the failures. The following steps are performed during a FMEA [163]:

- System Description: a detailed description of the process is initially required so that it can be broken down into either a functional, hierarchical or reliability block diagram. The assessment needs also a detailed description of the various actions performed by each section, including a description of their modes of success or failure;
- Establishment of Purpose and Scope of The Study: the purpose and scope of the assessment needs to be stated, along with how the results will be displayed, and whether there will be any interactions between the results and any further studies to be performed;
- Perform the Study: initially all items should be identified in the various system or sub-systems produced by the graphical breakdown of the system. All the possible failure modes of thees items should then be collated. the most likely failures caused by the failure modes can then be identified, and their effects on the process evaluated;
- Results Reporting: the report should include the diagram produced in the System Description, a detailed record of the analysis, a summary, recommendations indicating possible improvements, failures which have significant effect on the process, as well as those that produce none.

5.2.1.3 Failure Modes, Effects and Criticality Analysis (FMECA)

Failure Modes, Effects, and Criticality Analysis (FMECA) employs the same methodology as implemented for FMEA with the following additional steps to be performed:

- Determination of the severity of the effects caused by the failure: the severity of the failure is generally classified as being in the range of; complete loss of capability with loss of life, to negligible effect on success with no injuries;

- Evaluation of the frequency of the occurrence: an assessment is made to evaluate the frequency of the effect caused by the failure, this is usually done by an examination of previous data for similar systems to that under consideration, operating at similar conditions. In the event of no prior data being available then additional factors can be used to convert frequencies from unrelated systems. Once appropriate data is complied, the severity and frequency of the event caused by the failure can then be compared producing a criticality rating. The most critical failures would cause an event that produces very severe results very often, while the least critical would produce negligible results very infrequently.

5.2.1.4 Fault Tree Analysis

The Fault Tree is a graphical method of describing the combination of events leading to a defined system failure. FTA starts with all identifiable hazards and works backwards to determine their possible causes. Immediate events related to the top event are combined using logical operations, such as AND or OR, and the process is repeated, working back to the basic events, that are the root cause of the hazard. The logical operations are represented using a set of symbols defined in IEC 61025 [325] and shown in Table 5.2.

These symbols are very similar to those of conventional electronics logic. FTA provides a systematic description of the combinations of possible occurrences in a system that can result in failure or damage to the apparatus.

Table 5.2 Fault Tree Symbology

Logic gates	OR-gate	The OR-gate indicates that the output event occurs if any of the input events occur
	AND-gate	The AND-gate indicates that the output event occurs only if all the input events occur at the same time
Input events (states)		The basic event represents a basic equipment failure that requires no further development of failure causes
		The undeveloped event represents an event that is not examined further because information is unavailable or because its consequences are insignificant
Description of state		The comment rectangle is for supplementary information
Transfer symbols	Transfer out Transfer in	The transfer-out symbol indicates that the fault tree is developed further at the occurrence of the corresponding transfer-in symbol

Each fault tree deals with a specific event, the top event would be obtained from the hazard analysis; for EMI analysis these will be hardware failures or loss of functionality. The AND gate is used to represent a logic condition in which all inputs below the gate must be present for the output (at the top of the gate) to occur, while the OR gate is used to represent the situation in which any of the input events will lead to the output. A normal event is an event that is expected to occur during system operation; an undeveloped event is an event that has not been developed further due to lack of information or because it is not of sufficient consequence. A transfer symbol is used to represent a connection to another part of the fault tree within the same branch.

In fault tree construction a distinction is made between a fault and a failure. In FTA, *failure* is reserved for basic events, while a *fault* is more all-encompassing. The dissection of a system to determine what combinations of primary failures may lead to the top event is central to the construction of a fault tree. Failures may be classified as primary, secondary and command faults. A primary fault by definition occurs in an environment and under a loading for which the component is qualified. Primary faults are most often caused by defective design, manufacture, or construction and are therefore most clearly correlated to wearing failures, they may also be caused by excessive or anticipated wear, or may occur when the system is not properly maintained (i.e. maintenance of EMI gaskets) and parts are not replaced on time (i.e. timely replacement of a faulty surge protection device). Secondary faults occur in an environment or under loading for which the component is not qualified, for example an under rate surge protection device or filter, Such failure often occur randomly and are characterised by constant failure rates. Although a component fails when it has primary and secondary faults, it operates correctly when it has a command fault, but at the wrong time or place.

FTA is an important step in establishing the relationship between electromagnetic disturbances and a particular failure, it is useful in establishing a set of electromagnetic phenomena that needs special attention during detailed EMC and Functional Safety investigation. For a large system such as a distributed signalling system or a train, the fault trees are often many, large and complex. It is therefore important to limit the FTA to safety critical top-events. The evaluation of a fault tree proceeds in to steps: a qualitative analysis and a quantitative analysis; the qualitative analysis involves constructing a logical expression for the top event in terms of combinations of the basic events whilst the quantitative analysis involves using the logical expressions to give the probability of the top event in terms of the probabilities of the primary events.

The results from the FTA could be used to find different cases that ought to be tested in order to provoke a wanted top-event. There are some advantages when using fault tree analysis for evaluating EMC. The method can handle both common cause failures and time varying failure rates which is important when the behaviour during the presence of electromagnetic disturbances in a railway is analyzed. Another advantage is that the events in an FTA analysis are not limited only to faults, but could also involve degradation in performance. When a fault tree is constructed, it is possible to calculate probabilities using Boolean algebra.

5.2.1.5 Preliminary Hazard Analysis

Preliminary hazards analysis is used as an early means of hazard identification during the design and development of a process. The method is often used to follow-up on the hazards that have been identified during hazard analysis performed at the conceptual design phase. It follows an approach similar to HAZOP.

5.2.2 Risk Acceptability

The subject of risk has increasingly become a point of shared interest between many entities representing different sectors. This has given rise to different but converging definitions of risk [28, 33, 552, 555, 678]. Risk has been defined both qualitatively and quantitatively. Modares [552] defines risk qualitatively as the potential of loss or injury resulting from exposure to hazards. A hazard being considered as source of danger that is not associated to the likelihood with which that danger will actually lead to negative consequences. Quantitative definitions of risk associate hazards with their probability of nuisance to the people and the environment. For instance in [461], risk is defined to be a set of scenarios (S_i), each of which having a probability (or frequency P_i) and a consequence C_i. This quantitative definition to risk aims to estimate the degree or probability of loss related directly to the occurrence of hazards or potential failures of a system. The risk management process, as set out in International standards, consists mainly of five sequential stages, as illustrated in Figure 5.2, beginning with the establishment of the context within which risk has to be evaluated, in order to set both the objectives and scope of the system; this entails an exhaustive and detailed description of the system that is at risk. Having delimited the system, the potential hazards or sources of risk are identified, using one of the techniques described above; in this stage the list of initiating events, E_i, or scenarios of events leading to the undesired outcome is enumerated. Those events include essentially internal and/or external failures of both the technology used and the human force responsible for it. The next stage, usually referred to as *risk analysis* is reserved for estimating the likelihood, P_i, of the scenarios or events E_i, actually occurring and each scenario's consequence, C_i, is also estimated. The results of the risk analysis stage are thereafter used to compare and rank the various risk drivers and compute the total expected risk value, R, defined as: $R = R_i$ where $Ri = P_i C_i$ is the expected risk value associated with event E_i, the risk analysis is illustrated in Figure 5.2. In the evaluation stage, minor risks may be screened out and more attention paid towards risks with highest expected risk value. Risk treatment is the final stage, where action plans are determined in response to the identified risks and mechanism to control those risks are put in place. It should be noted that this risk management process may well require regular monitoring and review especially when applied with dynamic systems, which may evolve over time. Successful risk management requires that all parties who need to be involved at any stage are given adequate opportunity to do so and play an active role in the process and are kept informed of any developments and actions resulting from the process.

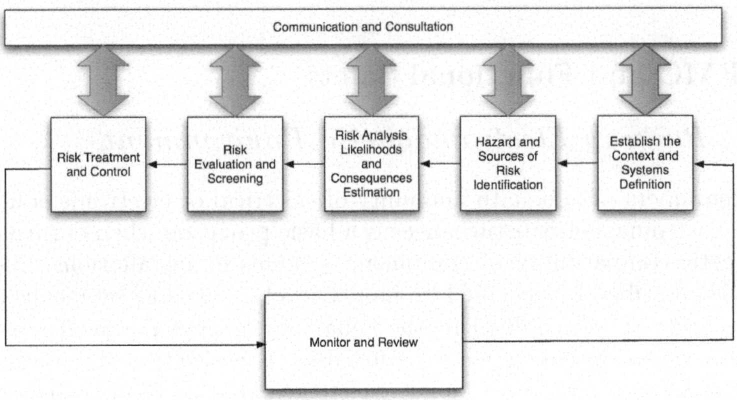

Fig. 5.2 A Typical Risk Management Process

With respect to EMC, it is clearly not satisfactory to have an EMI related hazard that could have catastrophic consequences and that could occur frequently; however it may be acceptable to have an EMI related hazard with negligible consequences that occur frequently. e.g. mobile phone interference on a VDU. The acceptability of a given level of risk is determined by the benefits, associated with that risk, and by the amount of effort that would be required to reduce the risk. The criterion for acceptance of a particular risk is based on a decision as to whether the risk is As Low As Reasonably Practicable (ALARP), bearing in mind the benefits of the system and the cost of any further risk reduction. Equally, the assessment of the risk is based on current industry knowledge and understanding, thus what was an acceptable risk yesterday may not be acceptable today. The assessment of risk is an imprecise process involving much judgment and intuition and a poorly defined database. This is more so in the railway industry, since electromagnetic immunity threshold of legacy equipment are not well known, and if known are out of date as the data may not contain reference to high frequency interference sources which are now abundant. Clearly, the determination of the risk acceptability should consider the life cycle of the system and also consider non-EMC issues which may influence our perception of the risk, for example Human Factors.

Within railway application, a technique normally applied to assess the potential for EMI risk is the implementation of a risk matrix. The risk matrix provides a simple form of determining the potential electromagnetic interface between known sensitive circuits and electromagnetic disturbance generators. The interface is classified as potentially of concern if marked (or labelled) with $\surd$ or "Yes". The risk matrix is usually generated at an early stage of the project, when sufficient information of the intended electromagnetic environment is acquired and just before a detailed EMC safety requirement specification is generated..

5.3 EMC and Functional Safety

5.3.1 Railway Electromagnetic Environment

The achievement of adequate immunity of electrical or electronic equipment against electromagnetic disturbances is a basic provision when ensuring electromagnetic compatibility of equipment, systems or installations. Such immunity is described by specified immunity levels as well as by means of performance criteria, which describe the behaviour of the equipment under test when exposed to electromagnetic disturbance. The selection of immunity levels and performance criteria should reflect the expected electromagnetic disturbance at the intended location of operation as well as the characteristics of the equipment. However, knowledge of the electromagnetic environment of the said location is a crucial factor when determining appropriate immunity levels. Obtaining such knowledge, whereas the resultant data need to have certain evidence, represents a very challenging task, directly deducible from the definition of the electromagnetic environment: totality of electromagnetic phenomena existing at a given location[ref]. For safety related system, the starting point is that the electromagnetic environment for a safety related system is specified in the SRS for the system concerned, as required by IEC 61508-2 [327]. This specification should be based on a competent assessment of the foreseeable EM threats. Two considerations regarding the electromagnetic environment and the potential severity of the electromagnetic disturbance can be made:

- Controlled Electromagnetic Environment: In this type of environment a compatibility level (per phenomena) may be chosen first, with immunity limits derived from it with the objective of ensuring an acceptable high probability of electromagnetic compatibility in that environment. The chosen compatibility level is selected based on technical and financial consideration with the objective that all equipment installed within the environment will be able to comply with the resultant emissions and immunity limits.

- Uncontrolled Electromagnetic Environment: In this environment, the compatibility level is chosen on the basis of existing or expected disturbance levels, and the immunity limits derived from it. However, there is a need to re-assess the disturbance levels taking into account the introduction of new equipment into the electromagnetic environment and the potential impact this may have on the emissions and immunity limits.

Both types of electromagnetic environment exist and play a crucial rule in the determination of immunity limits for the demonstration of functional safety. In a controlled electromagnetic environment, the infrastructure (or part of) is managed via a number of processes and requirements to ensure that the overall electromagnetic environment remains conducive for the operation of safety related systems. The compatibility level therefore is established based on the totality of electromagnetic disturbances obtainable within the infrastructure. The immunity limit is set and applicable to all equipment installed within the infrastructure. In this scenario, the immunity limits need not exceed those specified for normal operation; however the safety functions would need to be exercise to demonstrate the fail-safe capability of the system. In an uncontrolled electromagnetic environment, the immunity level would have to be derived based on the worst case electromagnetic disturbance expected; as the environment is not controlled emission level would need to be base lined against the worst case immunity level of equipment already installed in the environment being considered.

In general, the railway electromagnetic environment is considered a union of three basic electromagnetic environment archetypes, as shown in Figure 5.3.

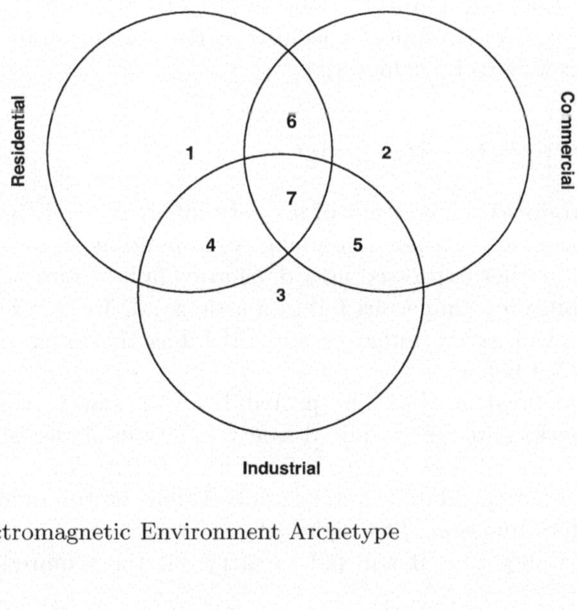

Fig. 5.3 Electromagnetic Environment Archetype

It is a spatial diagram, indicating the environments created due to the specification and delineation of separate areas as Residential, Commercial and Industrial:

- Any location in the centres of these separate areas will therefore experience the archetypical electromagnetic environments (areas 1, 2 and 3). However, locations at the boundary or boundaries between these separated areas (areas 4, 5, 6 and 7) will experience an electromagnetic environment that is a blend of the archetypical environments, containing phenomena from the adjacent archetypes
- As a functional diagram, demonstrating how changes in the activities performed in a designated area (i.e. the area 'function') can lead to the location's environment deviating from the archetype through the introduction of electrical or electronic equipment (and hence of phenomena or a severity of phenomena) not associated with the archetype.
- Each of the three main archetypical electromagnetic environments can be defined within a railway infrastructure,

While there are procedures implemented to control the electromagnetic environment in some parts of the railway (i.e, the use of cellular phones is restricted in signalling equipment rooms by some railway agencies - as an operational control), the electromagnetic environment within the railway can by no means be classified as controlled. The railway environment implicitly specified in IEC 62236 series of standards (or its European Norm equivalent, EN 50121 series of standards) therefore represents an attempt to provide a generic description of the railway environment. For existing railways, it is prudent to re-assess the electromagnetic environment prior to the introduction of new assets, such that statements of compatibility with existing assets (internal and external to the railway) can be made. For new railways, the electromagnetic environment specified in IEC 62236 series of standards represents a baseline to be achieved.

5.3.2 Safety Integrity Level

IEC 61508 introduces the concept of a safety integrity level (SIL). This is a discrete level (one out of a possible four), corresponding to a range of safety integrity values, either expressed as a dangerous failure rate per hour or an average probability of dangerous failure on demand, PFD, where SIL4 has the highest level of safety integrity and SIL1 has the lowest, as shown in Tables 5.3 and 5.4 below.

SIL can be defined as the "The probability of a safety related function satisfactorily performing the required safety functions under all stated conditions within a stated period of time." Safety Integrity levels relates to the probability of a safety related systems, sub-system or equipment achieving its required safety function. The higher the safety integrity of an device, the lower the probability that it will fail to carry out the required safety func-

Table 5.3 Safety Integrity Levels Classification - for on demand safety functions

SIL	Average PFD of the safety function, "on demand" or "in a year"	Equivalent mean time to dangerous failures, in years	Equivalent confidence factor required for each demand on the safety function
4	$\geq 10^{-5}$ to $< 10^{-4}$	$> 10^4$ to $\leq 10^5$	99.99% - 99.999%
3	$\geq 10^{-4}$ to $< 10^{-3}$	$> 10^3$ to $\leq 10^4$	99.9% - 99.99%
2	$\geq 10^{-3}$ to $< 10^{-2}$	$> 10^2$ to $\leq 10^3$	99% - 99.9%
1	$\geq 10^{-2}$ to $< 10^{-1}$	> 10 to $\leq 10^1$	90% - 99%

Table 5.4 Safety Integrity Levels Classification - for continuous safety functions

SIL	Average PFD of the safety function per hour	Equivalent mean time to dangerous failure, in hours	Equivalent confidence factor required for every 10,000 hours of continuous operation
4	$\geq 10^{-9}$ to $< 10^{-8}$	$> 10^8$ to $\leq 10^9$	99.99% - 99.999%
3	$\geq 10^{-8}$ to $< 10^{-7}$	$> 10^7$ to $\leq 10^8$	99.9% - 99.99%
2	$\geq 10^{-7}$ to $< 10^{-6}$	$> 10^6$ to $\leq 10^7$	99% - 99.9%
1	$\geq 10^{-6}$ to $< 10^{-5}$	$> 10^5$ to $\leq 10^6$	90% - 99%

tion. Environmental induced failures, such as electromagnetic disturbance, temperature and vibration are treated as part of the failure integrity.

Safety integrity is composed of two components:

1. Hardware Safety Integrity: the random hardware failures component that can normally be estimated to a reasonable level of accuracy;
2. Systematic Safety Integrity: the non-quantifiable part of the safety integrity and relates to hazardous systematic faults; systematic integrity is difficult to quantify due to the diversity of causes of failures; systematic failures may be introduced during the specification, design, implementation, operational and modification phase and may affect hardware as well as software.

IEC 61000-1-2 [320] states that the failure or malfunction of a safety related system due to an electromagnetic disturbance with a given strength is systematic as long as the duration of the disturbance is longer or equal to the operating cycle of the safety-related system. IEC 61000-1-2 assumes that the duration of the disturbance is such, moreover the influence of a given electromagnetic disturbances on a given safety-related system is considered the same for each occurrence. At system level, appropriate mitigation techniques increase immunity of the safety-related functions and must be considered as part of the systematic capability. Any equipment that has been developed to fully meet the requirements of IEC 61508 concerning systematic safety integrity for a given safety integrity level is said to have the corresponding systematic capability. Alternatively, the systematic capability of equipment can be demonstrated through *proven in use* evidence according to the strict requirements of IEC 61508-2. The estimation of systematic failure integrity is much harder than the estimation of random failure integrity. The concept of safety integrity level is used as an indicator of the required level of protection against systematic failure: each abstract function is allocated a safety integrity level at the early design phases, and this is inherited by the components that implement the function. IEC 61000-1-2 details that it is impossible to establish a correlation between SIL and the immunity level of a safety related system.

5.4 EMC Management and Functional Safety

5.4.1 Testing for Functional Safety

5.4.1.1 Severity Levels for Functional Safety

The process detailed in IEC 61000-1-2 [320] requires the inclusion of EMC safety requirements as part of the Safety Requirements Specification. These reflect the electromagnetic disturbances that are considered likely to have an influence on the correct functioning of the safety critical system. The associated electromagnetic phenomena, the applicable severity level per phenomena and the test configuration is one that requires detailed consideration.

Adverse levels of electromagnetic disturbances can occur at any location with a very low probability. The possibility of occurrence of higher electromagnetic disturbance levels is not considered in product (or product family) or generic standards. For the assessment of EMC Functional Safety, subjecting the equipment/system to an increased severity levels, as partial fulfillment of the safety life cycle requirements of IEC 61000-1-2, results in a level of confidence that dangerous failures caused by electromagnetic phenomena are kept to ALARP.

The derived severity levels are defined phenomenon by phenomenon, where necessary and are only related to the functional safety aspect of the equipment/system. The increased severity levels apply only to the safety-related functions having a specific performance criterion for functional safety (see sec. 5.4.3 below).

There is currently no product standard within the railway applications suit of standards, that deals with this subject matter, there is however a generic Industrial Standard, IEC 61000-6-7 [324] being developed by two working groups (IEC TC 77 WG 13 and MT 15) within the IEC. It is anticipated that this standard will be used for railway applications until such a time that a product specific standard covering EMC Functional Safety requirements is published. Clause 1.2 of IEC 61000-6-7 recognises the variety of the equipment that may be used for safety applications within Industrial locations and states that the severity levels specified within the Standard may underestimate (or over estimate) the totality of the electromagnetic environment. To mitigate this, the standard suggests that the applicable severity level for functional safety should be agreed with the end user where possible.

In [589], Ogunsola *et al* suggested a number of methods for the establishment of severity levels for functional safety; the suggested methods include:

- assuming the environment has been characterised - typically, emission measurement will be performed at 10m, then an approximation of the field strength at proximity to the apparatus can be estimated - a safety margin of 20 dB is added to this value to give the desired immunity level for functional safety test;
- immunity requirements are taken directly from product specific standards that include safety requirements;
- safety immunity levels are defined as levels specified in product specific standards (with no safety consideration) or generic immunity levels plus a 6 dB safety margin;
- no levels are defined: however apparatus is subjected to immunity test until failure occurs.

Armstrong in [18] proffers reasons why testing is not adequate for the demonstration of EMC Functional Safety; he informs that EMC testing alone ignores most of the issues that arise over the product's life cycle, that can affect how EMI increases safety risks, and that EMC testing applies a limited number of types of EM disturbance one at a time, ignoring the fact that electromagnetic disturbances, in real life, occur simultaneously. This position is valid, however EMC product design and risk assessment/management need to be universally robust for all systems performing safety related functions, to remove the dependency on testing. IEC 61000-6-7 and similar standards can be viewed therefore as a compromise between testing to demonstrate some level of increased confidence and no testing at all.

5.4.1.2 Safety Margin

The approaches suggested above, excluding the last, are based on the application of a safety margin on top of a baseline immunity level, specified in the product or generic standard or established through field measurement. Establishing the severity levels for functional safety via the application of a safety margin is also suggested by [21] and [624]. The concept of safety margin is not new and has been used extensively in Military EMC applications, MIL-STD-1541A [549], providing definitions for a point and integrated safety margin; a general definition of safety margin is the ratio between the susceptibility threshold and the maximum acceptable level of a disturbance at a test point or on a signal line. NASA System Guidelines for EMC [561] provides suggested approaches for the establishment of safety margins for each electromagnetic phenomena. Clause 4.10 of MIL-STD-1541A specifies the following applicable safety margin for safety critical system, the applicable safety margin is dependent on the category of the safety critical system:

- Category I: 12 dB for qualification; 6 dB for acceptance;
- Category II: 6 dB;
- Category III: 0 dB.

The standard further states that Category I and II safety margins shall be increased by 6 dB if the only practicable method of verification is entirely by analysis based on estimated emission or susceptibility characteristics. For railway application, a 6dB safety margin is deemed sufficient in demonstrating the fail-safe nature of the safety critical system. Thus taking the baseline levels from IEC 62236 Part 4, the severity levels specified in Table 5.5are suggested levels for functional safety.

Table 5.5 Suggested severity levels for functional safety based on a 6 dB safety margin

Electromagnetic Phenomena	Frequency range	Functional Safety Severity Levels
Radiated Immunity	80 MHz - 2.4 GHz; 80% AM	20 V/m
	800 MHz - 1 GHz, 80% AM	40 V/m
Conducted Immunity	0.15 MHz - 80 MHz, 80% AM	20 V

Table 5.5 reflects a view that some electromagnetic phenomena are non-linear by nature and thus the dependencies on the strength of the electromagnetic disturbance does not follow the inverse square law. Secondly, it is advisable to apply the limits and processes detailed in IEC 61000-6-7 or similar standards.

5.4.1.3 Test Configuration

Figure 2 of IEC 61000-6-7[1] [324] describes the decision and the process required with respect to EMC testing for normal compliance, as well as for Functional Safety. The process requires an assessment for normal performance prior to commencement of testing for EMC Functional Safety. It is therefore safe to assume that the test configuration used for the normal compliance is adequate for functional safety purposes provided the EUT configuration is such that the relevant safety functions can be exercised and observed during the test. Clause 8 of the IEC 61000-6-7 provides more details on this. It is

[1] The reader is cautioned that IEC 61000-6-7 is still being developed and thus the descriptive text and figures referred to herein may not make it to the final publication.

entirely possible that the EUT configuration for functional safety assessment is necessarily different from that for normal compliance, in which case the configuration of the EUT during the test would need to be reviewed against how it would typically be installed and used.

5.4.2 Aging Related Disturbance and Its Impact of Functional Safety

The above discussion on EMC and Functional Safety may give the impression that the evidence of compliance obtained during the qualification phase remains valid throughout the life of the device; however IEC 61000-1-2 recognises that the electromagnetic characteristics of a device is influenced by a number of other factors including aging. A crucial question in this regard is "how do we demonstrate the ability of an aged device to perform its safety function for applicable service conditions?". To answer this question, we need to understand the factors and impact of age related degradation.

Age related degradation is a significant issue which must be considered for the operational life of the railway; all signalling and communication devices, particularly those performing safety related functions must be evaluated to determine their susceptibility to age related degradation. The influence of environmental and operating conditions on device/equipment degradation, and the influence on the electromagnetic characteristics, must be assessed and understood. In practice, this might be difficult as aging cannot be usefully investigated in real time; and thus for experimental evaluation purposes accelerated aging testing must be performed.

Age related degradation is the effect of operational, environmental, and system conditions on equipment which lead to a degradation of performance of equipment over time [368]. Evidence of aging due the environmental effects (i.e. the synergistic effects of temperature, humidity, solar radiation etc) is well detailed in published research. Furthermore, backfits and design changes may introduce new aging mechanism or special aging concerns; for example, installation of additional signalling/communication devices in an air-condition control room may add to the heat load sufficiently to make local temperature control inadequate and thus contributing to thermal degradation of electronics. Thermal stress however has been demonstrated to impact the electromagnetic characteristics of electronics equipment [491, 29, 492, 479, 485].

The railway environment is characterised by high environmental stress; and age related degradation during normal operation may be the result of a variety of environmental stress mechanisms, acting either independently or synergistically. The environmental stresses may be constant, periodic, or intermittent. Signalling and Communication equipment stress resulting from normal operation may be classified into two general categories:

- External Environmental Stress: it is the environmental stress typically incurred from such influences, such as atmospheric conditions (pressure, temperature, humidity etc), mechanical shock and vibration;
- Internal Functional Load Stress: it is the stress resulting from power supply fluctuations (electrical stresses) and mechanical stresses.

Experimental research indicates that the electromagnetic emission of a device experiences a "drift" in both frequency and amplitude. The drift in frequency is due to changing clock frequency and harmonics. Thermal stress may result in amplitude changes of up to 10 dB or more and shifts in peaks and dips in the spectrum have also been observed due to changing resonance frequencies. Thermal cycles also have the ability to alter the electrical properties of a component, i.e. decrease in the permittivity of a capacitor and increase in leakage current. Thermal stress alone can result in a reduction of the electromagnetic immunity of a device by up to 10 dB; this therefore has serious implications for devices performing safety related function. The IEC, through its Technical Committee 77, is developing a generic immunity standard, IEC 61000-6-7 [324], that details the limits and process for the assessment of EMC and Functional Safety; it is expected that this standard will provide some guidelines on the assessment of aging influence on the electromagnetic characteristics of the device under test.

5.4.3 Performance Criteria for Functional Safety

The performance of a product, when tested for compliance, is described by set of performance criteria. The generalised performance criteria, as used in compliance testing, are specified in the railway application product standard and reflect those defined in the generic EMC Standards. These performance criteria, Performance Criteria A, B and C, are not specifically defined for functional safety, although there are occasions where Performance Criteria A might be acceptable.

For functional safety purposes a particular performance criterion FS is applied. It is defined as follows:

- the functions of the Equipment under Test (EUT) intended for safety applications are not affected outside their specification or may be affected temporarily or permanently if the EUT reacts to a disturbance in a way that a detectable, defined state or states of the EUT are maintained or, achieved within a stated time; also destruction of components is allowed if a defined state of the EUT is maintained or, achieved within a stated time;
- the functions not intended for safety applications may be disturbed temporarily or permanently.

5.5 System Compatibility

Numerous sources of electromagnetic disturbance exist in the railway; these can broadly be categorized as intra-system and inter-system sources of interference. Intra-system interference within the railway includes, for example, traction power supply and traction drives to signalling installation, and coupling between parallel DC and AC traction systems. Inter-system interference in this context is the electromagnetic interaction between the railway and its neighbour, and is exampled by the generation of earth current causing corrosion or interference to sensitive systems outside the railway environment, or the degradation of performance of railway telecommunication and signalling installations due to sources external to the railway. Electrified railways can generate electrical noise over a wide frequency band by a variety of mechanism giving rise to complex amplitude and time varying signature as a trainset travels along the installation. The electrical noise generated by a trainset will depend upon the operation of the vehicle and the noise can be enhanced by system resonance and interaction with other trainsets operating on the route.

The electromechanical interaction between the contact wire and the pantograph is a source of radio frequency interference; this interference is due to high voltage spark breakdown when the pantograph is raised or lowered and high current arcing during operation. Typically, high current low voltage sources, such as fields produced by traction currents and pantograph arcing, generate induction or near magnetic fields, whereas low current high voltage sources from switching, spark breakdown and arc formation create (far) electric fields. AC and dc fed traction drives are also a source of electromagnetic noise; the magnetic field emanating from a dc overhead Light Rail Transit (LRT) may be the source of mal-operation of nearby non-railway sensitive installations [228, 632, 520]. While signalling installation is generally considered to be a victim (in a railway context) it can also serve as a source of electromagnetic disturbance with reference to equipment located outside the railway environment and to other nearby signalling systems. This is exampled by:

- power and audio frequency track circuits with modulated carriers signals, which can form long loop transmitters;
- signalling transponders, which can function as localized radio frequency transmitter (e.g. Eurobalise and balise);
- the wheel – rail contact zone, which when diverting low power track circuit current can act as a transitory interference source.

There is a need therefore to demonstrate that the systems to be introduced within the railway environment are electromagnetically compatibility and can co-exist without mal-operation. Such compatibility activity may exceed those specified in IEC 62236 (or EN 50121) and may or may not involve measurements. Numerical analysis, such as those detailed in Chap. 7, may be necessary to demonstrate compatibility between such systems.

5.5.1 Signalling Compatibility

Railway signalling systems, by virtue of their functions, are systems that perform safety related functions and are considered sensitive installations. The frequency range of typical signalling systems, depending on technology range, starts from dc to a few hundreds kHz, with advanced signalling systems operating in the MHz range. The main source of electromagnetic interference within the railway is viewed as the train and this is particularly true if viewed from a railway signalling point of view, as both assets are in close proximity electromagnetically. Signalling compatibility, therefore was traditionally viewed as the process of demonstrating compatibility between a rail vehicle and signalling equipment, however with the increasing use of radio frequency technology for signalling applications, the definition and application of signalling compatibility is being extended to cover the compatibility between signalling and communication system, particularly where new signalling systems are being introduced into existing functional railway in which the new signalling/communication system is expected to be operational with the existing signaling/communication system before final switch over. As mentioned in Chap. 2, the operational frequencies of signalling systems are intentionally chosen to avoid the odd harmonics of the fundamental; the process of demonstrating compatibility between the train and signalling assets is referred to as signalling compatibility. Usually this process involves a detailed characterisation of the electromagnetic disturbance from the train, particularly at the frequency of operation of the said signalling asset, under all operational modes (i.e. braking, coasting. etc). Signalling compatibility in this context is well understood and detailed in public literature; for example, Holmstrom et al, in [313], investigated the compatibility between the audio frequency track circuits and a dc chopper propelled train - this work led the development of the UMTA recommended practices (refer to sec. 6.3.2.2 for a detailed discussion on UMTA standards) and Costamagna *et al.* in [229] and [230] investigated the compatibility between track circuits and four quadrant controller ac drive; Taufiq in [655] investigated the introduction of a quasi-square wave scheme in a traction voltage source inverter and its switching transients compatibility with reed relay type track circuits. Falvo *et al.* in [281] investigated the compatibility between an electric substation and a code track circuit. In all these, and many more references, the signalling system is viewed as the victim however interoperability between all subsystems of a rail network is a prerequisite for reliable and safe rail transport of passengers [621] and thus the need to consider signalling compatibility in a more general term to include compatibility between two signalling systems as demonstrated in [591].

In performing signalling compatibility, the permissible current limit(s) of the signalling asset(s) must be known. A number of railway agencies have published permissible current limits for core signalling assets used on their infrastructure, for example London Underground's ST00062A [493], Network Rail's suite of standards [564] - [580]; in the absence of published permissible limits, detailed analysis would need to be undertaken to demonstrate compatibility.

Signalling compatibility analysis therefore requires a detailed understanding of the operation of the equipment in question particularly that of its reaction to in-band electromagnetic interference. In performing signalling compatibility, three claims are usually made in support of the rationale detailed in a Safety Case:

- Compatibility Margin: in this case, a claim is made based on the immunity threshold of the victim circuit and the electromagnetic disturbance characteristics of the source;
- Frequency Separation: in this case, a claim is made that the victim and source circuits operate at considerable different frequencies and that it is inconceivable for in-band electromagnetic interference to occur - such a claim would include intermodulation etc.;
- Spatial Separation: in this case, a claim is made that the victim and source circuits are installed at a separation distance that limits coupling between the circuits - such consideration would include shielding (architectural and structural).

Chapter 6
Standards and Standardization

Standardisation is indispensable to life in this century. It is virtually indispensable as the air we breathe, and like the air we breathe it is invisible to all except its technicians. Standardisation impacts life dramatically, it shapes technology and defines the terms of commerce. It sustains our environment and makes us safe. It has given us the wherewithal to walk on the moon and yet it is invisible. Standards or the need for standardisation has been with us since the earliest societies; in simple terms setting standards sprang from the need to assign practical and fair value to space and objects.

6.1 What Is a Standard?

In the simplest of terms a *standard* can be defined as an agreed upon way of doing something. A more elaborate definition might denote a uniform set of measures, agreement, conditions, or specification between parties. Standards are something that most of us accept for granted; they have become such an integral part of our life, that the average person gives little or no thought to everyday products and services and how they work. Standards make modern conveniences possible. There are numerous versions of the definition of a standard and standardisation, but in this book the definition used by the International Organisation of Standardisation, ISO, is preferred. The ISO [412] defines a standard as "*a document, established by consensus and approved by a recognised body, that provides for common and repeated use, rules, guidelines or characteristics for activities or their results, aimed at the achievement of the optimum degree of order in a given context*" [ISO/IEC Guide 2: 2004, definition 3.2]. The ISO also offers this definition for Standardisation "*An activity of establishing with regards to actual or potential problems, provisions for common and repeated use, aimed at the achievement of the optimum degree of order in a given context*" [ISO/IEC Guide 2: 2004, definition 1.1]. The standardisation activity includes the process of formulating, issuing and implementing standards. Thus the term "standardisation" actually encompasses

A. Ogunsola and A. Mariscotti: Electromagn. Compat. in Railways, LNEE 168, pp. 217–314.
springerlink.com

a broad range of activities and ideas, from the actual development of a standard to its promulgation, acceptance and implementation. It is based on the consolidated results of science, techniques, and experience. It determines not only the basis for the present, but also for future development and it should keep pace with progress. It also includes the methods of evaluating whether products, processes, systems, services and personnel comply with a standard; this evaluating process is generally referred to as *conformity assessment.*

EMI is an electromagnetic coupling between circuits and systems that depends on the specific configuration of the systems and their neighborhood. Interface specifications, standards and regulations are protective measures aimed at reducing the risk of EMI within classes of configurations. They refer to, and tested according to, specified configurations and conditions, considered as references for each class. They do not guarantee immunity against EMI, but reduce the risk, to the extent that the system configuration within its environment resembles the reference. The term "standard" encompasses various types of publications, that have in common the following elements: they have been prepared by a technical committee or more generally by a group of persons, who are experts and represent the interests and the competence of various stakeholders, manufacturers, users, public bodies; they are subject to a mechanism of voting by the public or by a restricted group of readers and they are more or less frequently updated as a result of feedback by the readers, the members of the committee or similar committees working in parallel on other related subjects; they represent a useful reference containing technical elements coming from previous experience, analysis, discussions among the said experts. Depending on the years when the project of the standard was started, its scope width and its applicability, there are various standardization committees and bodies with a more or less articulated hierarchy and complexity; the same committees and bodies may be an expression of the specific needs of an industrial sector, a group of stakeholders, users and/or manufacturers, or an entire market, to discipline the quality and performances of the circulating products. Depending on the initial need and on the successive updates and activities, committees and groups may grow, reduce and disappear, merge together, change their mission and valence, for example from private to public interest.

A list follows of the various standardization entities that are more or less frequently referred to for railway applications, and for the most relevant a description will be given of their organization (a good reference for the structure of the most relevant standard making bodies is [679], sec. 2, and in general the websites of the same bodies). The list is preceded by a description of the life cycle of a standard, considering the formation of the working group, the issue of the various drafts for internal use and for public circulation, until the final release as approved standard. This life cycle represents the procedure normally followed by relevant standardization bodies of large dimensions; when standards are issued by a railway operator or a manufacturer, or they are represented by the technical elements and prescriptions contained in a law (as for EU Directives and some technical attachment) the procedure is usually simpler and with less iterations.

6.2 Types of Standards

Standards are not monolithic, nor are they of the same general type, acceptance, or function. Indeed much of the current work of standards expert is aimed at harmonising or rationalising the diversity among national, regional, and international standards. Standards come in different forms, they may be physical items or things, particularly standard units of weight and measure, or they may be written documents, published norms. The application of standards may be, but not limited to:

- Physical standards or units of measure;
- Terms, definition, classes, grades, ratings, or symbols;
- Test methods, recommended practices, guides, and other applications to products and processes;
- Standards for systems and services, in particular quality standardisations and related aspects of management system standards for quality and the environment;
- Standards for health, safety, consumers, and the environment.

In general, standards can be classified by function or origin, and into one of eight types:

- Basic Standards;
- Generic Standards;
- Product Standards;
- Design Standards;
- Process Standards;
- Specification and Codes;
- Management System Standards;
- Personnel Certification Standards.

In the following sections a large amount of standards are presented, that are at different extent applicable to a railway application or are more or less often used by the designers, suppliers and installers, who are involved in the realization of an electric transportation system, as far as EMC is concerned. This entire chapter is meant to represent on one side a complete reference of the EMC related standards, whatever their origin, applicability, and age, and on the other side a critical discussion of specific aspects related to particular phenomena, to harmonization and conflicting specifications, to references for good practice. For this reason a structured presentation of standards, starting from a high level classification and proceeding with a top-down fashion to the single publication, is not attainable, a different criteria and perspectives may be followed.

6.2.1 *Basic Standards*

These standards form the fundamental building blocks of product specific standards. They deal with the fundamental process of weighing and measuring a unit. With respect to EMC, the basic EMC standard is one that specifies the general conditions or rules necessary for achieving electromagnetic compatibility. Some basic EMC standards are normative international standards while others are non-normative technical reports. Within the IEC publication, basic standards are contained in the IEC/EN 61000 or CISPR 16 series of publication.

The IEC/EN 61000 family of standards is further subdivided into (as taken from the introduction appearing in the standards themselves):

- Part 1: General – General considerations (introduction, fundamental principles), Definitions, terminology
- Part 2: Environment – Description of the environment, Classification of the environment, Compatibility levels
- Part 3: Limits – Emission limits, Immunity limits
- Part 4: Testing and measurement techniques
- Part 5: Installation and mitigation guidelines – Installation guidelines, Mitigation methods and devices
- Part 6: Generic standards
- Part 9: Miscellaneous

None of these parts is strictly mandatory (except Part 6, that represents the latest harmonized Generic standards), in the sense that even the application to an equipment of Part 3, bearing emission limits and immunity levels, occurs through the product standard or a generic standard, as we will see in the next sections; in any other case the application of any of Part 3 standards is voluntary. Yet, Part 3 standards define a widely comprehensive, even if not really complete, list of test methods for the conducted and radiated phenomena, that will be briefly reviewed in sec. 6.4. The reason to say "not really complete" is that emission tests are limited to some conducted phenomena occurring at low frequency (harmonics and flicker), and they are complemented by the CISPR standards. CISPR standards were born as product standards and then some of them became *de facto* generic standards, being the most complete and useful reference for the treated electromagnetic phenomena. It is the case of CISPR 11 and CISPR 22 (now EN 55011 and EN 55022 respectively), dealing with the test methods and limits for radiofrequency conducted and radiated emissions for industrial, scientific and medical equipment and information technology equipment respectively. The limits have since then being applied to all other equipment, possibly with a restriction of the scanned frequency range and with a relaxation of the limit curves by a certain amount decided by the product committee, and the test method has been widely adopted as the reference method (scanning for maximum of the radiated emissions in one direction by a vertically movable antenna, use of the quasi peak detector, choice of the resolution bandwidths, and so on). It must be said that the same CISPR has through the years issued the "reference standard" for radiofrequency measurements, in terms of required performance of used equipment, test methods, processing methods, uncertainty, etc., and this has in turn supported and fostered the development of the two above standards at a degree of completeness to be accepted also for the issue of the ANSI/IEEE Std. C63.022 in 1996, following the first complete issue of CISPR 22.

The "reference standard" is the CISPR 16 publication, now published as EN 55016, is what has been being considered for years as the reference for compliant measurements and instrumentation at radiofrequency: this publication has evolved from two volumes in the '90s for a total number of pages of less than 400, to the present version split into a dozen standards, for more than 600 pages, issued starting on 2003 [134] - [143]. Its content is focused on the execution and the methods of measurement, on the performance of used instruments, on the adequacy of the test site and is thus out of the scope of this book; the reason of mentioning it here is to give an overall picture of the activities of CISPR and the level of standardization that is behind the execution of an EMC test. To complete the picture, it is worth citing the counterpart of this group of standards in North America on the United States side, that is the ANSI and ANSI/IEEE standards issued under the section C63, covering similar topics, in a more direct way, but not always with the same formal correctness [337, 338, 339, 340, 347, 350, 346].

Other CISPR publications cannot be identified as basic standards, since they are clearly targeted to a specific product (like EN 55014 for household appliances, EN 55015 for electrical lighting equipment, EN 55020 for sound and television broadcast receivers, EN 55024 for information technology equipment and EN 55025 for vehicles, boats and internal combustion engines), but they do not have the originality and the completeness that for EN 55011 and EN 55022 has justifyied their assumption as *de facto* basic standards.

6.2.2 Generic Standards

Within the EMC standardisation world, generic standards are for products operating in a particular EMC environment but which no specific standards yet exist. They are in effect general and can be viewed as simplified product standards. Generic EMC standards are important building blocks for the development of new product standards. Not only do they provide technical background, they also fill the gap during the time taken to develop a product standard. In some cases, they may be completely adequate in themselves for use with particular products and may be used for product certification. This in turn relieves a product committee from developing a new standard and is very much in line with the IEC's forward-looking approach to its EMC work. While referring to Basic EMC standards for detailed measurement and test methods, Generic EMC standards specify a limited number of essential emission and immunity tests, as well as minimum test levels. The aim is to ensure adequate compatibility at the same time as achieving a good balance between technical and economic considerations.

Generic standards cover different electromagnetic environments (sec. 6.5): “Residential, commercial and light industrial environment” (in brief, “Residential environment” or RE) and “Industrial environment” (or IE). The RE generic standards apply to “apparatus to be connected to a low-voltage public mains network or connected to a dedicated DC source which is intended to interface between the apparatus and the low-voltage public mains network. These standards apply also to apparatus which is battery operated or is powered by a non-public, but non-industrial, low voltage power distribution system.” The RE environment applies to houses and apartments (residential properties), shops and supermarkets (retail outlets), offices private and public (business premises), cinemas, bars and other entertainment (public entertainment), amusement parks and sport centers (outdoor locations), small workshops, laboratories and service centers (light industry).

The IE generic standards apply to “apparatus intended to be connected to a power network supplied from a high or medium voltage transformer dedicated to the supply of an installation feeding manufacturing or similar plant, and intended to operate in or in proximity to industrial locations.” The IE environment are characterized by the existence of industrial, scientific and medical (ISM) apparatus (see EN 55011 [132, 133]), heavy inductive or capacitive loads are frequently switched, currents and associated magnetic fields are high.

As stated in the EMC generic standards IEC 61000-6-X (and its European Normative EN equivalent) [194, 195, 196, 197], the object of these standards is to define, on one side, a sufficiently complete set of immunity test requirements . The severity specified do not necessarily cover all cases, since adverse electromagnetic disturbance may occur at any location, but with an extremely low probability; not all electromagnetic phenomena have been included for testing purposes, but only those considered as relevant for the equipment covered by these standards. On the other side, emission test requirements have been selected, so as to ensure that the generated electromagnetic disturbances do not exceed a level which could prevent other apparatus from operating as intended; not all disturbance phenomena have been included for testing purposes, but only those considered as relevant for the equipment covered by these standards. The severity levels and limits specified within the standards represents a tradeoff between the technical significance of each electromagnetic phenomenon and the related costs for testing and the adoption possibly of necessary countermeasures.

The technological advancements witnessed in the Industrial sector has produced a corresponding evolution of generic standards in terms of the number of tests and of test requirements. The old EN 50081 and EN 50082 group of standards [67]-[70], is now superseded by the new suite of IEC 61000-6-X standards, with amendments that take into consideration advancement of tracking technology at the time of publication availability. An example of an amendament update made to the generic standards is the change of the levels and duration of voltage dips applied for the voltage dips immunity test (as per basic standard IEC/EN 61000-4-11, and its European Norm equivalent, [182]), even if they remain differently specified for the two environments, Residential/Light Industrial and Industrial. Another example is the specification of different conducted emissions limits for the telecommunication ports (expressed in dBμA or in dBμV) to reflect the amendments to EN 55011 [133] and EN 55022 [145].

Also the IEC 61000-6-5 [323] is considered briefly here as an example of a very harsh environment other than the railway environment, and that in some cases may be part of a railway project if High Voltage intakes from the national grid are available and HV/MV stations must be constructed or even only equipped by the contractors. It is also possible that HV Transmission Lines share the railway right of way and thus represent an interference source in close proximity to the railway assets. This standard dates back to 2001 and has not yet been adopted by CENELEC; very synthetically it divides its application in three locations (H: HV substations; G: power generation and MV substations; P: protected areas) and five signal connections (l: local connections; f: field connections; h: connection to HV equipment; t: telecommunication; p: connections within a protected area). Simplifying a little, for all signal and supply connections other those of type l and p two immunity tests are prescribed: the "mains frequency voltage" (part of the IEC/EN 61000-4-16 [187]) and the "damped oscillatory wave" (as per IEC/EN 61000-4-12 [183]). Moreover, for low voltage dc input and output power ports also the IEC/EN 61000-4-29 [192] for the voltage dips, short interruptions and voltage variations immunity test finds its application. Finally, this standard is also to mention for the list of performance criteria, as it is presented in Table 7 of the said standard, for the various functions during the execution of an immunity test. The definitions go beyond the simple A, B and C performance criteria, and "short delay", "short delay with temporary wrong indication", "temporary degradation", and so on are added; the functions cover for example protection, regulation, metering, data transmission, command and control, alarm, monitoring, self-diagnosis.

6.2.3 Product Standards

These are standards that deal with the physical, chemical, electrical and mechanical characteristics of a product or material. Product standards establish conformity and interoperability by defining dimensions and qualities. EMC product standards may apply either to a particular product or to a group of products having general characteristic, that may operate in the same environment and have neighbouring fields of application. Product standards can be classified as either defining performance criteria or design criteria. Performance standards specify the level of expected performance for a product and may include test methods that simulate performance under actual conditions. All EMC Product standards are performance standards; an example of an EMC Product standard is the EMC standard for Uninterruptible Power Supplies [208].

In the last ten years the generic standards of the family IEC/EN 61000-6-X, with X covering Part 1 to 4 have undergone an intense transformation; several product standards on the contrary have remained unchanged since the '90s. When formally released, these standards made reference to generic standards in force at the time, such as EN 50081 and EN 50082, which have now been superseded. For this reason the evaluation of the equivalence between the set of tests prescribed by either a current generic standard or a product standard is a complex process; as product standards take precedence over generic standards, the blanket compliance to the requirements specified within a product standard containing superseded or withdrawn standards, may result in a product that is not necessarily fit for the intended environment. It should however be mentioned, that most EMC test laboratories would perform EMC test taking due cognisance of current standards and standards conflict only becomes an issue when non-compliance occurs.

Product standards, even if dating back to the '90s, are often supplied as reference in the EMC certificates and test reports, and thus represent a major issue for EMC management, particularly where EMC requirements (with dated references) are cited in contractual documentations. Another area of conflict, particularly within the European market, is the EMC Directive transition from the superseded 89/336/EEC [277] to the new 2004/104/EC [280], which requires an update of all EMC certificates for CE mark. The use of old product standards may be acceptable for products that have not been modified since manufactured: the solution is normally either an analysis of the real electromagnetic environment to understand if some tests may be relaxed or the execution/repetition of missing/inadequate tests. This usually requires a gap analysis to be performed.

6.2.3.1 Information Technology and Communication Equipment

Information Technology and Communication (ITC) equipment is widely used for a variety of tasks and applications within a modern railway system. ITC are generally used within the three archetypical electromagnetic environments, that constitute the railway electromagnetic environment; they are generally characterised by having a low immunity threshold and thus their use within the railway environment is usually accompanied by some form of assessment (EMI Risk assessment). Some ITC equipment are bespoken for a specific application and thus tested with due consideration of the application and intended environment of use; however in the majority of cases ITC equipment are procured as "commercial off-the-shelf" (COTS) equipment. In such instances the EMC qualification of the product is likely to be certified to the EN 55022 for emissions and the EN 55024 for immunity. The qualification in reality should depend on the assigned e.m. environment and on any specific e.m. phenomena that appear to manifest as a potential threat, particularly relevant for ITC equipment given that it is characterised as having a low immunity threshold (typical of a residential e.m. environment) and an intrinsic complexity. The latter refers to the use of high speed communication lines using low voltage signals, the tight specifications on external interference and eye diagrams of these communication protocols, the difficulty in establishing adequate performance tests to evaluate any influence on the performances themselves during immunity tests.

EN 55022 has been always treated as a reference for the measurement of conducted and radiated emissions, and as a result it is often cited as if it were a basic standard, whereas it is a product standard, the product standard for for ITC equipment. Regarding immunity, the environment considered by the EN 55024 (1998) [146] is that of a residential, commercial and light industrial environment, that generally covers the intended environment of use of the vast majority of ITC equipment. Particular attention is for ESD phenomena, which ITC equipment for residential and commercial use are exposed to, so that various coupling methods and points of applications are identified, as well as a high number of repeated discharges, in the order of several hundreds. The first critique is that in reality the evolution of the generic standard for this environment IEC/EN 61000-6-1 has been faster than the EN 55024, so that some tests don't share the same severity level, such as for instance, the magnetic field immunity tested at only 1 A/m, that is however a minor problem from a practical point of view, since magnetic field at supply frequency is not a relevant threat, especially nowadays that cathode ray tubes are no longer used for videos and video terminals; or, as another example, the adoption of immunity test levels for RF radiated field above 1 GHz, to cover new mobile telephones and wireless devices. The new release of EN 55024 [149] substantially incorporates the two amendments issued on 2001 [147] and 2003 [148], adding coupling-decoupling networks for the RF conducted immunity test of terminal equipment.

Since it is not so uncommon for ITC equipment to be installed within a railway application and used in e.m. environments other than the residential one, it is extremely important to identify and understand all the elements that may result in a electromagnetic threat close to the equipment and - where necessary - to specify additional tests for the identified phenomena and threats. Excluding the equipment that is installed in very harsh environments, such as trackside, heavy industrial environment or substations, there are many cases in which the industrial environment is applied, since a full residential environment cannot be ensured, such as in a technical room or in a control room. In this case an analysis is needed for the real sources and the elements that represent a "penetration" of the environment by external phenomena, the most common being represented by cables, possibly exposed to e.m. threats and in particular impulsive surges and fast transients.

With respect to voltage insulation and steady or impulse voltage withstanding capability, ITC equipment undergoes very severe tests for the equipment by itself and for all ports connected to telecommunication networks and cable distribution systems, as prescribed at sec. 5.2.2, 6 and 7 of IEC/EN 60950-1 [167], based also on several ITU-T Recommendations (see sec. 6.6.2.6). This means that when considering the susceptibility of ITC equipment connected to outdoor cables of significant length and thus exposed to the most severe transient over-voltages, the attention must be paid also to those tests prescribed by the product standard and not only to the sole IEC/EN 61000-4-5 surge test.

6.2.3.2 Audio and Video Products

Audio, video and audio-visual products are used extensively in passenger announcement information systems, both on-board and at stations, and also in some cases in passenger information and entertainment systems. When such systems are installed on-board, then EN 50121-3-2 for the on-board e.m. environment shall be used; at stations, residential, or industrial, or line-side e.m. environments may apply depending on the distance from the track and on the presence of nearby sources and a so-called "gap analysis" will be needed to verify if and how the limits and levels are fulfilled. The EN 55103 standards [150, 151] establish a set of five environments that span over various conditions, with reference to the IEC 61000-2-5 standard for the identification of the e.m. environments:

- E1: residential environment, including location types of class 1 and 2 as per IEC 61000-2-5;
- E2: commercial and light industrial;
- E3: urban outdoors, based on location of type 6 as per IEC 61000-2-5;
- E4: controlled EMC environment (e.g. recording or broadcasting studios) and rural outdoors environment (e.g. far away from railways, transmitters, overhead power lines, etc.);
- E5: heavy industrial, and locations close to broadcast transmitters.

The emission limits and test levels need in general to be established by the application standards, even if for emissions (EN 55103-1 for emissions) the audio, video and audio-visual products comply already with the lowest emission limits, those of Class B, and the widest range of conducted emission tests. Moreover, since these products are very sensitive to any type of disturbance produced by nearby equipment (e.g. stacked audio and video equipment), the EN 55103-1 specifies additional emission tests: magnetic field emissions from 50 Hz to 50 kHz measured at both 10 cm and 1 m distance from the enclosure.

For the immunity tests and related levels (EN 55103-2 for immunity) the product standard is kept as the reference for the evaluation of the performance of the DUT during the same immunity tests, in particular, for the evaluation of audio and video quality during the application of the disturbance. The specified tests are many and for the most sever environment (E5) they are almost equivalent with the industrial e.m. environment and IEC/EN 61000-6-2, except the extended frequency range for the RF radiated and the increased intensity for the magnetic field immunity tests. However, it is very uncommon that manufacturers test and certify their products at the most severe environment E5; the tendency is to certify equipment keeping on tests and levels approximately compatible with that of only the residential, commercial and light industrial environments.

6.2.3.3 Adjustable Speed Drives

The scope of IEC/EN 61800-3 comprises adjustable speed ac or dc motor drives for mains supply up to 1 kVrms, installed in industrial or residential environments, with the exception of traction applications and electric vehicles (these applications are treated by IEC/EN 60349-2 [161], but in the current version of the standard, there is no reference to EMC or supply conditions, other than the citation of the EN 50163). This standard in its 1996 version needs to be evaluated together with its A11 amendment on 2000, that brought substantial changes to the initial version. For ASD two environments are identified, the first one for domestic premises and the second one for all the others. More importantly, the standard defines two kinds of distributions (similar to that specified in EN 62040-2 for Uninterruptible Power Supplies), "restricted" and "unrestricted", where the former refers to applications where there is technical competence in the EMC requirements of drives applications, while the latter refers to all other applications. The standard can thus define for radio frequency emissions (both conducted and radiated) higher emission limits than those prescribed by CISPR 11 and 22 (corresponding to the EN 55011 and 55022), as it is done in the cited IEC/EN 62040-2 [208]. Besides radio frequency emissions, ASD are both a source and may be a victim of various power quality phenomena, related to harmonics, flicker and voltage fluctuations, but in particular they are quite exposed to voltage sags and short interruptions: first, if the supply shortage is sufficiently long the ASD performance worsen for a mere matter of power consumption (an energy storage system need to be set up to continue the prescribed function unaltered); second, voltage sags may be accounted for and immunity ensured by local dc link capacitors, that act as a momentary energy reservoir; however, when the voltage sag is terminated and the nominal voltage re-established, the dc link capacitor needs to be recharged and in the absence of any limiting devices it does so with a peak in the input current, that may cause damages or the tripping of some relay; third, for those ASD with the rectifier synchronized to the mains frequency (for example in order to set a time reference for the fire delay angle) any serious deformation of the supply voltage (such as during a voltage sag) may cause a commutation malfunction or even the block of the input rectifier.

6.2.3.4 Uninterruptible Power Systems (UPS)

UPS (Uninterruptible Power Systems) are particularly important and are treated here quite extensively, since they represent the e.m. environment for many information technology, communication, control, signalling equipment, as far as conducted phenomena at low and high frequency are concerned. The protection of an equipment from voltage variations, dips and interruptions by means of an UPS is sometimes mistakenly considered to extend to other conducted phenomena, such as harmonics and inter-harmonics and RF conducted disturbance, while they may represent an issue, because of the use of static converters to supply and regulate the output power of the protected line. This aspect is treated in the following, after a presentation of the classification scheme of UPS for EMC testing.

The IEC/EN 62040-2 standard is rather new with the CENELEC version on March 2006, preceded by a few months by the IEC version (Oct. 2005) [208]. This standard replaces the old IEC/EN 50091-2 (1995) by introducing in particular a new classification of UPS categories and related environments. There are two environments and four UPS categories:

- First environment (that may be abbreviated as E1), represents residential, commercial and light-industrial environments directly connected to a low voltage public mains without intermediate transformers.
- Second environment (that may be abbreviated as E2), represents all commercial, light-industrial and industrial environments other than those connected directly to a low-voltage mains that supplies building for residential purposes. It is interesting to note that the classification is not so straightforward, since for example E2 could include all commercial environments that are connected inside the same building to other commercial environments, thus classifiable as a residential environment in the sense of IEC/EN 61000-6-X standards, not as a first environment.
- Category 1 (abbreviated as C1) for UPS that can be used without any restrictions in E1, identified as corresponding to the residential environment.
- Category 2 (abbreviated as C2) for UPS with an output current ≤ 16 A and intended to be used without any restrictions in E2; these UPSs may be used also in E1 when connected through industrial or national plugs and sockets or permanently and in this case the following sentence shall be included in the instructions for use: "Warning. This is a category C2 UPS product. In a residential environment, this product may cause radio interference, in which case the user may be required to take additional measures."

- Category 3 (abbreviated as C3) for UPS with an output current > 16 A and intended to be used without any restrictions in E3; such UPSs are suitable for commercial and industrial installations with a minimum separation of 30 m from other buildings classified as E1 and in this case the following sentence shall be included in the instructions for use: "Warning. This is a product for commercial and industrial application in the second environment - installation restrictions may be needed to prevent disturbances."
- Category 4 (abbreviated as C4) for UPS used in complex environments, that need an agreement between the supplier and the customer regarding the applicable emission limits and immunity levels.

A final consideration is presented by the standard, concerning the relationship between environments and categories: if the environment is classified as E1, then UPS of category C1 and C2 may be used; if the environment is classified as E2, then UPS of category C2 and C3 may be used; finally, "if the environment is not covered exclusively either by the first or the second environment", then UPS of category C4 should be used. For the latter it is briefly commented that it seems to address environments different from those covered by IEC/EN 61000-6-X standards, such as those for specific applications, featuring large power installations, highly disturbed supply lines, etc., necessitating analysis and agreement between stakeholders; for such cases it is not uncommon that additional countermeasures are adopted in any case, consisting of supply filters, enforced cable segregation, etc..

The emission phenomena addressed by the standard are:

- radio frequency conducted emissions on the input port (mains port), with limits specified over the usual 0.15 - 30 MHz frequency interval, for the various categories, as shown in Table 6.1;
- radio frequency conducted emissions on the output port (protected line port), with limits specified again over the 0.15 - 30 MHz frequency interval, are identical to those specified for the input port above; a margin of +14 dB on the limits for C1, C2 and C3 (only below 100 A of phase input current) is allowed if the UPS is equipped with an output cable of ore than 10 m length;
- current harmonics on the input port, for which the IEC/EN 61000-3-2 limits apply, if the voltage and rated current are within its scope;
- radio frequency radiated emissions, specified at 10 m distance and shown in Table 6.2; concerning fluctuations of the recorded signals, the standard specifies that, if they are close to the limit, a 15 s observation time at each frequency must be applied, recording the maximum value, but the excluding "any brief isolated high reading", that leaves the door open to the occurrence of potentially interfering transients.

For immunity the standard divides among prescription contained in its sec. 7.3 (Basic immunity requirements - high frequency disturbance), sec. 7.4 (Immunity to low frequency signals), sec. 7.5 (Immunity to power frequency

Table 6.1 UPS limits in [dBμV] for radio frequency conducted emissions on the mains input ac port; for the output ac port the same limits apply with an allowance of +14 dB, except for C3 with input current greater than 100 A

Class	Frequency [MHz]					
	0.15 - 0.5 MHz		0.5 - 5 MHz		5 - 30 MHz	
	QP	AV	QP	AV	QP	AV
C1	66-56	56-46	56	46	60	50
C2	79	66	73	60	73	60
C3 $> 16, \leq 100$ A	100	90	86	76	90-70	80-60
C3 > 100 A	130	120	125	115	115	105
C4	unlimited					

Table 6.2 UPS limits in [dBμV/m] for radio frequency radiated emissions (QP values)

	30-230 MHz	230-1000 MHz
C1	30	37
C2	40	47
C3	50	60

magnetic field) and sec. 7.6 (Immunity to voltage dips, short interruptions and voltage variations), with some modifications with respect to the superseded EN 50091:

- for the basic immunity requirements, the tests specified for C1 are similar to those for residential environments (IEC/EN 61000-6-1 [194]), but with higher levels for the EFT and surge tests on dc, signal and control ports, and for C2/C3 the tests are again similar to the industrial environment (IEC/EN 61000-6-2 [195]), but with higher levels for the EFT tests on dc, signal and control ports; the upper frequency limit for the test of immunity on the enclosure port is for all categories 1 GHz;
- the immunity to magnetic field on the enclosure port is required to be performed at 10 and 30 A/m for C1 and C2/C3 respectively, so higher than the residential specifications, in line with the industrial specifications, but lower than the 100 A/m specified for railway applications;
- the test for immunity to power line frequency and inter-harmonics is specified at Annex D as a 10 V signal with frequency varying slowly between 140 and 360 Hz superimposed to the normal mains voltage at 50 or 60 Hz for single- and three-phase systems; specific reference to IEC/EN 61000-2-2 is made for low frequency conducted disturbance in the mains, but the intensity and frequency interval cannot be identified in the said standard, nor the general guidelines for harmonic and inter-harmonic disturbance have been considered in the said annex;

- supply voltage dips, short interruptions and voltage variations are part of the main functionality of UPS and the behavior is described in more detail in a dedicated standard, the IEC/EN 62040-3 [209], where the required performance of the UPS in terms of output voltage variation in amplitude and time is specified depending on the classification (the classification is different from the category and refers in general to UPS performance).

It is thus evident that the EMC of UPS is a complex issue that involves:

- the verification of the basic immunity requirements specified by the product standard with respect to those specified for the target environment;
- the integration of tests for radiated immunity above 1 GHz, if required by the application standard, such as the EN 50121-X (2006) and IEC/EN 61000-6-2 (2007);
- the verification of compatibility, that is not trivial if the higher levels of radiated and conductance disturbance (for UPS of category C3) are considered: radiated emissions may be as high as 50/60 dBμV/m and conducted emissions 100/130 dBμV. A particular concern is the fact that the latter is very close to the RF immunity test levels for connected equipment; it is evident that an equipment mistakenly classified as residential only because it is far from, but connected to, this UPS is critically exposed;
- the possibility for connected equipment to relax the requirements concerning immunity to supply variations, thus easing the installation of many equipment whose product standards are not in full agreement with the latest version of the generic standard for residential immunity EN 61000-6-1 [194].

6.2.3.5 Low Voltage Power Supplies - d.c. Output

Low voltage power supplies is the scope of EN 61204-3 [200]. There are many examples and variants of power supply units (PSU) used in industrial applications such as:

1. PSU intended as free-standing apparatus, representing a product *per se*; examples are power supplies for laboratory and domestic use, plug/wall type power supplies (e.g. for ITC equipment);
2. PSU that are components, subdivided into:
   a. those considered equivalent to an apparatus (such as power supplies with IT mains plug or for internal use in computer type hardware, 24 V power supplies intended for installation) and
   b. those for a professional installer/assembler (such as power supplies on PCB and rack sub-assembly power supplies).

The standard is fully applicable to PSU of type 1 and 2(a), but it is only for guidance for PSU of type 2(b).

Emission limits and immunity test levels are applied depending on the group the PSU belongs to; groups are defined based on the ac or dc input, on the conversion principle, on the method for output regulation. The emission phenomena addressed by the standard are:

- for radio frequency conducted emissions on the ac input port (mains port), the same limits for Class A and B equipment are applied as reported in standards EN 55011 [132] and EN 55022 [144], that have not been modified in the later versions [133, 145]; in case of a dc input port, the limits are no longer mandatory, but only recommended, with the Class A limits generally considered, and the Class B limits for input voltages of less than 60 V;
- for radio frequency radiated disturbance, analogously, the same limits for Class A and B equipment are applied as reported in standards EN 55011 [132] and EN 55022 [144]; it is possible to use the *clamp method* (described in the CISPR 16-1-2 (or EN 55016-1-2) [135] to convert a radiated measurement into a conducted measurement, under the assumption that the cables are the main source of radiated disturbance instead the enclosure itself, and this is true provided that the longest side of the PSU enclosure is shorter than a quarter of the wavelength of the highest measured frequency; each cable (laid such the maximum length does not exceed brought to 5 m, if longer than 25 cm) is stretched on a straight line over a non conducting plane and the clamp is moved along it by keeping the maximum recorded value, that is then converted to equivalent radiated field by means of the clamp conversion factor;
- current harmonics on the input port, for which the IEC/EN 61000-3-2 limits apply if the voltage and rated current are within its scope (as already treated in the previous section).

The immunity tests are subdivided into low severity and high severity levels, that are identical to those prescribed by IEC/EN 61000-6-1 [194] and IEC/EN 61000-6-2 [195], corresponding to the low severity (i.e. residential, commercial, light industry) and high severity (i.e. industrial) environments. An additional test of RF field immunity at 900 MHz on the enclosure port is specified in compliance with the first edition of IEC/EN 61000-4-3 issued on 1995.

6.2.3.6 Radio Equipment and Services

A considerable attention is dedicated to the standards covering radio equipments and radio services, because they represent an important element for a modern railway system and represent a potential victim or source of electromagnetic disturbance, as an inter and intra-railway noise source: for example, uni- or bi-directional communication systems and signalling systems are often implemented nowadays with wireless technology; the devices and equipment that are part of signalling, telecommunication, passenger information, public announcement, time and clock and alarm subsystems are interconnected by network backbones; new lines built within city centers must proof to be compatible with all public, personal, broadcast communication facilities nearby. The focus here is on radio equipment as a product and on the EMC relevant standards; for further analysis of interference, reference must be made to the technical characteristics of the single radio equipment and in particular transmitting power, receiver sensitivity, antenna characteristics, location and orientation (see sec. 7.3).

The technical and functional specifications and requirements in the form of standards, technical specifications and technical reports amount to an impressive number of items in the ETSI database, of which those associated with EMC number approximately 250. Moreover, over a decade across the turn of the century, the organization of the EMC standards has changed and several have been withdrawn or superseded. The general group of equipment under the definition of "radio equipment and services" of the former, now "historical", EN 300 339 has evolved in several sub-groups and classes of equipment, each with its own peculiarities; a new family of standards named EN 301 489-X has organized the EMC matter for all the radio equipment and services; many other standards issued in the early '90s are no longer active and were tagged as "historical" in 2005, after some years of "service" of the new EN 301 489. The EN 301 489-1 [264], entitled "common technical requirements", establishes performance criteria, methods of measurements and limits; specific conditions for various classes of equipment are defined in the other successive standards. The EN 301 489 family represents an evolution with respect to the previous EMC standards, that were developed independently for the various equipment and services once they became relevant.

Radio equipment is characterized by complex susceptibility and interference mechanisms, and the performance assessment during qualification tests is rather involved, so that the standard requires that the manufacturer, at the time of submission of the equipment for test, supplies the following information that will be recorded in the test report:

- the primary functions of the radio equipment to be assessed during and after the EMC exposure;
- the user settings and stored data that are required for normal operation and the method to be used to assess whether these have been lost after the EMC exposure;
- the type of modulation, the operating frequency bands, the bandwidth of the IF filter before the demodulator;
- the characteristics of the transmission used for testing;
- the necessary test equipment delivered to enable the assessment of the EUT;
- the ancillary equipment to be combined with the radio equipment for testing (where applicable);
- an exhaustive list of ports, with the maximum cable lengths allowed, classified as either power (ac or dc) or telecommunication/signal/control;
- the method to be used to verify that a communication link is established and maintained (if appropriate);

- any equipment thermal limitation which prevents continuous testing of the EUT;
- the environment(s) in which the equipment is intended to be used.

The RF exclusion band applies to radio equipment with an operating frequency up to 2.7 GHz, or for equipment operating above 2.7 GHz, but whose RF bandwidth extends to a frequency below 2.7 GHz. For equipment operating at frequencies above 2.7 GHz and whose RF bandwidth does not extend to a frequency below 2.7 GHz, there is no exclusion band.

The environment classification used by the EN 300 489-1 refers to the residential, commercial and light industrial environment (IEC/EN 61000-6-1 and -6-3), the telecommunication centre environment (TR 101 651) and the vehicular environment (ISO 7637-2), however without any reference to the (heavy) industrial environment.

Radiated emissions are extensively specified from 30 MHz up to 6 GHz; for the frequency range of 30 - 1000 MHz both Class A and Class B limits, as defined in EN 55022 standard, are specified for general application and for telecommunication centers; over the frequency range 1-3 GHz and 3-6 GHz, the limits are 50 (70) and 54 (74) dBμV/m at 3 m distance for the general application, increased to 56 (76) and 60 (80) dBμV/m always at 3 m distance, respectively, for telecommunication centers (the limit values are given as average values and as peak values between ellipses). The limits for conducted RF emissions on dc and ac ports are again those of Class B and Class A. Harmonic current emissions and voltage flicker on the ac input port are subject to the EN 61000-3-2 and -3-3 standards, that are applicable to equipment with input current $\leq$ 16 A; for larger input current the IEC/EN 61000-3-11 and -3-12 standards are indicated.

The immunity tests and test levels are approximately similar to those of the generic standard for residential environment IEC/EN 61000-6-1, with the exception of reduced surge levels for protected and/or indoor cables and the lack of magnetic field immunity test; in addition reference is made to ISO 7637-2 for vehicular environment when the equipment is powered from 12 V dc or 24 V dc lines.

In the following subsections, there are presented some radio equipment and devices that are particularly relevant for a modern railway or metro application, is discussed.

6.2.3.7 Short Range Devices

The definition of a Short Range Device (SRD) as given in EN 300 489-3 [265] is rather general. It is defined as a *"piece of apparatus which includes a transmitter, and/or a receiver and/ or parts thereof, used in alarm, telecommand and telemetry applications, etc, operating with analogue speech/music or data (analogue and/or digital) or with combined analogue speech/music and data, using any modulation type."* SRD are classified with respect to their primary function and performance as shown in Table 6.3.

Table 6.3 General classification of SRDs based on their primary function and performance

Primary Function Classification	
Type 1I	Transfer of messages (digital or analogue signals)
Type II	Transfer of audio (speech or music)
Type III	Other
Performance Classification	
Class I	Highly reliable SRD communications media (e.g. serving human life inherent systems, may result in a physical risk to a person)
Class II	Medium reliable SRD communication media (e.g. causing inconvenience to person, which cannot simply be overcome by other means)
Class III	Standard reliable SRD communication media (e.g. inconvience to persons, which can simply be overcome by other means

To clarify the performance classification, Annex B of the said standard details a list of SRDs, broadly divided into groups with similar application (Table 6.4).

Table 6.4 Performance classification of SRDs According to and Application examples

Performance Classification	Application Example
Class I	Telecommand/Telecontrol (such as remote control power and lightning, remote surveillance switching, remote control cranes etc)
	Wireless sensing/measuring (such as fire detection, process control, position locator, wireless data communication etc)
	Alarms (such as guard systems, personal security)
	Others (Identification/Access control, Medical telemetry)
Class II	Remote control appliance and lightning, etc, for residential use only
	Car alarms, radio call alert
	Video cordless terminals, Cordless local networks, Identification of rail wagons, Surface probing radar, Vehicle detection/monitoring
Class III	Garage door openers, Car Lock/Unlock devices, Radio remote control television

In Annex A, reference is made to three further standards, that deal with SRD depending on the applicable frequency range:

- EN 300 220 Part 1, “Electromagnetic compatibility and Radio spectrum Matters (ERM); Short Range Devices (SRD); Radio equipment to be used in the 25 MHz to 1 000 MHz frequency range with power levels ranging up to 500 mW; Part 1: Technical characteristics and test methods” [257];
- EN 300 330 Part 1, “Electromagnetic compatibility and Radio spectrum Matters (ERM); Short Range Devices (SRD); Radio equipment in the frequency range 9 kHz to 25 MHz and inductive loop systems in the frequency range 9 kHz to 30 MHz; Part 1: Technical characteristics and test methods” [258];
- EN 300 440 Part 1, “Electromagnetic compatibility and Radio spectrum Matters (ERM); Short range devices; Radio equipment to be used in the 1 GHz to 40 GHz frequency range; Part 1: Technical characteristics and test methods” [259].

In EN 300 489-3 it is clearly stated that “the frequencies on which the SRD transmitters are intended to operate shall be excluded from conducted and radiated emission measurements when performed in transmit mode of operation. There shall be no frequency exclusion band applied to emission measurements of SRD receivers, and/or associated ancillary equipment.”

Another standard in use that covers SRD is EN 300 683 [262]. EN 300 683 covers the assessment of SRDs and ancillary equipment with respect to EMC; the qualification details covered in the standard are mainly for residential, commercial and light industrial environment, excluding vehicular electromagnetic environment. Clause 7 of the standard details the electromagnetic phenomena that SRDs of the type detailed in the standard are likely to encounter against a number of reference EMC standards, whereas Clause 8 presents the test methods and limits per SRD port.

6.2.3.8 Digital Cellular Radio GSM/DCS

GSM and DCS cellular networks are very common in modern railway systems to support the customers mobile terminals. Many high speed lines use the modified GSM-R protocol for the radio signalling system (see ERTMS/ETCS in Chap. 2) and internal and external GSM/DCS systems are probably the most relevant victims for radio frequency emissions from vehicles and line. For this reason the GSM/DCS systems are considered both as products, where EMC testing is performed by the manufacturer for certification, and as devices, that may be sources or victims of electromagnetic emissions.

The assessment of GSM and DCS Phase 2 and Phase 2 + equipment is covered by requirements detailed in EN 301 489-1 and EN 301 489-7. These standards, however, do not cover the assessment of emissions from the enclosure port of radio equipment (base station (BTS), ancillary RF amplifiers and GSM repeaters, that is the remit of ETS 300 342-3. The test condition is as specified in Clause 4 of EN 301 489-1; prior to executing the test sequence, it is mandatory to record the reference level of the speech output signal on both the down-link and up-link on the test instrumentation.

6.2.3.9 Wideband Transmission Systems and Radio LANs at 2.4 and 5 GHz

For this category of equipment, no exclusion bands applies.

In Annex A reference is made to three further standards that address more closely the two categories of equipment operating at the 2.4 and 5 GHz bands:

- EN 300 328-1, "Electromagnetic compatibility and Radio Spectrum Matters (ERM); Wideband Transmission systems; Data transmission equipment operating in the 2.4 GHz ISM band and using spread spectrum modulation techniques; Part 1: Technical characteristics and test conditions" [260];
- EN 300 652, "Broadband Radio Access Networks (BRAN); HIgh Performance Radio Local Area Network (HIPERLAN) Type 1; Functional specification" [261];
- EN 301 893, "Broadband Radio Access Networks (BRAN); HIPERLAN Type 2; Harmonized EN covering essential requirements of article 3.2 of the R&TTE Directive" [263].

In addition to the standards referenced above, there are a number of new standards that have been developed, covering wireless transmission system application in the railway industry or have found specific use in the railway sector. For example, a number of train communication systems utilise wireless technology based on IEEE 802.11h [382]. Solutions based on 802.11h are implemented as an alternative to solutions where cabling is deemed too complex and time consuming, when a high degree of flexibility is required and/or when the environment of intended use is considered harsh and containment. Such systems have found ready applications in monorails and driverless transport systems. 802.11h was released in response to the non-applicability of 802-11a [380] in Europe, where the 802.11a intended frequencies was already allocated and in use by civil and military applications such as radar system. To ensure compatibility with other systems that operate at the same frequency as 802.11h devices, the device is equipped with two interference mitigation techniques:

- Dynamic Frequency Selection (DFS): It is the process implemented of detecting radar signals that must be protected against 802.11a interference, and upon detection, switching the 802.11a operating frequency to one that is not interfering with the radar systems. DFS, as a technology, was first detailed in EN 301 893 [263]; it specified the types of waveforms that systems operating in the 5250 – 5350 MHz and 5470 – 5725 MHz bands should detect and defined threshold and timing requirements;

- Transit Power Control (TPC): It is used to adapt the transmission power based on regulatory requirements and range information.

6.2.3.10 Terrestrial Trunked Radio (TETRA)

TETRA is a digital voice and data network that offers support to terminals in use to Police, security personnel, maintenance personnel. TETRA was specifically designed to meet the needs of Public Mobile Radio (PMR), and has a large number of features designed for Safety and emergency use. Within Europe, the operational frequencies of TETRA has been allocated based on agreement between the EU and NATO. As a result, European Public Safety and Security forces are using the radio frequency band 380-385 MHz and 390-395 MHz for TETRA operation. Most of the spectrum allocation within the TETRA band are reserved for Trunked Mode Operation (TMO). The Direct Mode Operation (DMO) frequencies are typically allocated at the lower end of the radio frequency band (from 380 MHz upwards) and specific frequencies for Air-Ground-Air (AGA) operation are allocated at the upper end of the band (from 385 and 395 MHz downwards) to allow international compatibility. For the European non-emergency services, TETRA frequencies are mainly allocated in the 410 to 430 MHz band with some countries allocating frequencies in the 450 to 470 MHz band. Countries that are not members of NATO can sometimes allocate radio frequencies in the 385-390 MHz and 395-399.9 MHz band for non-public safety TETRA users. In Asia Pacific and South America, the so-called 800 MHz band, i.e. frequencies 806-824 MHz and 851-869 MHz are available for Digital PMR systems like TETRA. The following standards are major references dealing with the EMC aspects of TETRA systems:

- Electromagnetic Compatibility (EMC) standard for radio equipment and services Part 1: Common technical requirements, [264]
- Electromagnetic Compatibility (EMC) standard for radio equipment and services; Part 18: Specific conditions for Terrestrial Trunked Radio (TETRA) equipment, [268]

6.2.3.11 Railway Applications - Electronic Equipment Used on Rolling Stock

Beginning with its 2001 version, EN 50155 is an environmental stress screening standard in the true sense, providing specifications and tests for electronic equipment used on rolling stock: it encompasses ambient temperature, shock and vibration, humidity, salt mist as polluting/corroding agent, but also some electrical stress conditions. Electrical stress conditions consist of the rated value of the battery voltage, the supply range (accounting for steady voltage changes), voltage fluctuations, voltage interruptions (due to many reasons, including supply changeover), dc ripple factor, overvoltages (intended as an increase by 40% lasting up to 1 s) and surges (with a peculiar waveform 5/50

μs and 1.8 kV amplitude, called by the standard "waveform A"). Then, the standard specifies other immunity tests with reference to the EN 50121-3-2 for their specification, such as ESD, EFT and RF radiated immunity; also for emission measurements reference is made to the EN 50121-3-2.

The 2007 version didn't contain any significant changes to the part dedicated to electrical testing conditions, and reference was kept to the old EN 50121-3-2 (2000), even if this version was released more than one year after the final issue of the new EN 50121-3-2 (2006): two tests were evidently different from those already defined in the set of harmonized basic standards, namely the surge voltage test and the voltage variations test, the latter justified by the specific arrangement and characteristics of the on-board supply system.

The relationship between the two standards is rather confusing, in the sense that EN 50155 makes reference to EN 50121-3-2 for the specification of several EMC tests; conversely the old version of EN 50121-3-2 referred to the EN 50155 for surge and voltage variation tests. However the current version of the EN 50121-3-2 includes a specification for a surge test as per the basic standard EN 61000-4-5, but still refers to the EN 50155 (2001) in the table of "normative references" without using it. Finally the current version EN 50155 (2007 amended by Corrigendum July 2010) also refers to the new EN 50121-3-2 (2006), and manufacturers tend to use it in place of the EN 50121-3-2 for their declarations of conformity: this is incorrect since, the correct standard for EMC compliance is EN 50121-3-2, and secondly, not all the EMC phenomena specified by the EN 50121-3-2 are included in the EN 50155, such as the magnetic field immunity test.

6.2.4 Design Standards

These specify the design or technical characteristic of a product in terms of how it is to be constructed. There are no EMC design standards, or at least any applicable to railway applications, although standard covering general design guidelines do exist.

6.2.5 Process Standards

These are concerned with the series of actions or operations used in making a product and provide the methodology to perform these processes in a consistent and reproducible way.

6.2.6 Specification and Codes

A specification is a set of conditions and requirements of precise and limited application that provides a detailed description of a procedure, process, material, product or service for use primarily in procurement and manufacturing. Standards may be referenced or included in specifications. A code is a collection of mandatory standards that has been codified by government authorities and become law.

6.2.7 Management Systems Standards

These types of standards specify the requirements that can be applied to any organisation, regardless of the product it makes or the service it performs. The two most widely known series of international standards that fall into this category are the ISO 9000 series and ISO 14000 series of standards. Within the EMC sector, ISO 17205 which governs the requirements for the competence of calibration and test laboratories falls within this category.

6.2.8 Personnel Certification Standards

Personnel certification programs help to verify that individuals have the necessary knowledge, skills and abilities to perform their work. Certification of personnel provides continual improvement in professional competence and leads to a highly skilled workforce, particularly within industries that have an impact on public safety. Within the EMC profession, the NARTE[1] EMC Engineer certification falls within this category.

6.3 Standards Classification by Origin and Scope

A standard has legal value if it is recalled by a law or regulation of the country where the equipment is sold or installed. In this case, it cannot be ignored and it shall be included in the list of applicable standards. Many standards in use nowadays are not really mandatory for an application, but they represent a precious technical or state of the art reference and they are indicated as applicable in the contract or in one of its attachments. Furthermore, a standard may be proposed later by one of the suppliers or by the technical board to solve a problem, to fill procedural lacks or as a support for some kind of test.

A railway application features various and different subsystems, that share the same complex environment, with various degrees of interaction, purposed or accidental. It is thus quite natural that there exist standards that are focused on a specific component or equipment, that may be sold separately and that may be installed in a variety of target applications and environments, among which our railway application and its subsystems; these standards are named "product standards" and, if they are closer to the specific product characteristics and they handle them in the proper way, they lose at once

[1] NARTE is the National Association of Radio and. Telecommunication Engineers.

the consciousness of the specific requirements that may be set forth by the various applications.

It may happen by the way that a component or equipment has no product standard dedicated to EMC and it was thus tested by the manufacturer using a generic standard; generic standards are generic application standards, so they assimilate various similar applications to the characteristics of a common electromagnetic environment. That is, the focus is on the application rather than the products, even if the application is generic, so covers a broad range of similar applications (the similarity being from the EMC viewpoint, that is regarding the intensity and the occurrence of specific electromagnetic phenomena, already considered in sec. 6.5).

If an application is believed to be particularly severe, then the details of its e.m. environment are described in specific application standards, such as the ones referred to as railway applications: the EN 50121 family covers the whole railway as a product, the rolling stock as a separate product, and the e.m. environments for embedded products on board, trackside and at substations, as seen above (sec. 6.5).

Standards can be classified into their origin or where they were developed. There are four primary sources for standards:

- Company Standards;
- Trade Associations and Professional Societies;
- Government;
- Consortia.

In addition, standards can be classified by their scope:

- National;
- Regional;
- International.

6.3.1 Company Standards

Most manufacturing companies (e.g. car manufacturers) or railway infrastructure owners (Network Rail, London Underground, etc.), develop internal documents for their own use, that define such activities as production processes, materials characteristics, and purchasing requirements. In addition to being used internally, suppliers to such companies must comply with company standards in order to do business with these companies. For example, any product that does not comply with London Underground Category 1 standard on EMC [494] would not be allowed to be installed on their infrastructure. Within the railway sector, company standards are quite common, since the railway systems in each country have evolved separately and railway operators or infrastructure owners have adopted their own rules for the testing and acceptance of equipment and components. Within Europe, attempts are being made by CENELEC through the EU Interoperability Directive to integrate and harmonise national and company specific railway standards, with the aim of driving several benefits: cross circulation of products, and in particular rolling stock; more fair rules in the homologation procedures, that can be applied by different manufacturers in each respective country, with an underlying agreement of equivalence; a better coverage of various technical details and issues based on the possible integration of different approaches. For example, the many testing procedures for track circuits adopted by the several EU countries (more than thirty approximately) have been described and included in the draft version of the EN 50238-2 [114] and EN 50238-3 [116] standards, following the EN 50238 [112], to support the effort to harmonization and interoperability.

In the past some of these company standards have been also partially adopted as national standards, but without the equivalence with the other neighboring countries standards, that is implicit in the derivation by a harmonised process.

6.3.1.1 Network Rail

Network Rail is the British mainline railway infrastructure owner, operator and maintainer. Network Rail's EMC strategy and assurance process is specified in [562] and [563]. These documents are mandatory requirement for the management and control of the electromagnetic environment within Network Rail's control. As part of its obligation to provide relevant data to Train Operators and system suppliers, Network Rail has produced a suite of EMC documents that detail the permissible limits for all signalling and communication assets in operation within their infrastructure. There are approximately 18 standards within this scope [564, 565, 566, 568, 573, 574, 575, 577, 578] with each addressing a particular asset with the exception of [579], which specifies generic compatibility rationale for all line-side equipment. Network Rail standards are designed to address the shortfalls of Railway Application standards, in that they typically cover in-band interference, at frequencies not normally covered by Railway Application standards.

6.3.1.2 RFI - Trenitalia - Ferrovie dello Stato

Ferrovie dello Stato is the name of the former major Italian infrastructure owner and railway operator; other minor owners and operators exist, but they do not represent a reference for standardization. Ferrovie dello Stato was split, in the year 2000, into the infrastructure owner RFI (Rete Ferroviaria Italiana, Italian Railway Network) and Trenitalia, both controlled by the holding Ferrovie dello Stato S.p.A.. The EMC relevant standards concern the testing and limits for electrical and electronic equipment for the signalling systems and for rolling stock conducted emissions in relationship with the interference to signalling systems. For other aspects reference is made to national standards (see sec. 6.3.5.4) or to European harmonized standards (see sec. 6.3.6.1).

The IS 402 standard [409] covers several environmental tests and for EMC testing it indicates a list of immunity tests covering ESD, EFT and surges, RF conducted and radiated, magnetic field at supply frequency and pulsed. The tests are specified by dividing the equipment in three groups and reference is made to the old EN 50082-2, to several EN 61000-4-X standards (for the test procedures and conditions), to the EN 50121-4 itself and in many cases the test values are directly specified in the text. The test levels are approximately correspondent to those of the EN 51021-4 standard.

The first standard for the limitation of the absorbed current distortion in terms of spectrum components was issued on 1996 and it is called "FS 96 mask": applicable to 3 kV dc railway lines for the current absorbed by the whole vehicle, it has developed rules for the apportioning of the limit for the whole vehicle to single on-board converters, by applying multiplicative coefficients, probably set taking also into account the superposition of the different emission components. The test conditions for the vehicle have always been established by selecting specific line routes for the test runs and by instructing the driver to keep an agreed driving style and driving conditions. For the single converters great care must be given to the selection of the electrical loads to connect (balanced or unbalanced, with significant current leakage to earth or not, etc.) and to the mix of loading conditions, that may statistically weight the emissions of the converter under test; a starting base for the load conditions and load mix is the UIC 550-2 standard [665]. The revision. 2 of the standard, changed some values and frequency limits and most of all it was complemented by a statistical evaluation of the distribution of relevant emission components close to the prescribed limits: the limits are thus specified as a hard limit $L_0(f)$ to be never violated and other limits taken as fractions of the hard limit $L_1(f) = k_1 L_0(f)$, $L_2(f) = k_2 L_0(f)$, $\ldots$, $k_2 < k_1 < 1$, which a probability limit is associated to: during the test runs the emission spectra are computed and stored for successive statistical processing, consisting of the calculation of the probabilities of occurrence for each frequency bin of the collected amplitudes with respect to the so determined amplitude limit bands. This standard has been included together with all the other national railway standards for the evaluation of interference to signalling circuits in the EN 50238-2 under development.

For the 2x25 kV 50 Hz new high speed lines, there is a limit mask dating back to 2006, that is still under study: for the low frequency range the limits are quite high, since this range is characterized by the low order harmonics of the supply frequency, with two exceptions for two small frequency intervals (75-95 Hz and 160-190 Hz) used by power frequency track circuits; above 350 Hz the curve profile is flat at 10 Arms up to 1800 Hz, where it goes down to 1.5 Arms up to 9 kHz and then 0.5 Arms up to 17 kHz (this maximum frequency represents the boundary for the last channel of the AFTCs in use). It is underlined that in general a unique curve for two different installed TCs is never really attainable, since the TCs have a different behavior with respect to interference, in terms of amplitude, time duration, spectral components and track conditions. The two circuits have been included in the last issue of the EN 50238-2 [115] as preferred circuits for interoperable lines, under a unique harmonize method of evaluation of interference (the band-pass filter method), that differs from the old Fourier based method for the evaluation of interference in use in Italy and also in France (see sec. **6.3.1.2**).

6.3.1.3 London Underground Limited, LUL

London Underground Limited, (LUL), is the infrastructure owner and maintainer of the oldest underground railway in world. LUL, like Network Rail, has a suite of EMC standards that specify requirements with respect to the demonstration of compatibility with their infrastructure assets. LUL requires that equipment, systems and installations comply with the EU EMC Directive (currently 2004/108/EC) and the UK implementation of the EMC Directive (currently Statutory Instrument No.3418) and that such installations meet the LUL defined set of specific EMC requirements, in order to help ensure interference-free operation in the underground's severe electromagnetic environment. The LUL mandatory Category 1 Standard 1-222 [494] describes the framework necessary to ensure that the EMC aspects of equipment for installation within LUL are adequately addressed.

The standards to be applied are listed in the LUL Manual of Good Practice G-222 [495], a document which goes into more technical detail on LUL's EMC requirements. The primary set of standards to be applied is the railway application EMC series of standards (i.e. EN50121), which define limits for the general railway environment. However, LUL has also identified additional EMC standards for their specific environment (and these should be complied with as it is), the responsibility of the parties involved with installation, design and commissioning, any specific electromagnetic environmental influences.

6.3.2 Trade Associations

It is possible for trade association to develop standards specific to their trade. Such standards represent the expressions of the interests and exigencies of a group of companies that operate in the same market sector or that use the same types of products. The respective members may be motivated by the need of protecting their products from excessive costs and overheads related to conservative and inadequate standards, or establishing rules, methods and limits for performances, suitable for a specific application, for example related to a specific electromagnetic environment, or safety and reliability levels, or ignition energy when there is risk of explosion. Such standards may be adopted as a company or national standard. For example, the Railway Industry Association (RIA) in the UK developed a set of standards including EMC specific standards, such as RIA 18, which became the starting point for the development of European EMC Railway Application Standard, EN 50121 (IEC 62236) series of standards.

6.3.2.1 American Railway Engineering and Maintenance-of-Way Association (AREMA)

AREMA is an association of railway and maintenance-of-way engineers in North America. The association details its common specification of the demonstration of compatibility with signalling equipment in AREMA Manual 11.5.1 Clause 6 [14] and a methodology of dealing with inductive interference in AREMA C&S Manual 20.1.4 [16]. Clause 6 of [14] does not specify requirements for immunity and refers to FCC Part 15 [283] for the demonstration of compliance for emissions. Appreciating the shortcomings of [14] with respect to EMC, AREMA has a project to develop detailed emission and immunity requirements for signalling equipment, the proposed standard is to be known as AREMA C&S Manual 11.5.2 [15]. The proposed AREMA Manual 11.5.2 is largely based on the requirements of IEC 62236 Part 4 [82]. The emissions requirements are based on IEC 61000-6-4 [197] while the immunity requirements are based on IEC 61000-4-2 [174], IEC 61000-4-3 [175], IEC 61000-4-4 [176], IEC 61000-4-5 [177], IEC 61000-4-6 [178], IEC 61000-4-8 [179] and IEC 61000-4.9 [180]. There are significant differences between AREMA Manual 11.5.2 and IEC 62236 Part 4, for the radiated immunity severity levels, specified in AREMA, between 1.4 GHz and 2 GHz is 20 V/m compared to 5 V/m in the IEC 62236 Part 4. Furthermore, for Fast Transients, Surges and Conducted CM immunity test, AREMA specifies different severity levels for internal and external signals. It is important to note that there might be EMC clauses specified in other AREMA standard, such as requirements for Electrostatic Discharge, ESD in AREMA 11.4.5 [13].

6.3.2.2 Urban Mass Transport Association / Urban Mass Transportation Administration (UMTA)

The UMTA series of recommended practice was developed, under the auspice of the US Department of Transportation, in the eighties, in response to an incompatibility issue observed during commissioning between a dc chopper train and a signalling system. The Technical Working Group that was set up to undertake this work, developed standardised methods of analysis and testing to qualify and resolve incompatibility issues observed with dc chopper systems, as well as addressed future electromagnetic compatibility issue of rail transit vehicles. In 1985 the recommendations were published as specifications and has thus taken the form as the US Rail Transit EMC standards. The publications consist of two parts (1) The Recommended Practice, and (2) The Suggested Test Procedures. These standards are still referenced in a number of US rail transit projects but there is increasing migration to Railway Application Standards (EN 50121 series of standards). Reference to UMTA

Specification and Test Procedures tends to be on upgrade/modernisation type projects where the EMC baseline has been underpinned on UMTA specifications and there is a need to maintain the same level of compatibility. The following figures highlight the test limits associated with UMTA specifications (the measurement limits are based on the MIL-STD-461 A and B; the limits shown below are based on a 100 ft, approximately 30 m, measurement distance for a single vehicle). The figure for conductive interference includes limits for a single vehicle, auxiliary power and propulsion inverter.

Conducted Emissions

The limits shown in Figure 6.1 are quite permissive in the low frequency range up to 320 Hz, if compared to corresponding limits in Europe, but are quite restrictive going to higher frequency: the few tens of mA are much lower than the corresponding limits of one or few hundreds mA for most track circuits at audiofrequency [130, 112], examined in sec. 3.1.1.6.

Limits
30 Hz – 50 Hz, 10 Arms
50 Hz – 70 Hz, 1 Arms
70 Hz – 90 Hz, 10 Arms
90 Hz – 110 Hz, 1 Arms
110 Hz to 320 Hz, 10 Arms
Emissions then follow a continuous curve through 10 A at 320 Hz, 0.08 A at 2 kHz, 0.016 A at 4 kHz, and 0.0046 A at 7 kHz

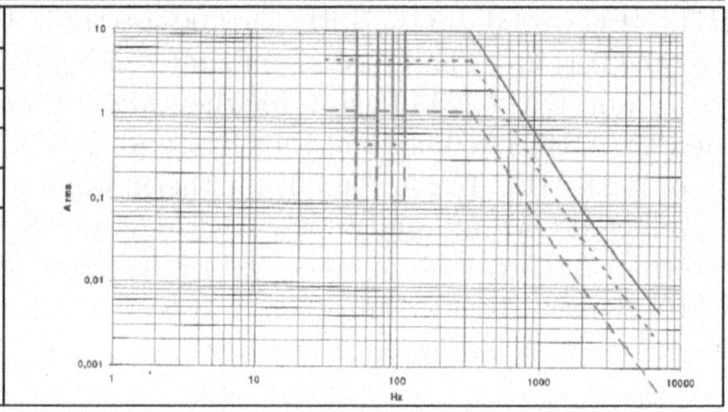

Fig. 6.1 UMTA Limits for Conducted Emissions (UMTA-MA-06-0153-85-6 [669]

Radiated Emissions

The test method focuses on broadband electromagnetic field emissions and even below 30 MHz emissions are measured as an equivalent electric field by specifying a rod antenna for measurement (opposite to the loop antenna suggested by EN 50121-2). The resolution bandwidth values for the EMI receiver or spectrum analyzer are 10 kHz below 30 MHz and 100 kHz above; these values are in accordance with those (9 and 120 kHz respectively) specified by CISPR for EMI receivers and can be easily adjusted by a corrective factor, to make the results comparable.

<table>
<tr><th colspan="2">Limits</th></tr>
<tr><td>0.01 MHz – 30 MHz, RE05 of MIL-STD-461 A</td><td rowspan="3">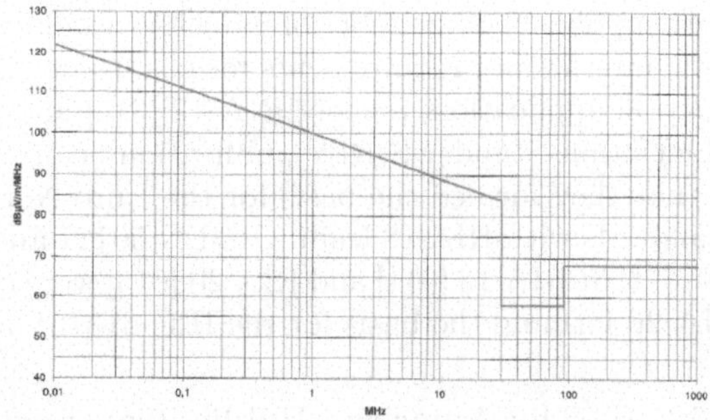
</td></tr>
<tr><td>30 MHz – 88 MHz 58 dB above 1μV/m</td></tr>
<tr><td>88 MHz – 1 GHz, 68 dB above 1μV/m</td></tr>
</table>

Fig. 6.2 Radiated Emissions UMTA Limits (UMTA-MA-06-0153-85-11 [670])

Inductive Emissions

Equipment can affect trackside equipment by generated AC fields or by the motion of the rail equipment past the trackside equipment. This causes a current to be induced into the trackside equipment through the track, by proximity to the trackside equipment, or to other equipment on board with parallel conductor runs. Inductive emissions limits are specified in terms of voltage coupled onto a reference circuit.

<table>
<tr><th colspan="2">Limits</th></tr>
<tr><td>30 Hz – 1 kHz, 20 mV</td><td rowspan="2">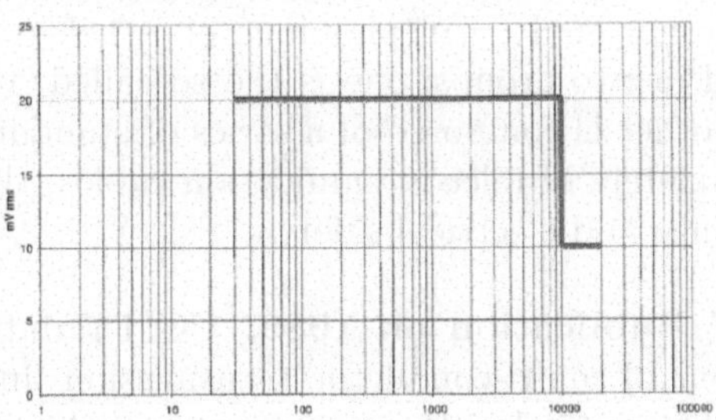
</td></tr>
<tr><td>1 kHz – 100 kHz, 10mV</td></tr>
</table>

Fig. 6.3 Inductive Emissions UMTA Limits (UMTA-MA-06-1053-85-8 [671])

6.3.2.3 Railway Industry Association (RIA)

The Railway Industry Association is the Trade Association for UK based suppliers of equipment and services to the railway industry. The trade association published a number of industry standards that dealt with electromagnetic interference requirements for electronic equipment used on trains or for traction application; namely RIA 12 [610] and RIA 18 [612]. As an Industry Association Standard, RIA publications can be considered very successful, in that they formed the basis for the Euro Norm EMC Railway Application Standards with RIA 18 leading to EN 50121 and RIA 12 leading to EN 50155 (Note that RIA 13 [611] and RIA 20 [613] were environmental standards, with RIA 20 forming the basis for the IEC 61373).

6.3.2.4 The American Public Transportation Association

The APTA is a non-profit international association of private and public member organisation, including transit systems and commuter rail operators among others. The APTA through its PRESS[2] task force published two standards of interest with respect to EMC:

- APTA SS-E-005-98 [12], a standard that provides minimum requirements for earthing and bonding on passenger rail vehicles, and
- APTA SS-E-010-98 [11], a standard that defines the requirements an EMC program for all rail equipment and track side equipment.

6.3.2.5 The ITU Telecommunication Standardization Sector (ITU-T, formerly CCITT)

The ITU Telecommunication Standardization Sector (ITU-T) is one of the three sectors (divisions or units) of the International Telecommunication Union (ITU); it coordinates standards for telecommunications, concerning the definitions of formats, protocols and physical characteristics for a wide set of telecommunication systems (TV, radio, telephone, etc.).

For EMC in railway systems, the International Telegraph and Telephone Consultative Committee (CCITT), (from the French name " Comité Consultatif International Téléphonique et Télégraphique") documents may be classified in two groups: one is the so called "CCITT publications of 1989", and the other is composed of a series of standards and guidelines concerning the parameters of telecommunication cables, their testing and the limits for interference and electrical safety.

CCITT Publication on 1989. CCITT did great work in writing down the theory of multi-conductor transmission line modeling, for the determination of per-unit-length parameters and the estimation of coupled voltages between power, traction and communication lines. This work, contained in

[2] Passenger Rail Equipment Safety Standard.

two volumes of the set of 1989 CCITT publication, [418, 419] is perhaps the most relevant reference available when analysing the interaction between traction and line-side (or buried) signalling and communication cables. The same publication in volume IV [420] contains a valuable collection of measurement results, that cover various phenomena, such as vehicle emissions, their propagation along the lines, the behaviour of traction lines versus frequency in terms of pantograph impedance, line resonance and line parameters values. The results contained within this volume should be viewed with caution for a number of reasons: the factors influencing line resonances and line parameters are numerous, but most importantly rail vehicles are continuously evolving and those in use at the time of these publications, which formed a core component of the study, are now old and/or not in service any longer. The structure of the volumes is very complex and only a brief outline is given below; most of the contents however may be traced back throughout this book.

Volume II [418] details, in its chapter 2, a simplified method for the calculation of induced voltage and currents, the method presented in this chapter has been critically reviewed in [527]; the publication then proceeds to deal with the modeling methods and the equations for the calculation of capacitive, inductive and conductive coupling in chapters 3, 4 and 5 respectively. The calculated parameters are then used with two-conductor and multi-conductor transmission line approaches for the determination of the induced (in the general sense of "coupled") voltages for power and railway systems.

While Volume II is oriented to the interpretation of real cases and the formulation of simplified expressions, Volume III [419] deals with the background electromagnetic theory and the development of the basic theory for the determination of the coupling terms and the per-unit-length parameters. Several configurations and conditions are covered in the presented formulations: air, earth and air-earth regions, presence and position of the return conductors, cable shield structures (thin, thick, solid, etc.) and its termination to earth, effect in terms of equivalent screening of interposed conductors.

Volume IV [420] is dedicated to real systems: chapter 2 reviews the main characteristics of railway systems, both dc and ac, for later use in chapters 3 and 4 where induction is evaluated and compared with measured results.

6.3.3 Professional Society Standards

Professional societies such as the Institute of Electrical and Electronics Engineers, IEEE (USA) and the Institution of Engineering and Technology, IET (UK) develop standards that support and advance their professional interest. These organisation, in particular the IEEE have an active and influential standards program.

6.3.3.1 IEEE (Institute of Electrical and Electronics Engineers)

IEEE is organized in societies, each one for a specific broad technical sector, such as "Electromagnetic Compatibility", "Antennas and Propagation", "Instrumentation and Measurement", and so on. A committee is formed, composed of members of the interested societies, to set up and develop the standard project; when a stable draft is available, it is submitted for inquiry between IEEE members, who subscribed to the Standards Society. There is a close relationship between the IEEE and the American National Standards Institute (ANSI), with many standards published under joint ownership as ANSI-IEEE Standards.

IEEE standards are numerous with approximately thirty standards dealing with EMC. In the following discourse, the classification is based on main topics that are to some extent relevant or applicable to the test and certification of equipment for emissions and immunity, on the analysis of the electromagnetic phenomena, on the characterization of instrumentation and on the appropriate design of power systems and electrical installations, and that thus may be referenced in railway construction contracts.

Emissions and Immunity of Equipment

Some standards cover the most relevant emission and immunity phenomena, in particular for equipment in an industrial environment:

- IEEE Std. 1346, "Evaluating Electric Power System Compatibility with Electronic Process Equipment" [389], that focuses on voltage sags and interruptions and their effects on connected equipment, considering in particular the immunity of Programmable Logical Counters, Adjustable Speed Drives, Personal Computers and control relays;
- IEEE Std. 519, "Recommended practices and requirements for Harmonic Control in Electrical Power Systems" [375], describing the most relevant sources of harmonic emissions (arc furnaces, various static converters including switch mode power supplies, etc.), the response of the supply network and its elements, and the effects of harmonics on various electrical equipment, to justify the given limits; limits are specified in terms of resulting supply voltage distortion and apportioned current distortion for each connected load, as a function of its peculiar characteristics and rated power; the standard also presents synthetically the analysis and simulation methods, as well as the measurement techniques, for the evaluation of harmonic distortion.

Analysis of Electromagnetic Phenomena and Installation Practices

The IEEE standards represent a repository of knowledge and guidelines for industrial installations, taking into considerations the most relevant source of disturbances and coupling mechanisms, and in particular reporting practical considerations and measurement results, describing typical cases and configurations. The cited standards span over more than 30 years and in some cases the first issue is even older, dating back to 1950 as for the IEEE Std. 140 [361].

IEEE Std. 518 [374] details good practice for the installation of controllers and control systems, but the content of this guide reflects good EMC practice that is applicable across many sectors. It is covered in this book, because like many "reference guide" it provides valuable background reading with respect to the electromagnetic interaction between noise sources and victim circuits, as well as mitigation techniques, to resolve common incompatibility problems. With respect to Controllers and Control systems, IEEE Std. 518 covers the subject of the interference model as follows:

- noise sources are recognized as switches and contactors operating inductive loads (showering arc accompanies current interruption with very large rate of rise, in the order of several kV/μs, and an even larger - by about an order of magnitude - rate of fall, that produce a spectrum extended up to several MHz), thyristors and other semiconductors used in static converters (characterized by very steep rise and fall profiles, up to a ten of kV/μs, triggering stray resonating circuits and thus followed by ringing phenomena, with the spectrum characterized by different modulation terms), welding machines, fluorescent and neon lights, heavy current conductors, thyratrons and ignitron tubes; the latter witness evidently the age of the standard as they have almost disappeared nowadays;
- coupling mechanisms are those normally recognized of common impedance, magnetic, electrostatic and electromagnetic radiation coupling; their description is accompanied by numerical examples;
- victims susceptibility is simply considered for continuous and transient electrical noise, and distinguishing between common mode and normal mode (differential mode) noise; the standard proposes different setups for the measurement of continuous and transient conducted susceptibility, radiated electric field susceptibility and magnetic field susceptibility, together with a discussion of the means and implementation details for disturbance coupling and hence to understand some mitigation techniques;

The major contribution of this standard is the part dedicated to "wiring type classification", that offers a systematic treatment of what it is also called "cable segregation criteria" (see Chap. 8). Different mitigation techniques are considered, such as separation, orientation, pairs twisting and shielding, coax cables, selection of grounding methods and termination impedance. The standard classifies the mitigation techniques in "suppressive wiring techniques" (reducing magnetic and electric fields close to the noise source), "barrier wiring techniques" (the obstruction of magnetic and electric field propagation is done at a larger distance from the source, in a region of space where far field condition may occur) and "compensatory wiring techniques" (it consists substantially in cancelling or counteracting the effects of the rate of change of the magnetic or electric field, without actually obstructing or altering the intensity and distribution of the fields). A comprehensive formulation is given for the determination of properties and performances of twisting and shielding and for grounding and bonding issues in an industrial environment.

A synthesis of the main mitigation techniques appear in a schematic and ordered way in the more modern IEEE Std. 1137 [386], devoted to assist readers in controlling or modifying the inductive environment and the susceptibility of victim telecommunications lines in order to operate within acceptable levels of steady-state or surge-induced voltages. This guide was thought to accompany the IEEE Std. 776 [379], concerning the "inductive coordination of electric supply and communication lines".

The IEEE Std. 776 identifies very clearly a series of effects related to inductively coupled voltage, that in turn determine a list of thresholds to be considered, while designing cableways and evaluating the effects of mutual inductive coupling.

- Safety. The standard identifies 50 Vrms continuous with respect to ground as the threshold normally considered by telecommunication facilities companies in North America, while 60 Vrms is recognized as a value used by electric power companies and railway telecommunications systems, as also supported by CCITT.
- Damage of installed cable. With a practical approach, using IEEE Std. 487 [373] as a reference, typical conductor-conductor dielectric strength for plastic insulated cables is 1.8 kVrms for 18 AWG, 1.4 kVrms for 22 AWG, 1.2 kVrms for 24 AWG, 0.96 kVrms for 26 AWG.
- Destruction of equipment. For line repeaters and other electronic equipment directly connected to telecommunication cables the limit energy dissipation is between 16 and 80 A^2s.
- Interference, referred to as "circuit degradation", in terms of noise with respect to ground is limited to 80 dBrnC.

Interestingly, the standard proposes a practical method to assess the level of exposure by using a reference probe wire of 100 ft (30.48 m) length and 22 AWG gauge, insulated and grounded at both ends. This conductor is placed at a distance from the culprit power line that is 50 or 75 ft depending on the voltage level, lower than or equal to/greater than 69 kV. No indication is given when the source of interference is an electric traction line, but we could assimilate it to a less than 69 kV power line; moreover, most of the telecommunication lines are located in the right-of-way of the traction line and so at a distance surely less than 50 ft.

Starting from the 1987 version a series of limits have been developed for the induced voltage on the probe wire, identifying two limiting harmonic distributions that form an envelope:

- for low order harmonics with $2 \leq n \leq 17$, corresponding to 120 and 1020 Hz on a 60 Hz system, a general limit corresponding to a $n^{-2.7}$ weighting factor of the induced voltage limit at the fundamental (e.g. 50 Vrms) is set, but if there are only three harmonic components or less above this limit, then the limit for them is increased to a factor of $n^{-2.0}$;

- for high order harmonics with $18 \leq n \leq 50$, corresponding to 1080 and 3000 Hz on a 60 Hz system,
  the limits above for low order harmonics are prolongued with an identical slope of $n^{-1.2}$ using the following expressions: $1/(17^2+n^{1.2})$ and $1/(17^{2.7}+n^{1.2})$ for the two limit curves adopted above for maximum three and for all harmonics;
- for higher order harmonics above the 50th the limit value for the 50th is applied;
- the same standard identifies an additional limiting curve applied to all harmonics with exponent $n^{-1.7}$ that has not to be overcome, that seems redundant and useless, but whose significance probably clarifies with the consideration about harmonic distributions in the following paragraphs.

IEEE Std. 1137 [386] integrates this first evaluation by means of preliminary measurements to be taken at potential victims (with reference to IEEE Std. 820, 1984):

- circuit loss, measuring the voltage drop on the selected line for a 1004 Hz 0 dBm test signal;
- circuit noise, also called "C Message Weighted Metallic Noise"; a $\leq$ 20 dBrnC value is recommended, a$>$ 20 dBrnC and $\leq$ 30 dBrnC value is acceptable, while a $>$ 30 dBrnC is not recommended;
- power influence, also called "C Message Weighted Longitudinal Noise" or "C Message Weighted Noise-to-Ground"; a $\leq$ 80 dBrnC value is recommended, a$>$ 80 dBrnC and $\leq$ 90 dBrnC value is acceptable, while a $>$ 90 dBrnC is not recommended;
- longitudinal balance, to be computed as the difference in dB of circuit noise and power influence; a $\geq$ 60 dBrnC value is recommended, a$<$ 60 dBrnC and $\geq$ 50 dBrnC value is acceptable, while a $<$ 50 dBrnC is not recommended.

A high power influence reading indicates one or more of the following: high influence of a nearby power line, insufficient separation between the power line and the communication cable, or poor shielding of the communication cable; poor shielding may be due to defective bonding and/or grounding of the communication cable.

A low longitudinal balance may indicate cable trouble (i.e. defective splices or wet cable), left-in service wire, grounded ringers, or operated protectors; note that the balance may be degraded by either a series unbalance, where the impedance of the two communication conductors is not equal, or by a shunt unbalance, where the impedance to ground of both communication conductors is not the same.

In order to assess the shielding effectiveness of a cable, the IEEE Std. 776 [379] suggests the comparison of the measured voltage-to-ground with that measured on the probe wire provided both are normalized to the respective values induced at the supply fundamental frequency. When considering the reason for longitudinal unbalance and poor shielding effectiveness, plots with a negligible slope with frequency indicate a resistive leakage problem, while a significant slope indicates a capacitive unbalance. Finally, in order to assess the consistency of the harmonic conditions on the inducing system, the plot of the statistical results of a survey of the circuit noise on several cables may help: the IEEE Std. 776 suggests to plot the mean at each harmonic of the supply fundamental, accompanied by the 1 dispersion limits around it, and two curves corresponding to the $n^{-1.7}$ and $n^{-2.7}$assumptions.

Grounding. Some standards exist that treat at a great extent the grounding of electronic equipment (IEEE Std. 1100 [385]), low voltage facilities (IEEE Std. 142 [363]) and at high voltage substations (IEEE Std. 80 [359], IEEE Std. 665 [378]), including installed measuring and control equipment (IEEE Std. 1050 [384]).

In all cases the exigencies of electronic equipment grounding are balanced with the objectives of the power circuit grounding; in particular, the IEEE Std. 1050 identifies for generating station grounding the following objectives:

- maintain safe voltages across the station area during high-voltage system transients (step and touch potentials);
- minimize the effects of lightning surges on equipment and structures;
- provide a low-impedance, ground fault current return path.

The standard proposes the modern approach to grounding, that is "all portions of the station ground system, whether for power, lightning, or signal reference, need to be interconnected and should not be isolated from each other for any reason; isolated grounding elements represent a safety and operational hazard". Going to the grounding of single equipment and subsystems, for electronic equipment used for instrumentation and control, a common ground or reference plane is suggested, claiming that the interconnections may be complex and the use of a separate return path for each signal would be impossible. Nowadays this approach is in reality followed if the application requires so and some extra cost and size are not an issue: modern high speed serial buses are routed with each signal next to its own return wire. The possibility of separating ground systems is proposed as a valid mean to reduce EMI coupled from the grounding connections of power equipment, in particular if the low frequency signals of instrumentation and control equipment to and from the field are considered. The single point grounding method is very effective and adequate when dealing with equipment operating at frequencies below 30 kHz and not exceeding 300 kHz (in particular circumstances the same philosophy can be extended up to some MHz). The signal ground reference point should be tied to the equipment safety ground point within the enclosure or local area and there should be no more than one (preferably filtered) ground lead exiting each equipment enclosure. This lead should be a stranded and insulated conductor sized to minimize the steady-state potential difference between devices, to meet the required mechanical strength and to cope with any corrosion issue. A disadvantage of the single-point grounding system is that it may fail to function as a signal reference and even may be detrimental at high operating or threat frequencies. The former are normally in relation to microprocessor systems and high speed connection buses, but are believed to be confined within a single equipment or a compact apparatus. Another issue is that "even an otherwise carefully designed and implemented single-point grounding system may unexpectedly evolve over time into unintentionally large, enclosed, EMI gathering and radiating loop areas once modifications and additions of cabinets and cabling are made that do not conform to the original philosophy." A special provision for enclosures is indicated for terminating the ac equipment grounding conductor that is run with the input power cable from the power distribution system; this connection is in addition to the solid bonding of any metal armor, conduit, or raceway provided with the incoming ac power conductors. A supplementary connection to the cabinet is also advised through a ground cable that is not run with any ac power circuit conductors and will connect to the closest local ground point, ground mat, or reference ground plane if one is available. If multiple enclosures are physically grouped together, first bond them by bolting them to one another and then use only one connection between the enclosure group and the supplementary ground.

When equipment and cabinets of the same system are widely separated or the frequencies are above 30 kHz and certainly above 300 kHz, cabinet ground points will be at relatively different potentials with respect to each other, and unwanted currents may then flow on any ground connection between the cabinets; separate single-point grounding connections to the common bonding reference network for each geographic grouping of equipment need to be created. The communications circuits between distant cabinets must have appropriate protection for the common-mode noise, which is likely to result from the impedance of the long, insulated signal grounds. Signals between systems should use either differentially connected signal isolation transformer coupling or ac capacitor coupling with transmitter/receiver circuits having a common-mode withstand voltage that safely exceeds the ground voltage under fault conditions. It may be even preferable to eliminate all metallic signal paths between the widely separated cabinets by the use of by fiber optics, optical-coupling means, or by using wireless communications at infra-red or radio frequency. Advantages of this system are that circuit construction is easier, and that standing wave effects in the ground system at high frequencies are avoided. A disadvantage of a multiple-point ground system is that the system may create multiple low-frequency ground loops that may cause inadvertent common-mode noise in low-frequency circuits.

A third grounding system, not advised by the standard for extended systems to be grounded within a power facility, is a floating ground system: a true floating system is difficult to achieve in practice and will commonly be found only inside otherwise grounded equipment and on physically very small subsystems. An example of its use would be on a printed circuit board within a cabinet, but only on circuits confined to the board's own area. This approach is also used for high performance sensitive measuring instruments, but it is not normally followed in the industrial practice, in particular because of the need of controlling and detecting leakage to earth or nearby systems.

Tray and conduit grounding is also considered with reference to IEEE Std. 665 [378]; in synthesis recommendations are: lubricate the locknuts and fixing parts before they are tightened; grounding locknuts must positively penetrate all paint or other non-conductive finishes; all metallic parts shall be continuous and adequately bonded together.

Concerning the shielding effectiveness of various conductor and cable arrangements (twisted pair, coaxial cable, influence of the terminating impedance) and the grounding of cable shields, the standard offers a concise treatment of shield grounding practices, in particular with respect to grounding at a single point or at both ends, for electrostatic and inductive coupling. The same subject is treated for power distribution cables in the IEEE Std. 575 [376]. The IEEE Std. 142 [363] deals at large with the problem of grounding of power distribution and supply systems for several industrial and commercial applications, considering the behavior in case of fault and thus closing the gap with the National Electric Code prescriptions and rules. At Section 1.15.9 it considers transit power systems as a special case of dc power systems: the running rails are used for the negative return to the substations and typically are not intentionally grounded in order to reduce stray ground current; yet, "several miles of track laid on ground does create a reference of usually low impedance to ground; the electric system designer should be aware of this and the effect of ground currents and of step and touch potentials in an isolated ground system." The problem of the stray currents is only marginally considered and the CENELEC EN 50122 [86] and EN 50162 [98] publications are still the best reference available.

The IEEE Std. 1050 [384] recommends a cable grouping (or classification), that can be implemented as part of a cable management system, through the definition of cable segregation criteria for railway applications (cable segregation is covered in Chap. 8 in this book), and that is focused on power generation applications: the group M stands for medium voltage cables in a single lay, the group P stands for low voltage cables in random lays, the group C stands for control non-shielded cables for digital input/output and control power, and the group L stands for instrumentation shielded cables for both digital and analog input-output, with particularly low voltage levels (thermocouples, RTDs, etc.). The separation distances briefly given here overlap with those more extensively treated by IEEE Std. 518 above, so that a consistency check must always be done between the different recommendations.

6.3.4 Government Standards

Standards are also written by governments for procurement and regulatory purposes. These types of standards are mandatory standards, as they are enforced through legislation and regulation or through contractual requirements for sale to government purchasers.

6.3.5 National Standards

National standards are either written by or adopted by a national standards body. Standards may be written by an organisation accredited by a national standards body, such standards are submitted to the national body for approval as a national standard. National standards may and may not be mandatory, depending upon whether they have or have not been adopted into regulation.

6.3.5.1 American National Standards Institute (ANSI)

The American National Standards Institute, ANSI, is the main standards body in the USA. There are quite a few organisations in the USA engaged in the standards development and standardisation activity; however ANSI is responsible for the accreditation of many of these organisations with the provision that such bodies implement ANSI's processes and procedures. With respect to standardisations in the field of EMC, ANSI accredited standards committee (ASC) C63 is responsible for the co-ordination of EMC standardisation activities in the USA. It has as its scope the remit to develop definitions and methods of measurements of electromagnetic noise and signal strengths - radiated and conducted -, to determine levels of signal strength, levels of unwanted signal, limiting ratio of noise (and/or unwanted sources) to signal and to develop methods of control of, and guideline for, influence, coupling and immunity.

ANSI ASC C63 has the following subcommittees reporting to it:

- SC 1: Measurement Techniques and Developments;
- SC 2: EMC Definitions;
- SC 3: International Standardization;
- SC 5: Immunity;
- SC 6: Test Laboratory Accreditation;
- SC 7: Unlicensed Personal Communication Services;
- SC 8: Medical Device EMC.

Subcommittee 1

Subcommittee 1 (SC 1) is perhaps is the one with the largest EMC activity. It is responsible for the provision of technical expertise and resource necessary for the development and maintenance of EMC standards, as well as those relating to measurements techniques, methods, associated instrumentation and limits. This subcommittee is therefore responsible for the following standards:

- C63.2 Emission Measurement Instrumentation [337]
- C63.4: Emission measurements [338]
- C63.5: Antenna calibration [339]
- C63.7: Open area test site construction [341]
- C63.10: Unlicensed Transmitter measurements [343]
- C63.22: Automated measurements [336]
- C63.23: Measurement Uncertainty [355]
- C63.25: Test Site validation time domain [357]
- C63.26: Licensed transmitter measurements [358]

Subcommittee 2

Subcommittee 2 (SC 2) is responsible for maintaining the definition of terms relating to electromagnetic compatibility up to date. The subcommittee is therefore only responsible for ANSI C63.14 [347].

Subcommittee 3

Subcommittee 3 (SC 3) is a special committee with a direct link to the work of CISPR. In fact it serves as the US technical advisory group (TAG) for the CISPR plenary. As such, it reviews CISPR policy and technical issues that affect US EMC standards interests. The subcommittee reviews key issues and progress in all the CISPR subcommittees, as they might affect the US and forms a US position on to issues. Due to the close association of IEC TC77 with CISPR, discussion of the work of TC77 and its subcommittees is also covered.

Subcommittee 5

This subcommittee is responsible for writing, updating, and changing existing and proposed C63 approved standards for new or existing immunity-testing techniques, associated instrumentation and test-methods development, as requested by the Main Committee ASC C63. This subcommittee is responsible for the following standards:

- C63.9: Office equipment immunity [342]
- C63.15: Immunity Measurement & Instrumentation [348]
- C63.16: ESD Test Methodology [349]
- C63.20: Nuclear Power Plant Immunity [353]
- C63.24: On-Site Generic Immunity testing [356]

Subcommittee 6

Subcommittee 6 is responsible for providing a resource for laboratory competency and assessment activity. It has a number of subgroups/working groups working on many aspects related to laboratories, regulators and accreditation bodies. Representatives of the accreditation bodies, laboratories, and regulators are invited to give a report on the status of EMC test and calibration lab activities. This subcommittee is responsible for the ANSI C63.11 Draft [344].

Subcommittee 7

This subcommittee provides the technical expertise and resources necessary for writing, updating, or changing existing and proposed spectrum etiquettes and related standards, to provide new or amended measurement techniques and associated instrumentation, measurement methods, and operational limits or protocols supporting dynamic spectrum access. This subcommittee is responsible for ANSI C63.17 [350].

Subcommittee 8

Subcommittee 8 is responsible for writing and maintaining existing and proposed C63 standards for medical devices. This subcommittee is responsible for the following standards:

- C63.18: On-Site Medical Immunity testing [351]
- C63.19: EMC for Hearing Aids [352]

6.3.5.2 Federal Communication Commission (FCC))

The FCC regulates interstate and international communications by radio, television, wire, satellite and cable in the USA. It was established by the Communications Act of 1934 and operates as an independent U.S. government agency overseen by Congress. Title 47 of the code of Federal Regulations covers Telecommunication and contains, in five volumes, the intentional and incidental use of the spectrum. The parts relevant to EMC are contained in Chapter 1, Part 15 which covers Radio Frequency Devices, and Part 18 which covers Industrial, Scientific and Medical Equipment, however it should be mentioned that the FCC Part 15 and 18 are not dedicated railway application standards and in general terms are deemed not applicable to the railways..

Part 15. This Part governs the emissions from intentional and unintentional radiators and sets out the regulations, technical specifications, administrative requirements and other conditions required to enable equipment to be placed on the market without an individual license. The requirements of FCC Part 15 are almost aligned with that of EN 55022, particular with respect the configuration and the exercising of the EUT. Part 15 is sub-divided into three subparts as follows:

- Subpart A: Covers general aspects, including definitions
- Subpart B: is concerned with unintentional radiators
- Subpart C: is concerned with intentional radiators; the unintentional emissions are generally limited to Class B for digital devices; this subpart also details the restriction that applies to bandwidth and rules for specific type of equipment such as tunnel radio systems, cable locating equipment to name a few.

Part 18. ISM equipment is defined as being designed to generate and locally use RF energy at frequencies of 9 kHz and above for industrial, scientific, medical, domestic or similar purposes, excluding applications in the field of telecommunications. Typical applications are listed as the production of physical, biological or chemical effects such as heating, mechanical vibration, etc.. ISM equipment may be operated on any frequency above 9 kHz except the frequency bands used for safety, search and rescue.

6.3.5.3 Verband Deutscher Electrotechniker (VDE)

The legal regulations for interference control in Germany are enforced by ZZF. The standards are developed by the Association of German Electrical Engineers, VDE, in co-ordination with Din Deutsches fur Normung, DIN, the Germany Institute for Standardisation. Standards produced by the VDE which are also adopted as National Standards are referenced as DIN-VDE. Standards prefixed "E" are draft standards. With respect railway application standards, the VDE series VDE 0115 or VDE V 0115 mirrors CENELEC or IEC railway application standards.

6.3.5.4 Comitato Elettrotecnico Italiano (CEI)

Even if the modern approach is in the preparation and adoption of harmonized standards at the European level, so that the national standards are only a translation of the original CENELEC text and they represent thus the formal adoption at the national level, there are some cases of national standards that have no counterpart on CENELEC side. An example is represented by the codes for electric installations and power lines and for the calculation of short circuit currents (CEI 11-28), that depend on the installation and grounding practices (CEI 11-37) and the neutral conditions, the classification of cables for their mechanical, electrical, chemical characteristics. The EN 50110-2 addresses the problem giving an overview of the national regulations concerning the operation of low voltage electrical networks.

A case is represented by CEI 11-1, a standard for electric installations with voltage level above 1 kV ac, that has an almost identical counterpart in the NF 013 [586], but with no official CENELEC standard: the common base is represented by a harmonized document HD 637, that never became a standard. The relevance of this standard is represented by the treatment of a horizontal buried through-earth conductor as a grounding element and its resistance to ground. Moreover, this standard introduces the concept of "global earth", that is used by the CEI 9-20/1, treating grounding and safety aspects for "railway d.c. fixed installations facing metallic structures or reinforced concrete" with nominal voltage limited to 1500 V. As a consequence of this limitation the standard covers tramways, metros and light railways, that are mostly excluded by harmonized standards, having given precedence to heavy traction railways and interoperability, and where grounding, segregation of grounding networks and the impact on surrounding installations are particularly relevant, taking into account several factors: the infrastructure and the traction circuit arrangement may be very different depending on economical choices, past or local best practice, etc.; these lines are all located in urban areas, where the expected impact is evident; the supply voltage is dc with not negligible absorbed current levels and stray currents and corrosion represent a major issue. The CENELEC counterpart is represented by the EN 50122-1 and EN 50162 standards and a wise design practice cannot ignore them.

CEI 306-2 is a guide dating back to 2003 on "cabling criteria for telecommunications and multimedia signals distribution in residential buildings" and is evidently overlapping with the scope of EN 50173 and EN 50174 standards, and together with CEI 306-10 (on "Generic cabling systems - Guide to the implementation and the standards") and CEI 306-11 (on "Information technology - Customer premises cabling for wireless access points") cite explicitly the EN 50173 and EN 50174 standards without an explicit harmonization.

CEI 103-1-12 is part of a set of standards related to private telephone exchanges and specifies the tests to apply to ascertain the withstanding voltage capability in relationship to both differential and common mode impulsive voltages and to the supply frequency inductive and conductive coupling. The tests are clearly derived by the ITU-T standards, but without an explicit reference for the definition of test levels and test circuits.

CEI 103-6 and 103-10 are another example of clear derivation from the ITU-T standards, in particular the set of publications under the name CCITT 1989 [417, 418, 419, 420, 421, 422] (see sec. 6.3.2.5).

6.3.6 Regional Standards

Regional organisation such the European Union, can and do develop standards for use within a particular geographic area or region. Member of such standardisation work are national standards bodies representing the interest of the countries in the region. CENELEC is a regional standardisation body for the European Union.

6.3.6.1 European Committee for Electrotechnical Standardisation (CENELEC)

CENELEC, as it is generally referred to, is the European standardisation body responsible for the development and maintenance of European electrical standards. CENELEC was created in 1973 as a result of the merger of two previous European organizations: CENELCOM and CENEL. Nowadays, CENELEC is a non-profit technical organization set up under Belgian law and composed of the National Electrotechnical Committees of 30 European countries. In addition, 9 National Committees from Central and Eastern Europe are participating in CENELEC work with an Affiliate status. Their ultimate goal as affiliates is gaining full membership to CENELEC standardization activities. CENELEC operates under the same internal regulations as CEN (the European Committee for Standardization); therefore, the process of developing European Standards within CENELEC is the same as within CEN.

CENELEC's mission is to prepare voluntary electrical standards that help develop the Single European Market/European Economic Area for electrical and electronic goods and services, removing barriers to trade, creating new markets and cutting compliance costs. CENELEC committees are not generally open to bodies outside of Europe. U S parties may formally request meetings with the chairs of CENELEC technical committees via ANSI's USNC/IEC. ISO and IEC may also nominate representatives to participate in CENELEC committees, when there are related working proceedings at both the European and the international levels.

The process by which CENELEC standards are offered to the International Electrotechnical Commission (IEC) for possible international standardization is established in the Dresden Agreement. This Agreement is similar to the CEN/ISO Vienna Agreement. It comprises a chapter on the common planning of work, whereby each new work item identified at the European level can be offered to IEC, provided the work can be completed in a defined time period, and a chapter on parallel voting procedures, providing for simultaneous publication of identical standards. This Agreement also guarantees that any standard already ratified by CENELEC will be automatically offered to the IEC for possible acceptance at the international level.

CENELEC, like other standardisation bodies, works on the basis of technical committees and sub-committees with specific focus on a technical aspect. With respect to EMC, CLC/TC 210 "EMC" was set up with the role not only to coordinate the EMC standardization activities undertaken by the product committees, but also to set up a system to encompass EMC activities. This led to the known hierarchical structure of basic, generic and product (family) standards, that form the basis of EMC compliance. Through the Dresden Agreement, the EMC generic and product requirements were taken over by IEC, that is now producing the substantial part of EMC standardization. This has resulted into the situation, where IEC Committees prepare most of the work, which is then submitted to parallel IEC/CENELEC procedures on the resulting Committee Draft for Vote (CDV) and the Final Draft International Standard (FDIS), leading to one major principle: to ensure that worldwide standards are transposed into European standards which, in their turn, are used in the legal European framework. With regards to the EMC railway application standards, it is fair to state that the CENELEC led the way, through its technical committee, CLC/TC 9X, in the development of EN 50121 series of standard with the publication of draft standard in 1996.

CLC/TC 9X is the technical body responsible for the standardisation of electrical and electronic systems, equipment and associated software for use in all railway applications, whether on vehicles or fixed installations, including urban transport. It consists of three sub-committees as follows:

- CLC/TC 9XA: Communication, Signalling and Processing Systems;
- CLC/TC 9XB: Rolling Stock;
- CLC/TC 9C: Fixed Installations.

CLC/TC 9X is responsible for the publication and maintenance of a significant number of railway application standards, below is a list of some of them:

- EN 50155: Electronics equipment used on rolling stock [93]
- EN 50261: Mounting of electronic equipment
- EN 50207: Electronics power converters for rolling stock
- EN 50121: Electromagnetic Compatibility [76, 78, 80, 82, 84, 74]
- EN 50126-1: The Specification and Demonstration of Reliability, Availability, Maintainability and Safety (RAMS)
- TR 50126-2: Application guide for safety
- EN 50128: Communication, signalling and processing systems - Software for railway control and protection systems
- EN 50129: Communication, signalling and processing systems - Safety related electronic systems for signalling
- EN 50163: Supply voltage of traction systems [99, 100]
- EN 50124-1: Insulation coordination – Basic requirements – Clearances and creepage distances for all electrical and electronic equipment
- EN 50124-2: Insulation coordination – Overvoltages and related protections [87]
- EN 50125-1: Environmental conditions for equipment – Equipment on board rolling stock [88]
- EN 50125-2: Environmental conditions for equipment – Equipment in fixed installations [89]
- EN 50125-3: Environmental conditions for equipment – Equipment for signalling and communications [90]
- EN 50500: Human exposure to electromagnetic fields [128]
- EN 50463: Energy measurement on board trains [125]

- EN 50159-1: Railway applications – Communication, signalling and processing systems – Safety-related communication in closed transmission systems [94]
- EN 50159-2: Railway applications – Communication, signalling and processing systems – Safety-related communication in open transmission systems [95]
- EN 50238-1: Communication, signalling and processing systems - Compatibility between rolling stock and train detection systems - General (corresponding to the former EN 50238 (2003) [112])
- EN 50238-2: Compatibility with track circuits [114]
- EN 50238-3: Compatibility with axle counters [116]
- CLC/TR 50507: Interference limits of existing track circuits used on European railways [130]
- EN 50239: Communication, signalling and processing systems - Radio remote control system of traction vehicle for freight traffic [117]

6.3.6.2 The European Telecommunication Standards Institute

The European Telecommunications Standards Institute (ETSI) is an independent, non-profit standardisation organisation with particular focus on telecommunications. It was founded in 1988, by the European Conference of Postal and Telecommunications Administration (CEPT) which it self was founded in 1959. Telecommunication systems are complex systems and standards are needed to specify the different interfaces, the necessary characteristics of ancillary equipment, the functional requirements, the target performance. The ETSI standardisation process is implemented through Technical Bodies: these are either Technical Committees, ETSI Project, or ETSI Partnership Project. In addition to these, ETSI has Special Committee and Industry Specification Groups. Each of these bodies may set up Working Group's as required. Due to ETSI's global and European specific outlook, standards prepared by ETSI are categorised as:

- European Standard, telecommunications series (EN): used when the document is intended to meet needs specific to Europe and requires transposition into national standards, or when the drafting of the document is required under an EC/EFTA mandate.
- ETSI Standard (ES): used when the document contains normative requirements and it is necessary to submit the document to the whole ETSI membership for approval.
- ETSI Guide (EG): used when the document contains guidance on handling of technical standardization activities, it is submitted to the whole ETSI membership for approval.
- Special Report (SR): used for various purposes, including giving public availability to information not produced within a technical committee. ETSI SRs are also used for "virtual" documents, e.g. documents that are

dynamically generated by a query to a database via the web. An SR is published by the technical committee in which it was produced.
- ETSI Technical Specification (TS): used when the document contains normative requirements and when short time-to-market, validation and maintenance are essential; it is approved by the technical committee that drafted it.
- ETSI Technical Report (TR): used when the document contains mainly informative elements; it is approved by the technical committee that drafted it.
- ETSI Group Specification (GS): used by Industry Specification Groups according to the decision making procedures defined in the group's Terms of Reference. This deliverable type is approved and adopted by the Industry Specification Group that drafted it.

The dramatic increase of telecommunication protocols and systems in the last two decades explains the massive production of standards and the changes to their organization. In particular, the EMC standards for telecommunication devices have been re-organized into the EN 301 489 series of standards with a total of 35 publications in the series.

6.3.7 International Standards

Voluntary international standards are one of the major categories of standards and standardisation work. International standards are developed by such organisations as the International Organisation of Standardisation, ISO, and the International Electrotechnical Commission, IEC, whose members represent national standards bodies, that participate in the development of international standards. ISO and IEC standards are consensus standards. International standards may be mandatory or voluntary, however within the scope of EMC, international standards are voluntary, although increasingly a growing number of countries are codifying IEC standards, making them mandatory.

6.3.7.1 The International Electrotechnical Commission (IEC)

Under the General Agreement on Tariffs and Trade (GATT) and its successor, the World Trade Organization (WTO), member countries are obliged to adopt international standards for national use, wherever possible. International standards concerning EMC are primarily developed by the International Electrotechnical Commission (IEC) and the International Special Committee on Radio Interference (CISPR). The IEC is perhaps the largest standardisation body in existence, that deals with electrotechnology. Founded in 1906, the IEC through the respective national committees has published some 3000 voluntary standards and has approximately 90 technical committee. The development and maintenance of EMC standards, within the IEC, is the remit of Technical Committee, TC, 77 and its sub-committees (TC 77A, TC 77B and TC 77C) and CISPR.

6.3.7.2 International Special Committee on Radio Interference (CISPR)

Under the General Agreement on Tariffs and Trade (GATT) and its successor, the World Trade Organization (WTO), member countries are obliged to adopt international standards for national use wherever possible. International standards concerning EMC are primarily developed by the International Electrotechnical Commission (IEC) and the International Special Committee on Radio Interference (CISPR). CISPR is responsible for the development of standardisation covering the frequency range 9 kHz to 400 GHz.

The remit of CISPR includes the following:

- Protection of radio reception in the range 9 kHz to 400 GHz from interference caused by operation of electrical or electronic appliances and systems in the electromagnetic environment.
- Measurement instrumentation, facilities, methods and statistical analysis for the measurement of disturbance.
- Limits for radio disturbances caused by electrical or electronic appliances and systems.
- Requirements for the immunity of electrical appliances, multimedia equipment, information technology equipment and sound and television broadcast receiving installations from interference.
- Liaison with IEC Technical Committees that maintain basic standards that apply the prescriptions of methods of measurement of such immunity. Test levels for such immunity tests will be set by CISPR in relevant product standards.
- The consideration jointly with other IEC and ISO committees of the emission and immunity requirements for devices and products where their standards cover EMC requirements which do not match to the respective requirements in CISPR standards.
- Taking into account the impact of safety issues on disturbance suppression and immunity of electrical equipment

6.3.8 Life Cycle of a Standard

The development of a standard begins as a response to an exigence expressed by a proposer, for example a specific company or an industry sector, another committee of the same standardization body or a different body, etc.

The project of a standard is discussed initially by the technical committee (TC) that best fits the foreseen standard scope, possibly identified by the directive committee of the standardization body, or because of direct acquaintance of one of its members with the proposer, or other; the preliminary discussion is needed to understand if the new standard is really needed, if that technical committee can develop it, if it could be of interest for or potentially interfere with the activities of another committee, if there are existing standards that overlap with the proposed topic, and if so, why they have not been adopted, and so on.

Then a working group (WG) is created by simply identifying the interested parties with a poll within the same TC and other potentially involved TCs, by topic affinity. The WG activity is started with a first meeting chaired by the secretary or convenor, who has been appointed in the meanwhile. The development of the activity is rather topic and WG dependent: some or several plenary meetings may be needed to clarify the matter, then the sub-activities may be delegated to volunteers within the WG, deemed expert for the specific sub-topic, even if a frequency activity of reporting and information exchange is needed to avoid overlapping and interference of the various sub-topics, in order to have a well ordered document made of the various contributions. This series of activities may iterate several times, depending on contingency (unavailability of some members, difficulty of some stakeholders in disclosing restricted information, disagreement in interpretations and definitions, and so on) and topic complexity (technical complexity, elements in common with other TCs and standards, and so on), and may end up in a stalemate situation, for which the standard project may be aborted.

If a working draft is developed and agreeable to all involved, the draft standard undergoes the phase of internal voting, that can be enlarged at a later stage to the entire standardization body. The vote is often influenced not only by technical opinions, but on the judgment of the opportunity of certain statements, on their consequences for products and installations, on the chance of need for further standardization to solve new issues. In some cases a standard with a valid technical content, but with arguable prescriptions and testing procedures may be transformed into a "reference guide", so a state of the art statement with no direct mandatory limitations and requirements. A positive outcome brings the draft standard into the status "draft standard" as it is known to the public and it is publicly circulated for voting. Two dates characterize the application of a standard: DOP (Date Of Publication) and DOW (Date Of Withdrawal); the former corresponds to the publication in the OJEC (Official Journal of the European Community), while the latter establishes the "date of cessation of presumption of conformity of the superseded standard", that is the date beyond which any superseded standards must be withdrawn and not applied any longer by users.

An IEC standard life cycle is similar in process to that of CENELEC; as with CENELEC, the IEC standard's life cycle starts with an idea and a work proposal. The idea is then developed into a specification and implemented in products or services. During the use of the standard, new needs arise or flaws are detected in the standard and this leads to a maintenance activity. ISO/IEC Directives, Part 1 [411] entitled the "*Procedures for the technical*

Table 6.5 ISO/IEC Standard Development Project Sequence

Project Stage	Associated document	
	Name	Abbreviation
Preliminary stage	Preliminary Work Item	PWI
Proposal stage	New Work Item Proposal 1)	NP
Preparatory stage	Working Drafts 1)	WD
Committee stage	Committee Draft 1)	CD
Enquiry stage	Enquiry Draft 2)	ISO/DIS; IEC/CDV
Approval stage	Final Draft International Standard 3)	FDIS
Publication stage	International Standard	ISO, IEC or ISO/IEC
1) These stages may be omitted as described in Annex F of ISO/IEC Directive		
2) Draft International Standard in ISO, committee draft for vote		
3) May be omitted		

work", details the ISO/IEC procedure to standard development and maintenance. The IEC standards development process consists of six phases as shown in Table 6.5.

Perhaps the significant difference between CENELEC and IEC standardisation development is in the approval process of a standard. In the IEC, an enquiry draft and approval draft are approved provided a two-thirds majority of the votes cast by permanent members of the technical committee or subcommittee are in favour, and not more than one-quarter of the total number of votes cast are negative. In the assessment of the votes for the enquiry draft, abstentions are excluded from the count as well as the negative votes not accompanied by a technical justification. For a final draft international standard, technical justifications that accompany negative votes are to be taken into consideration by the technical committee. Simply said, the IEC voting principle is based on one member one vote. In CENELEC however, standard approval voting principle is based on the concept of weighted votes. The weighted votes corresponds to the size of the country, larger countries like France, Germany, Italy and the UK have 29 votes each while smaller countries have 3 weighted votes. The requirements for a CENELEC standard approval are at least 71% of the weighted votes cast must be positive and majority of national committee's must support the publication of the said standard.

6.4 Classification of Electromagnetic Phenomena

Electromagnetic phenomena are reviewed and classified herein in order to introduce a hierarchy of the available standards and to better focus on some aspects related the standards and to the peculiarity of a railway application or

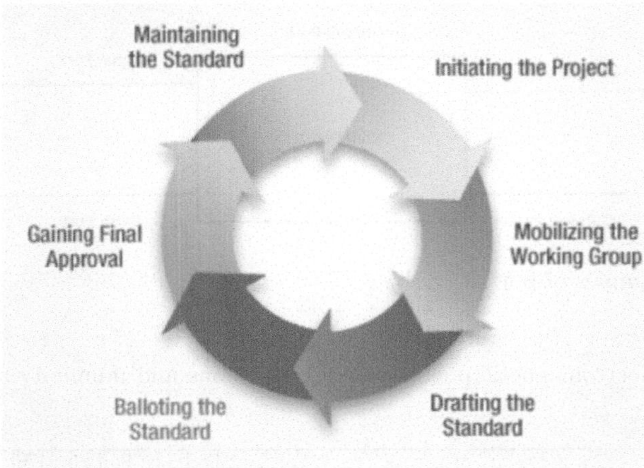

Fig. 6.4 Standards Life Cycle - IEEE

component: applicability, completeness, harmonization, to cite some. Thus, this section is purposely simplified and schematic and the electromagnetic phenomena are briefly reviewed, so that they are fully identified when cited in the following discussion. To do this it is looked forward to the list of the standards that will be developed in the next sections, as a justification and reference point for the chosen classification of the same phenomena.

The basic standards (that will be defined later as a part of the entire standard hierarchy) are those closer to the electromagnetic phenomena and to the definition of their properties and are thus used in this section as the starting point for the classification of the "standard" electromagnetic phenomena. The standards are continuously evolving and may occupy different levels of the hierarchy, with different visibility and treatment of the subject, also due to the maturity of the same standard, its scope and the competences of the standardization body and committee. For this reason we can say that there are various electromagnetic phenomena that are "covered" by standards at different extent, depending on the level of detail and complexity of the same standard, its applicability and purposed scope. Usually, following the classification performed by the standards, the phenomena are considered on a port-by-port basis, where a port is a "particular interface of the specified apparatus with the external electromagnetic environment"; examples of ports are given in Figure 6.5.

Focusing on the electromagnetic phenomena addressed by the IEC/CENELEC harmonized EMC standards (namely the generic standards and the railway applications standards, supported by the basic standards), that however find many equivalents in the ANSI/IEEE standards - to cite another relevant group of standards -, the electromagnetic phenomena are listed below in Table 6.6 per port or per port group (in case of similarities) and briefly described.

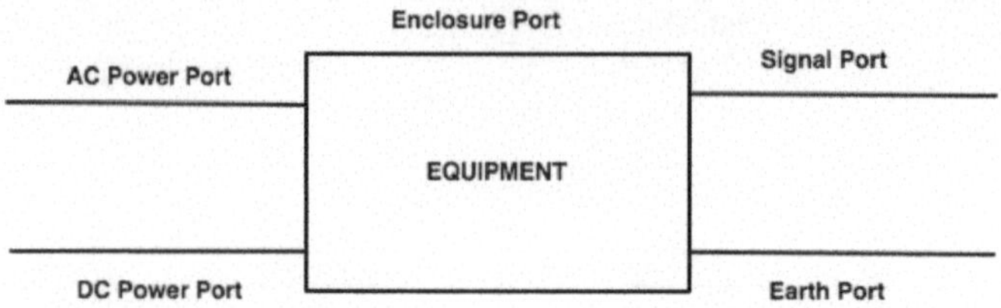

Fig. 6.5 Examples of ports

Table 6.6 Electromagnetic phenomena for emissions and immunity in relation to ports

Electromagnetic phenomenon	Enclosure	AC power in	DC power in	AC power out	DC power out	Telecom	Signal and control	Earth
Harmonic disturbance emissions		•	•	•	•			
Flicker disturbance emissions		•						
Radio frequency conducted emissions		•	•	•	•	•	•	
Radio frequency radiated emissions	•							
Magnetic field emissions	•							
Electric field emissions	•							
Electrical fast transients immunity		•	•	•	•	•	•	•
Surges immunity		•	•	•	•	•	•	•
Magnetic field immunity	•							
Harmonic disturbance immunity		•	•			•	•	
Flicker disturbance immunity		•						
Radio frequency conducted immunity		•	•	•	•	•	•	•
Radio frequency radiated immunity	•							
Electrostatic discharge immunity	•							
Supply voltage variations immunity		•	•					

6.4.1 Common Electromagnetic Phenomena

This section aims at introducing and describing the most relevant electromagnetic phenomena, as a basis for the foregoing discussion and to address the vastity of the problem of defining the rules an equipment is subject to and the EMC interfaces that are worth of being considered for a comprehensive analysis.

6.4.1.1 Harmonic Disturbance

Harmonics disturbance generally derives from equipment with non-linear voltage/current characteristics. The harmonic components are by definition sinusoidal components with a frequency integer multiple of a supply fundamental frequency either in terms of voltage or current. This definition is trivial for an ac supply network where the concept of fundamental is clear and the possible sources of harmonics can be easily figured out; any distorting load or electrical equipment is a source of harmonic components, in the sense that the absorbed current is no longer sinusoidal and harmonic terms appear: transformers with a non-linear magnetizing curve and saturation, static converters, any load under unsteady and transient loading conditions, etc. Non-linear loads represent a significant percentage of the total loads in a railway installation and thus the Total Harmonic Distortion (THD) maybe very high and pose a threat. Harmonics can be distinguished as voltage or current harmonics; for example triple harmonics is a term used to describe harmonics components whose orders are multiples of three. Two additional groups of harmonics are used to distinguish harmonics due to converters systems:

- characteristic harmonics, which occur in the converters current under idealised operation conditions;
- non-characteristic harmonics, occurring under actual operating conditions, whose presence is not related to the converter pulse number.

For the vast majority of non-linear loads the magnitude of the harmonic component decreases with its increasing order. With respect to the relationship between frequency of the analysed waveform components and the fundamental frequency, we can distinguish between harmonics and inter-harmonics. For example, in the case of variable speed drives (VSD), it is commonplace to observe spectrum components that are not strictly located on the frequency axis at integer multiples of the supply frequency, but are rather the combination of two harmonic patterns built around two different supply frequencies. These components are the so called "inter-harmonics", caused by the inter-modulation of two harmonic patterns related to the input supply frequency f_s and motor load stator frequency f_m; the input rectifier and the output inverter each produce its own harmonic pattern, and the output inverter pattern is seen on the internal dc link as modulating the instantaneous dc link current, thus influencing the input harmonic pattern.

Inter-harmonics are voltage or currents with a frequency that is a non-integer multiple of the fundamental frequency. The term sub-harmonics is to indicate inter-harmonics whose frequency is lower than the fundamental. With respect to the harmonics, the inter-harmonics order is given by the ratio between the inter-harmonics frequency and the fundamental frequency. A simplified mathematical representation is shown in Table 6.7.

Table 6.7 Mathematical representation of harmonics

Harmonics	$f = n \times f_0;\ n > 0$
DC	$F = 0\,\mathrm{Hz}$
Inter-harmonics	$f \neq n \times f_0;\ n > 0$
Sub-harmonics	$f > 0\,\mathrm{Hz}$ and $f < f_0$

6.4.1.2 Flicker Disturbance

The term flicker is used to refer to the subjective impression that is experienced by human beings when subjected to changes occurring in the illumination intensity of light sources. One of the reasons for the complexity in the evaluation of this phenomena is the human factor involved in its definition, since it forces one to take into account the characteristics of the physiological process of perception. From an electrical point of view, flicker is caused by rapid voltage variations with a amplitude which is generally much lower than the threshold of immunity for electrical equipment. Voltage variation on the order of only a few tenths of a percent can produce a very significant malaise, especially if the frequency of repetitive deviation is between a few Hz and a ten of Hz (the maximum susceptibility occurring between 8 and 10 Hz). Conventional voltage flicker has traditionally been a concern when connecting large fluctuating loads such as arc furnaces. In the railway environment, flicker may result in an unsafe operation if the susceptible load performs a safety critical function, but in general it may impact only on the quality of performance.

Flicker is basically characterised by two parameters obtained by means of the flicker measurement apparatus (the IEC flickermeter). The two parameters are:

- The parameter P_{st}, which is a measure of short term flicker severity obtained for a 10-minute interval. This basic observation period of 10 minutes is a good compromise that is long enough to prevent too much weight to isolated voltage changes and to allow observation of the persistence of the disturbance. At the same time, it is short enough to characterise the voltage fluctuations produced by equipment with short duty cycle. P_{st} is a dimensionless quantity; a unit of P_{st} corresponds to the threshold of irritability, that is to the severity limit that should not be exceeded in order not to create malaise for an observer to any type of flicker, independently from the source of the disturbance.

- The parameter P_{lt}, is a measure of long-term flicker severity obtained for a two-hour period. P_{lt} is deduced from 12 consecutive values of P_{st}, as given by

$$P_{lt} = \sqrt[3]{\frac{\sum_{i=1}^{12} P_{st,i}^3}{12}} \tag{6.1}$$

The parameter P_{lt} provides a criterion for the long term flicker assessment, when flicker sources with long and variable duty cycles have to be considered or when several disturbing loads operate simultaneously in a random way. The threshold value for P_{lt} is 0.8 units. The standard IEC 61000-4-15 [186] provides the functional and design specification for flicker measurement apparatus, intended to indicate the correct flicker perception level for all practical voltage fluctuation waveforms

6.4.1.3 Supply Voltage Variations (Interruptions, Swells, Sags, Dips)

A voltage dip[3] is a sudden reduction of the voltage at a particular point of an electricity supply system below a specified dip threshold - within a time period no shorter than 10 ms - followed by its recovery after a brief time interval. A voltage dip is in most of the regulations a two-dimensional electromagnetic disturbance, the level of which is determined by both voltage and time. The wave-shape is only in principle sinusoidal. A voltage dip is caused by a fault occurring somewhere on the MV feeders derived from the same HV/MV transformer or further away, for remote faults at the HV level; this implies that the fault impedance may be highly variable in amplitude and reactive part. Moreover, various types of faults may occur, single phase fault to ground, phase-to-phase and three-phase (possibly to ground) faults. Three examples of two phase voltages at LV level are shown in Figure 6.6, as recorded at the LV side: it is evident that at the occurring of a voltage dip the voltage waveform may be highly distorted and not all the phases experience the same voltage reduction.

The *depth of the voltage dip* is defined as the difference between the reference voltage and the residual voltage during the voltage dip, often expressed as a value in volts or as a percentage or per unit value of the reference voltage itself. The *reference voltage*, U_{ref}, is the value specified as the base on which depth, thresholds and other values, which characterise the disturbance, are expressed in per unit terms. The duration of a voltage dip is the time between the instant at which the voltage at a particular point on a supply system falls

[3] The term "dip" is used in IEC publications, whereas the term "sag" is used in North America publications.

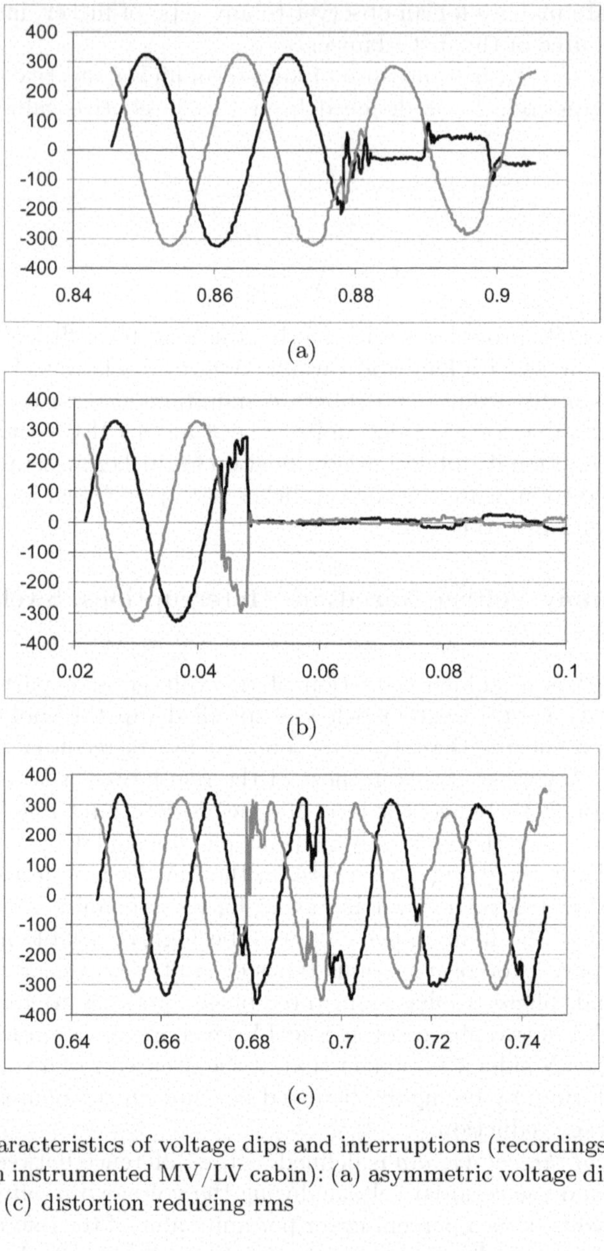

Fig. 6.6 Characteristics of voltage dips and interruptions (recordings of two phase voltages at an instrumented MV/LV cabin): (a) asymmetric voltage dip, (b) voltage interruption, (c) distortion reducing rms

below the dip start threshold and ends when voltages in all phases are equal to or above the dip end threshold,. The *duration of the voltage dip* depends on the specified threshold value.

A *short supply interruption* is a particular case of a voltage dip; it is a reduction of a voltage on all phases at a particular point of an electric supply system below a specified threshold (usually at 5% of the nominal voltage), followed by its restoration after a brief interval of time.

An overvoltage is a change of the instantaneous voltage beyond the threshold corresponding to the maximum steady variation of the supply voltage (usually taken as 10 or 20% of the nominal voltage, depending on the supply characteristics; the EN 50155 standard [92, 93] specifies larger tolerances, up to 25 or 40% for on-board equipment). Long overvoltages of limited amplitude are often referred to as voltage fluctuations or variations, intending a voltage change close to or slightly above the said maximum. Overvoltages are normally intended as transient overvoltages, that is "oscillatory or non-oscillatory overvoltages usually highly damped and with a duration of a few milliseconds or less" (from EN 50160 [97]). Depending on their cause and on the response of the supply network they can assume much different time durations and wave-shapes: lightning induced overvoltages may be very large depending on the level of exposure, and for EMC aspects reference is usually made to the standardized surge levels in CENELEC, IEEE and ITU-T standards; overvoltages coming from the switching of capacitor banks or other reactive loads are also of large amplitude and often characterized by an oscillatory nature, due to the supply network resonance; switching of smaller inductive loads is more frequent and is at the origin of the so called "electrical fast transients", characterized by a lower amplitude and again disciplined by the standardized EFT levels. The effects of overvoltages are various, from the destruction of equipment for energy dissipation or dielectric perforation down to disruption of its operation for less relevant faults of the weakest electronic components or even fuse blow out.

6.4.1.4 Radiofrequency Conducted Phenomena

These phenomena are relevant for both emissions and immunity, even if on a slightly different frequency range, according to standards.

Radiofrequency Conducted Emissions. Conducted emissions are the radio frequency noise present in the physical wiring or traces of an electrical system. This noise is generated by switching operations or harmonic resonances within a circuit. This results in unwanted common mode and differential mode currents within a system. Similarly to radiated emissions, the allowable conducted emissions from electronic devices are controlled by regulatory agencies. If a product passes all radiated emissions regulations but fails a conducted emissions test, the product cannot be legally sold. The primary reason that conducted emissions are regulated is that electromagnetic energy that is coupled to a product's power cord can find its way to the entire power distribution network the product is connected to and uses the larger network to radiate more efficiently than the product could by itself. Other electronic devices can then receive the electromagnetic interference through a radiated path (or, less frequently in case of good design practices, through a direct electrical connection). The frequency range where conducted emissions are regulated is typically lower than the one where radiated emissions are regulated. The longer wavelengths where conducted emissions are a problem need a much larger antenna to radiate and receive electromagnetic interference than the shorter wavelengths required for radiated emissions.

For the specific purpose of making conducted emissions voltage measurements on the mains input of apparatus, EN 55016-1 [135] defines a transducer known as an Artificial Mains Network (AMN)[4]. The AMN has three main purposes:

- to define the RF impedance seen by the EUT's mains port;
- to couple the interference signal from the EUT mains terminals to the measuring instrument in a defined fashion, and to prevent the mains voltage from being directly applied to the measuring instrument;
- to reduce extraneous ambient noise that might be present on the incoming mains circuit.

Several variants of AMN are specified in the EN 55016-1-2 [135], but one in particular (the "$50\,\Omega\,/\,50\,\mu\text{H} + 5\,\Omega$" version) has become established as the norm and is widely available from commercial suppliers. Other impedance stabilising networks are specified for military, aerospace or automotive tests or are defined in draft specifications for signal or control ports.

Radiofrequency Conducted Immunity. The source of disturbance assumed by the standard (EN 61000-4-6 [178]) is basically an electromagnetic field originating from intentional transmitters (and thus with the necessary intensity) acting on the whole length of the cables entering and exiting the equipment. The standard assumes that the equipment size is small with respect to the wavelength of the impinging field; the length of the cables is such that they can resonate in $\lambda/4$ and $\lambda/2$ conditions for the considered frequency range. The frequency range for which immunity tests take place is limited at low frequency by the maximum wavelength for which a significant coupling can occur and at high frequency by the intensity of the induced currents coupled onto and their propagation along the cables. It is underlined that the applied disturbance is a common mode disturbance and a relevant factor to ensure a reproducible and uniform test is the value of the terminating impedance, that the standard fixed at $150\,\Omega$; this value is in reality defined with a tolerance band that is $\pm 20\,\Omega$ for the first frequency interval from 0.15 to 26 MHz and $+60\,\Omega / -45\,\Omega$ for the second frequency interval from 16 to 80 MHz, the maximum applied frequency. The prescribed termination resistance value must be met also considering the combination of coupling and decoupling networks, that are needed to apply the test signal and to keep decoupled any additional equipment (called *auxiliary equipment*), needed to reproduce the correct setup for the equipment under test. The conducted disturbance is applied simulating the effect of the impinging electromagnetic field by means

[4] This is commonly known as the Line Impedance Stabilisation Network.

of various methods: direct injection may be applied if the conductors of the cable are accessible, but the preferred method is that of indirect coupling, obtained by means of the EM clamp or current clamp devices; the latter in fact allows the application of the test signal also onto cable bundles in the final installed configuration, without the need of dismounting the cable assemblies.

The intensity of the test signal is fixed at 1, 3 and 10 Vrms, plus an additional test level left open for future use and that also can be decided prior testing between the manufacturer and the customer.

6.4.1.5 Conducted Impulsive Disturbance (Surges, Electrical Fast Transients)

As an initial standpoint, surges and EFT may be differentiated and broadly classified as slower pulses (with a rise time between 1 and 10 μs and a duration time, or time to half-value, between 50 and 700 μs, depending on the type of test and recalled standard), but with a high energy content, the former, and faster pulses (with rise time and time duration equal to 5 and 50 ns), and thus featuring a much lower energy, even if a much broader bandwidth, the latter. The power bandwidth of the two types of impulsive test signals extends up to a few or a ten of kHz for surges, while it is much larger (around a ten of MHz) for the EFT; analogously the second corner frequency associated to the rise-time varies between 30 and 300 kHz for surges and about 50 MHz for EFT.

The CENELEC basic standards (EN 61000-4-5 [177] and EN 61000-4-4 [176], respectively) define the various types of pulse shapes, the methods of generation and application, and the parameters tolerance. In particular, even if the peak amplitude is of the same order (from 0.5 to 4 kV), the generator output impedance is much lower for surges, that thus apply a much larger current that together with the long time duration justifies the statement concerning the larger energy content.

The origin and the causes of such pulses to occur on power supply networks are to be found in internal and external reasons. The EN 61000-4-5 for surges identifies a broad range of cases subdivided into two classes, that indeed are able to lead to pulses with a widely variable rise and duration times and pulse shapes, and as a consequence bandwidths:

- Power system switching transients, that in turn can be separated into transients associated with
  - a) major power system switching disturbances, such as capacitor bank switching;
  - b) minor local switching activity or load changes in the power distribution system;
  - c) resonating circuits associated with switching devices, such as thyristors;

  - d) various system faults, such as short circuits and arcing faults to the grounding system of the installation;

- Lightning transients: the major mechanisms by which lightning produces surge voltages are the following:
  - a) direct lightning stroke to an external (outdoor) circuit injecting high currents producing voltages by either flowing through ground resistance or flowing through the impedance of the external circuit;
  - b) an indirect lightning stroke (i.e. a stroke between or within clouds or to nearby objects which produces electromagnetic fields) that induces voltages/currents on the conductors outside and/or inside a building;
  - c) lightning ground current flow resulting from nearby direct-to-earth discharges, coupling into the common ground paths of the grounding system of the installation; the rapid change of voltage and flow of current which can occur as a result of the operation of a lightning protection device can induce electromagnetic disturbances into adjacent equipment.

More concisely the EN 61000-4-4 identifies the origin of fast transients as "those originating from switching transients (interruption of inductive loads, relay contact bounce, etc.)" and thus partially overlapping with the phenomena listed above by the surges standard.

6.4.1.6 Radiofrequency Radiated Phenomena

Also in this case these phenomena are relevant for both emissions and immunity, again on slightly different frequency ranges according to standards.

Radiofrequency Radiated Emissions. This part is probably the most complex of those related to the measurement of EMC phenomena concerning emissions. The frequency range of the relevant radiated emissions was fixed by CISPR initially as 30-1000 MHz, as an overlap of the relevance of radiated emissions with respect to conducted ones (thus fixing a minimum lower frequency, decided taking into account the behavior of the power and I/O leads and cables as radiating elements) and of the technical feasibility in terms of reproducibility, repeatability and uncertainty, as well as of performance requirements for the used instrumentation (thus fixing the upper frequency to 1 GHz). There are signs to extend the upper frequency for emissions up to about 3 GHz, taking into account the increase in the clock speed of information technology equipment and data networks, as well as the widespread use of wireless networks operating in the GHz range. This extension has been already done for immunity testing, as it will be seen in the next paragraph, accounting for the presence of many transmitting devices of notable power in the GHz range.

The radiated electromagnetic emissions for the majority of applications are limited to the frequencies above 30 MHz and are represented by electric field emissions in horizontal and vertical polarizations with respect to the ground. Because of the unavoidable reflection of the emissions from the EUT on the floor, that is in general a partially conducting surface, the standardization committees chose to use a perfectly reflecting floor using a ground plane beneath the EUT and accounting for it by searching the point above it where the composition of direct and reflected rays gives the maximum amplitude. This search is done by vertical scanning, moving the antenna from a minimum to a maximum height above ground, and the point of maximum is located differently depending on frequency. This of course is a more complex procedure than a direct measurement and requires that the measuring antenna is mounted on a pole and moved vertically with an automatic mechanism.

This is far too complex and time consuming for the measurement of radiated emissions from the traction line or the rolling stock, and for both fixed height measurements are preferred, also taking into account that the soil surface is much less reflective than an industrial floor.

What is peculiar of the tests on the traction line (EN 50121-2 [75, 76] and 50121-3-1 [77, 78]) is the extension of the radiated emissions measurements down to 9 kHz, using a magnetic loop antenna for the magnetic field component alone. The antenna is placed at a fixed height above ground and is oriented so to have the direction of the measured magnetic field vertical and horizontal with respect to ground and orthogonal to the line.

Typically, such measurements are split to determine the amplitude of the electromagnetic fields emanating from the rail vehicle under all modes of operation and the amplitude of the electromagnetic fields emanating from the signalling and communication equipment to be installed on the vehicles. In particular, for measurements performed to characterise the electromagnetic fields emanating from the rail vehicle, the following factors usually govern the measurement setup:

- typical separation distance between rail systems and potentially sensitive circuits located wayside and in residences and industrial facilities;
- test procedures and limits - some railway agencies, particularly in the USA, specify test procedures and limits that are a combination of military and commercial standards;
- the effects on the rail vehicle emissions levels due to variation in vehicle speed, different modes of operation and infrastructure configuration (location of bonds, location of rail with respect to ground, etc.);
- measurement test equipment - availability, configuration and accuracy.

Being the emissions from rolling stock transient in nature the use of a peak detector is indicated by the standards for the dynamic tests, while the resolution bandwidths are the one normally used by CENELEC compliant tests, that is 200 Hz, 9 kHz and 120 kHz.

While the measurement of radiated emissions from a single equipment can be performed in a shielded semi-anechoic room, thus reducing external disturbance to a suitably low level, when measurements are done outdoor in front of a vehicle or a traction line, great care must be taken to completely describe and evaluate external disturbance. Both steady and transient disturbance may originate from nearby transmitters over the fairly wide frequency range between 9 kHz and 1 GHz, as well as being captured and re-radiated by the traction line itself, acting as a long wire antenna. The transient nature of many disturbing signals requires several measurements before, during and after the tests, so that a complete statistical evaluation can be carried on.

Radiofrequency Radiated Immunity. The straightforward approach in defining the correct frequency range for the generation of a synthetic disturbance to apply is the listing of all transmitting devices of relevant power, that may be thought to operate in the licensed and unlicensed bands: walkie-talkies and VHF radios in the 140-170 MHz band, FM broadcast transmitters in the 87-107 MHz band, TV broadcast transmitters in the upper VHF and UHF band, other portable short range radios operating between VHF and UHF bands, various cellular and mobile telephone systems operating around 860 MHz, 900 MHz, 1800 MHz, 1900 MHz, 2300 MHz (to cite the most known and widespread frequency bands). Other radio sources may be found in the HF band (between a few MHz and a ten of MHz), long wave, medium wave and short wave radios, etc. To these intentional transmitters, other unintentional sources (such as electrical and electronic equipment with strong periodic emissions from internal clock signals, data buses, switching operations) must be added on the frequency axis. The lower bound of the considered frequency interval was set by a consideration on the coupling efficiency of an external electromagnetic field onto the victim equipment cables and leads, and this minimum frequency was set to 80 MHz for commercial products not featuring significantly long cables. Of course, in particular circumstances, where cables entering an equipment are much longer or they are exposed to radio sources of particular intensity, it may be wise to lower this minimum frequency value, partially overlapping with the specified range for the radiofrequency conducted immunity test.

The test is performed with an amplitude modulated signal fed from a RF signal generator to an amplifier and then an antenna. For the normally used antennas (biconical and logperiodic antennas) the electromagnetic field is applied in both polarizations with the antenna aiming at the equipment under test and care must be taken in ensuring field uniformity over a vertical plane orthogonal to the aiming direction; a larger distance between the EUT and the antenna ensures a higher degree of uniformity, but requires a much larger amplification power.

In the last release of the generic standards for immunity of residential and industrial environments [194, 195], as well as in the last releases of the EN 50121 standards [80, 82, 84], the prescribed frequency range is extended to 2500/2700 MHz to include modern telecommunication systems. In brief, the intensity of the applied electromagnetic field is fixed at 1, 3 and 10 V/m for residential and industrial environments; railway application standards added a 20 V/m level for a portion of the frequency range, as well as a 5 V/m level for tests above 1 GHz.

6.4.1.7 Magnetic Field (Steady and Transient)

Relevant magnetic field emissions are expected by all current carrying conductors and by power equipment and apparatus featuring large size transformers, such as transformation cabins, power converters and drives, etc.. All these emissions are expected at the fundamental frequency, that for traction systems is fixed by the EN 50163 standards at 0, 16.7 and 50 Hz; 60 Hz could be added for other non-European countries and in rare legacy cases also 25 and 40 Hz (see sec. 2.1) should be considered. All surrounding equipment are thus subject to a steady magnetic field intensity that is proportional to the flowing current intensity, to the distance, to the conductors geometry, etc.; the EN 50121 standard normalized the amplitude of the magnetic field level for the test of equipment immunity to particularly high levels, that for trackside environment (EN 50121-4) and substation environment (EN 50121-5) was fixed at 100 and 300 A/m for ac and dc systems. In the generic immunity standards these values are lower and equal to 10 and 30 A/m for the residential and industrial environments. A problem of inconsistency is represented by the EN 55024, the product standard for immunity of information technology equipment, that specifies quite improvidently 3 A/m, for a test that is usually not critical even at much higher levels.

The basic standard that describes the test instrumentation and the metrological performances is the EN 61000-4-8 [179]: it is underlined that the generation of a uniform enough magnetic field at dc or supply frequency is relatively easy, by means of solenoids of various geometries and Helmholtz coils [223]; the accuracy and uniformity is normally better than 1% on a cubic space large about 20% the coil side at the center of the two coils. The magnetic field caused by supply currents is considered steady in the sense it is subject normally only to slow variations, due to changes in the load demand.

On the contrary transients may be produced by short circuits and consequent protections triggering and the phenomenon may affect equipment located both trackside and at substation; however the standardization committee judged relevant the phenomenon only for the EN 50121-4 environment and included it into the standard, by means of a dedicated test for transient magnetic field, again at a 300 A/m intensity, based on the basic standard EN 61000-4-9 [180]. In this case the coil used to apply the magnetic field, as well as the equipment used to generate the test signal, needs to be designed very carefully for the required extended frequency response.

6.4.1.8 Electrostatic Discharge

An Electrostatic Discharge, ESD, follows the loss of charge that can occur when a significantly high quantity of charge is accumulated on an object and consequently a high voltage difference is developed with respect to the surrounding environment and objects. For an ESD event to occurs, charge must be generated, accumulated and discharged. A net charge generates an electric field; this electric field, when sufficiently intense, can cause a discharge or breakdown of the medium surrounding the charged object. In electronics devices, as the distance between conductors and devices decreases, the sensitivity of the device to ESD increases. All devices have a maximum voltage threshold and when this threshold is exceeded, either immediate or latent damage occurs. ESD can also be a source of danger to personnel - the energy in an ESD can ignite vapour, gases and liquids.

6.4.2 Other Electromagnetic Phenomena

Some additional electromagnetic phenomena that are not fully covered by the standards (in particular those for certification of products), but likely to be relevant inside a modern transportation system, are briefly considered in the following.

6.4.2.1 Coupled Voltage on Cables

This phenomenon mainly deals with the induced voltage that couples on wayside cables, due to the traction current flowing in the traction circuit, both in steady and transient conditions. Steady conditions cover all the normal conditions of the system supply with acceleration, coasting and braking operations made by rolling stock; typical transients are for example supply short circuits and rolling stock inrush current at pantograph lift. In some exceptional infrastructure configurations (such as degraded rails insulation, grounded rails, faulty cables that become part of the return circuit), also conductive coupling may become relevant.

The problem is so widespread that is the main topic of the CCITT 1989 publications and there different approaches are found:

- one in particular, called often "CCITT simplified method" [418, 419], relates the induced voltage e on the victim cable to the inducing current I and its frequency f, to a mutual inductance term M measured in μH, to the length of parallelism l measured in km and to a screening coefficient k, that hides all the complexity of the traction line arrangement, earthed conductors, autotransformers, booster transformers, specific values of some of system parameters;

$$e = 2\pi f \, I \, l \, k \, M \tag{6.2}$$

$$M = 142.5 + 45.96\,x - 1.413\,x^2 - 198.4\,\ln x \qquad [\mu\text{H/km}] \tag{6.3}$$

- the "exact" method is based on a multiconductor transmission line (MTL) approach with the p.u.l. parameters that are determined with the expressions reported in sec. 7.5.5, 7.5.6, 7.5.7 and 7.5.8, and in some cases may be subject to a relevant variability, due to environmental conditions and worst case analysis.

The choice of the simulation or calculation method is a key issue: it must be well supported in terms of trustfulness and robustness, but it must be also accurate in responding even to little changes of system configurations and electrical parameters, resulting from modifications of system design. Since the amount of countermeasures and design modifications at the end of the study has a direct economic impact, the company supporting the costs has all the interest in selecting a method with the lowest implicit overestimating margins, to implement only really needed countermeasures. On the other side, the technical board supervising verification, validation and safety aspects very often prefers a conservative approach; the results are shown with associated uncertainty and accompanied by worst-case maximum values. From this the need for a simple, trustable and accurate method and the CCITT simplified method is widely recognized in these terms.

The study presented in [527] takes into account the accuracy of the simplified method taking the exact one as the reference, and identifies a set of system elements and properties and their relevance to induced voltage on victim cables:

- the expressions of the approximated method proposed by CCITT are first analyzed for consistency and slight inaccuracies are found that can be easily compensated for by correcting the expressions coefficients;
- the rail-to-earth conductance g_{Re}, the conductance to earth of the through-earth collector (or grounded return conductor, g.r.c.) $g_{re}$and the soil resistivity ρ_e are the most relevant parameters, that influence the sharing of the return current in the return circuit (as already pointed out in [508, 514]) and thus the amount of induced voltage on a trackside victim cable;
- the values of the screening factor k are collected and discussed for different configurations:
  - simplest Rail Return system and Return Conductor system, Booster and Autotransformer (called Negative Feeder) systems; number and separation of ATs;
  - number and separation of Impedance Bonds
  - number of return rails (one or both in a track) and number of tracks;
  - use of shielded cables and use of additional screening earthed conductors;
  - distance a of the victim cable with respect to the vertical projection of the catenary;
  - length of parallelism modifying the screening factor values above and establishing "point exposure" and "total feeder section exposure".

The results shown in [527] are reported synthetically in the following figures (in the cited reference the results are reported and commented extensively). The combined effect of the number of tracks and earth resistivity is shown in Figure 6.7, where the change of the screening factor for extreme values of the earth resistivity (respectively below and above the average value by a factor of ten and twenty) is between ±15% and ±25% for 50 Hz and 3000 Hz.

Rail return		only one rail			both rails		
rho [ohm m]		5	50	1000	5	50	1000
Freq. [Hz]	50	0.49	0.42	0.35	0.38	0.32	0.27
	200	0.525	0.435	0.355	0.42	0.335	0.275
	800	0.575	0.46	0.37	0.485	0.37	0.29
	2000	0.625	0.505	0.40	0.53	0.40	0.31
	3000	0.64	0.515	0.41	0.55	0.42	0.32
Aver. change		122%	ref.	80%	128%	ref.	77%
50 Hz change		117%	ref.	83%	119%	ref.	84%

Fig. 6.7 Screening factor for single and double rail return circuits vs. frequency and earth resistivity ρ_e

Considering an AT system (also called a Negative Feeder system), the amount of induced voltage on a victim cable is proportional to the circuit length (taking as reference length values 500 and 2000 m) while moving the victim circuit along the line, as shown in Figure 6.8. In the cited reference the results are reported in terms of the Normalized Induced Voltage calculated for 1 A of inducing current.

The same comparison is repeated for a line with 1 AT only (as shown in Figure 6.9) and the resulting screening coefficient is approximately 30% smaller, thus corresponding to the values of screening effect normally assigned to AT action.

Since the ratio of the length of the victim circuit and AT separation is relevant to the determination of the equivalent screening factor, the results of the analysis are presented also focusing on the linearity with respect to the ratio of the length of victim circuits.

The effect of ATs has been considered a few times with slightly different results reported in CCITT and other standards [587] and other values obtained by simulation with the exact method: all these values are reported in Figure 6.11 for comparison and give the immediate picture of the spread of suggested values, adopted for such analysis[5].

The values of the Normalized Induced Voltage (that correspond to the screening factor) show a spread of four to six over two decades of frequency, that cover the vast majority of the relevant induction effects, at the fundamental frequency (be it 16.7 Hz or 50 Hz) and its harmonics up to a relevant order.

5 For the meaning of the symbol k_{1b} used in the caption please make reference to [527].

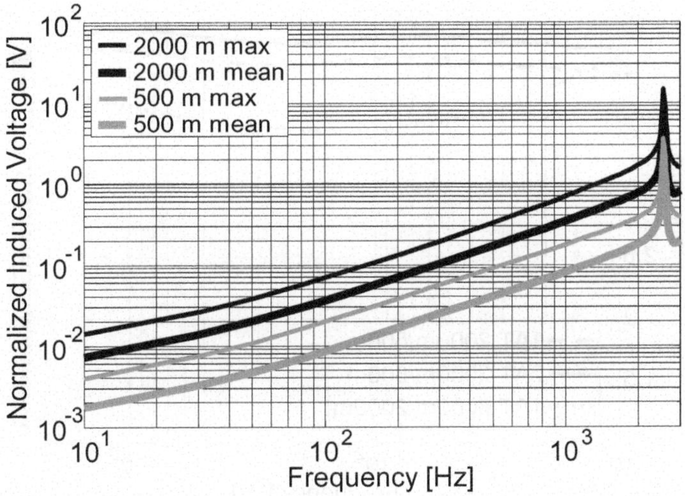

Fig. 6.8 Comparison of 2000 m (black) and 500 m (gray) victim circuits for 2 AT: mean value (thick line), envelope of maxima (thin line)

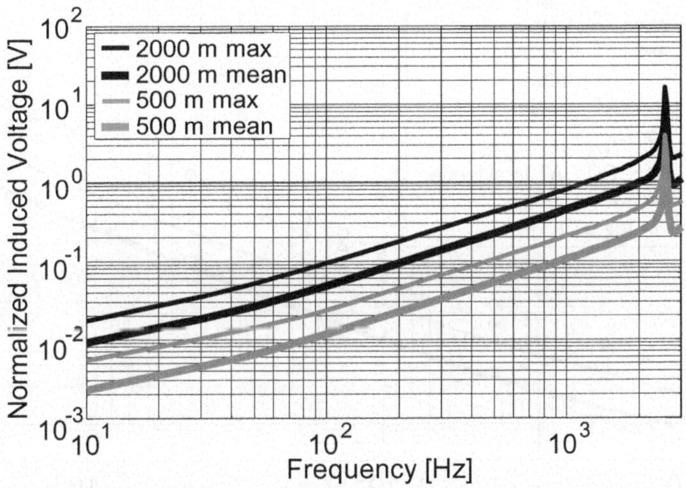

Fig. 6.9 Comparison of 2000 m (black) and 500 m (gray) victim circuits for 1 AT: mean value (thick line), envelope of maxima (thin line)

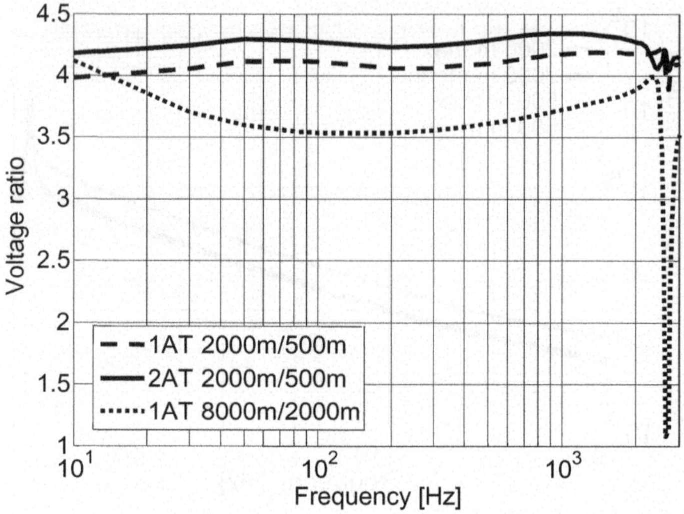

Fig. 6.10 NIV ratio for various victim circuit length ratios and AT number

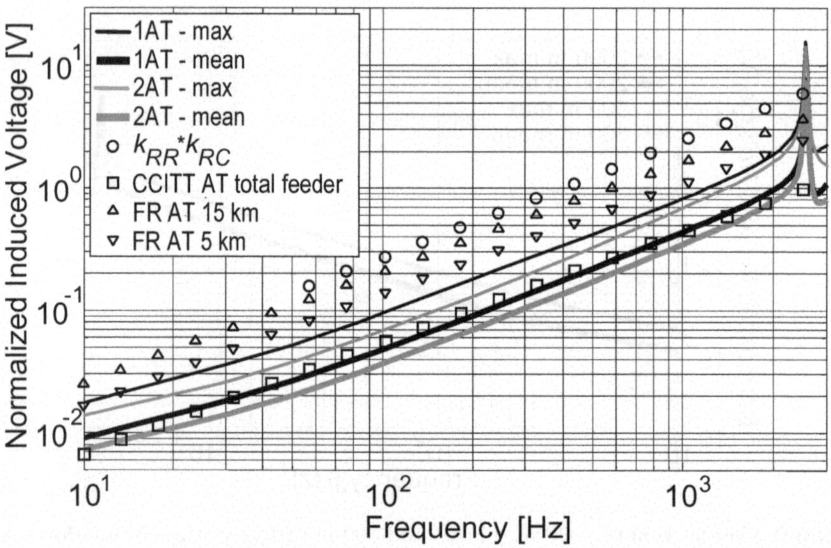

Fig. 6.11 AT and 2 AT for the 2000 m victim circuits: MTL results (mean value (thick line), envelope of maxima (thin line)) and CCITT results ($k_{1b} = k_{RR}k_{RC}$ (circle), French factor for AT every 15 km (upper triangle) and 5 km (lower triangle), CCITT factor for AT (square) for "total feeder length")

In Figure 6.12 the horizontal displacement a between the victim cable and the projection of the catenary is considered: even if the distance between the two is already accounted for in the calculation of the mutual inductance term M, the different horizontal position exerts some influence, leading to some reduction of the induced voltage by about five to ten % doubling and halving the horizontal distance with respect to the reference value of 6 m.

	Separation distance *a* [m]	
Freq. [Hz]	**3**	**12**
50	0.879	0.970
700	0.877 (0.90)	0.941 (0.96)
1100	0.884 (0.89)	0.927 (0.95)
1600	0.906 (0.89)	0.897 (0.95)
2000	0.956 (0.90)	0.844 (0.95)

Fig. 6.12 NIV values for a half (3 m) and double (12 m) of a^* are lower and they are reported as fraction of V* to allow the comparison of CCITT and MTLS values

In conclusion, in [527] the approximation errors and the uncertainty related to the estimation of the screening factors of the CCITT approximate method are reported and commented. The screening factors bear all the complexity of the underlying system, so that several factors exert their influence on and combine into a single screening value: traction power system topology and arrangement, geometrical and electrical parameters, operating conditions, type of victim circuit, quality of earthing. A critical search and analysis of published material was done and advised screening values are evaluated by means of the average value and their spread with respect to reference values calculated with a MTL method: the influence of the soil resistivity is about $\pm 10\%$ for a variation of one order of magnitude around an average reference value; the reduction of the rail-to-earth conductance may produce a significant increase of the induced voltage up to 300%, depending on the position of the victim cable; the number of ATs has a benign effect on the voltage reduction (about 30%). Experimental data confirm that the screening coefficients are overestimated by a factor of two to three, except for the lowest AT screening coefficient defined by CCITT "in total feeder section".

6.4.2.2 Electromagnetic Field in the Reactive Region

The radiation from a source transits three regions as shown in Figure . The transitions between these regions are not distinct and changes between them are gradual. The reactive near-field region is the region close to the source and up to about one wavelength (1λ) away from any radiating surface. In the reactive region, the energy decays very rapidly with distance. In the radiating near-field region, the average energy density remains fairly constant at different distances from the antenna, although there are localized energy fluctuations.

When sources of electromagnetic field, such as radio transmitters, are located at close distance from victims, the estimated E-field intensity by means of far field approximations (see sec. 7.2 and 7.3) can be much different from the real E-field intensity in some portions of the space around the transmitter, since the electromagnetic field is in its reactive region. The larger the reactive field region the lower the operating frequency, so the attention is for the HF and VHF frequency intervals. Various fixed and portable transmitters operate in the amateur radio band (140-170 MHz), TETRA band (380-420 MHz), 433 MHz band, etc., with a reactive field region that extends up to about 0.5 m around the antenna. When located on-board or in narrow spaces, a practical factor of 6 dB may be adopted with respect the far field estimation, in order to get an estimate of the real level of exposure; at lower frequencies, such as in the HF range, the antenna has normally a much larger size and a simple approximation is not possible. The practical factor is based on a Hertzian dipole approximation in the VHF range and assuming that in a small portion of space due to a particularly unlucky phase relationship the reactive terms are summing in phase to the far field one (see sec. 7.3.2 and 7.3.5).

A more accurate analysis can be made by either solving the Hertzian dipole equations for elementary sources representing the real sources, or by calculating the radiation pattern of an equivalent representation of the real sources based on calculable wire antennas (such as dipoles, long wire antennas, loops, all considered in sec. 7.3.3, 7.3.4, 7.3.7, 7.3.6), or last by using finite element methods, that can treat problems of any complexity, provided that the available simulator and computational capabilities are adequate.

6.5 Electromagnetic Environments

When thinking of a complete railway system the main elements that may be selected as the characterizing ones are the vehicle, the line and the supply substations. These elements are characterized by a range of operations, they feature different electromagnetic phenomena - or better they feature different intensities for various electromagnetic phenomena -, and may be subject to different coupling paths, if the distances and the presence of various radiating and conductive elements is considered. The overall picture is even more complex if other equipment and subsystems are considered, such as telecommunication, heating, air conditioning and ventilation, passenger announcements and information, alarm system, lighting, escalators and elevators, ticketing and fare collection, etc. It is evident that several parts of these subsystems are located far from the three main elements above, and thus experience more benign environments, that are classified as residential and industrial environments, for compatibility with the already existing prescriptions provided by the applicable generic standards .

It is observed that classification and environments identification are conceived in order to put in place a significant but sensible structure, to put order in the conformance assessment and certification activity and are the result of a tradeoff between the use of existing standards and environments (to maximize the re-use of test reports and certificates) and the faithful reproduction of electromagnetic phenomena (without underestimating real threats). For the effective management of the electromagnetic environment within the railway, the principles of electromagnetic zoning is applied, as illustrated in Figure 6.13. The railway boundary is defined as an area up to 10 m from the nearest running rail and extends 3 m beyond a sub-station boundary. The area is generally characterised as being equivalent to a heavy industrial environment. Equipment installed within this boundary falls within the remit of IEC 62236, with some specific requirements placed on equipments installed within the 3 m boundary from the centre of the nearest running rail. IEC 62236, does not specify requirements for in-band interference for signalling equipment and these are generally specified in company standards (see sec. 6.3.1). The EN 50328 [112] is an attempt to harmonise such for common signalling assets within Europe.

Beyond the railway boundary, the electromagnetic environment is generally classified as light industrial/commercial environment and thus the requirements IEC/EN 61000-6-1 and IEC/EN 61000-6-3 applies or any applicable product standard. Depending on the size of the infrastructure, the transition from heavy industrial to light industrial electromagnetic environment occurs gradually and the resulting interface zone - or buffer zone - is generally deemed to be that of a heavy industrial environment to account for the worst case electromagnetic threat. In this case, IEC 62236 equally applies to the buffer zone.

For those equipment and components that are located within the three electromagnetic environments typical of a railway application, there is the exigency of harmonizing the prescribed tests and the known e.m. threats with those accompanying them. In this case, the supplier must give an articulated description of the assigned e.m. environments for each equipment, the expected coupling paths (if relevant) and the prescribed tests; this is done in three phases that are treated in this and successive chapters by means of electromagnetic environments description, electromagnetic phenomena and interference analysis, and standards and tests assignment, almost automatic once the e.m. environments are selected. For these reasons the description of the electromagnetic environments is necessarily carried out in close agreement with the tests specified in the applicable standards and - where possible - with the environment standards, that for IEC/CENELEC are grouped under the family IEC/EN 61000-2-X. Yet, for each environment some hints are given of e.m. phenomena that are not covered by the same EMC standards and that are addressed instead by lightning, earthing and corrosion standards, in order to get a more complete description and to prepare the ground for the

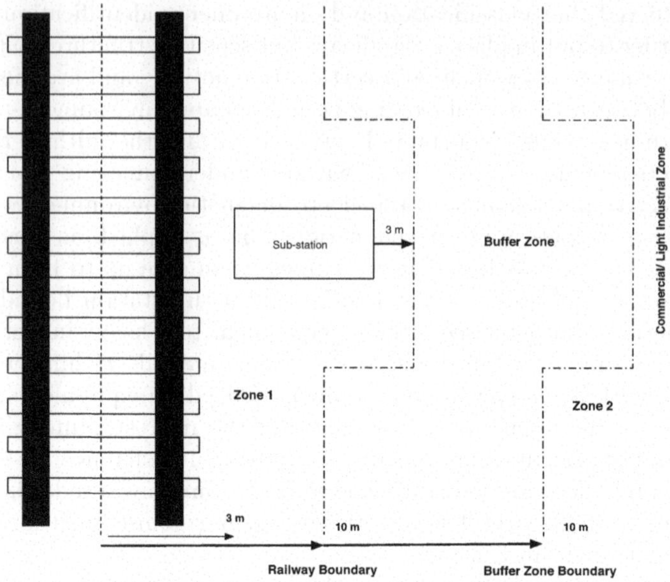

Fig. 6.13 Railway Electromagnetic Zones

cross-fertilization and to the harmonization of requirements, that are very common to complex systems and in particular to EMC matter.

6.5.1 Train or on Board e.m. Environment

The Train environment (as per EN 50121-3-2 [80]) refers to all the equipment installed on board, that may include signalling devices, communication apparatus, information technology equipment, etc.. It is characterized by a strong magnetic field at the traction frequency, produced by all the traction system apparatus and its connections (main transformers, filters, traction converters and motors, braking choppers, circuit breakers, etc.), and by a significant transient magnetic field in case, for example, of circuit breaker tripping. For the presence of several switching devices and radio frequency communication systems in close proximity, the expected field intensity over a wide frequency range is high; it is easy to see that over the entire MF and HF bands, ranging from tens of kHz to some or a ten of MHz, the electromagnetic emissions from static converters and other switching apparatus are in the reactive field region, so that magnetic and electric field components are both relevant and the direction of vectors as well as their ratio (the wave impedance) reduce the efficacy of electromagnetic shielding solutions. The presence of many capacitor banks and filters on the various supply lines makes the occurrence of surges (such as those produced by lightning phenomena) unlikely, but this doesn't mean that transient voltages may be excluded, since less intense transients, possibly accompanied by ringing and oscillations at various frequencies, may be put in relationship with the switching of capacitors, filters and reactive loads.

It is however possible that a specific equipment with a reduced number of e.m. interfaces and located in a particularly quiet portion of the on board environment experiences a more benign environment, at least for a restricted set of e.m. phenomena; in this case the use of less severe tests during product testing and certification could be allowed, provided that the real e.m. environment conditions are demonstrated. If the manufacturer wants to apply the tests of a more benign environment, for example I environment, on a portion of the on board environment, he focuses on the demonstration that the I environment is really isolated and protected, considering those elements that may represent an electromagnetic interface between the two: cable connections, shared power supply lines, insufficient shielding and so on.

6.5.2 Line and Track Electromagnetic Environment

The Line and track environment (as per EN 50121-4 [82]) refers to the equipment installed at a short distance from the line (set by the standard at 3 m from the track axis) or with a specific e.m. connection to the track and line, even at larger distance (cables connected to remote equipment that run in parallel with the line and track within the prescribed 3 m distance for a length longer than 30 m); examples are signalling devices and their cable connection to relays and electronic apparatus installed in protected rooms, but in connection, in terms of coupling path, to the L environment. A different approach is to be followed if the victim cable is in direct connection with the rails, since the overvoltages appearing in case of a lightning stroke hits the rails (directly or in close proximity) are well above the prescribed surge test levels of the L environment (the most severe environment anyway); special countermeasures must be set up by the manufacturer or installer to bring the residual overvoltage to a safe level, compatible with the prescribed EMC tests, thus avoiding an unnecessary increase of the testing severity for the connected equipment.

6.5.3 Substation Electromagnetic Environment

The Substation environment (as per EN 50121-5 [84]) refers to the equipment installed at fixed power supply installations, namely the electric substations and other few technical rooms; examples are programmable logical counters and terminal units located inside such environment or directly connected to sensors located in the same environment.

6.5.4 Residential and Industrial Electromagnetic Environment

The Residential and the Industrial environments contain and address all the equipment already CE marked by the manufacturer for other final applications and that may be transferred to the present one, given that they are not used and installed in critical zones, that fall under the other railway environments; the R environment is for office and protected environments, while the I environment covers all other more severe "industrial like" environments, not influenced by the vehicle, the substation or the line and track at close distance, such as escalators and elevators, power distribution facilities, etc.

The attention is focused on CE marked equipment, that often corresponds to COTS (Commercial Off-The-Shelf) equipment, for which an extended test report is not always or readily available, and only the compliance to either the Residential or Industrial environment is available, as stated in the respective Declarations of Conformity. With the ever more extensive use of COTS equipment for several applications (network connections and use of various kinds of information technology and communication equipment for access control, fare tolling, SCADA, environmental monitoring, passenger information and announcement, communication monitoring and support, etc.), this aspect is receiving increasing attention, since the DoC may refer to product standards that are not fully overlapping with the generic standards for Residential and Industrial environments. Such product standards have been considered extensively in various sections of this chapter (sec. 6.2.3.2, 6.2.3.1, 6.2.3.6, 6.3.6.2) and are here only briefly recalled: ITE standard for immunity (EN 55024), audio products standard (EN 55103), several ETSI standards for radio LANs, short range devices, radio equipment; for all these standards one or few immunity tests are missing or they are specified at a lower level with respect to the Residential Environment standard (EN 61000-6-1), or they may be subject to a slightly different specification, such as in the case of the application of exclusion bands for equipment covered by the ETSI standards. It is thus quite common the problem of addressing and demonstrating the equivalence between certified products with respect to their own product standards and the target environment, that for a railway application, following the EN 50121 standards, cannot be any different from the one prescribed by EN 61000-6-1 and EN 61000-6-3 for immunity in R and I environments respectively. In this case, the analysis shall cover the peculiar characteristics of the target environment and if the missing or less demanding immunity tests may be considered as a minor or non-existing problem. If the target environment is in principle an Industrial Environment, then several test are not adequately specified and an integration of immunity tests, as well as emission tests (going from Class A limits to Class B limits), is needed, supporting any analysis by additional tests. It is however possible that the Industrial Environment has been assigned with a conservative approach, in the absence of any clear indication of the real target electromagnetic environment; in this case, it cannot be excluded that a portion of the target environment, far from significant noise sources of conducted and radiated electromagnetic disturbance, may be re-classified as a Residential Environment or as a Light Industry Environment, with a few additional tests to address those phenomena, for which the analysis and the facts couldn't exclude particularly intense electromagnetic threats.

6.5.5 Technical Attachments to Laws and Regulations

This case is rather articulated in the sense that there are various approaches to the technical standardization in different countries and federations of countries. For example, in many countries, for the prescriptions concerning electrical safety and electric power systems design, there were issued laws containing minimum requirements for safety of humans and equipment: minimum conductor sizing, maximum earthing resistance, prohibition for toxic or flammable materials, and so on.

For EMC an example is represented by the EU Directives 95/54 [278], and successive 2004/104 [279], on motor vehicles, that specify limits and test procedures by themselves, with reference also to IEC standards (CISPR 12 and CISPR 25) and product standards (ISO 11451 and ISO 11452). These Directives specify both emission and immunity tests with respect to the radiated electromagnetic field only. The prescriptions for the site where the emission measurement shall be performed are particularly demanding for what concerns the area clear of obstacles and reflecting objects around the vehicle under test and the measuring antenna (the 95/54 Directive specifies a 30 m radius, reduced to 15 m in the 2004/104 Directive; compared to the specified minimum margin of 6 dB of any external disturbance with respect to the limits, even 15 m of clearance is too conservative a distance). For the immunity tests a 30 V/m reference level of the applied E-field intensity is specified, allowing lower levels at some specific critical frequencies that may occur when testing on-site, but never below 20 V/m (that represents the minimum E-field intensity) and for a number of frequency values less than 10% of the total frequency points.

6.6 Railway Application EMC Standards

6.6.1 Railway Applications: EN 50121[6]

As already pointed out, the EN 50121 family of standards defines and addresses an application environment, that of railway applications. The family

[6] In this book these standards are simply indicated as EN, even if there exist an IEC counterpart with a different code (IEC 62236). The European harmonised standard EN 50121 series is equivalent to the IEC 62236 series.

of standards is split into a part that deals with the emissions towards the external environment (named the "outside world") and those regulating the emissions and immunity for the various e.m. environments. The enforcement of EN 50121, in Europe was determined not by means of the EMC Directive (as for the other EMC standards), but by citing it in the Interoperability Directives, also called TSI (Technical Specifications for Interoperability).

6.6.1.1 EN 50121-2: Emission to the Outside Environment

The EN 50121-2 [76] addresses the radiated electromagnetic emissions between 9 kHz and 1 GHz from the whole railway line in full operating conditions, such as during trial runs after commissioning of the various subsystems. It distinguishes between emissions from the traction line and from the electric substation; in both cases the disturbance are mainly produced by the circulating rolling stock, for which operating conditions and speed values are specified, but the attention is on the separation of contributions, so that it is required that vehicles are far enough from the measurement location. Both measurements are performed with the same set of antennas: loop antenna for the lower part of the frequency interval, 9 kHz - 30 MHz, where magnetic field component is considered as the most relevant, and biconical and log periodic antennas, for the rest of the frequency interval, 30 - 1000 MHz, where the electric field is measured, being in almost far field conditions. The measurement methods are however different:

- the traction line emissions, because of the intermittent and dynamic nature of the electric arc during current collection at the pantograph or sliding shoe on the moving vehicle, are measured in peak mode, by sweeping the prescribed frequency intervals; the reason for the deviation from the quasi-peak method, as prescribed among others by the EN 55016-1-1 [134], is that "the method of EN 55016-1 is only designed to protect radio communication from interference and takes no account of electronic safety systems, such as those used beside the railway track or at airports, where short-time transients may cause interference" and this viewpoint characterizes also military applications and related standards, such as MIL STD 461 [548];
- the substation emissions are performed in quasi-peak mode, with no strict requirements on the substation load, prescribed above 15 or 30%, depending on the type of test; the required time to perform the tests would be very long, if they were not limited to a measurement only in the maximum orientation.

The statement on the required accuracy is rather peculiar: "the accuracy of the measuring equipment for the radio frequency tests shall not differ more than ± 4.0 dB from the requirements of EN 55016-1-1". So, it seems that the stated margin of 4 dB must be summed on top of the limits declared by the cited EN 55016-1-1 [134], without following a modern statement of accuracy based on uncertainty. Even if the discussion of modern uncertainty and error concept [414] is beyond the scope of this book, it is evident that the statistical evaluation by Type A or Type B approaches leads to the determination of the dispersion of an assumed normal distribution for the variable of interest, that characterizes the entire measurement chain, and this is not compatible with a more classical hard limiting of the "error" or deviation, that implicitly corresponds to an assumed worst case uniform probability distribution. When performing measurements of electromagnetic emissions from large systems exposed to a wide range of external sources of electromagnetic field, the standard procedures and techniques [414, 143, 355] must be integrated by a series of prescriptions and precautions, as discussed in [517].

6.6.1.2 EN 50121-3-1: Emissions from Rolling Stock

The radiated electromagnetic emissions from the rolling stock are again measured between 9 kHz and 1 GHz, with a measuring equipment similar to the one considered in the EN 50121-2. Two types of tests are specified, a stationary test (referred to as static testing) and a slow moving test (referred to as dynamic testing), with the prescribed speed that depends on the type of vehicle (urban with 20+/-5 km/h and main line vehicles 50+/-10 km/h). Measurements in the dynamic conditions are to be taken while the vehicle is passing the antenna, accelerating or decelerating with approximately one-third of its maximum tractive effort within the given speed range. This implies that a long enough track is available for starting from standstill accelerating to the rated speed, passing in front of the measurement location, and decelerating securely until a complete stop; for the next measurement the vehicle may turn back and repeat the complete maneuver facing the measurement location with the other side (in the standard there is no prescription concerning the selection of the vehicle side). Attention is given to the influence of the infrastructure: the line (or third rail) should be "infinite"; any discontinuity should be avoided, as well as substations, transformers, neutral sections, etc.; the distance from external potentially reflecting objects, such as trees, walls, bridges, tunnels or vehicles, shall be 30 m for main line vehicles and 10 m for urban vehicles; the measurement location shall be placed midway between masts or suspension poles, external to the line on the side of the track used for measurements (this last factor has impact on the way the vehicle is turned back at each run, since the same track must be used).

Another measurement prescribed by the EN 50121-3-1 regards the so called "psophometric current", a frequency weighted evaluation of the traction current, representing the interference to analog telephone lines (see also sec. 3.2.2.1); for this reason the frequency weighting is applied with a curve, named psophometric weight table or vector, that are specified in the ITU-T O.41 publication [455]. Other psophometric weighting tables exist (and the EN 50121-3-1 opens to this possibility in sec. 6.2.2), such as the one in the Network Rail standard 50016 [578], that extends from 16.7 Hz to 6000 Hz, with an additional curve and procedure for higher frequency testing. Concerning the psophometric current limit the EN 50121-3-1 doesn't specify any value and it is suggested that an agreement is reached between the stakeholders or reference to any applicable standard is made. Moreover, since the EN 50121-3-1 itself at sec. 6.2.2 states that "the relationship between the current absorbed or generated by the traction vehicle and the noise in the telephone line is neither under the total control of the vehicle manufacturer nor of the operator of the network", it is evident that measuring the psophometric current of a vehicle is a choice; other possibilities exist, such as measuring the interference noise voltage and its psophometric value on a sample trackside victim cable of adequate length and termination, as specified by [578].

In addition, it is underlined that the specifications on pshophometric current test are related to the measurement of a single unit (I_{pso}), or vehicle; to extend the measurement results to multiple units (in number of n) within a train consist ($I_{pso,tr}$), rules for the composition of the single emission terms are reported.

- For dc railways the traction line supply is normally derived by diode rectifiers from a three phase supply and for this reason, even if filters are applied, the ripple of the rectifier output contributes considerably to the psophometric current for the vehicles in the supply section, since the vehicles at the supply characteristic harmonics may be considered almost passive loads. To simplify the standard distinguishes between
  - dc systems with dominating rectifier ripple (vehicles with camshaft control; vehicles with chopper or inverter control, substation with 6-pulse rectifier without filtering): $I_{pso,tr} = n \times I_{pso}$
  - dc systems with converters on the vehicle and low rectifier ripple: $I_{pso,tr} < I_{pso}$ for choppers operating in interlaced mode, $I_{pso,tr} = \sqrt{n} \times I_{pso}$ for choppers operating without synchronisation or for inverters directly connected to the power supply.
- For ac systems the psophometric current generated by vehicles in the supply section depends mainly on the type of converter used on board the vehicle.
  - ac systems with phase controlled converters $I_{pso,tr} = \sqrt{n} \times I_{pso}$: this seems to be based on a statistical mix of vehicle types, speeds and actual current consumption, but recent experience with high power trainsets shows that this $\sqrt{n}$-law is not applicable in the case of equal speeds, equal power and equal vehicle types, when $I_{pso,tr} = n \times I_{pso}$ applies;
  - ac systems with 4 quadrant converters (4QC, pulse width modulated line converter): $I_{pso,tr} < I_{pso}$ if 4QC operate in interlaced mode (normal operating condition) $I_{pso,tr} = n \times I_{pso}$ if n equal units operate in non interlaced mode.

The matter of the correct summation rule for conducted emissions originating from different sources (i.e. different vehicles) is considered in sec. 2.2.5.2 and in the new draft of the EN 50238-2 [115]. In the latter three curves are indicated for the summation of conducted emissions from different sources: on one side a linear rule for sources "syncrhonized to a common reference", at the other end a square root rule for sources "completely uncorrelated", and an intermediate third curve for sources "synchronized to independent clocks", that are however loosely related to justify a weak correlation (the curve was determined by Monte Carlo simulation).

6.6.1.3 EN 50121-3-2: e.m. Environment for on Board Equipment

A specific standard for the on board equipment is needed, since the on board e.m. environment is rather peculiar, characterized by reduced clearance and short separation between installed equipment, by some particularly intense e.m. phenomena (magnetic field caused by traction currents, e.m. field in near field region produced by static converters and radio equipment, possible reflections and resonances both for conducted and for radiated phenomena) and by constraints in cable routing (due to space restrictions and cable ways often decided by the vehicle manufacturer, rather than the suppliers of third party equipment). The immunity tests are particularly severe with one exception:

- the radio frequency e.m. field immunity test on the enclosure port is prescribed at 20 V/m for the usual 30 - 1000 MHz frequency interval, and the additional tests appeared in the 2006 version are at the maximum level among the other EN 50121 standards and generic standards;
- the electrical fast transients immunity test is performed at ± 2 kV on several ports, dc and ac auxiliary ports, and signal, measurement and process control ports; the origin of intense fast transients on board is normally due to the use of ac or dc contactors for control and regulation of inductive loads, such as compressors and fans, for air conditioning or cooling, without adopting the necessary countermeasures, such as snubber circuits on the power electric contacts to extinguish any electric arc during turn off;
- magnetic field immunity on the contrary doesn't appear in the list of prescribed tests, even if it has been already considered and applied in the substation e.m. environment and past experience confirms that stray magnetic field at the traction frequency and its main harmonics may be relevant, in particular as a source of picked up noise for long cable lines.

Tests on the earth port are not specified in this standard, even if the generic industrial standard EN 61000-6-2, issued in the same year, specifies EFT and RF conducted disturbance tests. Even if it is not possible to put the two environments in close relationship, so that the tests specified in one standard can be readily transferred into the other, it is very common that interference on-board occurs with the chassis, the enclosure and supposed earthed "clean" cables as the propagation mean.

The EN 50121-3-2 is made mandatory also by the fact that it is recalled for the definition of EMC conditions by many product standards, such as EN 60077-1 "Railway applications - Electric equipment for rolling stock - Part 1: General service conditions and general rules", and by the close relationship with the EN 50155 [92, 93], that specifies a series of "environmental" tests and conditions for electronic equipment on-board (see sec. 6.2.3.11), including peculiar electromagnetic phenomena, now in agreement with EN 50121-3-2, but giving some inconsistencies at the beginning (see sec. 6.2.3.11).

6.6.1.4 EN 50121-4: e.m. Environment for Signalling and Telecommunications Equipment

The scope of the EN 50121-4 [82] is defined above as it appears in the title of the same standard. However, this standard is often referred to as the one for "trackside equipment", in contrast to the equipment located on-board and covered by the EN 50121-3-2; the reason for this identification of the standard with the trackside environment is to be found in the Note 1 of Table 1 and 2 of the same standard that says "the tests given apply to apparatus inside 3 m", but it is extremely important to underline that this note is first, not recalled for all tests of Table 1, second, accompanied by other considerations on the length of exposed cables connected to a given port for Table 2, and, third, does not appear in the other successive tables. The complete Note 1 for Table 1 reports also "zone and vital equipment such as interlocking or command and control which are mounted in areas where a high risk of interference from mobile radio telephones has been identified. For other apparatus within the railway environment, requirements of IEC/EN 61000-6-2 apply." This sentence has the effect of suggesting an overall classification as Industrial Environment of all locations and ports that do not fall under Note 1 scope; yet, mentioning "mobile radio telephones" indicates that Note 1, even if recalled by the tests on both radio frequency and magnetic field immunity tests, is in reality applicable only to the former. By comparison of the test levels and frequency intervals of the radio frequency electromagnetic field immunity tests appearing in EN 50121-4 and IEC/EN 61000-6-2 [195] it is observed that: the frequency intervals do not fully correspond, but partially overlap; the 800 - 1000 MHz interval in the EN 50121-4 is not present in the IEC/EN 61000-6-2; the test levels are higher for the EN 50121-4. It is thus highly recommended that any signalling and telecommunication equipment that may be subject either to EN 50121-4 or IEC/EN 61000-6-2 be in reality fully tested by the superposition of the two prescriptions, as shown in the following table.

Fast transients and surges are the most critical tests, and the interpretation for an industrial environment is that the most exposed port is the one of the ac power supply, that is in fact tested at the same levels specified by the

Table 6.8 Comparison of prescribed immunity tests and related levels as per EN 51021-4 and IEC/EN 61000-6-2

	Test description	EN 51021-4	IEC/EN 61000-6-2
Signal I/O ports	RF conducted immunity 0.15 - 80 MHz	10 Vrms	10 Vrms
	Fast transients 5/50 ns, 5 kHz rep. freq.	±2 kV, capacitive clamp	±1 kV, capacitive clamp
	Surges 1.2/50 μs	±2 kV, line-to-earth ±1 kV, line-to-line	±1 kV, line-to-earth
DC power ports	RF conducted immunity 0.15 - 80 MHz	10 Vrms	10 Vrms
	Fast transients 5/50 ns, 5 kHz rep. freq.	±2 kV, capacitive clamp	±2 kV, capacitive clamp
	Surges 1.2/50 μs	±2 kV, line-to-earth ±1 kV, line-to-line	±0.5 kV, line-to-earth ±0.5 kV, line-to-line
AC power ports	RF conducted immunity 0.15 - 80 MHz	10 Vrms	10 Vrms
	Fast transients 5/50 ns, 5 kHz rep. freq.	±2 kV, capacitive clamp	±2 kV, capacitive clamp
	Surges 1.2/50 μs	±2 kV, line-to-earth ±1 kV, line-to-line	±2 kV, line-to-earth ±1 kV, line-to-line
Earth port	RF conducted immunity 0.15 - 80 MHz	10 Vrms	no test specified
	Fast transients 5/50 ns, 5 kHz rep. freq.	±2 kV, capacitive clamp	no test specified

generally more severe EN 50121-4. On the contrary, signal I/O and dc power ports are tested at insufficient levels as per IEC/EN 61000-6-2, if an outdoor installation or an exposed configuration is to be considered. Going into greater detail of the Note 1 of Table 2 of EN 50121-4, we read "This test applies to I/O Port connected to cable inside 3 m - boundary or connected to cable longer than 30 m within 10 m boundary. I/O ports connected to cable other than above shall comply with the requirements of IEC/EN 61000-6-2 except that Note 2 of Table 3 of IEC/EN 61000-6-2 is not applicable." Note 1 of

Table 3 for DC power ports similarly indicates a 30 m length limit, above which the surge test is due, without specifying any distance from the track. These two notes aim at underlining the conditions that make these cables exposed and in particular at an extent to justify test levels more severe than those adopted for the industrial environment. In many cases the effort by the manufacturer to adopt suitable countermeasures also for increased severity is minimum, consisting very often in selecting the correct surge protective devices (SPDs) and insulation for the input components, located in the EMI filter.

6.6.1.5 EN 50121-5: e.m. Environment for Fixed Installations Equipment

The scope of the EN 50121-5 [84] is expressed as "railway fixed installations associated with power supply", and includes "the power feed to the apparatus, the apparatus itself with its protective control circuits, trackside items such as switching stations, power autotransformers, booster transformers, substation power switchgear and power switchgear to other longitudinal and local supplies." It's a long list of installations and locations that may be characterized by slightly different electromagnetic environment conditions: it is sufficient to observe that any installation located trackside may be characterized at a certain extent also by the electromagnetic phenomena considered for the trackside environment and its standard EN 50121-4 [82].

For the substation environment the most peculiar electromagnetic phenomena appearing in the list of the prescribed immunity tests of the EN 50121-5 [84] are: the steady magnetic field as per IEC/EN 61000-4-8 [179] procedure, but at traction frequencies, namely dc, 16.7 and 50 Hz; the damped oscillatory voltage wave induced during current interruption by ac contactors and circuit breakers on inductive circuits and applied to signal ports as per IEC/EN 61000-4-12 [183]. Due to the same mechanism, an additional oscillatory magnetic field test as per IEC/EN 61000-4-10 [181] could have been applied to the enclosure port, but it is not. Also the pulse magnetic field test, appearing in the EN 50121-4 [82] for trackside equipment, doesn't appear in the list of prescribed tests. The other phenomena are in common with the generic standards, even if with higher levels in some cases.

Concerning the equipment that are subject to the EN 50121-5 standard, the standard itself excludes substation filters, that are used for "harmonic suppression or power factor correction". The standard applies to "items of apparatus which are situated: a) within the boundary of a substation which delivers electric power to a railway; b) beside the track for the purpose of controlling or regulating the railway power supply, including power factor correction and filtering; c) along the track for the purpose of supplying electrical power to the railway other than by means of the conductors used for contact current collection, and associated return conductors (included are high voltage feeder systems within the boundary of the railway which supply substations); d) beside the track for controlling or regulating electric power supplies to ancillary railway uses (included are power supplies to marshalling yards, maintenance depots and stations); e) various other non-traction power supplies from railway sources which are shared with railway traction."

The list needs some comments: first, at point b) we read that items for power factor correction and filtering are included, but not the filters themselves, and this is somewhat curious, because it could be very difficult and subject to interpretation what is "filter" and what is instead its control or regulation; second, at point c), excluding the conductors and other passive elements of high voltage feeders, the only EMC relevant items are the high voltage protections located at the same substations and metering equipment. Protections and metering equipment are normally covered by their product standards recalled in the Declarations of Conformity, because this equipment is normally considered COTS or COTS-like, since it is widely adopted and sold in large numbers. The product standards are all less severe than the EN 50121-5 environment standard and in some cases EMC tests are completely excluded:

- EN 50263 [118] (for measuring relays and protection equipment);
- IEC/EN 60255-22 [153] - [160] (these standards deal with EMC assessment of relays);
- IEC/EN 60898 [165] that, at its sec. 7.14 and 7.15, excludes the relevance of low voltage circuit breakers for EMC purposes, so that they do not undergo any EMC test, even at low severity levels;
- IEC/EN 61543 [202], that makes reference to the IEC/EN 61008-1 [198] and IEC/EN 61009-1 [199] and EMC tests therein.

6.6.2 Environment Standards

Environment standards are those reporting the characteristics of electromagnetic phenomena for a given application or environment; examples of characteristics are the description of the sources, of the expected levels, of the occupied frequency interval or the coupling paths. So, environmental standards may address one, few or a whole set of electromagnetic phenomena in relationship to a particular application or operating environment.

6.6.2.1 Industrial Supply Systems

For the industrial environment, IEC/EN 61000-2-4 [169] presents the compatibility (or reference) levels for harmonics, inter-harmonics and voltage deviations, dips and short interruptions for three classes of supply networks: class 1, for protected supplies, with compatibility levels lower than public networks; class 2, for various industrial and non-public supplies, similar to the public network; class 3, for industrial supplies characterized by the presence of distorting loads (converters, large motors with frequent starts and stops, etc.). The reason for such a classification is that industrial networks are characterized by a variety of technical solutions, distorting loads and applications, more articulated than public networks, that are considered in the next paragraph. It is however important to underline that industrial environments and applications are quite diverse and have been designed, manufactured and installed over several decades. For this reason they are characterized by different technologies for the power semiconductor devices (bipolar junction transistors, MOSFETs, thyristors, GTOs, IGBTs, etc., with much different switching properties), for the passive components (capacitors and inductors, with different dielectric strengths, parasitics values, frequency performances), for the control circuits (electronic devices for I/O, microprocessors and related power supplies, with different clock frequencies, signal levels, switched currents, bus connections, etc.). All this leads to a wide variety of emissions and susceptibility profiles, in terms of relevant phenomena, amplitudes, distribution on the frequency axis, transients parameters, and thus the definition of the "industrial electromagnetic envrionment" is not easy nor straightforward and both new and old systems need a thorough analysis of relevant phenomena, existing levels and susceptibility of installed equipment (either by means of dedicated tests or collection of certificates and test reports).

6.6.2.2 Public Low and Medium Voltage Networks

Public Low Voltage and Medium Voltage networks are characterized by a series of conducted phenomena, that represent a potential disturbance for connected loads and that are subject to a general and statistical characterization contained in the IEC/EN 61000-2-2 [168] and EN 50160 [97] standards respectively. The considered conducted phenomena are voltage unbalance, voltage swells and sags, voltage interruptions, flicker, harmonics and interharmonics. The EN 61000-2-2 and the EN 50160 reports reference levels for the amplitude of these phenomena for the LV and for the MV and HV respectively, in terms of compatibility levels, so the level of the disturbance that is expected with a small probability (less than 5%) of being exceeded.

Supply Voltage Unbalance. Supply voltage unbalance in a poly-phase system is due to either a difference of amplitude or phase in the line-to-line voltages, that define the balanced positive sequence, it is defined in the IEC/EN 61000-2-2 as

$$\delta U = \sqrt[3]{\frac{6(U_{12}^2 + U_{23}^2 + U_{31}^2)}{U_{12} + U_{23} + U_{31}} - 2} \tag{6.4}$$

and it is quantified as less than 2% and 3% for LV and MV networks.

Voltage Flicker. Flicker is defined in terms of step voltage changes that are normally within a few % of the nominal voltage and is quantified by means of short term and long term severity levels, P_{st} and P_{lt} defined over 10 minutes and 2 hours time intervals, respectively. P_{st} is defined in IEC/EN 61000-4-15 [186] as a weighted average of various percentiles, obtained as time-amplitude statistics by the standard flickermeter. P_{lt} is in turn defined on the basis of 12 consecutive 10 min. P_{st} values as

$$P_{lt} = \sqrt{\frac{1}{12} + \sum_{i=1}^{12} P_{st,i}^3} \tag{6.5}$$

Harmonics. The harmonics of the supply voltage are defined by means of a standard Fourier approach, based on the spectrum computed over a suitable time-window of duration T_w. In the EN 61000-2-2 the Discrete Time Fourier spectrum is computed, on the one side, over one period of the fundamental, based on the definition of harmonics as integer multiples of a fundamental frequency, that is made correspond to the power supply frequency, and, on the other side, over a 200 ms time-window (that corresponds to 10 periods of a 50 Hz fundamental and 12 periods of a 60 Hz fundamental) with reference to the standard IEC/EN 61000-4-7 for inter-harmonics analysis. However, for inter-harmonics no prescriptions are given in any of the two EN 50160 and IEC/EN 61000-2-2..

The left column for LV networks refers to long-term (longer than 10 min.) harmonics and related effects and are characterized by a Total Harmonic Distortion value of 8%. For very short-term effects (lasting for less than 3 s) these values must be multiplied by a correcting factor $k = 1.3+0.7/45\times(h-5)$ and the corresponding THD is 11%.

6.6.2.3 Traction Line Supply Voltage

The characteristics of the traction supply voltage are addressed by identifying the rated values for European railways (750 V, 1500 V, 3000 V, 15 kV, 25 kV) and the admissible variations, described by limit values for maximum time duration and curves of voltage-time points; transients and fluctuations may be defined as almost steady, slow, fast, etc. depending on their time duration: EN 50163 in its Table 7 and Annex A specifies limit values of over-voltages starting from 20 ms of time duration (defining three time intervals, C, D and E, for long-term, non-permanent and permanent over-voltages respectively), together with special national conditions for European countries; shorter time intervals are treated by EN 50124-2 (recalled by the EN 50163 itself), by defining time intervals A and B, for short-term and medium-term over-voltages.

The fundamental frequency for ac supply is also subject to the specification of limits, for both 50 and 16.7 Hz systems, distinguishing for systems with and without synchronous connection. The limits of the ver. 2004 [99] are specified over 95%/99.5% of the time depending on the absence/presence of the synchronous connection and over 100% of the time as absolute limits. The amplitude values are specified as ±1% and ±2% for the two cases for

Table 6.9 Values of individual harmonic voltages for LV, MV and HV networks

Harm. order h	LV %	MV %	HV %
2	2	2	1.9
3	5	5	3
4	1	1	1
5	6	6	5
6	0.5	0.5	0.5
7	5	5	4
8	0.5	0.5	0.5
9	1.5	1.5	1.3
10	0.5	0.5	0.5
11	3.5	3.5	3
12	0.46	0.5	0.5
13	3	3	2.5
14	0.43	0.5	0.5
15	0.4	0.5	0.5
16	0.41	0.5	0.5
17	2	2	–
18	0.39	0.5	0.5
19	1.76	1.5	–
20	0.37	0.5	0.5
21	0.2	0.5	0.5
22	0.36	0.5	0.5
23	1.41	1.5	–
24	0.35	0.5	0.5
25	1.27	1.5	–

both 50 Hz and 16.7 Hz systems; the maximum absolute limits have been changed in the last 2007 amendment [100], so that now all synchronous systems are subject to a limitation of $+4/-6\%$ and non synchronous systems to a $\pm 15\%$ limitation. For 16.7 Hz systems interconnected by the 16.7 Hz grid the maximum limitation is reduced to $+2/-3\%$. In general the effect of frequency variations is a change in the position of harmonic components on the frequency axis with potentially harmful effects on signalling and for extreme variations a possible reduction of vehicle performance.

The EN 50163 considers traction supply voltage characteristics in a more ample way by addressing phenomena like:

- line voltage dips, lasting less than 1 s and with a depth of less than 50% of the nominal voltage;
- short interruptions, generally caused by tripping and auto-reclosing of circuit breakers, without an exact definition in terms of amplitude and

duration ("70 % of the short interruptions may be less than 10 s") and a quantification estimated from a ten to a hundred per year;
- long interruptions, caused by "external events and actions, that cannot be prevented by the infrastructure manager";
- voltage distortion, not further detailed.

An essential standard for the definition of the traction supply power quality is the EN 50388 [124], cross-citing with EN 50163, that goes beyond the definition of the phenomena and addresses also the the issue of interoperability between the power supply and the rolling stock. Harmonic distortion and dynamic effects resulting from the interaction between the traction current distortion, the on-board controller of traction converters and the traction line impedance and transfer function are considered. Overvoltages are caused by system instability, related to the interaction between the active controllers on-board residing on different vehicles through the traction line transfer function, giving rise possible dynamic instabilities; on the other hand over-voltages are simply due to the excitation of traction line resonances by the frequency components of the vehicle traction current. The interaction between the traction supply and the rolling stock is evaluated by the calculation of the quality index of the power supply: the "useful voltage" $U_{\mathrm{av},u}$ is defined as the average value of the RMS fundamental component of pantograph voltage over a well-defined geographical area that is part of the network, such as a whole line, and for one or several trains; thus, it is distinguished between $U_{\mathrm{av},u(\mathrm{zone})}$, for the average operation over all the circulating trains in a given zone, and $U_{\mathrm{av},u(\mathrm{train})}$, for the average operation for one train over a predetermined journey or, in general, its timetable.

Extensive measurements have been done in the recent past within the EU Project RAILCOM and the results concerning the three main supply systems (dc, 16.7 Hz and 50 Hz) published in a few works focusing on the description of the used instrumentation and its metrological characteristics [526], on the statistical characterization of power quality in dc railways [523, 521], on the useful voltage and transients in dc railways [524], on the frequency stability of 16.7 Hz systems [522] and 50 Hz systems [525].

6.6.2.4 On-Board Supply Voltage

There are two standards that specify the environmental conditions for on-board apparatus: the EN 50125-1 [111] and the EN 50155 [93] for the behaviour specified for on-board supply systems. The EN 50125-1 makes simply reference to the EN 50163 for the supply conditions. With reference to the EN 50155 the nominal values Un of the supply from accumulator batteries are 24, 48, 72, 96 and 110 Vdc. These voltage supply values are subject to a minimum/maximum value of 0.7/1.25 Un; voltage fluctuations are allowed between 0.6 Un and 1.4 Un with a time duration of less than 0.1 s; longer fluctuations up to 1 s are allowed and the standard specifies that they shall not cause damages, provided that the equipment may be not fully functioning.

In the Annex B (informative) of the EN 50125-2 an example is given of the expected levels of harmonic and inter-harmonic distortion, that characterize the traction line voltage.

The nominal input voltages on-board are specified in UIC 550 [664] for different frequencies as shown below:

Table 6.10 Nominal input voltage in accordance with UIC 550

Frequency [Hz]	Nominal Voltage [V]	U_{min1}	U_{max1}	U_{min2}	U_{max2}	U_{max2A}
15 - 52	1000	800	1150	700	1200	1250
dc	1500	1140	1650	1050	1740	1860
50	1500	1000	1800	900	1950	2050
dc	3000	2000	3600	1800	3900	4050

UIC 550 requires any electrical loads connected to the train line to be designed to operate satisfactorily within the voltage range of between U_{max2} and U_{min1}. Possible distortion of the waveform should be taken into account. In the case of alternating current, the ratio between the root-mean-square value and the peak value may be 1:2. In the case of a d.c. voltage, this is a unidirectional three phase current without smoothing. In Clause 4.3 of UIC 550, a requirement which can be interpreted as a voltage variation requirements states that equipment may stop working when the voltage is outside the a voltage range $U_{min2} - U_{max2A}$, it however requires such equipment to start up again once the voltage returns to the $U_{max2} - U_{min1}$. U_{min2} and U_{max2} are both duration dependent.

For hauled stock supplied with electrical power by diesel tractive units or generator vans through the train line, the conditions to be met regarding voltage and frequency are those set out in UIC 626 [667].

6.6.2.5 Fixed Installations Supply Voltage

It was chosen to consider the phenomenon of over-voltages of various kind (long term, medium term, short term, transient or impulsive) as an additional environmental condition, that takes into account environment and product standards for different locations and classes of equipment in a "horizontal" way. Thus, considering an hypothetical equipment for a railway application, the requirements contained in the various relevant standards are considered and compared, to evaluate the degree of coverage of over-voltages matter.

The EN 50125-2 [89] addresses the environmental conditions for equipment located at fixed electrical installations; the EN 50125-2 (contrarily to the EN 50125-1 and -3 standards of the same family) cites the phenomenon of lightning, with reference to the EN 50124-2, and EMC, with reference to the EN 50121-5.

6.6.2.6 Telecommunication and Signalling Installations Supply Voltage

The EN 50125-3 concerns the environmental conditions for telecommunication and signalling equipment in general; for lightning, reference is made again to the EN 50124-2 and for EMC to the EN 50121-4. For power supply, besides the specification agreed in the contract, reference is made to IEC/EN 61000-2-4 [169], to the harmonized document HD 472 S1 [214] (concerning the "Nominal voltages for low voltage public electricity supply systems") and to the product standard EN 60870-2-1 [164], with a scope covering equipment of teleprotection and distributed automation systems. The IEC/EN 60870-2-1 was issued at the beginning of 1996 and in its specifications for EMC tests it makes reference to old versions of the EN 61000-4-X standards and for some tests the specification is rather generic without a reference to a specific basic immunity test procedure or standard. On the contrary, noting that the IEC/EN 60870-2-1 [164] is a standard for an specific and uncommon class of products, there is no reference to other product standards that would be more suited for telecommunication equipment. To the latter the IEC/EN 60950-1 [167] applies, that considers information technology equipment with particular emphasis on cable connections to power supply and telecommunication networks; the over-voltage withstanding capability is more demanding than for other generic electrical equipment. The standard specifies steady-state and impulse tests for electric strength; the prescribed levels are at the highest levels, compatible with the over-voltage categories specified in IEC/EN 60947-1, with maximum 10 kV level for port connected to outdoor antennas; the impulse tests are carried on with two wave shapes, the 10/700 μs to simulate lightning current and the 1.2/50 μs to simulate induced impulsive over-voltages caused by lightning current.

The same IEC/EN 60950-1 makes reference to the ITU-T standards, that specify the "resistibility" to over-voltages and over-currents of telecommunication equipment in various conditions: K.20 [428] for equipment installed in a telecommunications centre, K.21 [429] for equipment installed in customer premises, K.45 [440] for equipment installed in the access and trunk networks, all using as the basic reference standard the K.44 [439], reporting definitions and testing methods; K.66 [447] for protection of electric supply networks, K.67 [448] for the expected surge level due to lightning.

6.7 Applicability of Standards

It has been already underlined that a standard is mandatory if it is cited as mandatory in a law, regulation or directive, that is in force in the country where the equipment, subsystem or apparatus is installed. Moreover, other standards may become mandatory, if they are referenced in the contract between parties, the customer on one side and the contractor, main contractor or the consortium on the other side; in this case, any type of standard, harmonized or not, may be adopted, provided it is not in contrast with the other standards.

Depending on the type of supply, the hierarchy of standards and requirements may be highly articulated. When a supplied subsystem must be embedded into an existing system, the compliance is specified with respect to not only the applicable standards, but also with the existing equipment, whose susceptibility determines the reference levels for emissions and the pre-existing electromagnetic field level in turn may determine specific immunity tests and levels.

For a vehicle, a loco or a coach the specifications usually consist of the superposition of the EMC standards for hosted equipment, such as the same EN 50121-3-2 and recalled standards, and of the EMC standards for the whole vehicle, such as the same EN 50121-3-1 for the phenomenon of radiated emissions, whilst the interface between the vehicle and the existing or designed signalling equipment is a complex combination of existing company standards and signalling specifications (become reference specifications, hopefully harmonized). In this case even a manufacturer's specification or a rule or practice internal to the infrastructure owner or railway operator obtains the dignity of a standard, since they are recalled in the contract and honestly are probably the best choice to define the existing scenario, thus becoming a "de-facto" standard, yet without the needed harmonization and iterated verifications.

The same happens when a piece of signalling is procured for an existing railway system, where the supplier must cope and comply with several installation rules, in particular for trackside equipment, for grounding, bonding and cable segregation, that convey the past experience on the specific installation, paying attention for example to specific electromagnetic phenomena (e.g. a particularly intense emission from existing rolling stock, or power supply system, or intense surge voltages due to lightning) or environmental conditions (e.g. temperature range, humidity, corrosion due to salty water, etc.).

For a whole transportation system, the selection of the standards is a critical aspect, since it influences the whole life-cycle of the project, the adequacy of some technical solutions, the degrees of freedom to reduce costs and to increase efficiency. Often a project is deemed by the standards originally referenced in the technical annexes of the contract, that are usually prepared by a board of experts appointed by the customer, and that reflect necessarily their point of view, their experience and their knowledge of the involved phenomena and technical aspects. The number of experts is however kept to a minimum and their competence is required to focus on more important topics, such as civil works, electric power systems, electric traction, reliability and safety, that have a more relevant impact on overall system cost, performance and safety. EMC is thus implicitly or explicitly considered as a secondary element and some deficiencies may arise: for example, the list of standards might be incomplete, provisions for cable routing, bonding or grounding might be partial or inconsistent, and in particular there might be confusion between several issues, such as EMC and lightning requirements for grounding system and for environments classification, distinction between application standards, generic standards, and product standards, use of compatible, harmonized and up-to-date standards. For these reasons it is highly advisable that the list of standards and the technical requirements are discussed by the contractor's EMC experts as soon as possible, in order to remove any inconsistency and to complete the set of applicable standards, taking into account all the different subsystems and installed equipment.

From a general point of view the applicability of a single standard per se is specified by the standard itself in its "scope" section, but it must be however harmonized and compatible with the overall system standardization, the environments classification, the assumed margins and immunity severity, and depends also on the features of the application and its environment:

- many are the EMC industrial standards for a generic industrial environment, that report testing methods and installation practices, derived in reality from various experiences, in different applications and environments (heavy industry, process industry, oil&gas, chemical, electrical energy production and distribution, etc.); in this case the problem is before the single product, but encompasses the infrastructure and the layout, in terms of minimum earthing resistance, structure of the bonding network (star, mesh, mixed, separated for different subsystems, its termination on the earthing points, in common with the power supply system), acceptable techniques and materials for bonding, overvoltages control (use of transient voltage suppressors, of what type, with status monitoring or not, located at major points or at all cables, etc.), and so on;
- the EMC standards for domestic environments apply to a large range of domestic appliances, low power equipment and information and communication technology equipment, for which few or no installation specifications are given and their applicability has been extended to office and light industrial environments (while increasing the severity of the test contained in the last releases of the residential environment standard EN 61000-6-1 [194]); however, all this poses a problem of compatibility with the single product standards still adopted to test the new products or that at least appear in the acquired test reports, and that may date back to the '90s, when the respective committees assumed a more or less close compatibility with the residential environment standard at that time, the EN 50082-1 [69].

6.8 Safety, Risk and Probability with Respect to EMC Standards

In principle, EMC tests and EMC compliance are based on certainty and leave no room for probabilistic reasoning related to risk analysis: the result of a test is "pass" or "fail", an equipment is compliant or not, and these results are fully applicable, provided the said equipment is installed in a compatible environment with the correct installation practices. This attitude is of course in favor of safety, as it establishes a deterministic framework where certainty replaces probability.

A closer look to the physical principles and to the common experience, however, may help in locating several cracks in this neat image:

- electrical noise is always present during measurements and tests, so that the results and the pass/fail answer must take this into the due consideration, by adopting margins with respect to the limits and conservative approaches (not too conservative as a tradeoff with economic and practical issues); the same margins must account also for instrumental and setup uncertainties, that, depending on the used approach, may describe a probabilistic or worst case deterministic approach;
- some of the electromagnetic phenomena considered for immunity testing are of natural origin or, if man-made, are not deterministic, so that a tiny residual probability cannot be excluded that the same phenomena show up with intensity or spectral distribution beyond the assumed values, even if these values have been determined with a conservative approach; lightning and lightning induced effects may vary depending on several environmental and technological factors, such as the height above sea level, the shape of the territory, the length and degree of exposure of victim cables, the soil resistivity, etc. and for this reason there are no definitive indications by environmental standards of the expected amplitude and number overvoltages and procured voltage dips and short supply interruptions;
- the position and orientation of sources and victims may vary with time, thus bringing them closer than what was assumed by the standard committees, while fixing the immunity test levels; examples may be found in the vehicle itself, in the antennas of mobile communication equipment, or in people moving with their personal devices (mobile terminals, or audio entertainment equipment, or hearing aids, pacemakers, etc.).

6.8.1 Target Electromagnetic Environment

The target electromagnetic environment is thus relevant not only because it automatically defines the applicable standards, but also because it partially determines the type of electromagnetic phenomena and the chance that they manifest with a particularly large intensity. When the electromagnetic environments are identified and circumscribed within a railway system, portions of space are identified together with the input/output "ports", that may represent an interface between internal and external electromagnetic phenomena and equipment. It is thus extremely important that ports, intended in a broad sense, are exhaustively identified, to understand and correctly evaluate coupling paths.

6.8.2 EN 61000-2-2: Harmonization of Emission and Immunity Levels

The EN 61000-2-2 [168] introduces and describes the rationale adopted to define adequate immunity levels, once the emission levels of the installed equipment and the general characteristics of the electromagnetic environment are known.

6.8.2.1 Relation between Compatibility Level and Emission Limits

It must first be noted that some disturbances have their sources in atmospheric phenomena, especially lightning, or in the normal and unavoidable response of a well-designed supply system to electrical faults or to the switching of load or of particular devices. The principal disturbances in this category are transient overvoltages, voltage dips and short supply interruptions. Emission limits cannot be assigned for these phenomena, since the emission sources are largely uncontrollable. In this case, the compatibility level is intended to reflect the level of severity which can be expected in practice. Many disturbances, however, have their sources in the equipment by which the public electricity supply is utilised, or, to a small extent, in equipment forming part of the supply system itself. The disturbance arises when such equipment draws a current which is not a regular or constant function of the voltage supplied, but contains abrupt variations or fails to follow the complete cycle of the voltage waveform. These irregular currents flow through the impedances of the supply networks and create corresponding irregularities in the voltage. Although reduction of some of the network impedances is sometimes considered in order to mitigate the effects of a specific source of disturbance, the general case is that they are fixed, largely on the basis of voltage regulation and other considerations not concerned with disturbance mitigation. The voltage irregularities, in turn, are conducted to other equipment, some of which they can disturb. The severity levels at which they reach the other equipment depend on the types of equipment which form the sources of the emissions, the number and location of such sources operating at any given time, and on how the emissions from these diverse sources combine together to yield particular levels of disturbance at particular locations. These levels should not exceed the compatibility level. Therefore, emission limits have a more complex relation with the compatibility level than immunity levels. Not only are the sources of emission highly diverse, but also, especially in the case of low-frequency disturbances, any source to which a limit is to be applied is only one of many sources combining together to produce the environmental disturbance level represented by the compatibility level.

Moreover, many emission limits are expressed in terms of current, although the compatibility levels are expressed in terms of voltage for most types of disturbances. This makes it necessary to consider network impedances.

Nevertheless, the objective of setting emission limits is to ensure that actual disturbance levels will not exceed the compatibility level, apart from the low-probability events that are accepted in EMC. This means that emission limits for equipment of any particular type cannot be established independently, but must, for each disturbance phenomenon, be co-ordinated with the limits set for all other sources of the same disturbance. The co-ordination must be such that when all sources comply with their individual limits, and are acting together to the degree that can be expected in the relevant environment, the resulting disturbance level is less than the compatibility level. The sources of emission are extremely diverse, but it is useful to divide them into two broad categories:

- Large equipment and installations: at one time these were almost the only significant sources of low-frequency emissions such as harmonics and voltage fluctuations. The important point relating to them is that they are always brought to the attention of the electricity supplier, who therefore has the opportunity, together with the operator or owner of the disturbing equipment, to devise an operating regime intended to maintain emissions within acceptable limits, and a method of supply which can ensure that emissions within those limits are unlikely to disturb other equipment connected to the supply network. This solution is specific to the location involved.
- Small equipment: to an ever increasing extent equipment of relatively low power, widely used in domestic, commercial and the smaller industrial premises, is the source of high levels of low frequency disturbances. This equipment is purchased on the open market and is generally installed and operated without reference to the electricity supplier. The emissions from any single piece of equipment are small in absolute terms, but the total number connected is very large and may account for 50% of system demand. Moreover, for much of this equipment the emissions are large relative to the rated power. Therefore this type of equipment has become a large and increasing source of low frequency disturbances. The only feasible method of controlling these emissions is to ensure that the equipment is designed and manufactured in compliance with appropriate emission limits.

Thus, in order to maintain the compatibility level as a true indication of the maximum probable level of disturbance in the electromagnetic environment, it is necessary to co-ordinate in a coherent manner the emission limits adopted for this wide range of products, including both the larger installations which are brought to the notice of the electricity supplier and the smaller equipment which the user installs at his own discretion.

6.8.2.2 Planning Levels

For large loads and installations, those responsible for the power supply system have a particular role. In determining the appropriate emission limits for such installations, they use the concept of planning level, as defined in Clause 3.1.5 of IEC 61000-2-2 [168]. Planning levels are relevant primarily to medium voltage and higher voltage networks. However, low frequency conducted disturbances pass in both directions between low voltage and the higher voltage networks. The co-ordination of emission limits must take account of all voltage levels. The use of planning levels is described in IEC/TR3 61000-3-6 and IEC/TR3 61000-3-7. The important points are:

- The planning level is a value adopted by the body responsible for planning and operating the power supply system in a particular area, and is used in setting emission limits for large loads and installations which are to be connected to the system in that area. It is used as an aid in distributing the emission limitation burden as equitably as possible.
- The planning level cannot be higher than the compatibility level. Generally, it is lower by a margin which depends on factors such as the disturbance phenomenon involved, the structure and electrical characteristics of the supply network (provided it is adequately designed and maintained), the background levels of disturbance, the possibility of resonance, and load profiles; it is, therefore, locally specific.
- Although the planning level is related mainly to large equipment and installations, account must be taken also of the many other sources of disturbance, notably numerous low-power equipment connected at low voltage. The margin available to accommodate emissions from large installations depends on how effectively limits are applied to the low power equipment. Any difficulty in this regard is an indication that a stricter approach to emissions from low power equipment is required. The over-riding objective is to ensure that the predicted level of disturbance does not exceed the compatibility level.

6.8.2.3 Illustration of Compatibility, Emission, Immunity and Planning Levels

The relationship between the various EMC levels and limits is shown in Figure 6.14. Although not mathematically exact, it illustrates the relationships between the values. The figure is intended to have schematic significance only. In particular, the relative positions of the two curves show that overlap can occur, but should not be interpreted as an accurate indication of the extent of the overlap. From the figure, a number of observations can be made:

1. It is assumed that only one electromagnetic disturbance is being considered
2. It is assumed that there is a verifiable (or reasonable verifiable) relationship between the process/manner that the emissions is measured and the

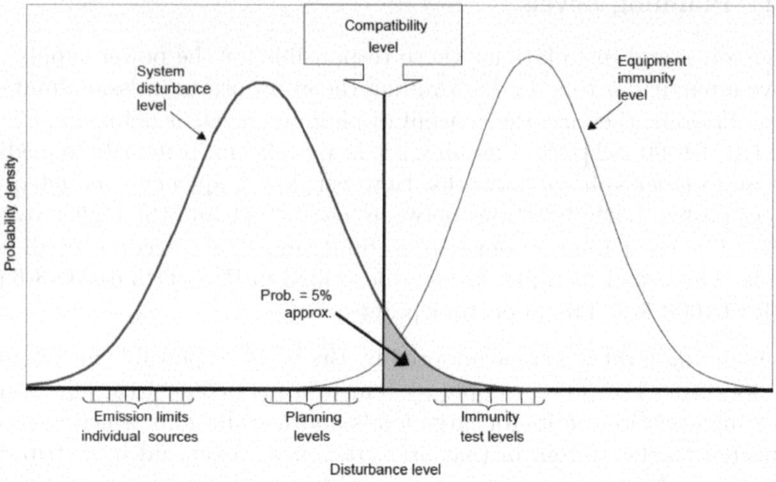

Fig. 6.14 Relation between Compatibility, planning, immunity and emissions levels

manner that the said electromagnetic disturbance is incident on another equipment

3. It is assumed that the equipment has been designed with an inherent margin to ensure compliance with the specified limits. The equipment design margin can not be specified or per-determined in standards as it is within the remit of the equipment manufacturer.

For a detailed understand of these terms and the inter-relation between and the influence they have on the derivation of specified limits in standards, please refer to IEC 61000-1-1 [318]

Chapter 7
Applied Electromagnetic Theory

Electromagnetic theory plays a central role to the field of electromagnetic compatibility, it is therefore important that a basic understanding of the electromagnetic theory and its application in railway electromagnetic compatibility be treated in context. It is not the intention of this chapter to cover in detail electromagnetic as this is covered in various classic textbooks; however in this chapter an overview of electromagnetic theory is presented with the view of enhancing the understanding of analytical and theoretical aspects of electromagnetic compatibility in railway applications.

7.1 Electromagnetic Phenomena

7.1.1 General Definitions

In this section basic definitions of common electromagnetic variables used in EMC are covered. The coverage includes their relationships and measuring units, applicable to radiated and conducted phenomena.

7.1.1.1 Ports and Matrix Representation

In several disciplines it is very useful and comfortable to use the concept of port, made of two terminal wires, where the input currents are bounded to sum to zero (or, in other words, are required to be equal in magnitude and of opposite polarity). Several electric circuits and equipment are suitable to be treated on a port basis (such as amplifiers, cables, transformers, etc.), which several matrix representations are related to. In the following a two-port network is considered, but n-port networks may be treated using the examples developed herein.

Depending on the choice of the input and output variables, starting from the four available variables, that is the voltages and currents of the two ports, V_1, I_1, V_2 and I_2, some matrix representations are possible.

A. Ogunsola and A. Mariscotti: Electromagn. Compat. in Railways, LNEE 168, pp. 315–428.
springerlink.com

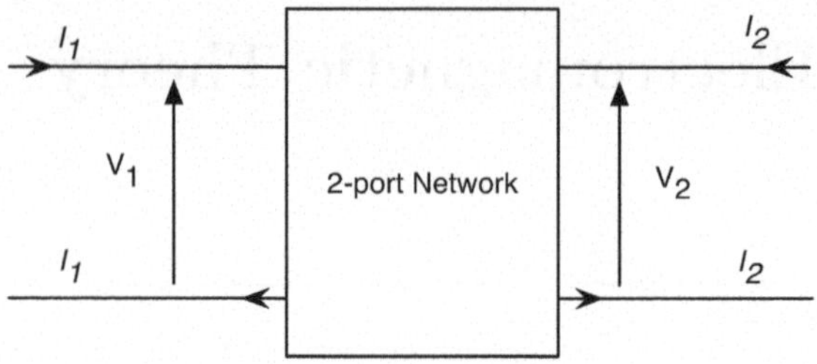

Fig. 7.1 Two-port network and related variables

Impedance Matrix. The input variables are the two currents I_1 and I_2. The output variables are the two voltages V_1 and V_2.

$$\begin{bmatrix} V_1 \\ V_2 \end{bmatrix} = \begin{bmatrix} Z_{11} \; Z_{12} \\ Z_{21} \; Z_{22} \end{bmatrix} \begin{bmatrix} I_2 \\ I_1 \end{bmatrix} = [Z] \begin{bmatrix} I_1 \\ I_2 \end{bmatrix} \tag{7.1}$$

The relationships that define the terms inside the $[Z]$ matrix are very straightforward and identify a volt-amperometric test on the two-port network, while holding a variable to a known condition.

$$Z_{11} = \left.\frac{V_1}{I_1}\right|_{I_2=0} \qquad Z_{12} = \left.\frac{V_1}{I_2}\right|_{I_1=0} \qquad Z_{21} = \left.\frac{V_2}{I_1}\right|_{I_2=0} \qquad Z_{22} = \left.\frac{V_2}{I_2}\right|_{I_1=0} \tag{7.2}$$

This representation is suitable for electrical supply networks and to express the concept of input impedance or loading impedance in a variety of circuits (probes, transformers, etc.).

Admittance Matrix. The input variables are the two voltages V_1 and V_2. The output variables are the two currents I_1 and I_2.

$$\begin{bmatrix} I_1 \\ I_2 \end{bmatrix} = \begin{bmatrix} Y_{11} \; Y_{12} \\ Y_{21} \; Y_{22} \end{bmatrix} \begin{bmatrix} V_2 \\ V_1 \end{bmatrix} = [Y] \begin{bmatrix} V_1 \\ V_2 \end{bmatrix} \tag{7.3}$$

The relationships that define the terms inside the $[Y]$ matrix are directly related to the one above for the impedance terms. The admittance matrix $[Y]$ is the inverse of the impedance matrix $[Z] = [Y]^{-1}$ and this can be easily demonstrated by substituting (7.1) into (7.3), or vice-versa.

The use of this representation is the same as that of the impedance matrix above; the choice between the two is based on the availability of either the two sets of variables during problem definition.

Transmission Matrix. The input variables are the current and voltage on one port I_1 and V_1. The output variables are the current and voltage on the other port $-I_2$ and V_2.

$$\begin{bmatrix} V_2 \\ I_2 \end{bmatrix} = \begin{bmatrix} T_{11} \; T_{12} \\ T_{21} \; T_{22} \end{bmatrix} \begin{bmatrix} V_1 \\ I_1 \end{bmatrix} = [T] \begin{bmatrix} V_1 \\ I_1 \end{bmatrix} \tag{7.4}$$

The $[T]$ matrix contains non homogeneous terms, that are either the ratio of two voltages or two currents (so dimensionless), or the ratio of one current and one voltage, thus taking the dimension of an impedance or admittance. In analogy with the impedance terms they are defined as

$$T_{11} = \left.\frac{V_2}{V_1}\right|_{I_1=0} \qquad T_{12} = \left.\frac{V_2}{I_1}\right|_{V_1=0} \qquad T_{21} = \left.\frac{I_2}{V_1}\right|_{I_1=0} \qquad T_{22} = \left.\frac{I_2}{I_1}\right|_{V_1=0} \tag{7.5}$$

The opposite relationships between port 2 variables on the right-hand side and port 1 variables on the left-hand side are not used since they would lead to expressions that cannot be implemented in order to measure the transmission parameters themselves: for example, $T_{11} = {}^{V_1}/{}_{V_2}|_{I_2=0}$ requires the application of a driving signal V_2 to a port where also the condition $I_2 = 0$ applies. This representation is suitable for the assemble of a series of building blocks to model more complex electrical networks, such as electric supply networks, the railway traction line, etc. The choice of the input and output variables allows the direct cascade connections of different blocks, resulting simply in the chain multiplication of the respective $[T]$ matrices; to this aim one of the two currents is reversed in polarity and marked positive while exiting the port.

7.1.1.2 Decibel

The decibel (abbreviated dB) is a logarithmic representation of ratios, so that values, ranging several orders of magnitude, may be kept on a handy scale. The use of dB is so widespread since any gain or attenuation (concepts so often used in electronics, signal processing and measurements) are translated directly into sum and difference operations.

The dB is defined as $10 \log(X_2/X_1)$, where X_1 and X_2 represent two homogeneous variables with a unit of power. It is evident that if X_2 is ten times X_1, then their ratio equals 10 dB. The variables X_1 and X_2 may represent two variables of a circuit or system (for example the input and the output power of a module, so that their ratio represents a gain or attenuation) or if X_1 is taken equal to a reference value that expresses the reference unit of measure, then X_2 becomes expressed in dB of this unit of measure: so, if $X_1 = 1\,\mathrm{W}$, then X_2 is expressed in dBW; if $X_1 = 1\,\mathrm{mW}$, then X_2 is expressed in dBmW or more commonly dBm; it is immediate to relate watts to milliwatts by observing that $1\,\mathrm{dBW} = 30\,\mathrm{dBmW}$.

By recalling the power law that for electric variables puts power in relationship with either voltage or current, the dB representation of the latter is obtained by a coefficient of 20 in front of the log operation: $20 \log(Y_2/Y_1)$. Again, the expression may represent a true ratio between two variables in a circuit or system, or the ratio of a variable with respect to a reference variable, thus obtaining dBV if $X = 1\,\text{V}$, dBmV if $X = 1\,\text{mV}$, dBA if $X = 1\,\text{A}$ and so on; the relationship between different powers of the same measuring unit is obtained with the criterion of 20 dB/decade, so that $1\,\text{dBV} = 60\,\text{dBmV}$.

In this transformation we have momentarily neglected the value of the resistance R, which the voltage or current is applied to; power may be related to voltage or current when the reference resistance value is known.

As an example, the very common transformation between X [dBm] and Y [dBV] commonly used in radio frequency measurements and calculations is developed for $R = 50\,\Omega$, the reference resistance for radio frequency equipment, such as antennas, cables, spectrum analyzers, amplifiers.

$P = 1\,\text{mW}$; $P = V^2/R$; $X = 10\log(P) = 10\log(V^2) - 10\log(R) = 20\log(V) - 10\log(50\,\Omega) = Y - 17$

Y [dBV] may be transformed to Y_2 [dBμV] by adding 120 dB.

If the reference resistance had been a different value, let's say $R_3 = 150\,\Omega$, then the new voltage value Y_2 [dBμV] would be larger by $K = 10\log(150\,\Omega) - 10\log(50\,\Omega) = 21.8 - 17\,\text{dB}$. Pay attention that the K so calculated is applied to a 20 dB/decade variable, Y_2 [dBμV], to obtain another 20 dB/decade variable, Y_3 [dBμV].

7.1.2 Transmission Lines

This subject is covered in many text books and thus only the main elements will be treated in this section, starting from an outline of the derivation of the telegraph equations.

Let's consider an infinitesimal section of line of length dz at the longitudinal position z. Starting from the p.u.l. parameters r, l, c and g, assumed constant over the entire transmission line, we may say that this section has a series resistance $r\,dz$, series inductance $l\,dz$, shunt capacitance $c\,dz$ and shunt conductance $g\,dz$.

It is known that for the most general approach, the equations for transmission lines must be derived in the time domain, to account for transients and arbitrary waveforms. Using the instantaneous quantities for the voltage and current at the two ports at the left and the right of the section, the Kirchhoff equations give directly

$$v(z + dz,\, t) - v(z,\, t) = -r\,dz\,i(z,\, t) - l\,dz\,\frac{\partial\, i(z,\, t)}{\partial\, t} \tag{7.6}$$

$$i(z + dz,\, t) - i(z,\, t) = -g\,dz\,v(z,\, t) - c\,dz\,\frac{\partial\, v(z,\, t)}{\partial\, t} \tag{7.7}$$

Since the infinitesimal section has the shunt elements on the right, one may argue that the voltage and its derivative that produce the current in the second equation is in reality $v(z + dz,\, t)$. The objection is first, that we couldn't do otherwise in order to derive a symmetric set of equations to solve, and, second, that the term $v(z+dz, t)$ may be developed in Taylor's series and all the higher order terms would be anyway negligible.

Dividing by dz and letting dz approach zero give the well known set of linear partial differential equations

$$\frac{\partial v(z,t)}{\partial t} = -r\, i(z,t) - l\,\frac{\partial i(z,t)}{\partial t} \tag{7.8}$$

$$\frac{\partial i(z,t)}{\partial t} = -g\, v(z,t) - c\,\frac{\partial v(z,t)}{\partial t} \tag{7.9}$$

whose solution will explicit v and i in terms of z and t, once the boundary conditions are known, that is the characteristics of the source at $z = 0$ and of the load at $z = L$.

If the line section is analysed in the frequency domain, then phasors $V(z)$ and $I(z)$ take place of the instantaneous quantities $v(z, t)$ and $i(z, t)$, and the solution of the set of equations lead to a similar set of linear differential equations.

$$\frac{d\,V(z)}{d\,z} = -(r + j\omega l)\, I(z) \tag{7.10}$$

$$\frac{d\,I(z)}{d\,z} = -(g + j\omega c)\, V(z) \tag{7.11}$$

The solution of the (7.10) and (7.11) leads to a second order differential equations that admit a time-harmonic integral of the form

$$V(z) = V_1 \exp(-\gamma z) + V_2 \exp(+\gamma z) \tag{7.12}$$

$$I(z) = I_1 \exp(-\gamma z) + I_2 \exp(+\gamma z) \tag{7.13}$$

The solution indicates that there exist two voltage waves (and two current waves) travelling along the two possible directions on the transmission line, the leftmost one in that of increasing z and the rightmost one in that of decreasing z, and thus they are usually named *progressive wave* and *regressive wave*, or *forward wave* and *backward wave*.

The quantity γ, named *propagation constant*, is derived from the p.u.l. parameters as

$$\gamma = \sqrt{(r + j\omega l)(g + j\omega c)} \tag{7.14}$$

and is in general a complex number. Its real and imaginary parts

$$\gamma = \alpha + j\beta \tag{7.15}$$

are named *attenuation constant* and *phase constant* respectively (or better, especially for α, attenuation coefficient and phase propagation coefficient, since the former is in reality far from being constant because of skin effect and other power loss effects), and their function is clarified if (7.15) is replaced in (7.12) for example

$$V(z) = V_1 \exp(-\alpha z) \exp(-j\beta z) + V_2 \exp(+\alpha z) \exp(+j\beta z) \tag{7.16}$$

Going to the instantaneous voltage by application of the Re{} operator

$$\begin{aligned} v(z,t) = {} & |\mathrm{V}_1| \exp(-\alpha z) \,\mathrm{Re}\{\exp(j(\omega t - \beta z + \varphi_1))\} + \\ & |\mathrm{V}_2| \exp(+\alpha z) \,\mathrm{Re}\{\exp(j(\omega t + \beta z + \varphi_2))\} \end{aligned} \tag{7.17}$$

the amplitude (given by the peak values $|\mathrm{V}_1|$ and $|\mathrm{V}_1|$ multiplied by the attenuation effects $\exp(-\alpha z)$ and $\exp(+\alpha z)$) and the orientation (given by the combination of the angular pulsation ωt, the initial phase φ_1 and φ_2, and the phase rotation $\pm\beta z$) are separated.

Let's consider a single "point" at the longitudinal position $z_0$and time t_0, corresponding to the voltage of the first forward wave; at a slightly later instant of time $t_0 + dt$ the point has moved from z_0 to $z_0 + dz$ and the voltage value in z_0 is no longer the previous one. By neglecting the attenuation of the amplitude, and thus equating the angular positions to make the "same points" in (z_0, t_0) and $(z_0 + dz, t_0 + dt)$ correspond as a simple horizontal shift, we may write $\omega t_0 - \beta z_0 + \varphi_1 = \omega(t_0 + dt) - \beta(z_0 + dz) + \varphi_1$, that leads to $\omega dt - \beta dz = 0$: the velocity, represented by the ratio dz/dt, with dz and dt approaching zero, is equal to ω/β,and it is called *phase velocity*. The same result is obtained for the backward wave, except for the sign of the direction of propagation. So, the wave looks still to an observer moving at its velocity; if the observer is fixed in space observing a longitudinal position, the time variation of the instantaneous voltage is sinusoidal.

There is another type of velocity, named *group velocity*, that refers to several frequency components (belonging to a radio transmission, for instance) that have not the same phase velocity; this case is however not of interest here and for our transmission lines phase velocity and group velocity will have the same value.

7.1.2.1 The Measuring Units of the Propagation Constant

The measuring unit of α and β is 1/m, since both appears multiplied by z in the exponential argument, that must be dimensionless. Yet, some distinction is needed. It has shown to be convenient to find a unit for the dimensionless argument αz, that expresses the attenuation: so, a voltage or current is said to undergo an attenuation of X *nepers*[1] when its magnitude changes by a factor

[1] The name neper derives from the latinization of Napier, the Scottish mathematician who invented the logarithms.

$\exp(-X)$. The measuring unit for α is thus neper/m. The phase constant β analogously is measured in radians/m; by definition of the wavelength λ, as the length in space travelled by a sinusoidal wave after a phase rotation of 2π, the known relationship $\beta = 2\pi/\lambda$ derives.

It is interesting to identify the relationship between nepers and dB, both used to express an attenuation where the two voltages (or currents) are taken as V_a and V_b, at two positions z_a and z_b. By definition of neper the attenuation between V_a and V_b is X neper, if $V_a/V_b = \exp(-X)$; then, by definition of dB $A = 20\log(V_a/V_b) = 20\log(\exp(-X)) = 8.686X$.

7.1.2.2 The Characteristic Impedance

The presence of a backward wave is related to the phenomenon of reflection, that is in turn related to the type of termination at the source and the load; so that, there exist transmission lines where there is no backward wave for specific termination conditions.

The characteristic impedance Z_0 is ideally calculated from a uniform line with no reflected waves ($V_2 = 0$ and $I_2 = 0$), by taking the ratio of the voltage and current waves, and it may be demonstrated that this ratio is constant at any point along the transmission line (depending only on frequency).

By taking (7.16) with only the forward wave for simplicity and by differentiating with respect to z

$$\frac{dV}{dz} = -(\alpha + j\beta)V_1 \exp(-\alpha z)\exp(-j\beta z) \tag{7.18}$$

the direct comparison with (7.10) gives

$$Z_0 = \frac{V}{I} = \frac{V_1}{I_1} = \frac{r + j\omega l}{\alpha + j\beta} = \sqrt{\frac{r + j\omega l}{g + j\omega c}} \tag{7.19}$$

The quantity has the dimension of an impedance and it is a characteristic of the line behaviour (neglecting reflections), so that it was named *characteristic impedance*. Its dependency on frequency is twofold: direct, since the numerator and denominator show the reactive terms led by a $j\omega$ operator, and, indirect, since the p.u.l. parameters may depend on frequency. If the resistance r and the conductance g are considered negligible (as it is, for example, for high quality coaxial cables), then the characteristic impedance becomes a pure resistance $Z_0 = \sqrt{l/c}$. It is underlined however that the characteristic impedance is a somewhat intangible entity, in that it cannot be measured directly by a measuring bridge and its resistive part is not dissipative. It may be determined indirectly by some impedance measurements (at least two), as it will be seen shortly. The line velocity v_0(that corresponds to the phase velocity) also can be expressed in p.u.l. parameters, and, again under the assumption of negligible r and g parameters, the known formula is derived $v_0 = 1/\sqrt{lc}$.

7.1.2.3 Input Impedance

The *input impedance* of a transmission line is the ratio of voltage and current at the input port, that is $Z_{in} = V_1/I_1$, so that for a non reflective transmission line the input impedance and the characteristic impedance coincide. Since it was said that the ratio V/I is constant over the entire line, at the far end $z = L$, the same relationship holds for the voltage and current phasors across the terminating load Z_t, so that $Z_t = Z_0$ becomes the condition for a non-reflecting transmission line termination (this condition is often referred to as *matched load condition*).

In general, the input impedance and the matching conditions at either end (source and/or load side) may be evaluated in the following way. By differentiation of (7.16) with respect to z, and writing the equality with (7.10), the following expression for the current I is derived:

$$I(z) = \frac{\gamma}{r + j\omega l} V_1 \exp(-\gamma z) - \frac{\gamma}{r + j\omega l} V_2 \exp(+\gamma z) \tag{7.20}$$

where the term $\gamma/(r + j\omega l)$ is equal to $1/Z_0$. The phasor current coefficients for the forward and backward waves are thus related to the corresponding coefficients of the phasor voltage by

$$I_1 = \frac{V_1}{Z_0} \quad I_2 = -\frac{V_2}{Z_0} \tag{7.21}$$

In the general case of possibly mismatched source and load impedances, the impedance at any point along the line is given by

$$Z(z) = Z_0 \frac{V_1 \exp(-\gamma z) + V_2 \exp(+\gamma z)}{V_1 \exp(-\gamma z) - V_2 \exp(+\gamma z)} \tag{7.22}$$

The attention is now on the reflection coefficient, in order to then derive the input impedance without the voltage terms V_1 and V_2.

7.1.2.4 The Reflection Coefficient and the Terminating Impedance

The *reflection coefficient*, either indicated by ρ or Γ, is defined as the ratio of the values of the reflected wave (going in the direction of decreasing z and associated to the term $\exp(+\gamma z)$) and of the incident wave (going in the direction of increasing z and associated to the term $\exp(-\gamma z)$). At the load port $z = L$, the reflection coefficient ρ_t is given by

$$\rho_t = (V_2/V_1) \exp(2\gamma L) \tag{7.23}$$

and is a complex number. Taking (7.22) at $z = L$ and dividing all the terms by the same quantity $(V_1 \exp(-\gamma z))$, it is possible to isolate the reflection coefficient at terminating load as

$$\frac{Z_t}{Z_0} = \frac{1+\rho_t}{1-\rho_t} \tag{7.24}$$

that solved for ρ_t gives

$$\rho_t = \frac{Z_t - Z_0}{Z_t + Z_0} \tag{7.25}$$

It may be noted that for some extreme terminating conditions, physical reasoning helps as well in finding the value of the reflection coefficient. If $Z_t = 0$, that is a short circuit, then the incident and reflected voltage wave must be identically equal at $z = L$, since their sum must be zero; the reflection coefficient is thus $\rho_t = -1$. On the contrary, if the line is terminated on an open circuit, $Z_t = \infty$, the same reasoning should be applied to the phasor current waves; the equation above confirms that the reflection coefficient is $\rho_t = -1$. If Z_t is a generic reactive impedance the modulus of the reflection coefficient is still unity, but the phase is variable; this is reasonable observing that a reactive impedance is not dissipative, cannot absorb power from the incident wave and is thus totally reflecting.

Dividing (7.22) by V_1 and substituting the ratio V_2/V_1 by manipulation of (7.23), the line impedance $Z(z)$ is

$$Z(z) = Z_0 \frac{\exp(-\gamma z) + \rho_t \exp(-2\gamma L)\exp(+\gamma z)}{\exp(-\gamma z) - \rho_t \exp(-2\gamma L)\exp(+\gamma z)} \tag{7.26}$$

Multiplying all the terms at numerator and denominator by $\exp(\gamma L)$, the expression is put in a form where the characteristic impedance, the reflection coefficient at the terminating load and the distance x between the point z and the load at $z = L$ are the sole parameters used.

$$Z(x) = Z_0 \frac{\frac{Z_t}{Z_0}[\exp(+\gamma x) + \exp(-\gamma x)] + [\exp(+\gamma x) - \exp(-\gamma x)]}{[\exp(+\gamma x) + \exp(-\gamma x)] + \frac{Z_t}{Z_0}[\exp(+\gamma x) - \exp(-\gamma x)]} \tag{7.27}$$

The input impedance at $z = 0$, that is $x = L$, may be readily obtained, where the complex exponential have been replaced by the hyperbolic tangent:

$$Z_{in} = Z_0 \frac{\frac{Z_t}{Z_0} + \tanh(\gamma L)}{1 + \frac{Z_t}{Z_0}\tanh(\gamma L)} = Z_0 \frac{Z_t + Z_0 \tanh(\gamma L)}{Z_0 + Z_t \tanh(\gamma L)} \tag{7.28}$$

It is now possible to treat the problem of the determination of the line characteristics on the basis of impedance measurements.

7.1.2.5 Measured Input Impedance

Under the two extreme conditions of short circuit and open circuit termination, the measured input impedance is expressed as

$$Z_{in,sc} = Z_0 \tanh(\gamma L) \tag{7.29}$$

$$Z_{in,oc} = \frac{Z_0}{\tanh(\gamma L)} \tag{7.30}$$

and the multiplication of the two equations gives

$$Z_0 = \sqrt{Z_{in,sc} Z_{in,oc}} \tag{7.31}$$

Besides reasoning on the accuracy of the measurement method and the propagation of uncertainty in measurements, it is evident that both $Z_{in,sc}$ and $Z_{in,oc}$ must be in the scale of the used instrument, so neither too small nor too large; it is possible to state that the appropriate length L is close to an odd number of eighths of the test wavelength, or, put in another more practical way, that the measurements are to be taken at somewhat different frequency values to avoid singularities, based on the assumption that the line p.u.l. parameters are not varying appreciably for changes of the test frequency within less than few % (let's say 10%).

Given an uncertainty of $u(Z_{in,sc})$ and $u(Z_{in,oc})$ in the measurement of $Z_{in,sc}$ and $Z_{in,oc}$ the estimation of Z_0 is characterized by with independently distributed errors on $Z_{in,sc}$ and $Z_{in,oc}$, the calculated Z_0 is characterized by an uncertainty $u(Z_0) = \sqrt{\frac{1}{2}\left[u(Z_{in,sc})^2 + u(Z_{in,oc})^2\right]}$.

Also the attenuation factor α and the phase propagation factor β may be estimated from the said measurements. Taking the square root of the ratio of the $Z_{in,sc}$ and $Z_{in,oc}$ expressions and replacing with the $\tanh(\gamma L)$ term

$$\sqrt{Z_{in,sc}/Z_{in,oc}} = \tanh(\gamma L) = \frac{1 - \exp(-2\gamma L)}{1 + \exp(-2\gamma L)} \tag{7.32}$$

it is possible to isolate the $\exp(-2\gamma L)$ term

$$\exp(2\gamma L) = \frac{1 + \sqrt{Z_{in,sc}/Z_{in,oc}}}{1 - \sqrt{Z_{in,sc}/Z_{in,oc}}} \tag{7.33}$$

Then, taking the log and separating the real and imaginary part of γ,

$$\alpha = \frac{1}{2L} \ln \left| \frac{1 + \sqrt{Z_{in,sc}/Z_{in,oc}}}{1 - \sqrt{Z_{in,sc}/Z_{in,oc}}} \right| \tag{7.34}$$

$$\beta = \frac{1}{2L}\left\{\arg\left[\frac{1+\sqrt{Z_{in,sc}/Z_{in,oc}}}{1-\sqrt{Z_{in,sc}/Z_{in,oc}}}\right] + 2n\pi\right\} \tag{7.35}$$

where the $2n\pi$, n integer, impedes to determine β uniquely, but rather a series of values that differ by π/L.

It appears that for the correct determination of α a piece of line is needed, with a length that causes a loss of some dB at minimum; on the contrary the unambiguous determination of β requires a piece of line so short that the discrimination of π/L is possible (put in another way, since β may be independently determined by means of the simple relationship $\beta = 2\pi/\lambda$, the value of λ in the transmission line, and not in free space, needs to be known with enough accuracy to discriminate which value of n above is the most appropriate).

7.1.2.6 Standing Wave Pattern and VSWR

Let's consider again the phasor voltage equation (7.16), that may be reformulated as

$$\begin{aligned} V(z) &= V_1\left[\exp(-\gamma z) + \frac{V_2}{V_1}\exp(+\gamma z)\right] = \\ &= V_1\left[\exp(-\gamma z) + \rho_t\exp(-2\gamma L)\exp(+\gamma z)\right] = \\ &= V_1\exp(-\gamma L)\left[\exp(+\gamma x) + \rho_t\exp(-\gamma x)\right] \end{aligned} \tag{7.36}$$

where $x = L - z$. The phasor current equation may be immediately determined dividing by Z_0.

When two waves of identical frequency travel in opposite directions the interference phenomenon called *standing waves* occurs. In the extreme case of a lossless ($\alpha = 0$) line with maximum reflection at the load termination (open circuited, thus $\rho_t = 1$), the magnitude of the voltage produced by the composition of the two waves is

$$|V(x)| = |2V_1\exp(-\gamma L)|\,|\cos(\beta x)| \tag{7.37}$$

The magnitude oscillates thus between minima (zero valued) and maxima; the separation of consecutive minima (or maxima) is $\beta(x_{i+1} - x_i) = \pi$ or $x_{i+1} - x_i = \lambda/2$; the minima are located at $\beta x_i = \pi/2 + n\pi$, n integer, and the maxima $\pi/2$ apart.

The instantaneous voltage at the points of minimum is always zero, while at the points of maximum oscillates in time at the same frequency of the signal and spans to the maximum level.

The most common characterization of the standing wave voltage pattern is the VSWR (Voltage Standing Wave Ratio), defined as

$$VSWR = \frac{|V(x)|_{\max}}{|V(x)|_{\min}} = \frac{1+|\rho_t|}{1-|\rho_t|} \tag{7.38}$$

If the line has losses ($\alpha > 0$) the expression above is modified in

$$VSWR = \frac{|V(x)|_{\max}}{|V(x)|_{\min}} = \frac{1+|\rho_t|\exp(-2\alpha x)}{1-|\rho_t|\exp(-2\alpha x)} \tag{7.39}$$

Several problems and adverse effects may be put in relationship with the observation of a standing wave pattern and large VSWR values:

- the voltage at the points of maximum may exceed or be too close to the voltage ratings of some of the connected equipment or the cable itself, thus causing voltage breakdown;
- similarly the current at the points of maximum may exceed or be too close to the voltage ratings of the same cable, thus causing local heating;
- because of the presence of an impedance mismatch (at the output and then proceeding backward producing a variation of the input impedance with frequency) the power transfer efficiency from the source will vary with frequency.

The reflection coefficient ρ_t may be calculated back from the VSWR as

$$\rho_t = \frac{VSWR-1}{VSWR+1} \tag{7.40}$$

Determination of the Reflection Properties of a Sample Coaxial Cable. Let's consider a piece of RG8 coax cable of length $L = 3.70\,\mathrm{m}$ and examine the measurement of the *return loss*, that can be performed with any spectrum analyzer. The return loss RL is a one-port measurement and is defined as the reflection coefficient at the input port with respect to the $Z_0 = 50\,\Omega$ reference impedance, usually expressed in dB. The other end of the cable is terminated on either an open circuit or a short circuit. Since the coax cable is terminated with two "N" connectors to plug into the input port of the spectrum analyzer and to fit the replaceable termination on the other end, the length is measured between the two reference planes at the connectors electrical contact points.

$$RL = 10\log\left[\frac{Z_{in}-50}{Z_{in}+50}\right] \tag{7.41}$$

By replacing (7.28) a relationship between the return loss RL, the terminating impedance Z_t, the cable length L and the propagation constant γ is obtained

$$\begin{aligned} RL &= 10\log\left[\frac{\tanh(\gamma L)-1}{\tanh(\gamma L)+1}\right] = 10\log\left[\exp(-\gamma L)\right] = \\ &= 10\log\left[\exp(-\alpha L)\right] + 10\log\left[\cos(-\beta L)+j\sin(-\beta L)\right] \end{aligned} \tag{7.42}$$

The first term represents the attenuation and with constant α it would be a horizontal line, constant with frequency; since α increases with frequency it exhibits a downward slope and $\alpha(f)$ may be estimated by curve fitting. The second term represents a ripple superimposed to the first term, oscillating by the same mechanism shown for (7.37), where the distance between two adjacent peaks is such that $2\pi L/\lambda = \pi$ and thus $L = \lambda/2$. By observing the distance on the frequency axis between successive local minima or maxima the resonant frequency f_0 is derived and the wave velocity is $v = 2f_0L$.

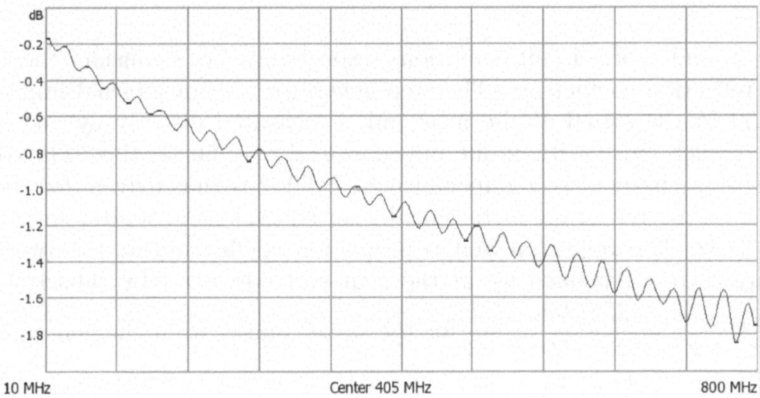

Fig. 7.2 Return Loss measurement in open-circuit condition

The results of the measurements on the cable sample are shown in Figure 7.2: 17.5 periods are counted between the fourth and the ninth division (a more accurate estimation than counting three periods per division) and each division is equal to $(800 - 10)/10 = 79\,\text{MHz}$: $f_0 = 6 \times 79/17.5 = 27.086\,\text{MHz}$. The cable velocity is $v = 2f_0L = 200.44\,10^6\,\text{m/s}$ and the velocity factor v/c is exactly 0.667 (as it is known from cable datasheet).

For the attenuation constant term only (7.42) may be simplified as $RL = -\alpha L\,10\log(e) = -\alpha L\,4.3429$.

Let's imagine an ideal line drawn amid the repeated oscillations, the decay is about 0.2 dB @ 10 MHz and increase to -0.83 dB @ 250 MHz, so that $\alpha = 0.0461/L = 0.0124\,\text{np/m}$ and $\alpha = 0.1911/L = 0.0517\,\text{np/m}$ at the two frequency values respectively.

The same cable is now tested in the time domain by means of a large bandwidth (6 GHz) large sampling rate (40 GS/s) Digital Storage Oscilloscope (DSO), as shown in Figure 7.3; the screen dumps are reported in Figure 7.4.

The two voltage waveforms captured at the near end of the transmission line made of RG8 cable match qualitatively the known behavior due to reflections from the far end of the line with and without 180° phase rotation for

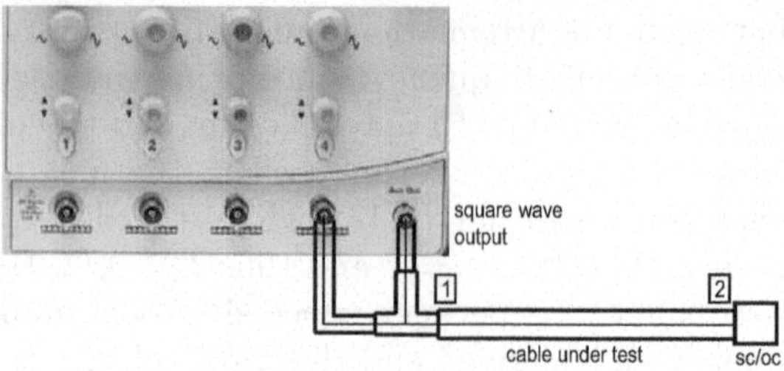

Fig. 7.3 RG8 test arrangement in time domain

the open- and short-circuit conditions respectively. Let's consider the values of the reflection coefficients. The step generator has an internal impedance $Z_g = 50\ \Omega$; the signal at the near end is measured directly by one of the DSO channels, also with input impedance $Z_{in} = 50\ \Omega$; the RG8 coaxial cable with a characteristic impedance $Z_c = 50\ \Omega$ is thus driven, for the calculation of the reflection coefficient, by an equivalent generator impedance $Z'_g = 25\ \Omega$. The reflection and transmission coefficients at the two cable ends are then determined as: at the near end, identified by subscript 1, is $\rho'_1 = \dfrac{Z'_g - Z_0}{Z'_g + Z_0} = \dfrac{25-50}{25+50} = -\dfrac{1}{3}$ for the wave coming from the generator external to the cable under test, while for the successive reflections inside the cable a change in sign is made, so that $\rho_1 = \dfrac{1}{3}$; the far end is terminated on either a short circuit with $\rho_{2,sc} = 1$ or on an open circuit with $\rho_{2,oc} = -1$. The voltage applied to the coaxial cable is thus given by the multiple reflections inside the cable after the first application of the square wave generator voltage at port 1, each characterized by a 33% reflection coefficient; for open circuit termination at port 2 there is a voltage phase reversal so that the portions of wave given by the odd order reflections subtract from the original square wave, as shown in Figure 7.4(a); on the contrary, for a short circuit termination on port 2, the reflected voltage waves always add to the first square wave as shown in Figure 7.4(b), again with the amplitude of each new term reduced by 33% reflection coefficient.

Transfer Impedance. Transfer impedance is in general a parameter that identifies the goodness of a shielded cable (e.g. a coaxial cable *in primis*, but also any other shielded pair) with respect to external disturbance in the form of an electromagnetic field hitting the external surface and voltage appearing on the inner conductor(s). The parameter links this voltage to the current that is flowing in the cable shield, from this the name and the unity of measure of impedance. For a fraction dx of the total cable length the voltage term is

$$dV = Z_t I_{sh} dx \tag{7.43}$$

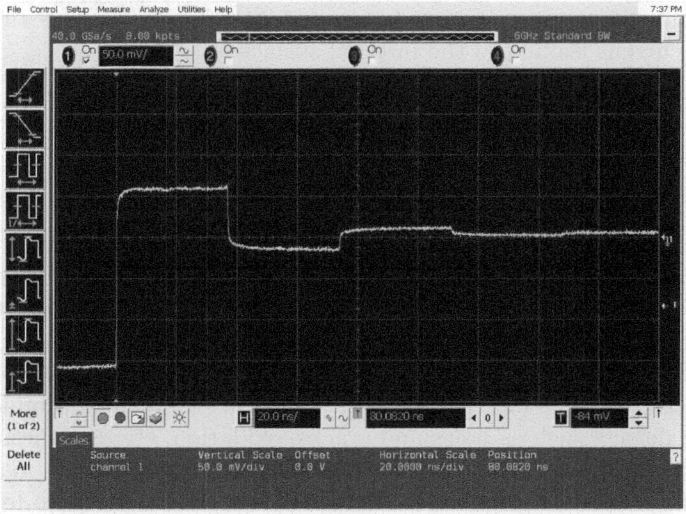

(a)

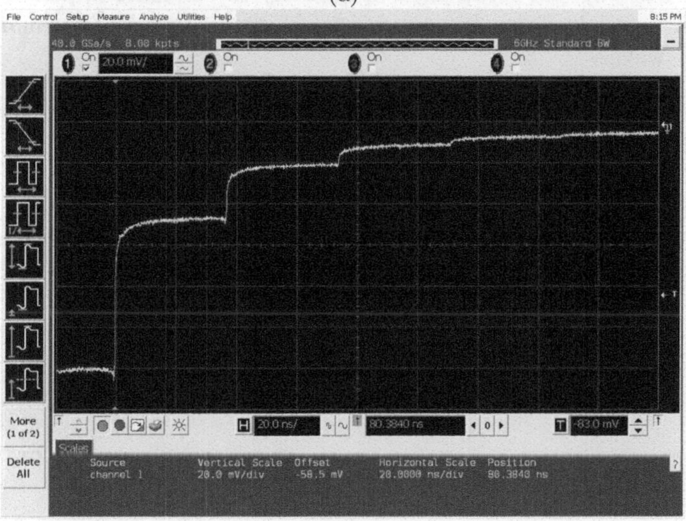

(b)

Fig. 7.4 Time domain measurement of step signal reflections: (a) open circuit condition, (b) short circuit condition

where Z_t is the transfer impedance and I_{sh} is the current flowing in the shield on the internal surface, so originating from the external electromagnetic field and passing through the shield. It is evident that for a perfect shield at a conveniently high frequency the current is confined completely on the external surface, because of a relevant skin effect, and the corresponding transfer impedance term is zero. In general, by diffusion even a solid shield will have current leaking to the internal surface given by

$$Z_t = R_{dc} \frac{\gamma t_{sh}}{\sinh(\gamma t_{sh})} \tag{7.44}$$

where γ is related to the skin effect and to the skin depth δ by $\gamma = (1+j)/\delta$, and t_{sh} is the shield thickness.

When a braided shield or a non-ideal foil shield are considered, the Z_t term ideally increases because of a poorer dc resistance with respect to a solid shield, but this would explain only a small increase uniform that can be compensated for by increasing the shield thickness; at higher frequency the larger Z_t values are on the contrary due to various non-idealities of the wire pattern or of the foil characteristics: optical coverage, non uniform distribution, defects such as holes and cracks, oxidation, etc. For these reasons a real plot of the transfer impedance versus frequency can show many deviations from an initial approximate inductive behavior linear with frequency; moreover, also the test setup used to quantify the transfer impedance exerts its influence at very high frequency, especially in terms of resonances and stationary waves.

7.2 Electromagnetic Field

This section provides a review of the electromagnetic theory and of the electromagnetic field fundamentals, in view of the successive treatment of antennas and of sources and victims of electromagnetic coupling at different distances and frequency values. The concepts of static electric and magnetic fields are first considered, as a preliminary basis for wave equation and electromagnetic wave propagation. The treatment is purposely not complete and only the main findings are recalled here.

7.2.1 Electric Field

Electric field is generated by the electric charge, that is measured in coulomb; it is indicated by **E** and defined as the vector force exerted on a unitary test charge. The measuring unit of the electric field is V/m from the definition

$$\mathrm{N/C} = \mathrm{J/m/(A\,s)} = \mathrm{W\,s/m/(A\,s)} = \mathrm{V/m}.$$

The electric field is in relationship with the flux density **D** and the permittivity ε, that for an isotropic medium is a scalar, so that the flux density is a vector with the same direction as **E** and with strength proportional to the quantity of charge. The **E** field originates at positive charges and terminates on negative charges or at infinity.

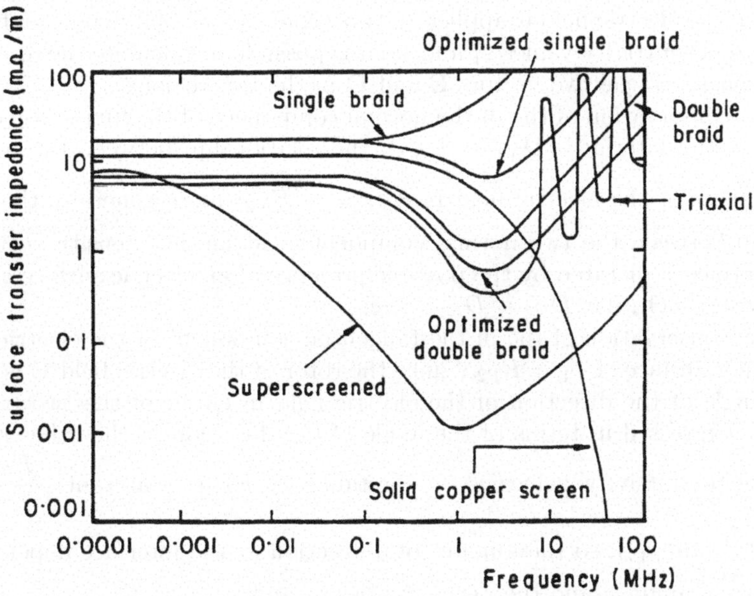

(a)

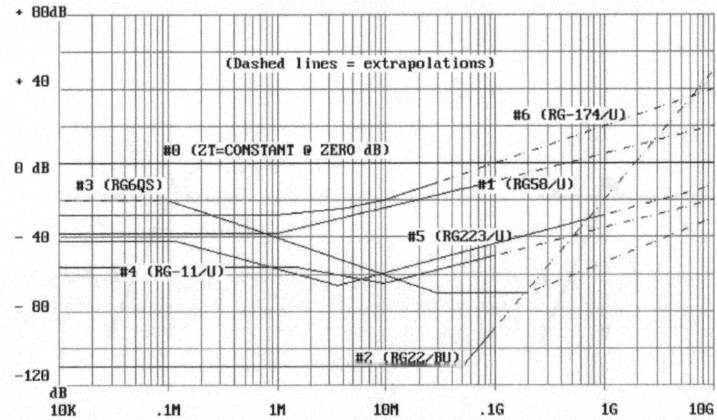

(b)

Fig. 7.5 Example of cable transfer impedance: (a) reference curves indicated by DEF STAN 59-41 [239], (b) exemplification of curves for most common values [289]

Gauss's Law states that the integral of the flux density over a closed surface is equal to the total charge Q therein. The permittivity is assumed for our purposes to be a constant scalar and for a given material is expressed as the relative permittivity ε_r (a real positive number) times the permittivity of free space $\varepsilon_0 = 8.854\,\mathrm{pF/m}$.

When the electric field is applied to two regions made of different materials with two permittivity values ε_1 and ε_2, it is possible to calculate the direction and intensity of the two vectors **E** and **D** in the two regions:

1) by conservation of the of the normal component of the flux density vector **D** at the interface ($D_{N1} = D_{N2}$), the relationship between the normal components of the electric field is $E_{N1} = \dfrac{\varepsilon_2}{\varepsilon_1} E_{N2}$; more generally the relationship between the two normal components of the flux density vector at the interface is dictated by the possible presence of an electric surface charge density ρ_S, such that $D_{N1} - D_{N2} = \rho_S$;

2) by conservation of the of the tangential component of the electric field **E** at the interface ($E_{T1} = E_{T2}$, since the rotor of the electric field **E** is zero) , the angle of the direction of the electric field in either of the two regions may be expressed in terms of the angle of the direction in the other region and the permittivity values: $\phi_2 = \arctan(E_{N2}/E_{T2}) = \arctan(\dfrac{\varepsilon_2}{\varepsilon_1}\dfrac{E_{N1}}{E_{T1}}) = \arctan(\dfrac{\varepsilon_2}{\varepsilon_1}\tan(\phi_1))$; so the amount of diffraction at the interface depends on the grazing angle $\phi_1$and the permittivities ε_1 and ε_2.

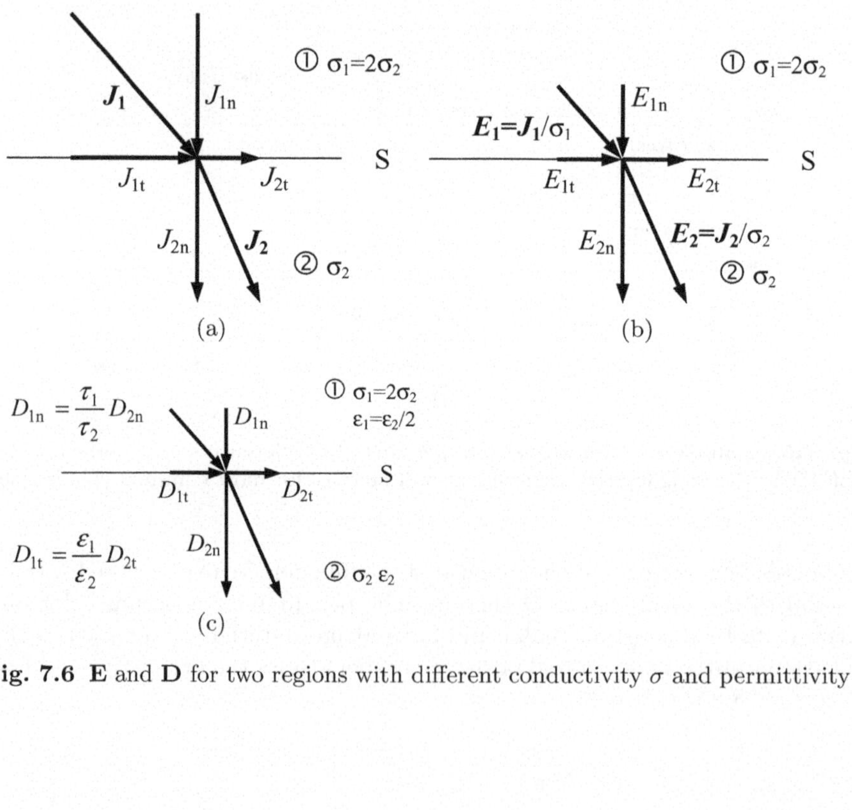

Fig. 7.6 **E** and **D** for two regions with different conductivity σ and permittivity ε

7.2.2 Magnetic Field

Magnetic field is generated by steady current flow or by magnetic materials; as it was done for the electric field there is a magnetic field strength (or simply magnetic field) **E** and a magnetic flux density (or induction field vector) **B**. For linear materials (those with no hysteresis behavior) they are related by the magnetic permeability, a scalar, so that $\mathbf{B} = \mu\mathbf{H}$, so that the two vectors have the same direction. The Biot-Savart Law (a more complete formulation of the Ampère's Law) in its integral form

$$\mathbf{H} = \frac{1}{4\pi}\int \frac{I\mathbf{dl} \times \mathbf{r}}{r^3} \tag{7.45}$$

says that the magnetic field vector **H** at a point is given by the integral of the current elements $I\mathbf{dl}$ in vector product with the vector between each current element and the point itself, normalized by the same distance amplitude to the third power. From this it is evident that the unity of measure of **H** is A/m. The unity of measure of the magnetic permeability is H/m, so that the unity of measure of the induction field vector **B** is $\mathrm{Wb/m^2} = \mathrm{T}$. Again the magnetic permeability μ may be expressed as a relative magnetic permeability μ_r times the magnetic permeability of free space $\mu_0 = 4\pi\, 10^{-7}\mathrm{H/m}$.

When the magnetic field is applied to two regions made of different materials with two permeability values μ_1 and μ_2, it is possible to calculate the direction and intensity of the two vectors **H** and **B** in the two regions:

1) by conservation of the of the normal component of the induction magnetic field vector **B** at the interface ($B_{N1} = B_{N2}$, since $\mathrm{div}\,\mathbf{B} = 0$), the relationship between the normal components of the magnetic field is $H_{N1} = \frac{\mu_2}{\mu_1} H_{N2}$; more generally the relationship between the two normal components of the induction field vector at the interface is dictated by the possible presence of a magnetic surface charge density τ_S, such that $B_{N1} - B_{N2} = \tau_S$;

2) by the relationship between the tangential component of the magnetic field **H** at the interface ($H_{T1} - H_{T2} = J_T$, since $\mathrm{rot}\,\mathbf{H} = J_T$) , the angle of the direction.

7.2.3 Electromagnetic Wave

We know that the essence of Maxwell's equations is that a time varying electric field produces a magnetic field, possibly related to a circulating current, and that a time varying magnetic field produces an electric field, related to the derivative of the linked magnetic flux.

The cross relation of the two rotor equations leads to the harmonic expression of the wave equation, that under the assumption of orthogonality

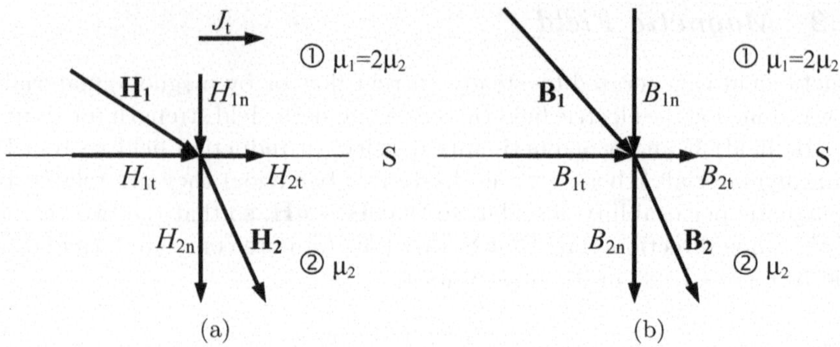

Fig. 7.7 **H** and **B** for two regions with different permeability μ

between the electric field, the magnetic field and the direction of propagation, takes the form

$$\frac{\partial^2 E_x}{\partial t^2} = \mu\varepsilon \frac{\partial^2 E_x}{\partial z^2} \tag{7.46}$$

The solution is the wave equation of the electric field component along the x axis, that under the assumption of orthogonality is the only non-zero electric field components, that propagates along the z axis. The solution is a wave equation, function of z and t, composed of a progressive (indicated by "f", forward) and a regressive (indicated by "b", backward)component, of the form

$$E_x(z,t) = E_f \exp(\omega t - \gamma z) + E_b \exp(\omega t + \gamma z) \tag{7.47}$$

where the arguments of the exponential $(\omega t - \gamma z)$ and $(\omega t + \gamma z)$ describe the propagation of the wave in time and space, as it was for transmission lines in the previous section.

7.2.3.1 Impedance - Wave and Medium

The impedance Z_w of a wave impinging normally to a surface is $Z_w = E_t/H_t$ where the tangential fields E_t and H_t are mutually perpendicular to each other. If we consider a conducting surface of thickness t,with a uniform linear current density $\hat{J}$ equal to J_T above, the conduction equation based on the conductivity σ is $J = \sigma E$. If t approaches zero and σ increases so that the product $G = \sigma t$ remains constant,

$$\begin{aligned} \hat{J} &= GE, \\ E &= R\hat{J} \end{aligned} \tag{7.48}$$

where G and R are called the surface conductance and surface resistance of the sheet respectively. More generally, the surface admittance Y_s and the surface impedance Z_s (both complex quantities) are defined by the equations:

$$\begin{aligned} \hat{J} &= Y_s E \\ E &= Z_s \hat{J} \end{aligned} \tag{7.49}$$

The primary electromagnetic properties of a medium are its conductivity σ, dielectric constant ε, and permeability μ. In transmission line theory, two secondary constants are (see 7.1.2) introduced, the propagation constant Γ and the characteristics impedance η. In three dimensional theory, the important constants are the intrinsic propagation constant γ and the intrinsic impedance η is defined by:

$$\begin{aligned} \gamma &= \sqrt{j\omega\mu\,(\sigma + j\omega\varepsilon)} \\ \eta &= \sqrt{\frac{j\omega\mu}{\sigma + j\omega\varepsilon}} \end{aligned} \tag{7.50}$$

These constants are independent of the geometry of the wave, and in general both are complex quantities:

$$\begin{aligned} \gamma &= \alpha + j\beta \\ \eta &= R + jX \end{aligned} \tag{7.51}$$

where

α is the attenuation constant;
β is the phase constant;
R is the intrinsic resistance;
X is the intrinsic reactance.
For a perfect dielectric:

$$\gamma = j\beta \qquad \beta = \frac{2\pi}{\lambda} \qquad \eta = \sqrt{\frac{\mu}{\varepsilon}}$$
$$\nu = \frac{1}{\sqrt{\mu\varepsilon}} \qquad \mu = \frac{\eta}{\nu} \qquad \varepsilon = \frac{1}{\eta\nu}$$

where the terms have their usual meanings.

7.2.3.2 Polarization

Polarization of an antenna in a given direction is defined as "the polarization of the wave transmitted (radiated) by the antenna. Note: When the direction is not stated, the polarization is taken to be the polarization in the direction of maximum gain." In practice, polarization of the radiated energy varies with the direction from the center of the antenna, so that different parts of the pattern may have different polarizations. Polarization of a radiated wave is defined as "that property of an electromagnetic wave describing the time-varying direction and relative magnitude of the electric-field vector; specifically, the figure traced as a function of time by the extremity of the vector at a fixed location in space, and the sense in which it is traced, as observed along the direction of propagation." Polarization then is the curve traced by the end point of the arrow (vector) representing the instantaneous electric field.

The polarization of a wave can be defined in terms of a wave radiated or received by an antenna in a given direction. The polarization of a wave radiated by an antenna in a specified direction at a point in the far field is defined as "the polarization of the (locally) plane wave which is used to represent the radiated wave at that point. At any point in the far field of an antenna the radiated wave can be represented by a plane wave whose electric-field strength is the same as that of the wave and whose direction of propagation is in the radial direction from the antenna. As the radial distance approaches infinity, the radius of curvature of the radiated wave's phase front also approaches infinity and thus in any specified direction the wave appears locally as a plane wave." This is a far-field characteristic of waves radiated by all practical antennas. The polarization of a wave received by an antenna is defined as the "polarization of a plane wave, incident from a given direction and having a given power flux density, which results in maximum available power at the antenna terminals." Polarization may be classified as linear, circular, or elliptical.

7.3 Antennas and Electromagnetic Field Radiation

An antenna may be defined as a structure associated with the region of transition between a guided wave and a space wave (transmission), or vice-versa (reception); the term "space wave" is used, and not "free space wave" in the sense that in many cases there is always a certain degree of interaction of the antenna with the surroundings and the propagation is not that of free space. For a broad class of linear antennas the Reciprocity Theorem [23] says that the radiation pattern (describing the distribution of electromagnetic field around the antenna itself) is the same for a transmitting or receiving antenna.

As an introductory concept, the *radiation pattern* is defined as a mathematical function or representation of the radiation properties of the antenna in the space coordinates. The radiation properties may include the radiation intensity, the field strength or field polarization.

Based on this it is possible to define:

- *isotropic radiator*, as a lossless antenna having equal radiation in all directions; although not really existing, it is taken as a reference radiator to quantify the distribution of the radiation intensity around a particular antenna;

- *directional antenna*, with the property of receiving or radiating the electromagnetic field in some directions than others; in particular, this definition, even if of general value, is applied to those antennas that show a significant directivity in a particular direction;
- *omnidirectional antenna*, that is a particular type of directional antenna with a non-directional pattern in a given plane, as it will be a dipole (see sec. 7.3.3) or a loop antenna (see sec. 7.3.6); it is anticipated here that the radiation pattern of these antennas (see sec. 7.3.3) is often approximated by a power of $\sin(\theta)$, that is $|\sin^n(\theta)|$, and independent on φ, so omnidirectional in the $0 \leq \varphi \leq 2\pi$ plane.

Since an antenna is a source of electromagnetic field, composed in turn in the far field region by the E-field component and by the orthogonal H-field component, there are two main planes at least, each containing the E- and H-field vectors and used usually as reference planes for the antenna placement and orientation in drawings.

To distinguish between portions of space with intense radiation and portions of space with weak radiation, the term lobe is often used, since these regions of relatively higher radiation have all the shape of a balloon. In a radiation pattern it is possible to distinguish between the main lobe, minor lobes, side lobes and back lobes (or back radiation lobes).

The main variables that are used in the following to describe antenna parameters are, for an antenna in transmission mode: P_g, the power generated by an external source (e.g. radio frequency amplifier or generator), P_a the "accepted" power, entering the antenna connector, and P_r, the power effectively radiated outside the antenna into the surrounding space.

We define the antenna efficiency as $\eta = P_r/P_a$.

If we consider the distribution of power around the antenna, given a spherical coordinate system with the antenna as the center, we assume the radiation intensity is independent of distance r, then we are under far-field assumption.

7.3.1 Antenna Parameters

7.3.1.1 Radiated Power and Radiation Intensity

The radiated power may be defined as the integral over the entire volume of the radiation intensity $\Phi(\theta, \varphi)$, that under the far field assumption depends only on the coordinates $\theta$and φ. The radiation intensity has the unity of measure of a power density, so $\mathrm{W/m^2}$.

$$P_r = \int\int \Phi(\theta, \varphi) \sin\theta d\theta d\varphi \tag{7.52}$$

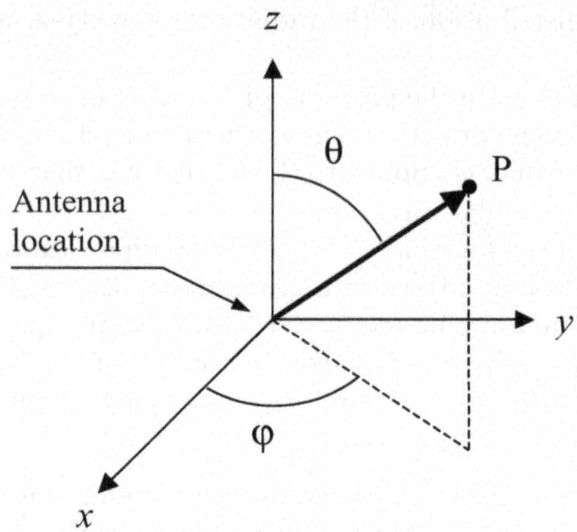

Fig. 7.8 Coordinate system for antenna properties

The average radiation intensity is then defined as

$$\Phi_{avg} = \frac{P_r}{4\pi} \tag{7.53}$$

The radiation intensity $\Phi(\theta, \varphi)$ is derived from the radiation density and then from the application of the Poynting Theorem to the E-field and H-field components. The radiation density $W_{rad}(\theta, \varphi)$ is given by

$$W_{rad}(\theta, \varphi) = \frac{1}{2}\mathrm{Re}[\mathbf{E} \times \mathbf{H}] \tag{7.54}$$

and the product of it and the square of the distance gives the radiation intensity, that is the radiation density per unit solid angle (1 srad).

$$\Phi(\theta, \varphi) = r^2 W_{rad}(\theta, \varphi) \tag{7.55}$$

7.3.1.2 Gain and Directivity

The directivity is then defined as the ability of an antenna to concentrate the emitted power in a particular portion of space; the directivity function is the ratio of the radiation intensity with the average radiation intensity

$$D(\theta, \varphi) = \frac{\Phi(\theta, \varphi)}{\Phi_{avg}} \tag{7.56}$$

so that its maximum is called simply directivity using the symbol D or D_0.

There is a practical formula for antennas with only one narrow major lobe and negligible minor lobes, that gives an estimation of the maximum directivity with respect to the half-power beam-widths (HPBW) in two orthogonal planes (usually the horizontal and vertical plane)

$$D_0 = \frac{32400}{\Theta_h \Theta_v} \tag{7.57}$$

The HPBW angles Θ_h and Θ_v are measured in degrees and correspond to the angle in a plane containing the direction of maximum that intercepts the half-power points (-3 dB) of the power radiation pattern or, more generally, the points where the value is half the maximum. This formula is very useful for directive antennas, such as radar, radio link and some access point antennas, for which, given the plots of the radiation patterns, it is possible to estimate the directivity, and compare it to the gain, for greater confidence.

The "directivity" is often confused with the "gain": the difference is that the latter takes into account also antenna efficiency, so that in analogy we have a gain function

$$G(\theta, \varphi) = \eta D(\theta, \varphi) \tag{7.58}$$

and the maximum (that corresponds to the same direction in (θ, φ)) is called gain G.

Balanis proposes a practical formula related to the HPBW angles, similar to the one for directivity, where the numerator is only 30000; from this an implicit estimation of the antenna efficiency of 92.5%, that is a power loss of 0.33 dB.

By putting in relationship the input power P_a and the radiation intensity in one direction $\Phi(\theta, \varphi)$, it is possible to obtain an operative relationship that involves the gain function

$$P(\theta, \varphi) = \frac{\Phi(\theta, \varphi)}{r^2} = G(\theta, \varphi) \frac{P_a}{4\pi r^2} \tag{7.59}$$

Practically, the gain function may be found immediately by taking the radiation pattern used for the directivity and by assigning the gain G to the point of maximum of the plot. The factor $P_a/4\pi r^2$ represents the power density that would result if the power accepted by the antenna were radiated by a lossless isotropic antenna.

7.3.1.3 Antenna Factor

This parameter is a transducer factor that relates the intensity of the incident field strength and the voltage across a matched load at antenna input. The antenna factor K can be related to the antenna gain by the following formula

$$K = \frac{9.73}{\lambda \sqrt{G}} \tag{7.60}$$

where the parameters G and K are in linear scale; the same expression in dB takes the form

$$K = -29.8\,\mathrm{dB} + 20\log(f/1\,\mathrm{MHz}) - G \tag{7.61}$$

7.3.1.4 Bandwidth

Until now the antenna characteristics have been considered independent on the frequency. Yet, all these characteristics vary with the frequency and not in the same way; for this reason the concept of antenna bandwidth is ambiguous, if not related to a specific characteristic, such as the gain, the input impedance, the half-power beam-width, etc.: it may be said simply that the bandwidth is the range of frequency values within which the performance of the antenna with respect to one of the characteristics meet a given requirement. The requirement may be to stay within a given percentage of the maximum value and usually the bandwidth is approximately centered around a central frequency which the maxima of some characteristics correspond to.

7.3.1.5 Input Impedance

The input impedance of an antenna Z_A is simply given by the ratio of voltage and current at the input connector of the antenna itself. The input impedance may be thought composed of a real resistive part R_A and an imaginary reactive part jX_A; the resistive part is in turn composed of the radiation resistance R_r and the antenna loss resistance R_l.

A simple circuit is considered, where the antenna is supplied by a generator with amplitude E_g and series impedance $Z_g = R_g + jX_g$; the antenna is thus operated in transmitting mode. The calculations that follow aim at estimating the power losses and the conditions for the best transmission of power from the generator to the antenna and then to the space around; the antenna is a simple antenna for which R_r and R_l may be considered in series.

The current supplied by the generator to the antenna input port is

$$I_g = \frac{E_g}{Z_g + Z_A} = \frac{E_g}{(R_g + R_r + R_l) + j(X_g + X_A)} \tag{7.62}$$

The power delivered for radiation P_r and dissipated as losses P_l are

$$P_r = R_r\,|I_g|^2 \qquad P_l = R_l\,|I_g|^2 \tag{7.63}$$

The remaining power is dissipated on the generator resistance R_g or exchanged by the reactive components.

The maximum power transfer occurs when the generator and the antenna have conjugate matching of their impedances:

$$R_g = R_r + R_l \qquad X_g = -X_A \tag{7.64}$$

and the power terms become:

$$P_r = |V_g|^2 \frac{R_r}{4(R_r + R_l)^2} \qquad P_l = |V_g|^2 \frac{R_l}{4(R_r + R_l)^2}$$

$$P_g = |V_g|^2 \frac{1}{4R_g} = |V_g|^2 \frac{R_r + R_l}{4(R_r + R_l)^2} \tag{7.65}$$

If the antenna is lossless ($R_l = 0$) then half of the power supplied by the generator reaches the antenna and is radiated, the other half being dissipated in the generator resistance; in all other cases the fraction of radiated power is even less.

Analogously, if the antenna is operated in receiving mode, the terminating load (represented by the measuring or receiving apparatus) is characterized by an impedance $Z_t = R_t + jX_t$, that replaces the generator impedance in the expressions above. The same expressions apply, where Z_t. P_t and V_t replace Z_g, P_g and V_g respectively. In particular, since the objective in receiving mode is to deliver power to the terminating load, and we know that under conjugate it reaches half of the received power, the power delivered to R_r and R_l sum to the other half of the received power; the first term across R_r is called scattered (or re-radiated) power.

7.3.1.6 Effective Aperture and Directivity

When considering for simplicity a receiving antenna, the effective aperture (or effective area) is that antenna parameter that puts in relationship the power density W_{inc} of the incident wave and the power P_t appearing at antenna terminals on the terminating load and has thus the unity of measure of a surface:

$$A_e = \frac{P_t}{W_{inc}} \tag{7.66}$$

The maximum effective aperture A_{em} is obtained for the conjugate matching conditions seen above. For a transmitting antenna the effective aperture puts in relationship the radiation density W_{rad} and the accepted power P_a.

Other areas may be defined that correspond to other two terms of the power balance above, that is scattering area and loss area. The aperture efficiency of an antenna is the ratio between the maximum effective aperture and the physical area. There are types of antennas for which the effective area can be even equal to the physical area (so reaching a 100% efficiency), but this doesn't imply that the received power is larger than one half, since also the scattering area (neglecting any power loss for simplicity) can be as large, so reaching a scattering efficiency of 100%, thus keeping the balance between the two area terms at 50% (as it should be for conjugate matching).

The relationship between the (maximum) effective aperture A_e and the (maximum) directivity D is (without demonstration)

$$A_e = \frac{\lambda^2}{4\pi} D \tag{7.67}$$

7.3.2 Elementary Antenna - The Hertzian Dipole

An infinitesimal current element of length dz is considered at the centre of the coordinate system, normally referred to as "Hertzian Dipole"; with reference to Figure 7.8 the current element is positioned along the z axis. The three components of the electric field E and magnetic field H along r, θ and φ are determined as follows:

$$E_r = \frac{\eta}{2\pi} k^2 I\, dz \left[\frac{1}{(kr)^2} - j\frac{1}{(kr)^3} \right] \cos\theta\, e^{-jkr} \tag{7.68}$$

$$E_\theta = j\frac{\eta}{4\pi} k^2 I\, dz \left[\frac{1}{(kr)} - j\frac{1}{(kr)^2} - \frac{1}{(kr)^3} \right] \sin\theta\, e^{-j\beta r} \tag{7.69}$$

$$H_\varphi = j\frac{k^2}{4\pi} I\, dz \left[\frac{1}{(kr)} - j\frac{1}{(kr)^2} \right] \sin\theta\, e^{-j\beta r} \tag{7.70}$$

$$E_\varphi = H_r = H_\theta = 0 \tag{7.71}$$

As a note, it is underlined that the parameter k is called the "wave number", and is equivalent to the phase constant $\beta = 2\pi/\lambda$. It is interesting to observe that the ratio of the first order components of the electric and magnetic field, that are projected along θ and φ axes (with propagation occurring along the r axis), give the free space impedance, that is $Z_0 = E_\theta^{(1)}/H_\varphi^{(1)} = 120\pi = 376.7$, the unity of measure being ohm.

The other components of higher order form the reactive field. A general definition of the three regions function of distance from the antenna is: that region in which the reactive component predominates is said the *reactive near-field region* (some call it simply "near-field region"); the outer region where the radiating component predominates, is further subdivided into two regions, *radiating near-field region* (where the relative phases and amplitudes of contributions from various elements of the antenna are functions of the distance from the antenna), and *radiating far-field region* (where, at larger distance, straight lines from the antenna elements to the observation point are essentially parallel and the angular distribution of radiated energy is independent on distance). The radiating near field-region is sometimes called intermediate region, since it separates the near-field region *tout-court* from the far-field region.

Following the development in [25], the Hertzian dipole equations above may be computed for the three regions, where $kr \ll 1$, $kr > 1$ and $kr \gg 1$ respectively.

Reactive near-field region, $\beta r \ll 1$:

$$E_r = -j\frac{\eta}{2\pi} I\,dz \frac{1}{kr^3} \cos\theta\, e^{-jkr} \tag{7.72}$$

$$E_\theta = j\frac{\eta}{4\pi} kI\,dz \frac{1}{r} \sin\theta\, e^{-jkr} \tag{7.73}$$

$$H_\varphi = \frac{1}{4\pi} I\,dz \frac{1}{r^2} \sin\theta\, e^{-jkr} \tag{7.74}$$

It is interesting to observe the two E-field components have both a $-j$ coefficient with respect to the H-field component, so that they are in time-quadrature; this implies that their time-averaged product is zero and so it is the associated power flow. Furthermore, the two E-field components have similar expressions and are in an in-phase condition.

Radiating near-field region, $kr > 1$:

$$E_r = \frac{\eta}{2\pi} I\,dz \frac{1}{r^2} \cos\theta\, e^{-jkr} \tag{7.75}$$

$$E_\theta = j\frac{\eta}{4\pi} kI\,dz \frac{1}{r} \sin\theta\, e^{-jkr} \tag{7.76}$$

$$H_\varphi = j\frac{1}{4\pi} kI\,dz \frac{1}{r} \sin\theta\, e^{-jkr} \tag{7.77}$$

The two E-field components lose their in-phase condition, so that the resulting E-field vector is a rotating vector with the extremity tracing an ellipse. However, the two components E_θ and H_φ are in time-phase condition and this indicates the formation of an average power flow directed radially in the outward direction, thus starting the formation of the radiation phenomenon, typical of the far-field region; from this the correctness of naming this region *radiating* near-field region.

Far-field region, $kr \gg 1$:

$$E_r = 0 \tag{7.78}$$

$$E_\theta = j\frac{1}{4\pi} kI\,dz \frac{1}{r} \sin\theta\, e^{-jkr} \tag{7.79}$$

$$H_\varphi = j\frac{1}{4\pi} kI\,dz \frac{1}{r} \sin\theta\, e^{-jkr} \tag{7.80}$$

Finally we reach the known condition of orthogonal propagation of the E-field and H-field components projected only on the θ and φ axes, with a 120π ratio (that corresponds to the said free space intrinsic impedance) and with a linear dependence on distance; the associated power density, given by time-average product of the E-field and H-field, shows a dependence on distance of the type $1/r^2$.

The distance $r*$ at which there is the transition from the reactive field to the radiating field region is conventionally taken as $r* = \lambda/2\pi$, that in our treatment above corresponds to the condition $\beta r = 1$. It is important to observe that at this distance the amplitude of the far field and reactive field components is only equal, so the meaning of this distance is that above it the far field components start prevailing, but in no way it indicates a distance at which the reactive field components are negligible.

7.3.3 Dipole Antenna

A dipole antenna is a very common antenna, since several sources may be approximated by this antenna with a satisfactory and accurate enough treatment: the biconical and log-periodic measuring antennas stem from a basic design built around a dipole; a slot in an enclosure may be modeled by an orthogonal dipole; the overhead contact wire excited by the electric arc at pantograph may be seen as a dipole with variable length, depending on frequency, as well as a long wire antenna; another very useful antenna, the monopole antenna, may be and is calculated by halving a dipole antenna.

By assuming a finite length dipole with negligible wire diameter, the current distribution may be assumed sinusoidal, going to zero at the two extremes, with the dipole center-fed. Even if it can be shown that a closed form solution for this current distribution is available, that is valid in all the regions (and all points in space), the followed approach is that of limiting the validity to the far-field region, so that it is applicable in general to a wide variety of distributions and leads to simpler formulations.

The dipole is subdivided in several infinitesimal dipoles of length dz, whose far-field E-field and H-field components can be computed by using (7.79) and (7.80). The sum of all the contributions from each infinitesimal element reduces, in the limit $dz \to 0$, to an integral. The integral gives an expression for the E-field (and in turn for the H-field, by simply dividing by the free space impedance) that is composed of a first term that is very similar to the one obtained for the infinitesimal dipole (7.79) multiplied by the integral of the current distribution. By assuming the sinusoidal current distribution above the mathematical expression of the E-field component E_θ is

$$E_\theta = j60kI_0 \frac{1}{r} e^{-jkr} \left[\frac{\cos(kl/2 \cos\theta) - \cos(kl/2)}{\sin\theta} \right] \tag{7.81}$$

The power distribution around the dipole antenna is given by the real part of the Poynting vector, that corresponds to the multiplication of $E_\theta$and H_φ; the radiation intensity is readily obtained by multiplying it by the unit solid angle, that is r^2.

$$U = \frac{30}{2\pi} k I_0^2 \left[\frac{\cos(kl/2\ \cos\theta) - \cos(kl/2)}{\sin\theta} \right]^2 \tag{7.82}$$

It is instructive to look at the radiation intensity plot as a function of θ for different length/wavelength ratios; the plots shown in Figure 7.9 are normalized plots (with the maximum value corresponding to 0 dB) for $l = \lambda/4$, $l = \lambda/2$, $l = 3\lambda/4$ and $l = \lambda$ cases, together with the reference case of the infinitesimal dipole, for which $l \ll \lambda$.

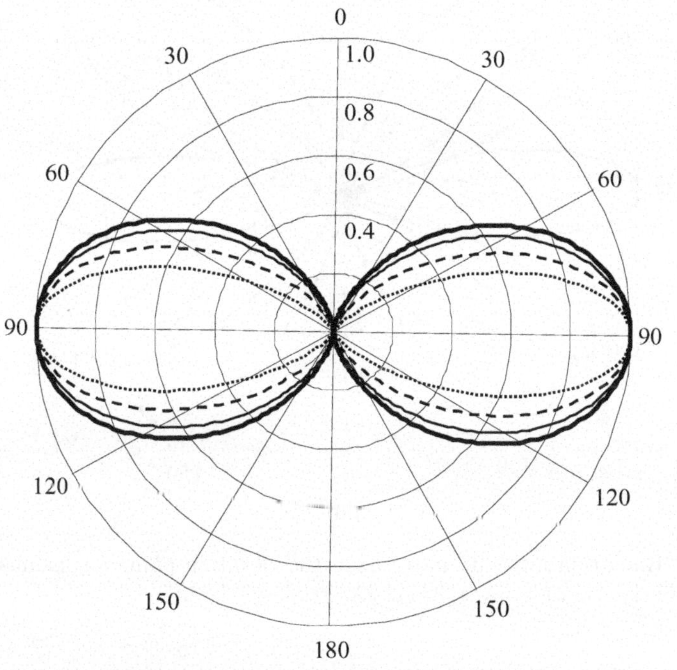

Fig. 7.9 Radiation intensity patterns in the elevation plane for a sinusoidal current distribution and $l = \lambda/4$, $\lambda/2$, $3\lambda/4$, λ.

All the patterns have one lobe with different width: as the length of the dipole increases the lobe becomes narrower and thus the dipole directivity increases; the half-power beam-width of the examined cases is reported in Table 7.1.

If the l/λ ratio is further increased, the number of lobes begins to increase. By using the expression (7.82) of the radiation intensity (that defines the

Table 7.1 Half-power beam-width of the dipole for $l \leq \lambda$

Case	HPBW
$l \ll \lambda$	90°
$l \ll \lambda/4$	87°
$l \ll \lambda/2$	78°
$l \ll 3\lambda/4$	64°
$l \ll \lambda$	47.8°

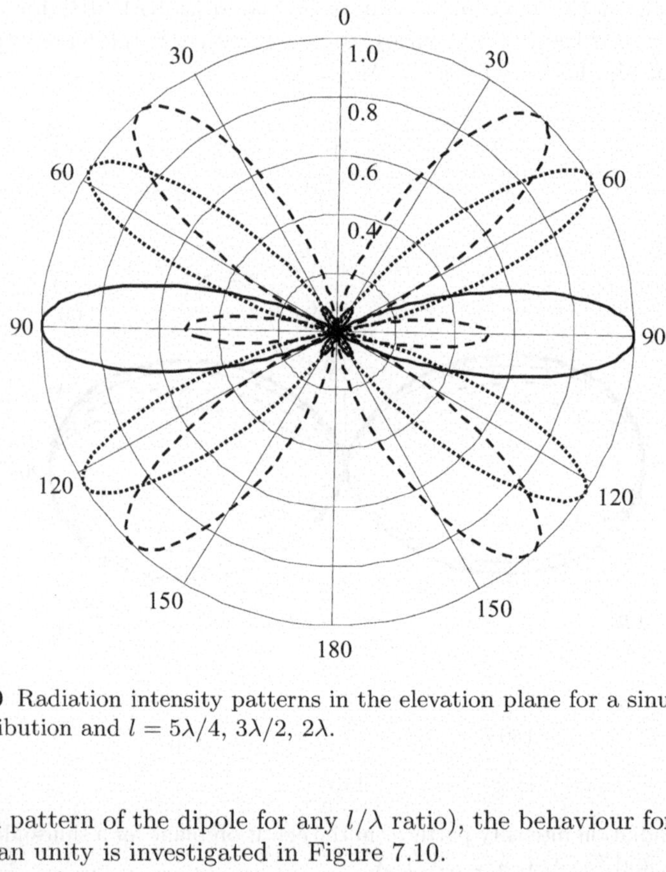

Fig. 7.10 Radiation intensity patterns in the elevation plane for a sinusoidal current distribution and $l = 5\lambda/4$, $3\lambda/2$, 2λ.

radiation pattern of the dipole for any l/λ ratio), the behaviour for l/λ ratio larger than unity is investigated in Figure 7.10.

The general concepts of radiation resistance and input resistance given in sec. 7.1.2.3 may be explicated here for the dipole; some of the calculations and simplifications are omitted, but the complete development may be found in [25].

The calculation and integration over a sphere of radius r of the Poynting vector give the radiated power

$$P_{rad} = \eta \frac{|I_0^2|}{4\pi} \left\{ C + \ln(kl) - C_i(kl) + \frac{1}{2}\sin(kl)\left[S_i(2kl) - 2S_i(kl)\right] + \right.$$
$$\left. + \frac{1}{2}\sin(kl)\left[C + \ln(kl/2) + C_i(2kl) - 2C_i(kl)\right] + \right\} \tag{7.83}$$

where

$C = 0.5572$ is the Euler's constant;

$C_i = -\int_x^\infty \frac{\cos y}{y} dy$ is the cosine integral function;

$S_i = -\int_0^x \frac{\sin y}{y} dy$ is the sine integral function;

I_0 is the peak value of the current distribution.

From the radiated power the radiation resistance R_r is calculated by simply dividing by the rms value of the current $\frac{1}{2}|I_0|^2$:

$$R_{rad} = \frac{2P_{rad}}{|I_0|^2} \tag{7.84}$$

This expressions will be used again in sec. 7.3.3.4 below.

7.3.3.1 Half-Wavelength Dipole

This is a special case of the formulas above with length $l = \lambda/2$ and it is very useful in practice, since several transmitting and receiving antennas are or may be modelled with the help of the half-wavelength dipoles: measuring antennas such as calibrated dipoles and biconical antennas; amateur radio antennas made of a Marconi semi-dipole in the 27 and 144 MHz bands; FM radio and TV broadcasting towers, equipped with arrays of tuned vertical dipoles.

The electric and magnetic field equations are taken from (7.81) and repeated here for $l = \lambda/2$:

$$E_\theta = j\eta I_0 \frac{1}{2\pi r} e^{-jkr} \left[\frac{\cos(\pi/2\ \cos\theta)}{\sin\theta}\right] \tag{7.85}$$

$$H_\varphi = jI_0 \frac{1}{2\pi r} e^{-jkr} \left[\frac{\cos(\pi/2\ \cos\theta)}{\sin\theta}\right] \tag{7.86}$$

The radiation intensity is again defined starting from the time-average power density, given by the multiplication of E_θ and H_φ

$$U = \eta \frac{I_0^2}{8\pi^2} \sin^3\theta \tag{7.87}$$

and its two-dimensional and three-dimensional patterns are plotted in Figure 7.11.

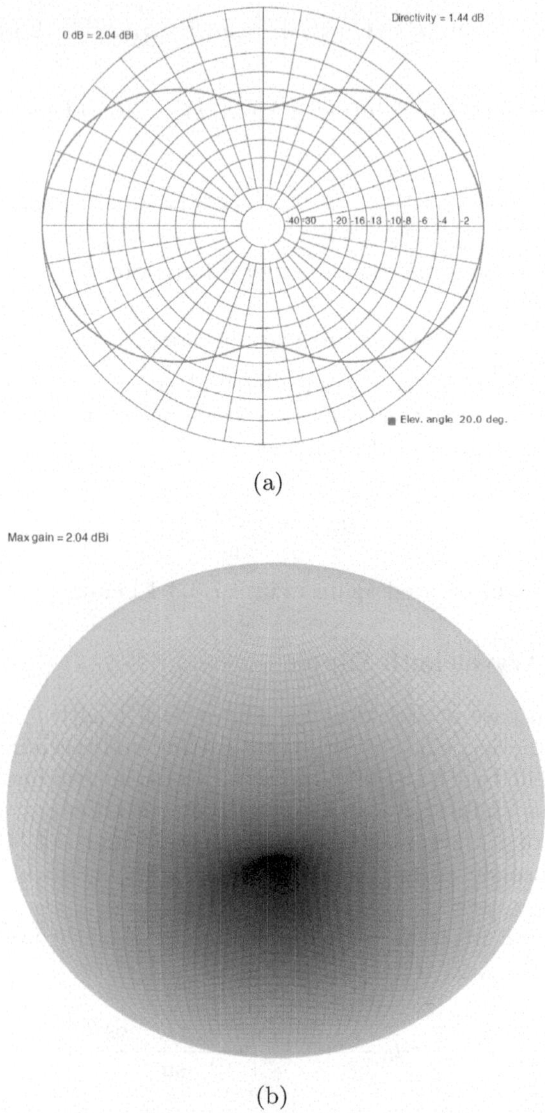

Fig. 7.11 Two- and three-dimensional patterns of the half-wave dipole

The total radiated power is calculated in [25] and is found to be

$$P_{rad} = \eta \frac{I_0^2}{8\pi} \mathrm{C_{in}}(2\pi) \simeq \eta \frac{I_0^2}{8\pi} 2.435 \tag{7.88}$$

(where $\mathrm{C_{in}}(\cdot)$ indicates the cos-integral function). The directivity, by definition, is the ratio of the maximum radiation intensity over the average power density, that is $P_{rad}/4\pi$: $D_0 \simeq 4/2.435 = 1.643$; the maximum effective area is in force of (7.67) $A_{em} \simeq 0.13\lambda^2$.

The radiation resistance R_r may be calculated from the radiated power P_{rad}, by equating (7.88) to the simple definition (7.63): $R_r = \frac{\eta}{4\pi}\mathrm{C_{in}}(2\pi) \simeq 73\,\Omega$.

7.3.3.2 Vertical and Horizontal Dipoles Near a Perfect Ground

The complete treatment may be again found in [25], here only some considerations and graphical results are reported, to give the basic elements to understand and evaluate the effect of ground.

The analysis of an antenna (or more generally a conductor) near the ground, considered as an infinite conductive plane, is made by means of the virtual images of the antenna itself, to account for reflections. These are not real sources, but the application of standard theory for real sources anyway allows to calculate the field components, that combined with those from the real sources above ground take into account reflections and give the final field. The composition of direct and reflected field component is eased by assuming a unity reflection coefficient and equal angles for the incident and reflected field.

The geometries that are considered here are those of infinitesimal vertical and horizontal dipoles (Hertzian dipoles) at a height h above ground. The total field above the ground plane (below it is zero, since the ground plane is a perfect conductor) and some of the performance parameters are given by the following relationships.

Vertical dipole:

$$E_\theta^v = j\eta k I_0 l \frac{1}{4\pi r} e^{-jkr} \sin\theta \left[2\cos(kh\cos\theta)\right] \tag{7.89}$$

The factor between square brackets is a function of the antenna height h and of the observation angle θ and it is known as pattern multiplication factor, or array factor. It creates multiple lobes as a function of the value of kh, that is the ratio of the height over the wavelength.

Without demonstration the maximum directivity and the radiation resistance are

$$R_r = 2\pi\eta \left(\frac{l}{\lambda}\right)^2 \left[\frac{1}{3} - \frac{\cos(2kh)}{(2kh)^2} + \frac{\sin(2kh)}{(2kh)^3}\right] \tag{7.90}$$

$$D_0 = \frac{2}{\left[\frac{1}{3} - \frac{\cos(2kh)}{(2kh)^2} + \frac{\sin(2kh)}{(2kh)^3}\right]} \tag{7.91}$$

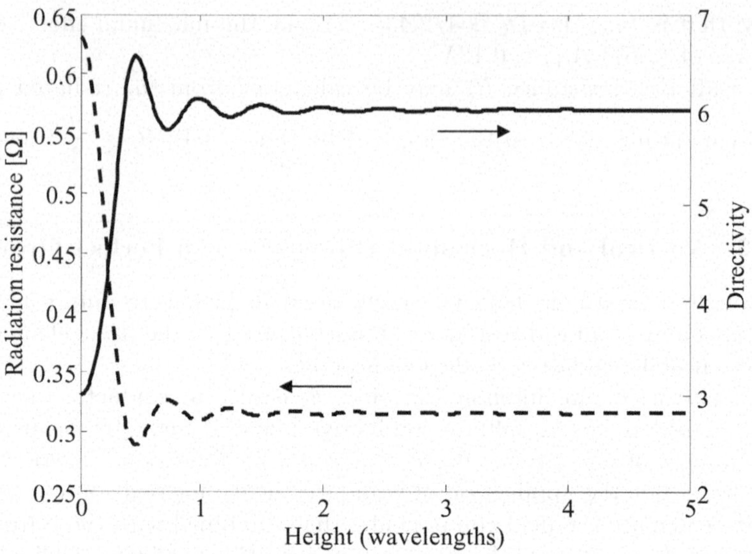

Fig. 7.12 Directivity and radiation resistance of an infinitesimal vertical dipole at height h above perfect ground

and they are plotted versus height (measured in multiples of the wavelength) in Figure 7.12.

Finally, the radiation patterns (that are of course influenced by the effects of the ideal ground and the relative height above it) are shown below, in Figure 7.13, for various height values, and by dividing the figure in two quadrants for display purposes.

Horizontal dipole:

$$E_{\psi}^{h} = j\eta k I_0 l \frac{1}{4\pi r} e^{-jkr} \sqrt{1 - \sin^2\theta \sin^2\varphi}\, [2j \sin(kh\cos\theta)] \tag{7.92}$$

The factor between square brackets is again the array factor.

Without demonstration the maximum directivity and the radiation resistance are

$$R_r = \pi\eta \left(\frac{l}{\lambda}\right)^2 \left[\frac{2}{3} - \frac{\sin(2kh)}{(2kh)} - \frac{\cos(2kh)}{(2kh)^2} + \frac{\sin(2kh)}{(2kh)^3}\right] \tag{7.93}$$

$$D_0 = \begin{cases} 4\sin^2(kh)/R_r & h \le \lambda/4 \\ 4/R_r & h > \lambda/4 \end{cases} \tag{7.94}$$

and they plotted versus height (measured in multiples of the wavelength) in Figure 7.14.

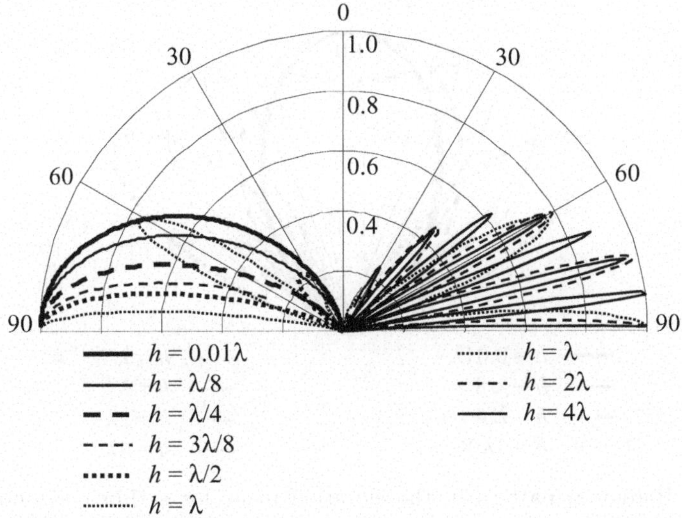

Fig. 7.13 Radiation pattern in the elevation plane for a thin vertical dipole above ground at different height values

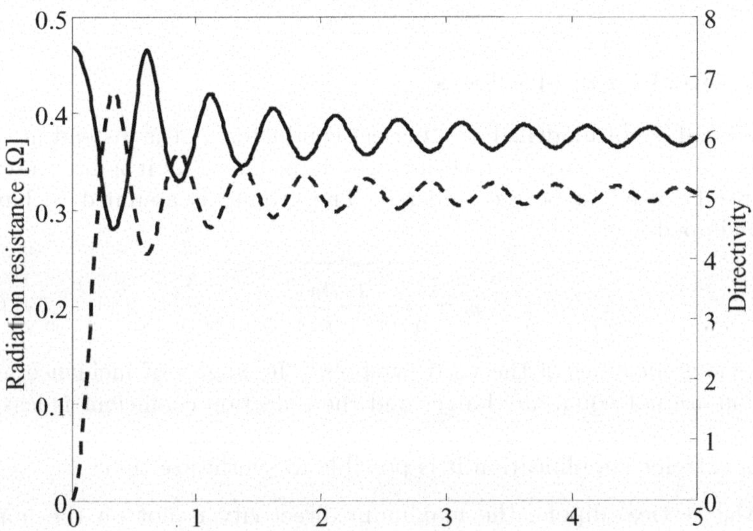

Fig. 7.14 Directivity and radiation resistance of an infinitesimal horizontal dipole at height h above perfect ground

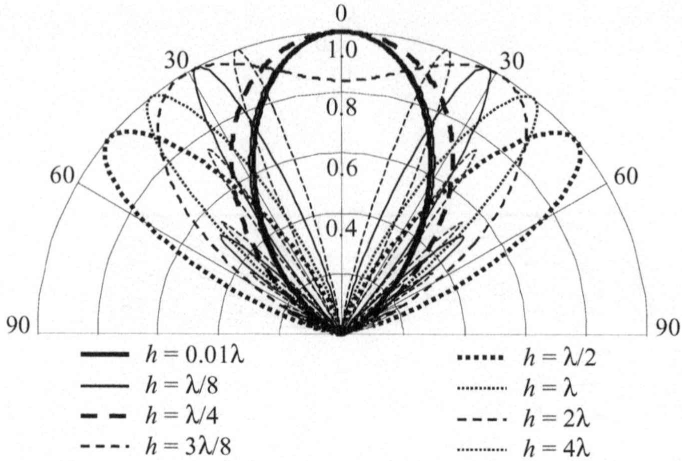

Fig. 7.15 Radiation pattern in the elevation plane for a thin horizontal dipole above ground at different height values

Finally, again the radiation patterns are shown below, in Figure 7.15, for various height values, and by dividing the figure in two quadrants for display purposes.

7.3.3.3 Real Ground Effects

If the ground is not modelled as a perfect conductor, but is instead modelled as a lossy dielectric with its own intrinsic impedance, as it is in reality, the above results need to be reconsidered. The ground is assumed to have an intrinsic impedance

$$\eta_g = \sqrt{\frac{j\omega\mu_g}{\sigma_g + j\omega\varepsilon_g}} \tag{7.95}$$

with obvious meaning of the used symbols. The angles of incidence and of refraction are not equal any longer and the reflection coefficient is less than unity.

With extreme simplification it is possible to synthesize that:

- for the vertical dipole, the maximum directivity is not on the horizontal plane any longer, as the main lobes of the radiation pattern are tilted upward by approximately up to 20 degrees, depending on the ground properties; the input impedance (a parameter that need more complex calculations and is normally the less accurate in numeric simulations) is always larger than that of perfect ground by up to a factor of two, depending on the height and ground properties;
- for the horizontal dipole, the changes are not so relevant and may be neglected.

7.3.3.4 Effects of Finite Diameter

If the assumption of infinitesimal wire radius is relaxed and the dipole radius is a, with a length over radius ratio l/a that is not infinite, then the calculations become more complex. If two antennas are close to each other or the antenna is close to ground, then the current distribution over the antenna surface is no longer uniform; for the derivation of the self impedance and radiation properties the current density is assumed however uniform. Central feeding is also assumed, so that the profile of the current is symmetrical.

There are several formulations available for the input impedance of cylindrical antennas, depending on the series expansion and the approximations used. The Moment Method with either Pocklington's or Hallén's integral equations is the most accurate for the calculation of the input impedance, but as a numerical method it doesn't give closed form expressions. Balanis proposes closed form expressions using the Induced EMF Method [[25], p. 405], where the radiation impedance and the input impedance are computed referred to "at the current maximum" and "at the current at the input terminals", respectively; it is noted that depending on the length/wavelength ratio the current maximum I_m may not occur at the antenna input terminals, where I_{in} is applied.

The relationships for the real and imaginary part of the radiation impedance are

$$R_r = \frac{\eta}{2\pi}\left\{C + \ln(kl) - C_i(kl) + \frac{1}{2}sin(kl)\left[S_i(2kl) - 2S_i(kl)\right] + \frac{1}{2}cos(kl)\left[C + \ln\left(\frac{kl}{2}\right) + C_i(2kl) - 2C_i(kl)\right]\right\} \tag{7.96}$$

$$X_r = \frac{\eta}{4\pi}\left\{2S_i(kl) - \cos(kl)\left[S_i(2kl) - 2S_i(kl)\right] + \sin(kl)\left[C_i(2kl) - 2C_i(kl) - C_i(2ka^2/l)\right]\right\} \tag{7.97}$$

while

$$R_{in} = \left(\frac{I_m}{I_{in}}\right)^2 R_r = \frac{R_r}{\sin^2(kl/2)} \tag{7.98}$$

$$X_{in} = \left(\frac{I_m}{I_{in}}\right)^2 X_r = \frac{X_r}{\sin^2(kl/2)} \tag{7.99}$$

where

$C = 0.5572$ is the Euler's constant;

$C_i = -\int_x^{\infty} \frac{\cos y}{y} dy$ is the cosine integral function;

$S_i = -\int_0^x \frac{\sin y}{y} dy$ is the sine integral function;

$C_{in}(x) = 0.5572 + \ln(x) - C_i(x)$

These expressions are very similar to those also reported by [[481], sec. 13-5] for the thin dipole, where a distinction is made between simpler formulae for dipoles long an odd multiple of $\lambda/2$ and a more general solution.

As a proof R_r and X_r are calculated for the $\lambda/2$ dipole by using (7.96) and (7.97): by replacing $kl = \pi$ the known $R_r = 73.13\,\Omega$ is obtained, together with $X_r = 42.545\,\Omega$ that indicates that for a perfectly resonant dipole (with $X_r = 0$) the length needs to be shorten by a few %; by successive approximation using the same expressions X_r has been brought to nearly zero for a 0.4883 length, so shorter by 2.34%, and the corresponding resistance becomes $R_r = 68.16\,\Omega$.

Balanis also presents a simplified formula for the input reactance (that becomes equal to the radiation reactance) of short dipoles with not negligible radius, taken from one of his references [297]

$$X_{in} = X_r = -120 \frac{\ln(l/a) - 1}{\tan(kl)} \tag{7.100}$$

while the input resistance (and the radiation resistance) is still given by

$$R_{in} = R_r = 20\pi^2 \left(\frac{l}{\lambda}\right)^2 \tag{7.101}$$

The above formulas (7.96) and (7.97) are plotted and shown in Figure 7.16, using the l/λ as the normalized length. For those who wish to recalculate the expressions above, it is underlined that, since $C_i(\cdot)$, $S_i(\cdot)$ and $C_{in}(\cdot)$ are computed by series expansions, the number of terms required to reach convergence of the series and get a meaningful and stable result may be large, depending on the value of the argument.

Krause, citing a work by Hallén [294], reports solutions in series expansion based on the factor $\Omega' = 2\ln(2l/a)$; first-order and second-order approximations are presented and the values of the coefficients are to be found in the original work by Hallén himself, who gives tabular data for the coefficients.

7.3.4 The Monopole

When a monopole is mounted on an ideal infinite ground plane, its impedance and radiation characteristics can be deduced from that of a dipole of twice its length in free space. Also known as $\lambda/4$ stub antenna, a base-driven monopole antenna can be modelled starting from an electric dipole and using a transversal ideal plane of symmetry represented by a perfect conductor sheet. In this ideal configuration the radiation pattern is that of the $\lambda/2$ dipole taken in the upper half space, being zero the electric field at and below the refer-

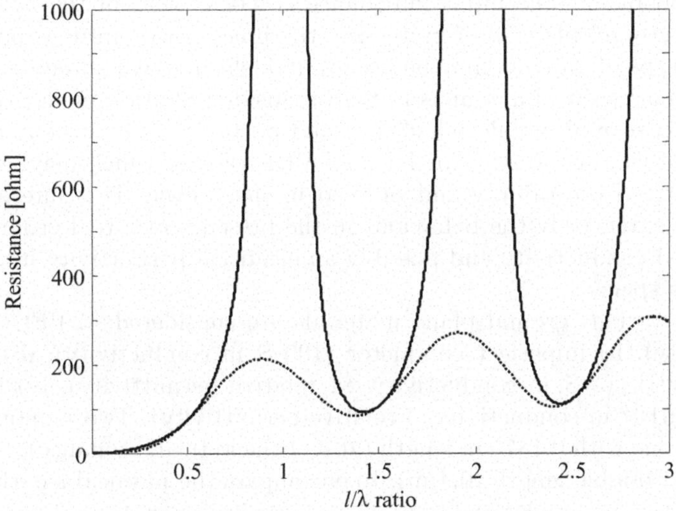

(a) Self resistance of a thin dipole

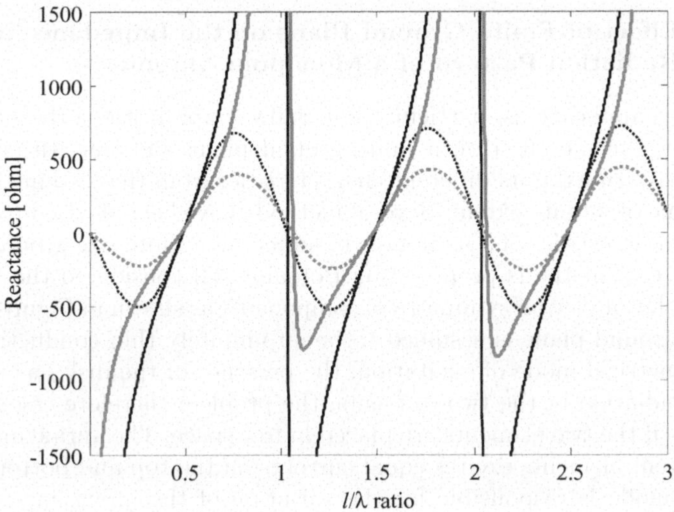

(b) Self reactance of a thin dipole

Fig. 7.16 Self-resistance and self-reactance for a dipole antenna with wire radius of $10^{-5}\lambda$ (black) and $10^{-3}\lambda$ (gray); solid lines represent the input resistance and reactance, dotted lines the radiation resistance and reactance ence ground plane. The input impedance is half of that of the $\lambda/2$ dipole. When the ground plane is of finite size, the image theorem does not apply. Since this type of antenna may be used to model some real antennas, such as whip antennas (helical antennas in the broadside radiation mode) that may be found on several mobile telephones and portable transceivers, or receiving/transmitting monopoles for HF and VHF ranges, which maybe located in proximity to the railway and/or used in the railway, it is interesting to treat more extensively the behaviour in the presence of a real ground plane with limited conductivity and possibly an electrical permittivity larger than that of free space.

For this various ground plane materials are considered: i) PEC (infinite conductivity), ii) imperfect conductor (100 S/m conductivity), iii) heavily lossy dielectric (5 S/m conductivity, 81 relative permittivity), iv) lossy dielectric (0.01 S/m conductivity, 5 relative permittivity). Two monopoles are simulated, one with 0.1875 m length (at $\lambda/4$ theoretical resonance), the other with almost double length (0.4 m), to encompass the physical length of normally adopted monopoles and to perform a sensitivity analysis. The radiation pattern obtained is shown in Figure 7.17.

7.3.4.1 Effect of Finite Ground Plane on the Impedance and Radiation Pattern of a Monopole Antenna

In numerous applications, particularly in railway applications the monopole antenna may be mounted on a finite ground plane and thus the resulting radiation pattern exhibits different characteristics from that mounted over a ground plane of infinite extent. Several methods have been devised to investigate the characteristics of a monopole mounted on a finite size ground plane. One of the first methods proposed by Bolljahn [34] considered the problem from the point of view of symmetrical components as shown in Figure 7.18, in which the ground plane is assumed to be an infinitely thin conducting disk. For a symmetrical mode of excitation, the presence of the disk has no effect upon the radiation of the two elements, the problem therefore essentially is the same as if the two elements are placed in free space. The anti-symmetrical pair of current elements excites equal currents on the top and bottom of the disk. This mode is responsible for the variation of the input impedance of the antenna as a function of the disk diameter and for the resultant radiation pattern with respect to the ground plane.

The change of the input impedance of a base-driven monopole mounted on a large circular ground plane can be expressed as [646]

$$\Delta Z = Z - Z_0 = \frac{-j\eta e^{(j2ka)}}{4\pi ka}\left[\frac{1-\cos(kh)}{\sin(kh)}\right]^2\left\{1+\frac{e^{[j(2ka+3\pi/4)]}}{\sqrt{(4\pi ka)}}\right\}^{-1}, \quad ka \gtrsim 30 \tag{7.102}$$

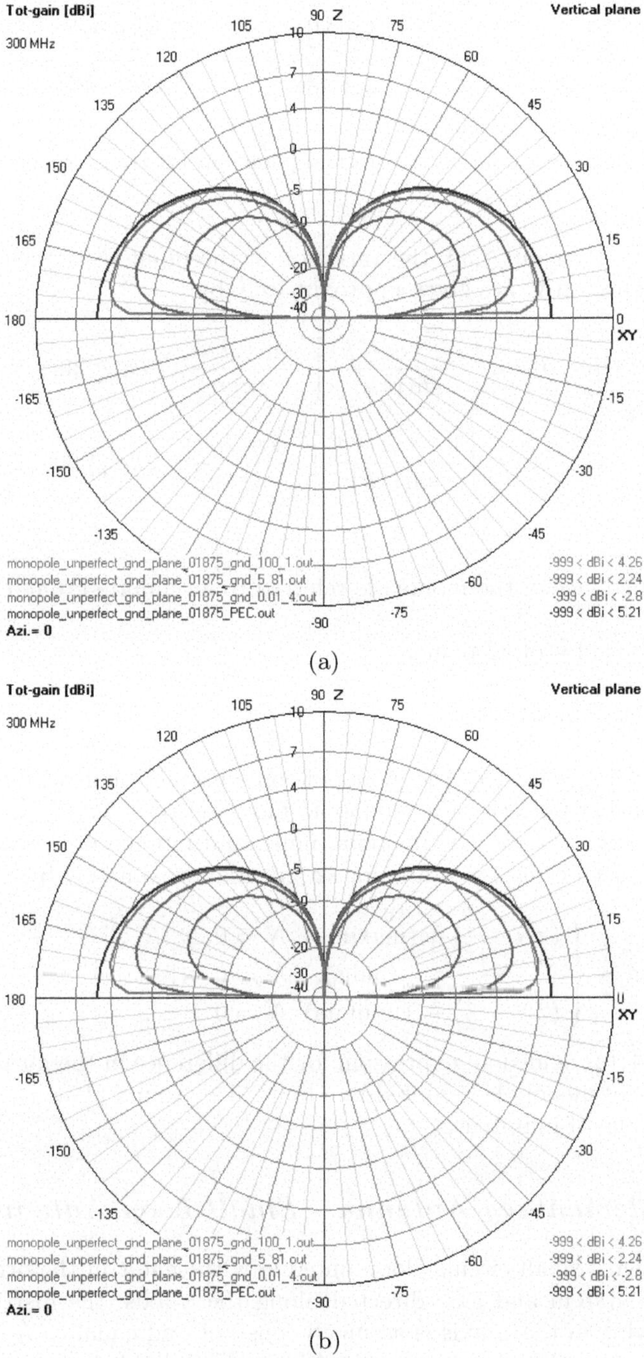

Fig. 7.17 Radiation pattern of a vertical monopole above real ground at 300 MHz: (a) 0.1875 m length at resonance, (b) 0.4 m length at resonance

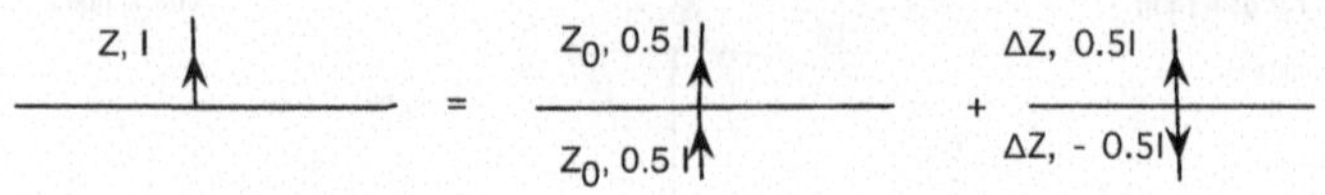

Fig. 7.18 Monopole mounted on a ground plane and its transformation

As $\sqrt{4\pi ka} \gg 1$ for $ka \gtrsim 30$, the variation of the input resistance, ΔR, and input reactance, ΔX, are given approximately by:

$$\Delta R = R - R_0 \approx \left[\frac{1-\cos(kh)}{\sin(kh)}\right]^2 \frac{\sin(2ka)}{4\pi ka}, \quad ka \gtrsim 30 \tag{7.103}$$

$$\Delta X = X - X_0 \approx \left[\frac{1-\cos(kh)}{\sin(kh)}\right]^2 \frac{\cos(2ka)}{4\pi ka}, \quad ka \gtrsim 30 \tag{7.104}$$

where

Z_0 = impedance of the monopole referred to an infinite ground plane;

ka = radius of ground plane;

h = height of monopole;

Z = input impedance of the monopole on a ground plane of radius ka;
The radiation pattern of a monopole mounted on a finite ground plane is shown in Figure 7.19. The maxima, nulls, and minima of the variation of input resistance occurs for values for ka given approximately by:

$$ka = \begin{cases} \left(2N + \frac{1}{2}\right)\frac{\pi}{2} \text{ (maxima)}, & N = 0, 1, 2, \ldots \\ N\frac{\pi}{2} \qquad \text{(nulls)}, & N = 0, 1, 2, \ldots \\ \left(2N + \frac{3}{2}\right)\frac{\pi}{2} \text{ (minima)} & N = 0, 1, 2, \ldots \end{cases} \tag{7.105}$$

Storer in [647] provided an expression for the difference in the far field radiation pattern of monopole mounted in a finite size ground plane from that for a ground plane of infinite size.

7.3.5 Elementary Antenna - Small Loop Antenna

Let's consider a small circular loop antenna with center in the origin of the coordinates system and axis directed along the z axis; the distribution of current flowing in the φ axis is assumed constant and equal to I_0. As it was done for the Hertzian dipole, the small loop is subdivided into infinitesimal elements of length dl; with reference to Figure 7.8 the current elements are

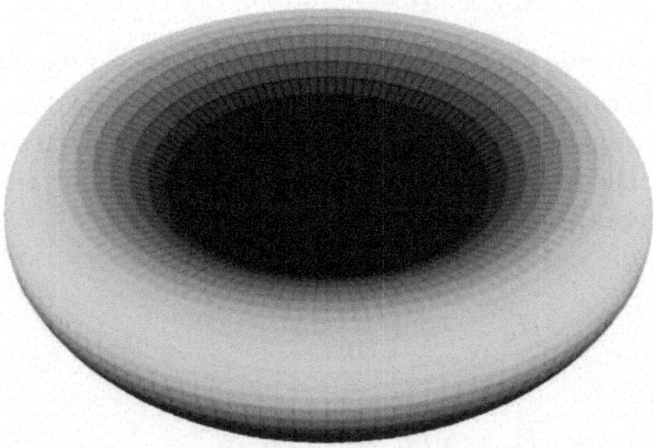

Fig. 7.19 A 1/4 wavelength monopole mounted on a finite ground plane of diameter 3λ

positioned along the φ axis. The three components of the electric field E and magnetic field H along r, θ and φ are determined as follows:

$$E_r = \frac{\eta}{2\pi} k^2 I\, dz \left[\frac{1}{(kr)^2} - j \frac{1}{(kr)^3} \right] \cos\theta\, e^{-jkr} \tag{7.106}$$

$$E_\theta = j \frac{\eta}{4\pi} k^2 I\, dz \left[\frac{1}{(kr)} - j \frac{1}{(kr)^2} - \frac{1}{(kr)^3} \right] \sin\theta\, e^{-jkr} \tag{7.107}$$

$$H_\varphi = j \frac{k^2}{4\pi} I\, dz \left[\frac{1}{(kr)} - j \frac{1}{(kr)^2} \right] \sin\theta\, e^{-jkr} \tag{7.108}$$

$$E_\varphi = H_r = H_\theta = 0 \tag{7.109}$$

Again, it is interesting to observe that the ratio of the first order components of the electric and magnetic field, that are projected along θ and φ axes (with propagation occurring along the r axis), give the free space impedance.

Following the development in [25], the small loop equations above may be computed for the three regions, where $kr \ll 1$, $kr > 1$ and $kr \gg 1$ respectively.

Reactive near-field region, $kr \ll 1$:

$$H_r = I\, a^2 \frac{1}{r^3} \cos\theta\, e^{-jkr} \tag{7.110}$$

$$H_\theta = I\, a^2 \frac{1}{4r^3} \sin\theta\, e^{-jkr} \tag{7.111}$$

$$E_\varphi = -jI\,a^2\frac{k}{r^2}\sin\theta\, e^{-jkr} \tag{7.112}$$

It is interesting to observe the E-field component has a $-j$ coefficient with respect to the H-field components, so that it is in time-quadrature; this implies that their time-averaged product is zero and so it is the associated power flow. Furthermore, the two H-field components have similar expressions and are in an in-phase condition.

Far-field region, $kr \gg 1$:

$$H_\theta = -I\,a^2\frac{k}{4r}\sin\theta\, e^{-jkr} \tag{7.113}$$

$$E_\varphi = \eta I\,a^2\frac{k}{4r}\sin\theta\, e^{-jkr} \tag{7.114}$$

This is the known condition of orthogonal propagation of the E-field and H-field components projected only on the θ and φ axes, with a $\eta = 120\pi$ ratio (that corresponds to the said free space intrinsic impedance) and with a linear dependence on distance; the associated power density, given by time-average product of the E-field and H-field, shows a dependence on distance of the type $1/r^2$.

The radiation resistance (and radiation efficiency) may be improved by increasing the number of turns; the same of course cannot be done with the Hertzian dipole.

7.3.6 Loop Antenna

A loop antenna is again a very common antenna, that is applied in several applications, particularly in railway signalling and communication systems, where loop antennas may form the core of a uni or bi-directional communication system; in the field of measuring antennas, the loop is used as a directive search antenna or, on a smaller scale, as a field probe (*sniffer*) to diagnose the operation of electronic circuits and equipment; many sources may be modelled by means of a loop antenna, such as large and extended circuits where the structure of the ground connection is "circular" (the so called *ground loops*) or parts of commutation circuits (for example the same *snubber circuits*) inside power static converters, where large currents may flow during switching transients; any re-closure of the absorbed traction current and the traction return current forms loops of large extension, that involve the aerial and the grounded conductors. The loop has a very low radiation resistance that makes it an inefficient radiator; so, when used for radio communication applications it is used as a receiving, and not a transmitting, antenna. It is possible to increase the radiation resistance either by increasing the number of turns or by inserting ferrite beads (with high magnetic permeability) along the loop wire. The shape of the field pattern of electrically small loop antennas resembles that of a dipole, with a null on the loop axis (perpendicular to its plane) and the maxima around the loop circumference on the plane itself. When the loop circumference is close to one free-space wavelength the loop approaches its resonance and the maximum of the field pattern moves towards the loop axis.

7.3.7 Long Wire Antenna

The long wire antenna is very useful to model the radiation of long bare conductors, such as the catenary or the third rail, excited by a source of radio frequency conducted emissions (such as an electric arc at the sliding contact of the pantograph or sliding shoe), at a frequency such that multiple wavelengths are contained in the conductor length. The long wire antenna model is also mentioned by the ERTMS standard for Euroloop considered in sec. 3.1.6.3. Detailed treatment of the physics of long wire antennas is available in open literature, however the solution detailed in [487] is detailed here.

The far field pattern for a long straight wire in free space with pure sinusoidal (standing wave) current distribution is expressed by:

$$E_\theta = \underbrace{\frac{60I}{r}}_{A} \left[\underbrace{\frac{1}{sin\theta}}_{B} \underbrace{\begin{matrix}\cos\\ \sin\end{matrix} \left(\frac{\pi L}{\lambda} \cos\theta \right)}_{C} \right] \mathrm{V/m} \tag{7.115}$$

The coefficient A relates the field strength to the anti-node current I in amperes and the distance r in meters from the center of the wire. The coefficient B gives the envelope for the pattern lobes, θ being the angle to the wire axis. The coefficient C oscillates from $+1$ to -1 and contains the information about the angles of the zeros in the pattern. The Cosine or sine in the coefficient is used when the number of half wavelengths is odd or even respectively. Coefficients B and C describe the shape of the pattern. The power flow at any angle θ is given by:

$$P = \frac{E\theta^2}{377} \mathrm{W/m^2} \tag{7.116}$$

For a perfect unattenuated travelling wave current distribution of any length the far field pattern is given as:

$$E_\theta = \underbrace{\frac{60I}{r}}_{A} \underbrace{\frac{\sin\theta}{1-\cos\theta}}_{B} \underbrace{\sin\left(\frac{\pi L}{\lambda} (1-\cos\theta) \right)}_{C} \tag{7.117}$$

For the angles of the series of maxima of diminishing amplitudes starting with the largest at θ_{m1} nearest to the forward axis of the wire, the angle of maxima can be determined from:

$$\theta_{mm} = \text{hav}^{-1}\frac{K}{2\frac{L}{\lambda}} \tag{7.118}$$

where Haversines $\equiv (1 - cos\theta)/2$ and K has successive values of 0.371, 1.466, 2.480, 3.486, and 4.495. For the angles θ_{0m} for the mth zero with respect to the axis, from θ_{01} to 180°

$$\theta_{0m} = \text{hav}^{-1}\frac{m}{2L/\lambda} \tag{7.119}$$

7.4 Shielding

Putting a barrier in the way of unwanted electromagnetic radiation is generally referred to by the general term *"shielding"*. This barrier is a conductive container, the material and construction of which determines its properties as a shield. Shields are used to contain fields generated in a system as a by-product of the system's operation, thus avoiding interference to nearby systems and meeting emission control requirements of national and/or international regulations. The shield is also used to protect components of a system against internally or externally generated fields.

Shields function on the basis of two major electromagnetic phenomena, reflection from a conducting surface and absorption in a conductive volume. Reflection from a shield occurs when there is a substantial difference between the impedance of the field in air and to the field of the shield material. High impedance fields, such as those from an electric dipole, high impedance data lines, etc., will be reflected well from a highly conductive shield such as copper or aluminum. Absorption is the transformation of the wave energy in the shield to heat. This property is frequently defined by the term skin depth which in turn, is a function of the conductivity and permeability of the shield.

The performance of a shield in attenuating electromagnetic energy is quantified through a figure of merit referred to as *shielding effectiveness*. The shielding effectiveness of any configuration is defined as a ratio, usually expressed in decibel, between two suitable electromagnetic parametric values such as power, electric field or magnetic field intensity. The ratio is numerically coincident for the different chosen quantities explicitly below only under very special circumstances [59]. The shielding effectiveness of metallic enclosures is not infinite because the conductivity of all metals is finite.

IEEE Std. 299 [366] defines shielding effectiveness as the ratio between the absolute value of the electric (or magnetic) field $\mathbf{E}(\mathbf{r})$ (or $\mathbf{H}(\mathrm{r})$) that is present at a given point r beyond the shield and the absolute value of the electric (or magnetic) field that would have been present at the same point in the absence of the shield itself. By definition, the latter is the incident field $\mathbf{E}^{inc}(\mathrm{r})$ (or $\mathbf{H}^{inc}(\mathrm{r})$).

$$SE_E = 20 \log \frac{\left|\mathbf{E}^{inc}(\mathbf{r})\right|}{\left|\mathbf{E}(\mathbf{r})\right|} \tag{7.120}$$

for electric shielding effectiveness,

$$SE_H = 20 \log \frac{\left|\mathbf{H}^{inc}(\mathbf{r})\right|}{\left|\mathbf{H}(\mathbf{r})\right|} \tag{7.121}$$

for magnetic shielding effectiveness and

$$SE_P = 10 \log \frac{\left|\mathrm{Re}\left\{\mathbf{P}^{\mathrm{inc}}(\mathbf{r})\right\}\right|}{\left|\mathrm{Re}\left\{\mathbf{P}(\mathbf{r})\right\}\right|} \tag{7.122}$$

for power shielding effectiveness.

Equations (7.120), (7.121) and (7.122) are defined as a loss, so that the shielding effectiveness is always positive. It is important to note that (7.120) and (7.121) would give identical results in the far field ($r \gg \lambda/2\pi$), but do not result in the same SE_{dB} in the near field since, the wave impedance are different. Figure 7.20 shows the conceptual mechanism for determining shielding effectiveness.

When a shield is sufficiently far from an electromagnetic source, (in the far field), the field incident to the shield can be modelled as a plane wave. The electric and magnetic fields are perpendicular to each other in the far field, and these fields are perpendicular to the direction to propagation. When an incident wave impinges on the shield, the electrons and other charged

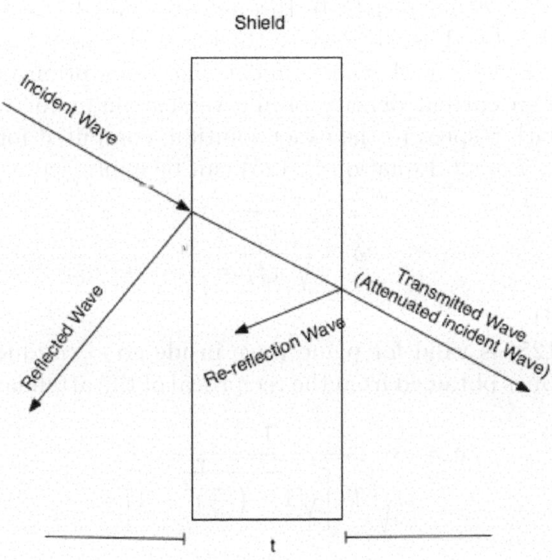

Fig. 7.20 Conceptual Mechanism of Shielding Effectiveness

particles in the shield respond to the electric and magnetic fields according to Lorentz's force law:

$$F = q(E + \upsilon \times \mu_0 H) \tag{7.123}$$

where q is the charge of each particle of velocity υ. The charges in the shield move in response to the electric and magnetic fields of the incident wave. These moving charges also generate a scattered magnetic and an electric field. The scattered fields change the total field inside the shield, which affects the charge motion. Eventually a steady state condition is reached inside the shield, assuming the incident field is in steady state. For a good shield the sum of the fields inside the shield is much less than the initial incident field. The power of the shield to generate a counter field is a function of the electrical parameters of the material, which includes the conductivity since the induced current density, J, is a function of the shield's conductivity, σ :

$$J = \sigma E \tag{7.124}$$

For a perfect conductor (where $\sigma \to \infty$), the induced current is entirely along the shield's surface however for real materials, the current and field penetrate a non-zero distance into the material. The depth of penetration is described by a parameter known as the *skin depth*, which is expressed as:

$$\delta = \frac{1}{\sqrt{\pi f \mu \sigma}} \tag{7.125}$$

If the material is thin compared to the skin depth, then a significant fraction of the incident field could penetrate through the shield. At any frequency, the skin depth is defined as the surface thickness of a metal in which $1 - \frac{1}{e}$ or 63.2% of the current is flowing , under the assumption of an approximately exponential current density profile versus conductor radius, that is quite accurate with respect to the exact solution, computed for a cylindrical conductor in sec. 7.5.5.2. Equation (7.125) can be expressed as:

$$\delta = \sqrt{\frac{10^{-3}}{\pi f \mu \sigma}} \tag{7.126}$$

where f is in MHz.

Equation (7.125) is valid for plane wave inside good conductors. A more general expression is obtained from the reciprocal of the attenuation constant:

$$\delta = \frac{1}{\omega\sqrt{\frac{\mu\varepsilon}{2}\left[\sqrt{1 + \left(\frac{\sigma}{\omega\varepsilon}\right)^2} - 1\right]}} \tag{7.127}$$

7.4.1 Impedance of Materials

The intrinsic impedance of the material is defined as the ratio of the electric field to the magnetic field for a wave propagating in one direction. This simple ratio is helpful in determining the degree of reflection of an impinging wave on a shield. The general expression for intrinsic impedance of a shield is given as:

$$\eta = \sqrt{\frac{j\omega\mu}{\sigma + j\omega\varepsilon}} \tag{7.128}$$

where σ is the conductivity, μ is the permeability, and ε is the permittivity of the shield.

7.4.2 Reflection and Transmission Coefficients

With reference to Figure 7.20, if the impinging wave is normally incident on the shield, the reflection coefficient has the same form as the voltage reflection coefficient for transmission lines. For a transmission line, the voltage reflection coefficient is:

$$\rho_L = \frac{Z_L - Z_0}{Z_L + Z_0} \tag{7.129}$$

For normally incident plane wave propagation, the electric field reflection coefficient is:

$$\rho_L = \frac{\eta_L - \eta_0}{\eta_L + \eta_0} \tag{7.130}$$

where η_L is the intrinsic impedance of the shield and η_0 is the intrinsic wave impedance or the intrinsic wave impedance of the medium proceeding the shield. The corresponding electric field transmission coefficient is:

$$T = 1 + \rho_L = \frac{2\eta_L}{\eta_L + \eta_0} \tag{7.131}$$

The magnitude of the reflection coefficient is approximately unity when there is a large impedance mismatch between the mediums:

$$|\rho_L| = \left|\frac{\eta_L - \eta_0}{\eta_L + \eta_0}\right| \approx 1 \quad \text{if } \eta_L \ll \eta_0 \left(\frac{\eta_L}{\eta_0} \ll 1\right) \quad \text{or } \eta_L \gg \eta_0 \left(\frac{\eta_L}{\eta_0} \gg 1\right) \tag{7.132}$$

Conversely, the magnitude of the reflection coefficient is approximately zero when the mediums are matched at the interface

$$|\rho_L| = \left|\frac{\eta_L - \eta_0}{\eta_L + \eta_0}\right| \approx 0 \qquad if \quad \eta_L \approx \eta_0 \quad \left(\frac{\eta_L}{\eta_0} \approx 1\right) \tag{7.133}$$

For a good conductor, the general shielding expression is given as:

$$SE = \frac{E^{inc}}{E^{trans}} = \frac{(\eta_0 + \eta_L)^2}{4\eta_0\eta_L}\left[1 - \left(\frac{\eta_0 - \eta_L}{\eta_0 + \eta_L}\right)^2 e^{-2\frac{t}{\delta}} e^{-j2\frac{t}{\delta}}\right] e^{\frac{t}{\delta}} e^{j\frac{t}{\delta}} e^{-j\beta_0 t} \tag{7.134}$$

where t is the thickness of the shield.

7.4.3 Absorption

The attenuation that occurs when a wave propagates through a shield is mainly due to the skin depth. The effect is generally described by an exponential decaying expression given below:

$$A_{dB} = 8.68\frac{t}{\delta} \tag{7.135}$$

Absorption depends on the conductivity of the material, the permeability of the material, the frequency and thickness of the material and does not depends on the near field/far field conditions.

7.4.4 Influence of Shield Thickness

For a thin shield, $(\frac{t}{\delta} \ll 1)$, whose impedance is much less than that of air, the general shielding expression of (7.134) reduces to:

$$SE = \frac{E^{inc}}{E^{trans}} = \frac{\eta_0}{4\eta_L}\left(1 - e^{-j2\frac{t}{\delta}}\right) e^{\frac{t}{\delta}} e^{j\frac{t}{\delta}} e^{-j\beta_0 t} \tag{7.136}$$

Equation (7.136) can be simplified further by reducing the complex exponential to a simpler form such that

$$\left|e^{-jx}\right| = \sqrt{\cos^2 x + \sin^2 x} = 1$$

Such that the ratio of the incident electric field to transmitted electric field can be written as:

$$\left|\frac{E^{inc}}{E^{trans}}\right| = \frac{\eta_0 \sigma t}{2\sqrt{2}} \tag{7.137}$$

For a thick shield, the reflected wave off the shield to air interface that returns to the air to shield interface (refer to Figure 7.20) may be neglected. Thick shields are so absorbent that the amplitude of this reflected wave is negligible after travelling a round trip distance of $2t$. The transmission coefficient may be used at the first interface to determine the amplitude of the electric field on the shield. The electric field within the shield is:

$$E_t^{inc} = E^{inc} T_{1,2} = E^{inc} \frac{2\eta_L}{\eta_L + \eta_0} \approx E^{inc} \frac{2\eta_L}{\eta_0} \tag{7.138}$$

where

$$\eta_L = \sqrt{\frac{\omega\mu}{2\sigma}} (1 + j)$$

7.4.5 Re-reflection Loss

When the impinging wave arrives at the second interface (shield to air interface), a reflection inside the shield is generated. This reflected wave travels towards the source, and generates a new reflection at the first interface. This second reflection propagates within the shield, creating at the second interface a new reflection and this process repeats itself. Each time the wave is reflected, part of it is also transmitted in the second medium. This implies that each time a small part of the wave is transmitted from the shield in the air. In this manner, the level of the resulting field increases slightly suggesting that the re-reflection loss has a negative impact of the total shielding. The phenomena of re-reflection can be modelled by a series expansion of the multiple reflections, which can be written in closed form.

$$B_{dB} = 20 \log \left| \left[1 - \left(\frac{\eta_L - \eta_0}{\eta_L + \eta_0} \right)^2 e^{-2\frac{t}{\delta}} \right] \right| \tag{7.139}$$

Shielding by reflection and absorption; transmission at the other interface; formulations for the various contributions and shielding effectiveness.

7.4.6 Global Shielding Effectiveness

Schelkunoff's transmission line model [620] to shielding informs that when a wave impinges on a shield, a reflection will be generated and that par of the wave will be transmitted through the shield. During the transmission, the wave may also experience further attenuation and multiple reflections may occur due to the second transition from shield to air. Thus the total shielding effectiveness is a combination of all these phenomena:

- Shielding effect of the first reflection, R
- Shielding effect due to absorption, A,
- Shielding effect due to multiple reflections, B

The global shielding effectiveness can therefore be expressed as:

$$SE = R + A + B \tag{7.140}$$

with reference to the influence of shield thickness, two main conclusions may be drawn from (7.140):

1. For a thick shield where absorption occurs during the transmission of the wave, the global shielding effectiveness simply reduces to $SE = R + A$
2. For a thin shield where no absorption occurs in the shield during the transmission of the wave, the global shielding effectiveness reduces to $SE = R + B$

7.4.7 Multilayer Shielding Materials

Multiple (or laminated) shields are formed by the stratification of two or more layers of different materials. A dielectric gap may or may not exist between the layers of the different materials - the presence of an air gap may impact the overall performance of the shields. Figure 7.21 below illustrates a schematic of a multi-layer shield.

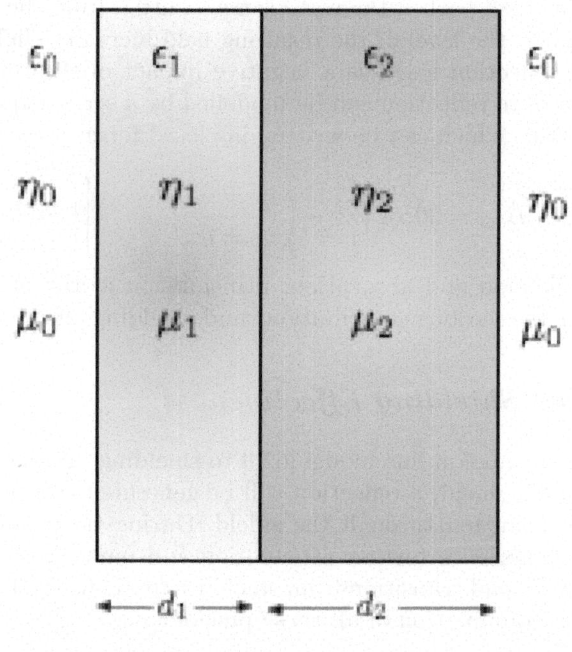

Fig. 7.21 Illustration of a multiple shield

For the two layer shield shown in Figure 7.21, assuming both materials are good conductors, the reflection loss consist of three terms shown below:

$$R_{dB} = 20 \log \left(\frac{1}{2} \left| 1 + \frac{\eta_1}{\eta_0} \right| \right) + 20 \log \left(\frac{1}{2} \left| 1 + \frac{\eta_2}{\eta_1} \right| \right) + 20 \log \left(\frac{1}{2} \left| 1 + \frac{\eta_0}{\eta_2} \right| \right) \tag{7.141}$$

The absorption term is

$$A_{dB} = 8.69 \left(\frac{d_1}{\delta_1} + \frac{d_2}{\delta_2} \right) \tag{7.142}$$

and the multiple reflection loss term is given as:

$$\begin{aligned} MR_{dB} = & 20 \log \left| 1 - \left[\frac{(\eta_1 - \eta_0)(\eta_1 - \eta_{in2})}{(\eta_1 + \eta_0)(\eta_1 + \eta_{in2})} \right] e^{-2\left(\frac{1}{\delta_1} + \frac{1}{\delta_2}\right) d_1} \right| + \\ & 20 \log \left| \left[\frac{(\eta_2 - \eta_1)(\eta_2 - \eta_0)}{(\eta_2 + \eta_1)(\eta_2 + \eta_0)} \right] e^{-2\left(\frac{1}{\delta_1} + \frac{1}{\delta_2}\right) d_2} \right| \end{aligned} \tag{7.143}$$

where

$$\eta_{in2} = \eta_2 \frac{\eta_0 \cosh X d_2 + \eta_2 \sinh X}{\eta_2 \cosh X + \eta_0 \sinh X}$$

and $X = \left(\frac{1}{\delta_1} + \frac{1}{\delta_2} \right) d_2$.

The above formulation assumes normal incidences, for consideration of angles of incidence other than normal incidence refer to [557]. The analysis of multiple shield could be simplified through the application of transmission line analogy as was performed in [236]. It is useful to highlight the equivalence between each shielding layer and a two port network for which:

$$\begin{bmatrix} V_1 \\ I_1 \end{bmatrix} = \begin{bmatrix} A & B \\ C & D \end{bmatrix} \begin{bmatrix} V_2 \\ I_2 \end{bmatrix} = [T] \begin{bmatrix} V_2 \\ I_2 \end{bmatrix} \tag{7.144}$$

where $[T]$ is the transmission matrix (refer to sec. 7.1.1.1 for a discussion on Transmission matrix). For an N layer stratified planar shield (7.144) becomes:

$$\begin{bmatrix} V_{i-1} \\ I_{i\ 1} \end{bmatrix} = \begin{bmatrix} \cos(k_{xi} d_i) & -jZ_i \sin(k_{xi} d_i) \\ -\frac{j}{Z_i} \sin(k_{xi} d_i) & \cos(k_{xi} d_i) \end{bmatrix} \begin{bmatrix} V_i \\ I_i \end{bmatrix} = [T_i] \begin{bmatrix} V_i \\ I_i \end{bmatrix} \tag{7.145}$$

where d_i is the thickness of the $i-$th layer, while V_i and I_i are the equivalent voltage and current (corresponding to the transverse electric and magnetic field, respectively) at the interface between the $i-$th and the $i+1$th layer. Equation (7.145) can be re-written as:

$$\begin{bmatrix} V_1 \\ I_I \end{bmatrix} = \prod_{i=2 \ldots N-1} [T_i] \begin{bmatrix} V_{N-1} \\ I_{N-1} \end{bmatrix} \tag{7.146}$$

An shielding effectiveness expression for a multi layer shield for which the region 1 and N are air is given as:

$$SE = 20 \log \left| \frac{1}{p} \prod_{n=2}^{N-1} e^{jk_{xn} d_n} \left(1 - q_n e^{-j2k_{xn} d_n}\right) \right| \tag{7.147}$$

where

$$p = \frac{2Z_0 \prod_{i=2}^{N-1} 2Z_i}{(Z_0 + Z_2)(Z_{N-1} + Z_0) \prod_{i=2}^{N-2} (Z_i + Z_{i+1})} \tag{7.148}$$

$$q_i = \frac{(Z_i - Z_{i-1})(Z_i - Z(d_i))}{(Z_i + Z_{i-1})(Z_i + Z(d_i))} \tag{7.149}$$

$$Z(d_{i-1}) = Z_i \frac{Z(d_i)\cos(k_{xi}d_i) + jZ_i \sin(k_{xi}d_i)}{Z_{i+1}\cos(k_{xi}d_i) + jZ(d_i)\sin(k_{xi}d_i)} \tag{7.150}$$

$$Z(d_{N-1}) = Z_0 \tag{7.151}$$

7.4.8 Enclosures and Cavities

Enclosures form an integral part of railway equipment design, they are designed to provided environmental (IP protection, vibration etc) as well as protection against electromagnetic disturbance (or to inhibit the transmission of electromagnetic emission from equipments housed within the enclosure which might impact neighbouring equipment. As such an analysis of the interaction between an electromagnetic field and an enclosure is provided in the section.

7.4.8.1 Ideal Source Free Enclosure

An ideal enclosure consist of a cavity with perfectly conducting walls and filled with a linear, isotropic, homogeneous, stationary, and lossless medium. The volume of the enclosure is denoted by V, while the boundary of the enclosure delimiting surface can be represented by S, as shown in Figure 7.22. In the absence of sources, electromagnetic fields can exist inside the enclosure if, at certain frequencies $f_m = \omega_m/2\pi$,, the source free Maxwell equations and boundary conditions are satisfied:

$$\begin{aligned} \nabla \times \mathbf{E}(r, \omega_m) &= -j\omega_m \mu \mathbf{H}(r, \omega_m) && \text{in } V \\ \nabla \times \mathbf{H}(r, \omega_m) &= j\omega_m \varepsilon \mathbf{E}(r, \omega_m) && in\ V \\ \mathbf{u}_n \times \mathbf{E}(r, \omega_m) &= 0 && \text{on } S \end{aligned} \tag{7.152}$$

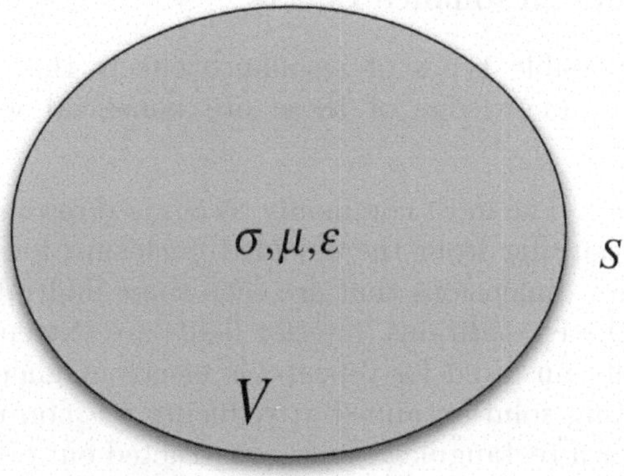

Fig. 7.22 An Arbitrary Metal Cavity

where $\mathbf{u}_n$ is the unit outward normal to the boundary S, and all other notations have their usual meaning (7.152) can be simplified as follows:

$$\begin{aligned}\nabla \times \mathbf{E}(\mathbf{r},\omega_m) &= k_m\hat{\mathrm{H}}(\mathrm{r},\omega_m) \text{ in } V\\ \nabla \times \hat{\mathbf{H}}(\mathrm{r},\omega_m) &= k_m\mathrm{E}(\mathrm{r},\omega_m) \; in \; V\end{aligned} \tag{7.153}$$

where $k_m = \omega_m\sqrt{\mu\varepsilon}$ is positive unknown parameter and $\hat{\mathbf{H}} = -j\eta\mathbf{H}$. The existence of oscillating fields is associated with the existence of positive values of $k_m$for which (7.153)admits nontrivial solutions when appropriate boundary conditions are applied. Such non trivial solutions are solenoidal eigenfunctions of the Laplace operator $\nabla^2[.]$. The solution of the electromagnetic problem inside the source free enclosure, at $f = f_m$ is given by:

$$\begin{aligned}\mathbf{E}(\mathrm{r}) &= A_m\mathbf{E}_m(\mathrm{r})\\ \mathbf{H}(\mathrm{r}) &= j\frac{A_m}{\eta}\mathbf{H}_m(\mathbf{r})\end{aligned} \tag{7.154}$$

This represents a resonant mode of the cavity. In the time domain, the electromagnetic fields. In the time domain, the electromagnetic field is given by:

$$\begin{aligned}\mathbf{e}(\mathbf{r},t) &= |A_m|\,\mathbf{E}_m(\mathbf{r})\cos(\omega_m t - \arg[A_m])\\ \mathrm{h}(\mathrm{r},t) &= -\frac{|A_m|}{\eta}\mathbf{H}_m(\mathbf{r})\sin(\omega_m t + \arg[A_m])\end{aligned} \tag{7.155}$$

The above equation inform that both the electric and magnetic fields of a resonant mode depends only on the solenoidal eigenfunctions and does not vary with time.

7.4.8.2 Enclosure Resonance Effects

There are three possible types of resonance effects that may occur in a shielded enclosures, knowledge of these are beneficial when designing a shielded enclosure.

1. *Cavity Resonance*: The most commonly recognised resonance effect is cavity resonance resulting from the shielded enclosure forming a reflecting cubicle of interior dimensions that are each some multiple of a half wavelength. Under these conditions, interior fields are extremely non-uniform and are generally unsuited for repeatable electromagnetic measurements without mitigating solution aimed at reducing the non-uniformity of the interior field. For a rectangular cavity constructed out of conducting walls and filled with a dielectric material having the following .electrical parameters ε_r and μ_r, the resonant frequency is given as

$$f_r\,(MHz) = \frac{300}{\sqrt{\mu_r \varepsilon_r}} \sqrt{\left[\left(\frac{m}{2a}\right)^2 + \left(\frac{n}{2b}\right)^2 + \left(\frac{p}{2c}\right)^2\right]}$$

   A rectangular cavity resonator can also support any degenerate TE_{mnp} and TM_{mnp} modes. The cutoff frequencies of the TE_{mn} and TM_{mn} modes can be determined from:

$$f_c\,(MHz) = \frac{300}{\sqrt{\mu_r \varepsilon_r}} \sqrt{\left[\left(\frac{m}{2a}\right)^2 + \left(\frac{n}{2b}\right)^2\right]}$$

2. *Inter-wall Resonance*: Another type of resonance is that that can occur at the higher frequencies due to reflections between the conducting sheets of a double shielded enclosure when the frequency is such that the spacing between sheets is some odd multiple of quarter wavelengths
3. *Penetration Resonance*: Practical enclosures that have seams joining metal sheets may exhibit a resonance effect at extremely low frequencies. The effect is due o wave transmission through a seam in parallel with transmission through the metal itself. The wavelength in metal is many orders of magnitude greater than in air. If the transmission loss through metal is of the same order of magnitude as through a tiny slit, the difference in phase could cause either a cancellation or an enhancement of the total transmitted wave.

7.4.8.3 The Quality Factor

The quality factor, Q_m of the enclosure for the mth resonant mode is defined as:

$$\frac{1}{Q_m} = \frac{P_{loss}}{2\omega_m W_m} = \frac{P_{dL} + P_{cL}}{2\omega_m W_m} = \frac{1}{Q_d} + \frac{1}{Q_c} \tag{7.156}$$

where P_{loss} is the power loss in resonant circuit, ω_m is the resonant frequency at mode m, W_m is the energy stored in the electric field, P_{dL} is the power loss in the dielectric material that fills the cavity, and P_{cL} is the power loss in the conducting walls. If the cavity is made of a perfect conductor and filled with a perfect dielectric, it will have infinite Q - although in practice it is not possible to achieve an infinite value of Q. To determine the value of Q, we will therefore need to determine Q_d and Q_c.

For a rectangular enclosure, operating in the TE_{101} , the power loss in the dielectric medium can be determined from:

$$P_{dL} = \frac{\varepsilon_0 \varepsilon''}{8} \left|H_{101}\right|^2 \omega_m^3 \mu^2 \left(\frac{\pi}{a}\right)^2 abc \tag{7.157}$$

$$W_m = \frac{\varepsilon_0 \varepsilon'}{16} \left|H_{101}\right|^2 \omega_m^2 \mu^2 \left(\frac{\pi}{a}\right)^2 abc \tag{7.158}$$

The Quality factor of the dielectric, Q_d, can therefore be determined from (7.157) and (7.158) to give:

$$Q_d \frac{2\omega_m W_m}{P_{dL}} = \frac{\varepsilon'}{\varepsilon''} = \frac{1}{\tan\delta} \tag{7.159}$$

Similarly, the power loss in the conducting walls is determined from:

$$P_{cL} = R_s \left|H_{101}\right|^2 \left(\frac{\pi}{a}\right)^4 \frac{1}{4} \left[2bc + \frac{2a^3 b}{c^2} + \frac{a^3}{c} + ac\right] \tag{7.160}$$

where $R_s = \sqrt{\frac{\omega\mu_0}{2\sigma}}$ and all other notations have their usual meanings.. The Quality Factor due to the conducting walls is given as:

$$Q_c = \frac{2\omega_m W_m}{P_{cL}} = \frac{\omega_m^3 \mu^2 a^3 \varepsilon_0 \varepsilon'}{2R_s \pi^2 \left[2\left\{1 + \left(\frac{a}{c}\right)^3\right\} + \left(\frac{a}{b}\right)\left\{1 + \left(\frac{a}{c}\right)^2\right\}\right]} \tag{7.161}$$

7.4.8.4 Slots and Apertures

An aperture is a hole or an opening in an enclosure of an equipment through which electromagnetic fields may enter or leak out. As shielding enclosures are usually employed to protect against radiation from external fields and leakage effects from interior components, any holes or aperture in the shielding may result in mal-operation of the said equipment. Apertures and slots may be an intentional design parameter such as those required for ventilation, signal and communication cable entry functions, or may results from design defects such as plate joints. The theoretical analysis of apertures depends on a number of factors: the shape and size of the aperture, the frequency of operation and the characteristics of the incident electromagnetic field. Since aperture can not be avoided it is paramount that they are understood, particularly where shielding as a technique is applied to safety critical apparatus such as those used in the railway.

There are several methods that have been employed to solve the transmission of electromagnetic waves through an aperture. These methods can be broadly classed into three using frequency as the determinant.

Low Frequency Region. In this region, the aperture dimensions are small with respect to wavelength. Lord Rayleigh in [608] proposed a solution based on a series expansion in ascending powers of the wave number of the certain quantities which leads to a sequence of simple integral equations with a kernel of the electrostatic type. In [31], Bethe studied the low frequency electromagnetic scattering by a small circular hole cut in an infinite perfectly conducting plane, he derived the leading terms of Rayleigh series by applying the scalar potential approach to the problem. Bouwkamp [37] furthered this solution by applying integro-differential equations.

Resonant Frequency Region. In this region, the wavelength is of the same order as the aperture dimensions, and a full wave analysis is usually employed. The most common approach being the integral equation method.

High Frequency Region. In this region, the wavelength is small compared with the aperture dimensions. Kirchhoff in [472] applied Huygens's principles to solve the aperture diffraction problem, however Kirchhoff's treatment is limited in that it can not account for the polarisation of the electromagnetic field. A vector equivalent of the scalar Kirchhoff's formulation was introduced by Stratton and Chu [648] and Schelknoff [620]

Small Circular Aperture. The penetration of electromagnetic field through a small circular aperture in a planar screen can be solved by applying electrostatic and magneto-static approximations of the fields at the aperture [25]. In this approach, the field scattered by the aperture is represented by the superposition of an electric dipole field and a magnetic dipole field due to electric and magnetic dipole sources located at the center of the aperture.

The electric dipole, $\mathbf{p}_\mathrm{e}$ is proportional to the component normal to the screen of the electric field, while the magnetic dipole $\mathbf{p}_m$ is proportional to the component parallel to the screen of the magnetic field through the electric and magnetic aperture polarizabilities. Table (7.2) provide the electric and magnetic polarizabilities for a few symmetrical shaped aperture.

Simple expressions exist for the two components of the electric field and magnetic field for a circular aperture under consideration is given below:

$$E_z\left(0,z\right) = \frac{aE_0}{\pi}\left[\frac{1}{a}\arctan\left(\frac{a}{z}\right) - \frac{1}{z\left(1+\frac{a^2}{z^2}\right)}\right] \qquad if \quad 0 \le z \tag{7.162}$$

Table 7.2 Electric and magnetic polarizabilities for simple symmetrical aperture

	$\alpha_{e,zz}$	$\alpha_{m,xx}$	$\alpha_{m,yy}$
Circle	$\frac{2}{3}R^3$	$\frac{4}{3}R^3$	$\frac{4}{3}R^3$
Ellipse	$\frac{\pi}{24}\frac{w^2 l}{E(e)}$	$\frac{\pi}{24}\frac{l^3}{K(e)-E(e)}$	$\frac{\pi}{24}\frac{l^3 e^2}{\left(\frac{l}{w}\right)^2 E(e)-K(e)}$
Narrow Ellipse	$\frac{\pi}{24}w^2 l$	$\frac{\pi}{24}\frac{l^3}{\ln\left(\frac{4l}{w}\right)-1}$	$\frac{\pi}{24}w^2 l$
Narrow Slot	$\frac{\pi}{16}w^2 l$	$\frac{\pi}{24}\frac{l^3}{\ln\left(\frac{4l}{w}\right)-1}$	$\frac{\pi}{16}w^2 l$

$$E_p(\rho, 0) = \begin{cases} \frac{E_0 \rho}{\pi\sqrt{a^2-\rho^2}}, & if \quad 0 \le \rho < a \\ 0 & if \quad a \le \rho \end{cases} \tag{7.163}$$

$$H_z(\rho, \phi, 0) = \frac{2H_0}{\pi}\frac{\rho}{\sqrt{a^2-\rho^2}} \qquad if \quad 0 \le \rho < a \tag{7.164}$$

$$H_\rho(\rho, \phi, 0) = \frac{H_0}{2} \qquad if \quad 0 \le \rho < a \tag{7.165}$$

The power transmitted through the aperture due to an incident uniform plane wave, is given as $P_T = P_I G$,, where P_I is the incident power to the aperture and G, is the gain of the aperture. The incident power is the incident power density, P_d, multiplied by the area of the aperture:

$$P_I = P_d\left(\pi a^2\right) \tag{7.166}$$

The gain of the circular aperture when uniformly illuminated by a plane wave is:

$$G - \left(\frac{2\pi a}{\lambda}\right)^2 \tag{7.167}$$

Therefore the power transmitted through the aperture is given as:

$$P_T = P_d\left(\pi a^2\right)\left(\frac{2\pi a}{\lambda}\right)^2 \tag{7.168}$$

If the aperture is subdivided into N smaller circular apertures, the radius of these smaller apertures should be

$$a_N = \frac{a}{\sqrt{N}} \tag{7.169}$$

if the sum of each of the N areas is equal to the area of the original aperture:

$$\pi a^2 = N\left(\pi a_N^2\right) \tag{7.170}$$

Thus the transmitted power through one of such smaller apertures can be expressed by inserting (7.169) into (7.168):

$$P_{TN} = P_d \left[\pi \left(\frac{a}{\sqrt{N}} \right)^2 \left(\frac{2\pi \frac{a}{\sqrt{N}}}{\lambda} \right) \right]^2 = \frac{1}{N^2} P_d \left(\pi a^2 \right) \left(\frac{2\pi a}{\lambda} \right)^2 \tag{7.171}$$

Since the wavelength of the incident signal is large compared to a, the total power transmitted through all of the smaller aperture is then

$$NP_{TN} = \frac{1}{N} P_d \left(\pi a^2 \right) \left(\frac{2\pi a}{\lambda} \right)^2 = \frac{1}{N} P_T \tag{7.172}$$

A measure of the reduction of the transmitted power due to N smaller aperture when compared to that of a single aperture is given as:

$$10 \log \left(\frac{P_T}{NP_{TN}} \right) = 10 \log N \tag{7.173}$$

In other words, the smaller the aperture the less the electromagnetic fields penetrate inside the shield. The decrease of the shielding effectiveness due to N identical apertures is thus given as $20 \log N$.

The above treatment does not take into account the distance between the aperture and other geometrical properties which may influence the shielding effectiveness due the aperture. In [648], shielding effectiveness expressions for various multiple aperture of different shapes is provided, The shielding effectiveness formulation for the aperture shapes given in Figure 7.23 below are:

For Figure 7.23a:

$$SE_{E-Field}\,(dB) = 20 \log \frac{c^2 L}{b^3} + 41.8 \frac{t}{b} + 2.08 \tag{7.174}$$

$$SE_{H-Field}\,(dB) = 20 \log \frac{c^2 L}{b^3} + 32 \frac{t}{b} + 2.01 \tag{7.175}$$

For Figure 7.23b :

$$SE_{E-Field}\,(dB) = 20 \log \frac{c^2 L}{b^3} + 41.8 \frac{t}{b} + 2.68 \tag{7.176}$$

$$SE_{H-Field}\,(dB) = 20 \log \frac{c^2 L}{b^3} + 32 \frac{t}{b} + 3.84 \tag{7.177}$$

For Figure 7.23c:

$$SE_{E-Field}\,(dB) = 20 \log \frac{c^2 L}{b^3} + 38.6 \frac{t}{b} - 0.6 \tag{7.178}$$

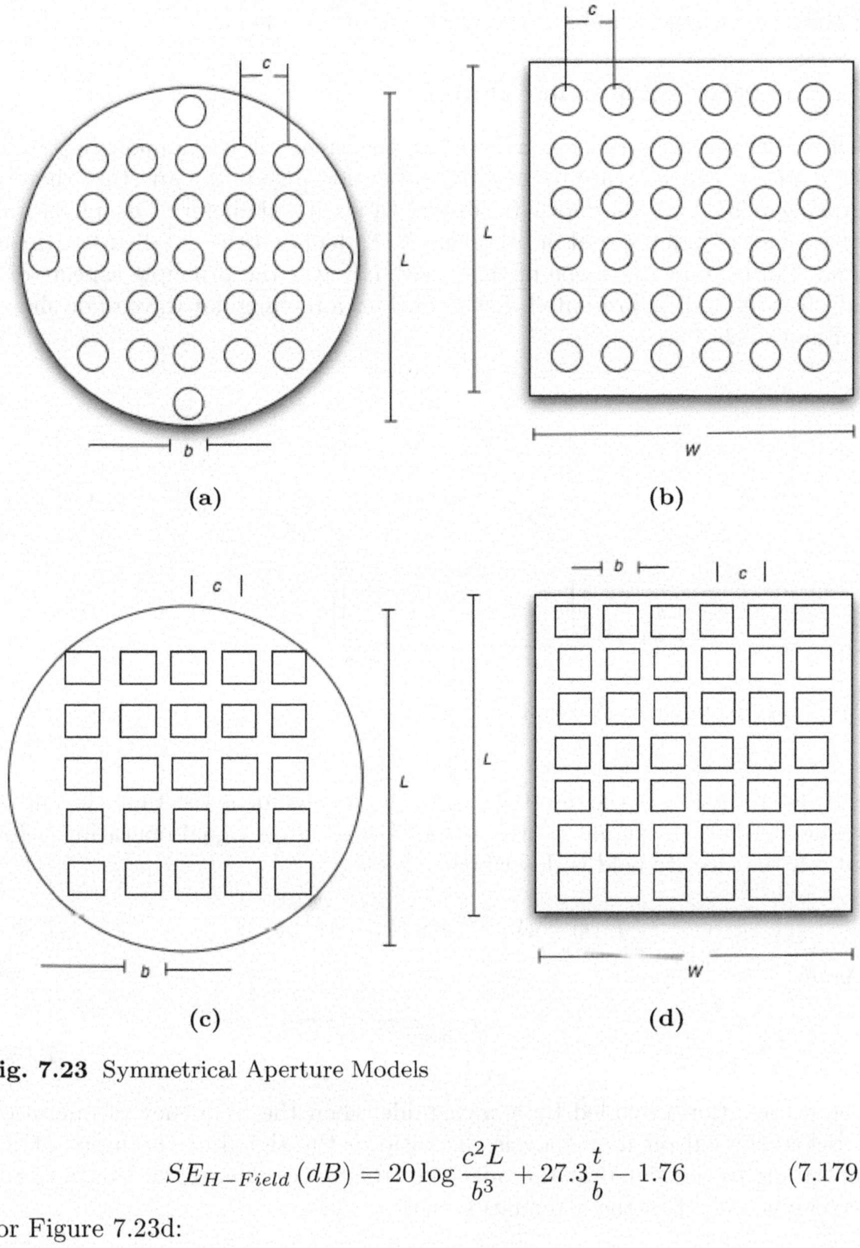

Fig. 7.23 Symmetrical Aperture Models

$$SE_{H-Field}\,(dB) = 20\log\frac{c^2L}{b^3} + 27.3\frac{t}{b} - 1.76 \tag{7.179}$$

For Figure 7.23d:

$$SE_{E-Field}\,(dB) = 20\log\frac{c^2L}{b^3} + 38.6\frac{t}{b} \tag{7.180}$$

$$SE_{H-Field}\,(dB) = 20\log\frac{c^2L}{b^3} + 27.3\frac{t}{b} \tag{7.181}$$

In the above formulation $L = W$ for rectangular shields, and L is the diameter of the circular shield and t is the thickness of the shield.

7.4.8.5 Waveguide below Cutoff

When the aperture dimension becomes too small, it is common to extend the depth of the aperture by placing waveguides inside the aperture thereby creating a hole whose depth is several times its diameter. The physics of waveguides is well covered in literature [218, 460] and the detailed treatment of such in beyond the scope of this book; however the principle aspects of a cut-off waveguide is presented. Let's consider a rectangular waveguide shown in Figure 7.24.

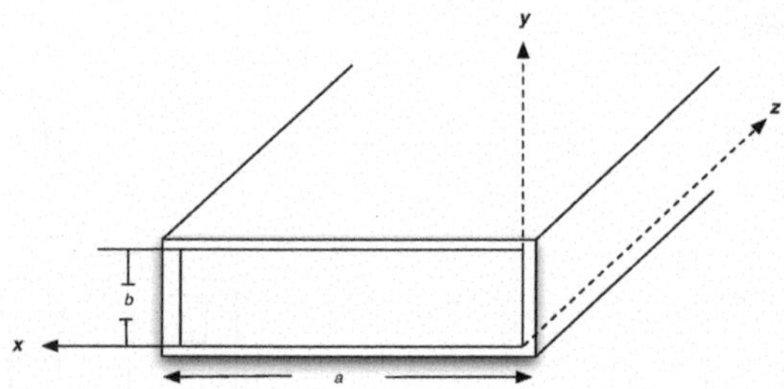

Fig. 7.24 Rectangular Waveguide

If the frequency of the input signal to the waveguide is less than the cut-off frequency then the signal is attenuated. For a TE_{m0} signal operating below cut-off, the electric field is described by:

$$E\left(t\right) = Ae^{-\beta_0 z} \sin\left(kx\right) \cos\left(\omega t\right) \hat{\mathbf{a}_y} \tag{7.182}$$

where

$$\beta_0 = \sqrt{\left(\frac{m\pi}{a}\right)^2 - \left(\frac{\omega}{c}\right)} \tag{7.183}$$

The attenuation provided by a waveguide when the frequency of operation is below the cut-off frequency is the ratio of the signal at the input of the waveguide to the signal at the output of the waveguide. If the length of the waveguide is l, then the attenuation is given as:

$$Attenuation = \frac{1}{e^{-\beta_0 l}} \tag{7.184}$$

or

$$Attenuation_{dB} = 20\log\left(\frac{1}{e^{-\beta_0 l}}\right) = 20l\beta_0 \log e^1 = 8.685\beta_0 l \quad (7.185)$$

$$= 8.685l\sqrt{\left(\frac{\omega_c}{c}\right)^2 - \left(\frac{\omega}{c}\right)^2} = \frac{8.685\omega}{c}\sqrt{\left(\frac{\omega_c}{\omega}\right)^2 - 1} \quad (7.186)$$

where

$$\omega_c = \frac{m\pi}{a}c$$

Two conclusions can be drawn the equations:

- Attenuation would increase as the length of the waveguide increases;
- Attenuation would increase as the ratio of the cut-off frequency to the signal frequency increases.

If the square of the ratio of the cutoff frequency to the signal frequency is much greater than unity, then

$$Attenuation_{dB} = \frac{8.686}{c}l\omega_c = \frac{8.686}{a}m\pi \quad (7.187)$$

$$\backsimeq 27m\frac{l}{a} \qquad if \qquad \left(\frac{\omega_c}{\omega}\right) \gg 1 \quad (7.188)$$

When the length of the waveguide is five times its width, the attenuation is given as:

$$Attenuation_{dB} = \frac{8.686}{a}(5a)\,m\pi = 136m \quad (7.189)$$

For circular waveguides, the mode with the lowest cut-off frequency is TE_{11}, and its cut-off frequency is given as:

$$\omega_c = \frac{0.585\pi c}{r} = \frac{1.837c}{r} \quad (7.190)$$

where r is the radius of the circular waveguide. The minimum attenuation for signal frequencies much less than the dominant mode cutoff frequency is

$$Attenuation_{dB} = \frac{8.686l\omega}{c}\sqrt{\left(\frac{0.585\pi c}{r\omega}\right)^2 - 1} \approx \frac{16l}{r} \qquad if \quad \left(\frac{\omega_c}{\omega}\right)^2 \gg 1 \quad (7.191)$$

For an aperture, having the geometrical dimensions shown in Figure 7.25, the attenuation of the waveguide as a function of frequency is given as:

$$SE = 27.2\frac{h}{g}\sqrt{1 - \left(\frac{f}{f_c}\right)^2} \quad (7.192)$$

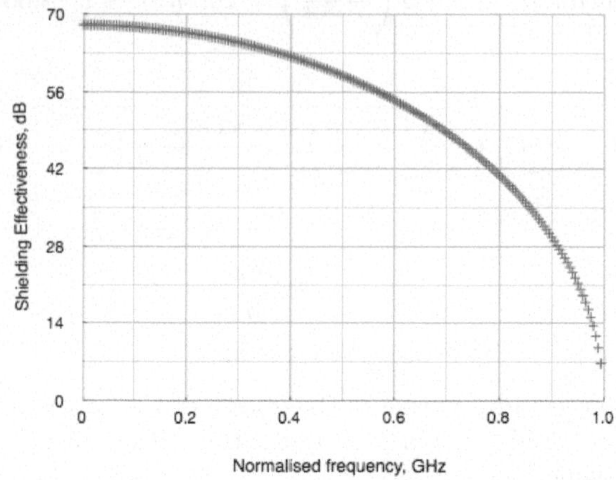

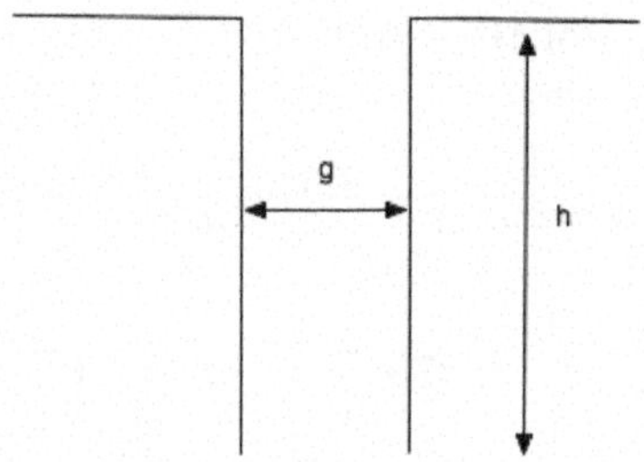

Fig. 7.25 Shielding effectiveness of a waveguide below cut-off with $h/g = 2.5$

Figure 7.25 shows the shielding effectiveness for a waveguide for a dimensional ratio of $\frac{h}{g} = 2.5$.

7.4.8.6 Babinets' Principle - The Law of Reciprocity

In the proceeding text we covered slots and apertures and provided approximate expressions for determining the attenuation from these slots. Here we further our understanding of the electromagnetic field distribution of slots and aperture through the application of Babinets' principle [23].

The electromagnetic field distribution in slots and aperture may be obtained by using Babinet's Principle as applied by Booker [35, 36] to slot antennas and complimentary wire antennas. In optics, Babinets' Principles states that the sum of the fields beyond any two complimentary absorbing screens add to produce the field that would exit there without any screen. If A_1 is a thin absorbing screen with apertures of any shape or size, then the screen A_2 obtained by interchanging the region of absorbing screen space and aperture space is called the complementary screen. If B_1 is the ratio at every point of the field on the right of the screen A_1 to the field without the screen, and if B_2 is the similar ratio for screen A_2, then Babinet's Principle in optic states that

$$B_1 + B_2 = 1$$

An extension of Babinet's Principle to conducting screens and polarised fields was provided by Booker. The generalised Babinets' Principle as applied to electromagnetic is as follows:

Let a source a_1 to the left of an infinite screen A_1 produce a field on the right of A_1, and let B_1 be the ratio of this field to the field that would exist there in the absence of the screen. Similarly, let B_2 be the ratio of the field to the right of the complimentary screen A_2 produced by a source a_2 which is conjugate to a_1 to the left of screen A_2 to the field that would exist there int he absence of the screen. Then

$$B_1 + B_2 = 1 \tag{7.193}$$

A source a_2 which is conjugate to source a_1 is defined as one for which the distribution of electric and magnetic source currents and charges is replaced by the corresponding distribution of magnetic and electric currents and charges respectively. This has the effect of interchanging **E** and **H** for the incident fields.

The basis of Babinet's principle is the duality property of the electromagnetic field. Maxwell's equation in a source free region are:

$$\nabla \times \mathbf{E} = -j\omega\mu \mathrm{H} \tag{7.194}$$

$$\nabla \times \mathbf{H} = j\omega\varepsilon \mathbf{E} \tag{7.195}$$

Equations (7.194) and (7.195) exhibit a high degree of symmetry in the expressions for **E** and **H**. If a field ($\mathbf{E}_1$,$\mathbf{H}_1$) satisfies these equations, then a second field ($\mathbf{E}_2$,$\mathbf{H}_2$) is obtained by the transformations given below:

$$\mathbf{E}_2 = \pm\sqrt{\frac{\mu}{\varepsilon}}\mathbf{H}_1 \tag{7.196}$$

$$\mathbf{H}_2 = \pm\sqrt{\frac{\varepsilon}{\mu}}\mathbf{E}_1 \tag{7.197}$$

This is a statement of the duality property of the electromagnetic field. The transformations given in (7.196) and (7.197) also affect the boundary conditions to be satisfied by the new field. As a results of the change in role of **E** and **H**, all electric walls have to be interchanged with magnetic walls and vice-versa. the new boundary conditions are still the continuity of the tangential components, and this changes the relative amplitudes of the fields on the two sides of the discontinuity surface.

The connection between slot admittance and dipole impedance was discovered by Booker. Booker's relations, $\frac{Z_1}{Y_2} = \frac{\eta^2}{4}$, often written in the form

$$Z_1 Z_2 = \frac{\eta^2}{4}, \tag{7.198}$$

says that $\eta/2 = 188.5\Omega$ is the geometric mean between the input impedance of a slender strip dipole and the input impedance of the complementary slender slot

7.5 Cable Coupling

Cables form an integral component of any railway installation, they are used for a variety of interfaces and purpose but it's use can generally be summed as for power, signalling and communications. Various types of cables are used within the railway infrastructure and it is not the intention to treat the subject in detail in this book. In this section, we limit the treatment on cables to a generic discussion on cable coupling.

7.5.1 Differential and Common Mode Signal

When electromagnetic disturbance is impressed on a circuit two possible modes describe the disturbance current or voltage, namely differential mode and common mode. The differential mode currents are equal In amplitude but in opposite polarity on each conductor, whereas common mode currents are equal in magnitude and polarity. The differential mode signal is the desired signal, the one containing the relevant information. It is easily predicted by the transmission line model or lumped circuit approximation model of the line. Common mode signals is referred to as the unwanted signal - disturbance currents. and is the main source of radiated noise. It is sometimes referred to as antenna mode currents, the common mode currents are difficult to predict and their existence depends on non-ideal aspects of the circuit such as asymmetries.

Two conductors carrying currents I_1 and I_2 in the same direction are shown in Figure 7.26. Total current on each conductor can be broken into a differential mode term, I_D, and a common mode term, I_C:

$$\begin{aligned} I_1 &= I_D + I_C \\ I_2 &= -I_D + I_C \end{aligned} \tag{7.199}$$

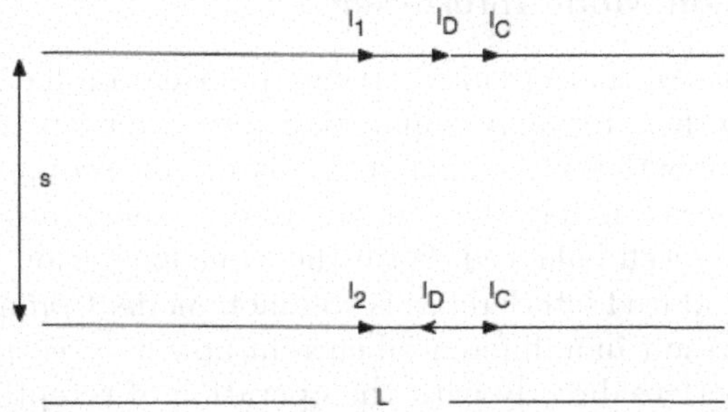

Fig. 7.26 Two current carrying conductors

From (7.199), the differential and common mode currents can be expressed as a function of the two total currents on the line as follows:

$$I_D = \frac{I_1 - I_2}{2}$$
$$I_C = \frac{I_1 + I_2}{2} \tag{7.200}$$

If the current measured on each conductor are equal but opposite, $I_1 = -I_2$, and the common mode current is zero. The differential mode current is the difference between the two currents divided by two. The common mode current can also be defined as the total current on both wires:

$$I_D = \frac{I_1 - I_2}{2}$$
$$I_C = I_1 + I_2 \tag{7.201}$$

Since the differential component is merely the difference of the two currents and the common mode is the average of the two currents, (7.201) can be written as:

$$I_D = I_1 - I_2,$$
$$I_C = \frac{I_1 + I_2}{2} \tag{7.202}$$

7.5.2 Common and Differential Mode of Interference

Following a discussion on differential and common mode signals, a brief discussion is presented below on differential mode and common mode interference, and how they impact electromagnetic compatibility.

7.5.2.1 Common Mode Interference

It is common practice that circuits carrying information have their two conductors close together. In many application they are even twisted together. This practice ensures that electromagnetic disturbance coupled into one conductor will be approximately the same as that coupled into the other, provided the circuit is well balanced. With the same excitation applied to both conductors, this induced interference is common mode. Common mode interference does not result in a differential current flow in a receptor circuit, and thus does not interfere directly with the operation of receptor circuit. Common mode interference can however result in other issues such as failure of the receptor component or circuit due to insulation breakdown or may result in exposure to hazardous touch voltages to railway staff and passengers.

Considering the circuit of Figure 7.27, the radiated field due to the common mode current is given as:

$$\hat{E}_{\theta,i} = MI_i \frac{e^{-j\beta r_i}}{r_i} F(\theta) \tag{7.203}$$

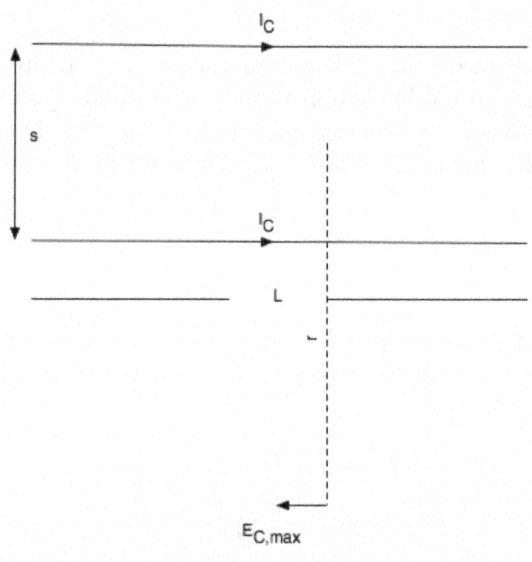

Fig. 7.27 Radiated emission from common mode currents

where $\hat{M}$ is the proportionality constant that depends on the electrical length of the conductor, $\hat{I}_i$ is the magnitude and phase of the current at the center of the array, and $F(\theta)$ is the array factor of the particular current distribution. For an assumed infinitesimal current element current distribution form:

$$\hat{I}_i(z) = \hat{I}_i \tag{7.204}$$

$$\hat{M} = j\frac{\eta\beta}{4\pi}L \tag{7.205}$$

$$F(\theta) = \sin\theta \tag{7.206}$$

The generalised expression of (7.203) can be simplified for a worst case where the fields are in the plane of the conductors, and perpendicular to the conductors as shown in Figure 7.27, as given below:

$$\left|\hat{E}_{C,max}\right| = 1.257 \times 10^{-6} fL\frac{\left|\hat{I}_C\right|}{r}\ V/m \tag{7.207}$$

The above expression informs that the radiated electric field due to common mode currents varies as a function of frequency, the line length, the magnitude of the common mode current and inversely with distance to the observation point.

7.5.2.2 Differential Mode Interference

Differential mode interference occurs when undesired current (or voltage) on the two conductors is not equal. The difference can be in magnitude, phase or frequency or waveform. Differential mode interference has the ability to disturb the operation of the railway signalling equipment at interference voltage/current levels far below those required for common mode interference to cause mal-operation of signalling equipment. Differential mode interference may occur in two ways,:

1. If the conductors are separated by some distance, then they can be exposed to different fields and have different voltages and currents induced - this method of differential mode interference may occur inside a poorly installed enclosure as well as on the track. However, in railway direct induced differential mode interference is usually insignificant considering the rail spacings; differential mode interference may become significant when the track is in close proximity to high current power lines.
2. The second means is via the conversion of common mode interference to differential mode interference due to imbalance of the receptor circuits. This is best explained via Figure 7.28 below, in which a common mode interference of $x\ V\ ac$ rms exist on both conductors. The differential mode interference is $0\,V$ if there is no connection to ground, however if one conductor is grounded, creating an unbalance, then one conductor would be at $x\ V\ ac$ and the other at $0\,V$ resulting in a differential mode current of $x\ V\ ac$ rms. If the source impedance is $0\,\Omega$ then there would be a dead short circuit across the voltage source. Current would actually be limited by the impedance of the current loop. With varying degree of receptor circuit unbalance due to either source impedance differences to the rails, or differing resistance to ground, the differential mode interference can vary

anywhere from $0\,V$ to 100% of the common mode interference amplitude. Usually railway signalling interference problems due to differential mode interference tends to be a combination of common mode coupling and receptor circuit unbalance. Different leakage resistance to ground from the rails through the ballast is a form of unbalanced impedance that can result in differential mode interference

Fig. 7.28 Common mode conversion to differential mode interference

A differential mode interference source may also result in a radiated electromagnetic disturbance impacting other railway systems. Such radiated field can be modeled from (7.203) with $I_1 = I_D$ and $I_2 = -I_D$; and solving for a worst case assuming that the field are in the plane of the conductors, and perpendicular to the conductors as shown in Figure 7.29

$$\left|\hat{E}_{D,max}\right| = 1.326 \times 10^{-14} f^2 Ls \frac{\left|\hat{I_D}\right|}{r} \; V/m \tag{7.208}$$

The above expression informs that the radiated electric field due to differential mode currents varies as a function of the area of the loop, $A = Ls$, the square of the frequency, the magnitude of the differential mode current and inversely with the distance to the observation point.

7.5.3 Impedance and Admittance of Conductors

This section is intended to collect, analyse and compare the various formulations for the calculation or estimation of series and shunt parameters of conductors, taking into account their shape, position and electrical and magnetic characteristics.

By recalling the mathematical formulation used for the simple two conductors transmission lines, it is possible to extend the model to n dimensional problems, where the number of the conductors forming a complex transmission line is n. While the mathematics of the modelling and solution of this type of transmission line will be treated later in sec. 7.6.2, it is anticipated here that the series and shunt parameters will take the form of matrices, with

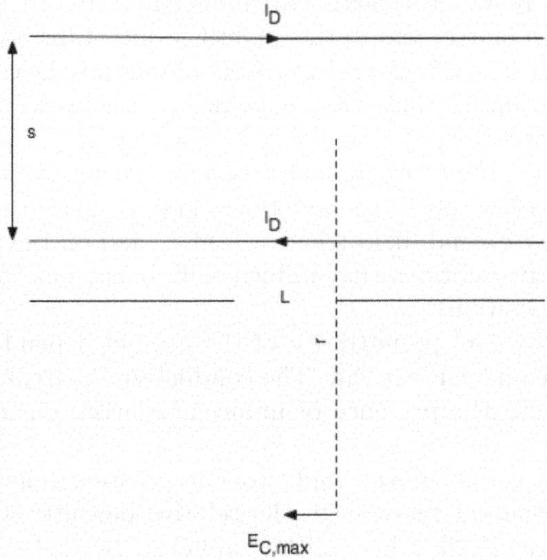

Fig. 7.29 Radiated emissions from differential mode currents

size $n \times n$, where the elements on the diagonal are referred to as *self* terms, linking one conductor to the reference located at infinite, and the elements off-diagonal are referred to as *mutual* terms, linking two conductors; oppositely to the mono-dimensional transmission line, where the driving signal is always applied as a differential signal between the two conductors that form the line, and there is thus no need to distinguish between self and mutual terms, for a more general viewpoint each conductor will be characterized by a series on n parameters for each property (resistance, inductance, capacitance and conductance), linking it to the other $n-1$ conductors and the infinite. Considering two conductors i and j, the self terms will bear the repetition of the same index i, i, and the mutual terms will indicate the two conductor indexes i, j. The series and shunt parameters matrices will be indicated by capital letters R, L, C and G.

Since the earth is one of the conductors of the whole railway system and influences the value of the parameters of the other conductors, it is considered first in the next subsection, analysing the behaviour versus frequency, the factor that influence its electrical properties and reference measurement results.

7.5.4 Earth and Materials

7.5.4.1 Earth and Track Construction Properties

We will see in the next subsections that the earth influences the self impedance of conductors (both resistance and inductance), in particular at low frequency. Moreover, the earth contributes directly to the determination of the conductance matrix and to the quantification of the stray current phenomenon. Last, the earth is the reference conductor for the calculation of the capacitance terms and its dielectric constant is considered for the estimation of the corrective terms.

In this subsection the attention is focused on the general modelling of the earth behaviour with respect to conduction and polarization, on the definition of the conductivity σ_e and dielectric constant ε_e and on the identification of the quantities and conditions that influence its behaviour (frequency, water content, soil composition).

The conductivity and permittivity of the ground depend on granularity, porosity, moisture and salt content. The conductivity is frequency dependent and complex due to the presence of induced polarisation in heterogeneous material.

Capacitance measurements of conductors near the earth indicate that there is a very large apparent increase in the relative permittivity of soil as the frequency decreases [532]. This can be explained by dielectric polarisation effects from inhomogeneous particles. The phenomenon is exacerbated by the migration of positive and negative ions through the contact zones.

The behaviour of the soil permittivity may be explained as the soil is a multiphase mixture of solid particles, water and air voids; also, the ionic conducting water is dispersed, yet acts as the dispersing medium. Further, the frequency variation of permittivity and the upper limit are largely independent of the soil type, and the increase in value with volume fraction of water is similar for different soil types, as predicted by the dielectric theories of mixtures. The semi-disperse soil model of Wobschall [680] was based on the ionic conductivity of water and its dispersion in the soil. They reported a very high dielectric constant and conductivity decreasing with frequency if the low dielectric phase is the solid and it is dispersed as well as conducting.

The physical reason for the increase in dielectric constant at low frequency is the accumulation of charge at the conductor-dielectric interface. Pure water has a relative dielectric constant $\varepsilon = 87.8 - 0.3T$, T the temperature in °C and zero conductivity, dry soil has a relative dielectric constant of 3.5 and a conductivity of 10 mS/m. Soil acts as a conductor and not as a dielectric if the relaxation time is much faster than the period ($\varepsilon/\sigma \ll 1/f$), holding for the frequency values of interest. In this case water acts as a pure dielectric and the particles become capacitor electrodes; the electric field is then concentrated across thin high-dielectric material regions, accentuating the accumulation of charge.

7.5.4.2 Influence of Water Content and Frequency

Scott *et al* [623] collected several experimental data that allow the definition of a relationship between the relative dielectric constant (also called permittivity) or the conductivity and the percentage of water w within the sample and the applied frequency f. The data have been interpolated fitting a power law in the two variables obtaining

$$\varepsilon_e = h_1 w^{h_2} f^{h_3} \qquad h_1 = 28370,\ h_2 = 0.774,\ h_3 = -0.592 \tag{7.209}$$

$$\sigma_e = k_1 w^{k_2} f^{k_3} \qquad k_1 = 0.082,\ k_2 = 1.435,\ k_3 = 0.072 \tag{7.210}$$

The two equations are plotted in the following figures as a function of either w or f, and parametrized on the other.

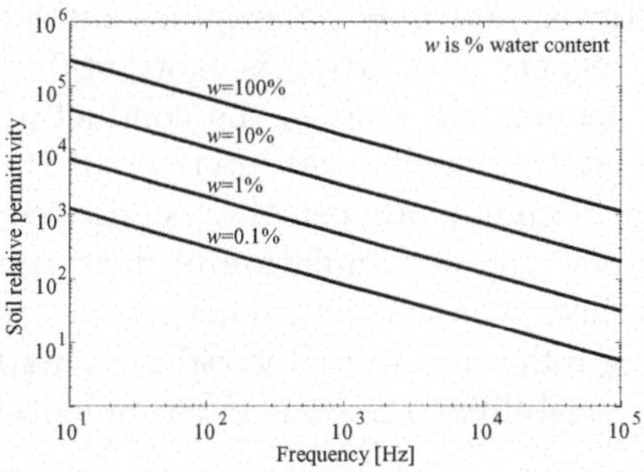

(a) Permittivity of soil vs. frequency as a function of water %

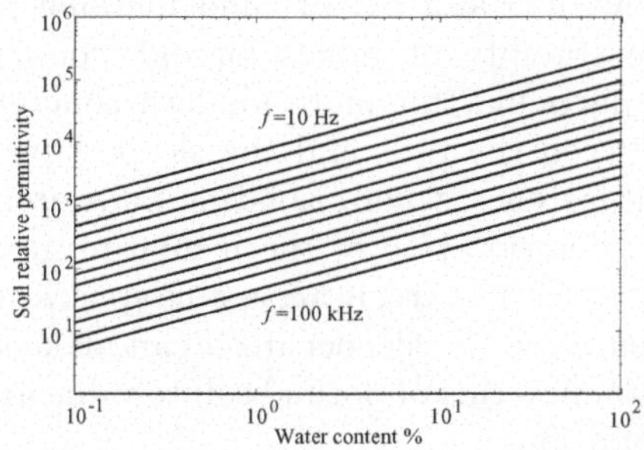

(b) Permittivity of soil vs. water % at specified frequencies

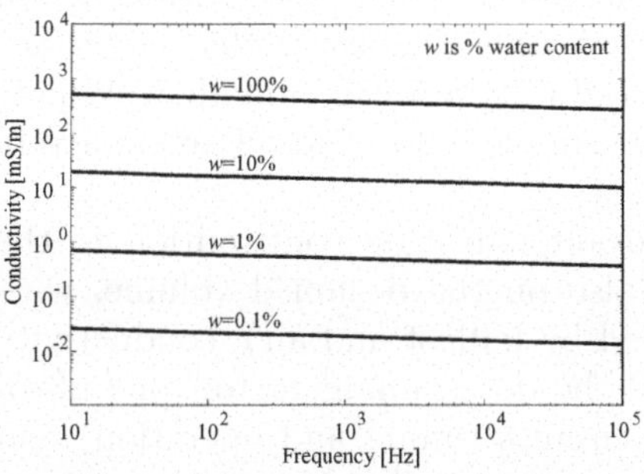

(c) Conductivity of soil vs. frequency as a function of water %

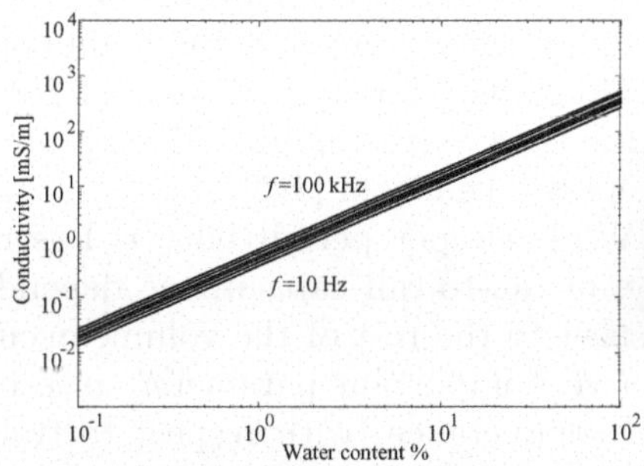

(d) Conductivity of soil vs. water % at specified frequencies

Fig. 7.30 Permittivity and conductivity of soil as a function of water content and frequency

Some values for the soil relative permittivity derived from experimental values are reported in [305]. The used model is a three-layer model by which the permittivity values are derived from the experimental data. At 50 Hz the values are quite dispersed and distributed on the assumed three layers with the largest values in the top layer (ranging between 200 and 800 approximately) and the smallest values in the bottom layer (ranging between

70 and 300 approximately); the average value is about 200. At 10 kHz on the contrary the values are smaller and less dispersed, so that the overall span is between 20 and 40.

7.5.4.3 Sleepers

Sleepers supporting the running rails are the first conductive element with which the rails come into contact and largely determine the rail conductance terms. Two main conduction phenomena are to be considered, one along the surface and one across the volume of the sleeper: the first phenomenon is influenced by the environmental conditions (in particular humidity, rain, snow, ice) and by dirt and metallic particles, and may be troublesome for the stability of values for old wooden sleepers; the latter is more stable, in particular for more modern concrete sleepers. Of course, the conduction between the rails and the sleepers occur at the mechanical interface of the rail fastener, where isolating pads and cushioning is almost always adopted: the deterioration of the insulating materials, the accumulation of dust and moisture in scarcely accessible spaces and the way in which fastener screws are secured to the sleeper itself can make a big difference in real world measured data, in terms of both absolute value and variability to the said environmental conditions.

Laboratory tests were performed on whole sleepers and slices both of the wooden and of the concrete type [305]. It is noted that conductivity for whole sleepers is larger even by an order of magnitude than that measured on samples, confirming that the main conduction mechanism is that of surface conductivity. Moreover, there is a large variability between individual sleepers of the same type. Figure 7.31 presents the simple data of wood and concrete sleepers taken from [305].

While sleeper permittivity is less relevant (since its contribution to the overall rail-to-rail capacitance depends also on the occupied volume, with respect to the rest of the volume occupied by ballast and air), conductivity is a very important parameter since it is the determinant factor for rail-to-rail conductance (with respect to track circuits tuning and operation) and rail-to-earth conductance (with respect to stray currents phenomenon).

7.5.4.4 Ballast

Track ballast provides mechanical support to the sleepers and prevents movement of the track under the dynamic action of train passages. They are made up of crushed hard rock or, sometimes, crushed gravel material, where smaller mineral particles have been sieved away. The ballast is contained and leans on a variety of structures, from concrete beds to simple trenches in the soil, with the use of asphalt or other impermeabilizing materials. Ballast performs a variety of functions principally from the reduction of applied stress to at acting as a weather interfaces, the maintenance of track position by the

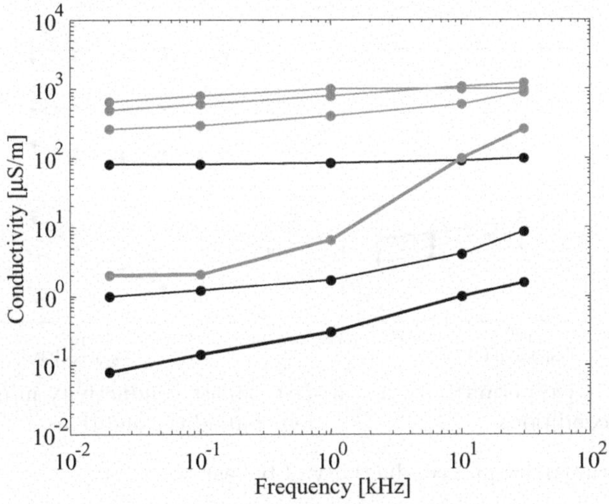

Fig. 7.31 Sleeper conductivity in µS/m for wood (black) and concrete (gray) sleepers; the thick lines represent the dry condition

resistance of applied vertical, lateral and longitudinal forces to sleepers, and the provision of drainage for water from track structures. In modern railways, but in particular for metros and light railways, sleepers and ballast are replaced by a concrete slab that greatly improves the electrical properties and drastically reduces maintenance. It is thus evident that absolute reliable figures of ballast conductivity and permittivity are hardly available, and in particular they are bounded to vary (in particular conductance) if the pollution of various agents is considered: dust, metallic particles, oil and grease, soil leaking from below and from flanks.

The test results reported in [305] were obtained by filling a container with clean ballast material (crashed stones of some cm size) and taking the measurement of capacitance and losses between two parallel metallic plates. The ballast under test was progressively drying, starting from wet conditions. The authors say that the test results were interpreted using three-dimensional finite element modelling. The results consist of a set of curves for different drying conditions that bracket the curve that refers to "damp exterior conditions". As shown in Figure 7.32 for some frequency values taken as reference.

The dielectric value of the ballast is a good indicator of the quality of the ballast and several investigators have conducted measurement using non-destructive techniques correlating the quality of ballast with its dielectric constant. In [225], Clark *et al* used ground penetrating techniques to estimate the dielectric properties of ballast. The dielectric constant was derived from experimental data using the low loss equation given below:

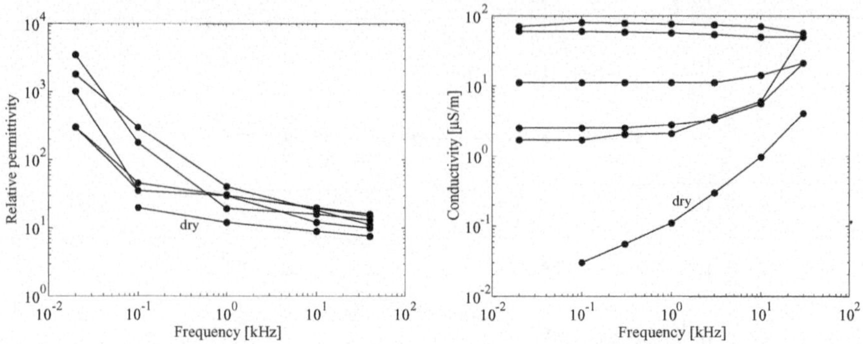

(a) Ballast relative permittivity in various humidity conditions
(b) Ballast conductivity in µS/m in various humidity conditions

Fig. 7.32 Permittivity and conductivity of ballast

$$\varepsilon_r = \left(\frac{2ct}{d}\right)^2 \tag{7.211}$$

where c is the speed of light; t is the time taken for electromagnetic wave to travel distance d; and d is the depth of the material layer. Sussmann *et al* in [652], performed a similar investigation and reached similar conclusion with respect to the relative dielectric values for ballast. In [225], Clark *et al* presented the dielectric values of good and poor quality ballast materials as a function of moisture content. Table 7.3 details the relative dielectric values published by Clark and Sussmann.

Table 7.3 Dielectric Values of Ballast - reproduced from Clark *et al* [226]

Material	Dielectric Value (Clark *et al* [225])	Dielectric Value (Sussmann *et al* [652])
Dry Clean Ballast	3	3.6
Moist Clean Ballast	3.5	4
Dry Spent Ballast	4.3	3.7
Moist Spent Ballast	7.8	5.1
Wet Spent Ballast	38.5	7.2

7.5.4.5 Concrete

As most of the supporting structures (sleepers, slab, viaducts, tunnels, etc.) are nowadays made of concrete, the electrical properties of concrete are important as they effect a variety of applications. The real part of the complex permittivity and the effective conductivity of concrete are fundamental parameters required for a detailed understanding of the electromagnetic wave propagation through concrete; high frequency behaviour is relevant when the attenuation offered to electromagnetic emissions or the influence on per-unit-length parameters of the traction line are considered. Soutsos *et al* [640] used a co-axial transmission line test rig to measure the dielectric properties of concrete. Measurements were performed on 19 samples at various levels of water volume as a percentage of concrete. All samples were made using Ordinary Portland cement and the concrete mixes were made with Staffordshire Pink 10 mm gravel aggregate with Hope Quarry sand. Soutsos *et al* presented experimental data of the relative permittivity and conductivity of concrete as function of frequency and moisture content covering the frequency range of 500 MHz to 1 GHz. Robert [614] performed a similar measurement on five samples of concrete; the maximum size of aggregate used was 30 mm. Plots showing the variation of permittivity with frequency were presented. The data showing the influence of moisture content, in particular the effects of chloride on the permittivity were also presented. Five material models with frequency-dependent permittivity were compared with the experimental data. The Cole-Cole model worked better than Debye's in fitting the measurement values, showing the best agreement. Ogunsola *et al* [590] presented a dispersive model for concrete that accounts for the complex permittivity dependence on frequency. The authors achieved this by modeling concrete as a Debye material with a dc electrical conductivity. The model parameters are fitted to published experimental data, particularly those of Soutsos and Roberts. The authors also explored the effects of moisture content on the electrical parameters of concrete, as well as on the shielding effectiveness of a concrete structure. They modeled the frequency-dependent complex relative permittivity of concrete as:

$$\hat{\varepsilon}_r = \varepsilon_\infty + \frac{\triangle\varepsilon}{1+\omega^2\tau^2} - j\frac{\omega\tau\triangle\varepsilon}{1+\omega^2\tau^2} \tag{7.212}$$

where $\triangle\varepsilon = \varepsilon_{static} - \varepsilon_\infty$ is the difference between the values of the real part of the complex relative permittivity at low and high frequency, respectively, and τ is the relaxation time. Equation (7.212) can be expressed in terms of the real and imaginary component of the complex relative permittivity of concrete which reflects the ability of concrete to store electromagnetic energy and accounts for the energy losses due to dielectric relaxation respectively. These are shown in (7.213) and (7.214) below:

$$\varepsilon_r^{'}(\omega) = \varepsilon_r + \frac{\triangle\varepsilon}{1+\left(\frac{\omega}{\omega_r}\right)} \tag{7.213}$$

$$\varepsilon_{r,eff}^{''}(\omega) = \frac{(\omega/\omega_r)\triangle\varepsilon}{1+(\omega/\omega_r)^2} + \frac{\sigma_{dc}}{\omega\varepsilon_0} = \frac{\sigma_{eff}(\omega)}{\omega\varepsilon_0} \tag{7.214}$$

Sandrolini *et al* in [619] extended the work detailed above, providing extended parameters for the electrical properties of concrete as a function of moisture content, as shown in Table 7.4.

Table 7.4 Fitted parameters for concrete from measured data

Moisture Content	ε_{static}	ε_{∞}	τ (ns)	σ_{dc} $(\Omega^{-1}\mathrm{m}^{-1})$
0.2%[640]	4.814 ± 0.002	4.507 ± 0.002	0.82 ± 0.01	$6.06 \times 10^{-4} \pm 0.06 \times 10^{-4}$
2.8%[640]	6.75 ± 0.03	5.503 ± 0.005	2.28 ± 0.08	$2.03 \times 10^{-3} \pm 0.04 \times 10^{-3}$
5.5%[640]	8.63 ± 0.02	6.023 ± 0.009	1.00 ± 0.02	$5.15 \times 10^{-3} \pm 0.06 \times 10^{-3}$
6.2%[627]	9.14 ± 0.06	5.93 ± 0.02	0.80 ± 0.03	$6.7 \times 10^{-3} \pm 0.3 \times 10^{-3}$
9.3%[627]	11.19 ± 0.05	7.20 ± 0.02	0.73 ± 0.02	$23 \times 10^{-3} \pm 2 \times 10^{-3}$
12.0%[640]	12.84 ± 0.03	7.42 ± 0.02	0.611 ± 0.006	$20.6 \times 10^{-3} \pm 0.2 \times 10^{-3}$

7.5.5 Self Impedance of Conductors

7.5.5.1 Elementary Formulations

The impedance of a straight circular conductor may be determined as a first approximation by its longitudinal resistance at dc R_{dc} and the ideal longitudinal inductance L, by assuming a re-closure at infinity, with the following expressions [511]:

$$R_{dc} = \rho \frac{l}{S} \tag{7.215}$$

$$L = \frac{\mu}{8\pi} \tag{7.216}$$

Several improvements to the inductance formulation exist to account for the geometrical properties of the conductor, such as its diameter d, or non-cylindrical shape [615]:

- for a cylindrical straight wire of non-magnetic material with length l and radius r

$$L = 2l\left[\ln\left(\frac{2l}{r}\right) - \frac{3}{4}\right] \tag{7.217}$$

- if the cylindrical wire is made of material with relative magnetic permeability μ_r the expression above becomes

$$L = 2l \left[\ln \left(\frac{2l}{r} \right) - 1 + \frac{\mu_r}{4} \right] \tag{7.218}$$

and the term $\mu_r/4$ disappears at very high frequency where the magnetic permeability vanishes reaching the free space value;

- if the conductor is complemented by a return circuit made of a similar parallel conductor at distance d, by neglecting the end effects, the self inductance is

$$L = 4l \left[\ln \left(\frac{d}{r} \right) - \frac{d}{l} + \frac{\mu_r}{4} \right] \tag{7.219}$$

- for a square or rectangular conductor (e.g. a bar)

$$L = 2l \left[\ln \left(\frac{2l}{R} \right) - 1 + \frac{R}{l} \right] \tag{7.220}$$

where R is the equivalent radius and is obtained as the Geometric Mean Distance (GMD) of the conductor cross section: for a rectangle of sides a and b, the equivalent radius is

$$\ln R = \sqrt{a^2 + b^2} - \frac{1}{6}\frac{a^2}{b^2} \ln \sqrt{1 + \frac{b^2}{a^2}} + \\ - \frac{1}{6}\frac{b^2}{a^2} \ln \sqrt{1 + \frac{a^2}{b^2}} + \frac{2}{3}\frac{a}{b} \arctan \left(\frac{b}{a} \right) + \frac{2}{3}\frac{b}{a} \arctan \left(\frac{a}{b} \right) - \frac{25}{12} \tag{7.221}$$

that for a square of side a simplifies to

$$\ln R = \ln a + \frac{1}{3} \ln 2 + \frac{\pi}{3} - \frac{25}{12} \tag{7.222}$$

that gives $R = 0.447a$;

- the self inductance of a circular conductor having a radius r and positioned at a height h above a ground plane is

$$L \approx l \frac{\mu_r \mu_0}{2\pi} \ln \left(\frac{2h}{r} \right) \text{ H} \tag{7.223}$$

- the mutual inductance between two conductors separated by a distance d and positioned at a height h above a ground plane is

$$M = l\frac{\mu_r \mu_0}{4\pi} \ln \left[1 + \left(\frac{2h}{d} \right)^2 \right] \text{ H} \tag{7.224}$$

- if the ground plane in (7.224) is replaced by conductors located at a distance h below the ground plane then the following expressions apply

$$M = l\frac{\mu_r \mu_0}{2\pi} \ln \left(\frac{d_{14} d_{23}}{d_{13} d_{24}} \right) \text{ H} \tag{7.225}$$

where $d_{14}, d_{13}, d_{23}, d_{24}$ are the geometrical mean distances between the conductors.

The ac longitudinal resistance R_{ac} is bounded to increase with increasing frequency for two concomitant mechanisms that are considered in more detail in the next paragraphs: the skin effect and, possibly, the ferromagnetic losses, due to the hysteretic behaviour of the magnetic material.

For any conductor of not infinitesimal cross section, the longitudinal inductance may be expressed as the sum of two terms, the external inductance L_{ext} (related to the geometry of the circuit, and thus at a first approximation equal to L used in the expression above) and the internal inductance L_{int} (related to the real distribution of the magnetic field inside the conductor itself and the partial linkage to the infinitesimal current elements distributed over the cross section). The internal impedance of a conductor is defined in [511] as the ratio of the surface voltage drop V_s and the total current in the conductor I. The internal impedance consists of a real part (which gives the ac resistance of the conductor) and an imaginary part (which gives the internal reactance of the conductor, arising from the magnetic flux within the conductor itself). The choice of the circuit path along the surface (on which the surface voltage drop is computed) is convenient because it does give the separation between "internal inductance" and "external inductance"; any other path internal to the conductor may be chosen as well.

$$Z_{int} = V_s / I \qquad R_{int} = \text{Re}\left\{ Z_{int} \right\} \qquad L_{int} = \text{Im}\left\{ Z_{int} \right\} / 2\pi f \tag{7.226}$$

Skin effect must be considered for the analysis of the internal resistance and inductance behaviour with respect to frequency. The fields and current magnitudes decrease with an approximate exponential decay with penetration into conductors, so, the skin effect may be defined as the phenomenon which tends to concentrate currents on the surface of conductors closer to the field sources producing them: with increasing frequency the current is concentrated over a smaller portion of the cross section of the conductor and the resistance increases; also, if penetration of fields decreases, then the internal magnetic flux and the internal inductance decrease. The variability of the phenomenon makes also the longitudinal inductance change with frequency because of skin effect; moreover, the magnetic permeability of the material changes (and normally decreases) with frequency and this has again a direct influence on the longitudinal inductance.

A first attempt to consider the variability vs. frequency of resistance R_{ac} including the skin effect is to replace a conductor of arbitrary shape with a circular conductor of radius r, with low frequency R_{ac} made corresponding to the former R_{dc}, that is proportional to the surface S ($S = \pi r_S^2$), and high frequency R_{ac} estimated on the equivalence of the perimeter ($p = 2\pi r_p$) [30, 594]. For the rails considered in [505] r_S is about 50% of r_p: the calculated radii for a UNI60 heavy traction rail are $r_S = 49.4$ mm and $r_p = 107.7$ mm; the two values bracket the equivalent radius that is obtained in [30] by Geometrical Mean Radius (GMR) formulation [615] (62.9 mm).

A second approach is that of sub-division of conductors proposed in [6, 5], where for the chosen elementary conductors a uniform current density may be assumed and then the entire set of coupled differential equations is solved with suitable boundary conditions, defined by the external circuit; this approach is numerical rather than analytical and the discretization of the conductor under analysis is a direct method that requires no simplifying assumption. The minimum number of sub-conductors is not set a-priori, but it is to be determined by increasing it progressively until the calculated resistance/inductance is no longer changing; in the proposed implementation only a constant value of the relative magnetic permeability may be used, to be determined by practical data or a sensitivity analysis, thus eliminating any explicit dependency on frequency and current amplitude.

7.5.5.2 Analytical Solution for Cylindrical Conductors

The analytical solution of the internal part of the series impedance vs. frequency may be obtained for a circular conductor with an expression that involves Bessel functions and its derivative, and that lends itself to a simplification that justifies the appearance of exponential terms.

The complete development may be found in [221, 609], where the problem is solved for a circular conductor, that lends itself to a simplification by expressing all the variables in the radial coordinate r thanks to the circular symmetry.

The time-varying magnetic flux produced by the unknown current density $J_z(r)$ oriented in the z direction is put in relationship with the electric field $E_z(r)$ (simply $J_z(r) = \sigma E_z(r)$ and replaced in the Faraday law in integral form; the obtained relationship is finally a modified Bessel's equation of order zero

$$\frac{\partial^2 J_z(r)}{\partial r^2} + \frac{1}{r}\frac{\partial J_z(r)}{\partial r} - j\omega\mu\sigma J_z(r) = 0 \tag{7.227}$$

The general solution

$$J_z(r) = A\, J_0(\sqrt{-j\omega\mu\sigma} r) + B\, Y_0(\sqrt{-j\omega\mu\sigma} r) \tag{7.228}$$

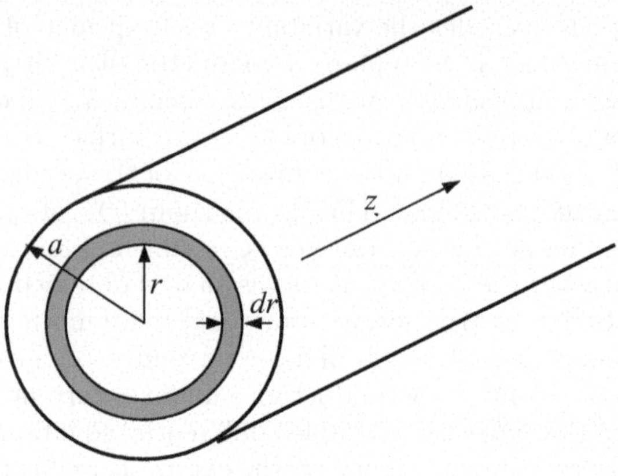

Fig. 7.33 Cylindrical conductor for the analytical calculation of the current density and series impedance

where $J_0(\cdot)$ and $Y_0(\cdot)$ are the Bessel functions of the first and second kind of order zero.

For a real argument these functions are calculated by infinite series and can be found in many mathematical tools and tables. If the argument is complex the series contain real and imaginary terms; the two Bessel functions are thus replaced by their real and imaginary parts, indicated as $\mathrm{ber}(x) = \mathrm{Re}\left\{J_0(\sqrt{-j}x)\right\}$, $\mathrm{bei}(x) = \mathrm{Im}\left\{J_0(\sqrt{-j}x)\right\}$, $\mathrm{ker}(x) = \mathrm{Re}\left\{J_0'(\sqrt{-j}x)\right\}$, $\mathrm{bei}(x) = \mathrm{Im}\left\{J_0'(\sqrt{-j}x)\right\}$. The arguments have been indicated as $\sqrt{-j}x$, since $\sqrt{-j\omega\mu\sigma}r$ may be transformed into $\sqrt{-j}\sqrt{2}r/\delta$, where $\delta = \sqrt{2/\omega\mu\sigma} = 1/\sqrt{\pi f\mu\sigma}$, that is the expression of skin depth.

Equation (7.228) may be then written as

$$J_z(r) = A\left[\mathrm{ber}(\sqrt{2}r/\delta) + j\,\mathrm{bei}(\sqrt{2}r/\delta)\right] + B\left[\mathrm{ker}(\sqrt{2}r/\delta) + j\,\mathrm{kei}(\sqrt{2}r/\delta)\right] \tag{7.229}$$

Since $\mathrm{ker}(x)$ tends to infinite when $x \to 0$ (that is $r \to 0$, at the center of the conductor), the coefficient B must be necessarily 0 and the rightmost term vanishes. The current density expression for cylindrical conductors is then

$$J_z(r) = A\left[\mathrm{ber}(\sqrt{2}r/\delta) + j\,\mathrm{bei}(\sqrt{2}r/\delta)\right] \tag{7.230}$$

If the radius a of the conductor is considered, then

$$J_z(a) = A\left[\mathrm{ber}(\sqrt{2}a/\delta) + j\,\mathrm{bei}(\sqrt{2}a/\delta)\right] \tag{7.231}$$

The p.u.l. impedance may be expressed as the ratio of the longitudinal voltage drop (that corresponds to the electric field $E_z(r)$ multiplied by 1 m) and the circulating current I_z. This current is put in relationship with the tangential magnetic field $H_\varphi(r)$ by the so called Ampère's law

$$I_z = 2\pi a H_\varphi(a) \tag{7.232}$$

$H_\varphi(r)$ in turn is put in relationship with the same electric field $E_z(r)$ by recalling the second Maxwell curl equation for the non-null components

$$\frac{\partial E_z(r)}{\partial r} = j\omega\mu H_\varphi(r) \tag{7.233}$$

Combining the above equations it is possible to write

$$I_z = \frac{2\pi a}{j\omega\mu\sigma}\frac{A\sqrt{2}}{\delta}\left[\mathrm{ber}'(\sqrt{2}a/\delta) + j\,\mathrm{bei}'(\sqrt{2}a/\delta)\right] \tag{7.234}$$

that, replacing A, becomes

$$I_z = \frac{2\pi a}{j\omega\mu\sigma}\frac{\sqrt{2}}{\delta}J_z(a)\frac{\left[\mathrm{ber}'(\sqrt{2}a/\delta) + j\,\mathrm{bei}'(\sqrt{2}a/\delta)\right]}{\left[\mathrm{ber}(\sqrt{2}a/\delta) + j\,\mathrm{bei}(\sqrt{2}a/\delta)\right]} \tag{7.235}$$

where the prime at the right of ber( ) and bei( ) terms indicates the first derivative.

The impedance is thus

$$\begin{aligned} Z &= \frac{j\omega\mu}{2\pi a}\frac{\delta}{\sqrt{2}}\frac{\left[\mathrm{ber}(\sqrt{2}a/\delta) + j\,\mathrm{bei}(\sqrt{2}a/\delta)\right]}{\left[\mathrm{ber}'(\sqrt{2}a/\delta) + j\,\mathrm{bei}'(\sqrt{2}a/\delta)\right]} \\ &= \frac{j}{2\pi a}\sqrt{\frac{\omega\mu}{\sigma}}\frac{\left[\mathrm{ber}(\sqrt{2}a/\delta) + j\,\mathrm{bei}(\sqrt{2}a/\delta)\right]}{\left[\mathrm{ber}'(\sqrt{2}a/\delta) + j\,\mathrm{bei}'(\sqrt{2}a/\delta)\right]} \end{aligned} \tag{7.236}$$

The formulation may be considered exact for round linear conductors, so taking into account also ferromagnetic conductors with assigned constant magnetic permeability μ, if any non-linearity related to hysteretic behaviour and the so called *iron losses* may be neglected. Practically the latter may be included as an additional equivalent resistance term on top of the skin effect term, but its treatment requires a more complex modelling (accounting for non-linear behaviour and hysteresis cycles) and/or the use of measured results.

In [487] the macroscopic properties of attenuation and phase constant, based on an equivalent skin depth, are put in relationship with the hysteresis properties of the material. Given the B-H relationship based on a complex permeability

$$B = (\mu' - j\mu'')H = \mu H = |\mu|\, H \exp(j\theta_\mu) \tag{7.237}$$

the complex skin depth is correspondingly defined

$$\delta = \sqrt{\frac{1}{\pi f \sigma |\mu|}} \exp(-j\theta_\mu/2) \tag{7.238}$$

and equivalent attenuation and phase constants are

$$\alpha = \sqrt{\pi f \sigma |\mu|}\,(1-\theta_\mu/2) \qquad \beta = \sqrt{\pi f \sigma |\mu|}\,(1+\theta_\mu/2) \tag{7.239}$$

7.5.5.3 Proximity Effect

Proximity effect describes the influence of neighbouring conductors carrying ac current and thus modifying the current distribution of the conductor itself. The effect of the proximity effect is that of reducing the cross section for current conduction due to the proximity of other current-carrying conductors, thus causing an increase of the measurable resistance from R to R'. There exist analytical solutions for some configurations of conductors with circular symmetry [418]; for other conductor shapes the solution is obtained numerically. An elegant solution is presented in [250] for tubular conductors in terms of series expansion and the reported results . Formulas of the ac resistance increase $p = R'/R$ are given for different s/d ratios, where s is the separation between centers of conductors and d is the conductor diameter:

$$\begin{aligned} p = {} & 1 + 0.12701 l^4 - 0.10965 l^8 + 0.09599 l^{12} + \\ & -0.8435 l^{16} + 0.0742 l^{20} - 0.0653 l^{24} \end{aligned} \qquad s/d = 2.0 \tag{7.240}$$

$$\begin{aligned} p = {} & 1 + 0.22872 l^4 - 0.17337 l^8 + 0.13619 l^{12} + \\ & -0.10804 l^{16} + 0.0860 l^{20} - 0.0684 l^{24} \end{aligned} \qquad s/d = 1.5 \tag{7.241}$$

$$\begin{aligned} p = {} & 1 + 0.53530 l^4 - 0.25686 l^8 + 0.15408 l^{12} + \\ & -0.0961 l^{16} + 0.0604 l^{20} - 0.0381 l^{24} \end{aligned} \qquad s/d = 1.0 \tag{7.242}$$

where the parameter l is in relationship with the skin effect and is given by $l = \sqrt{2\pi f/R}$, with R measured in absolute unit. Theoretical and measured values are in optimal agreement and show that for $s/d = 2.0$ the resistance increase due to proximity effect is limited to 10% while increasing the frequency by two orders of magnitude; on the contrary when the two conductors are side by side with $s/d = 1.0$ the percent increase in resistance may be as high as 70% for the same frequency change.

The more modern approach in [30] consists of a discretization of the domain occupied by the conductors and solution of the field equations in matrix form; this approach of course can treat conductors of any shape and the accuracy of the calculation method is not questioned, since the comparison with measurement results [604, 631] shows an almost exact match between the two.

7.5.5.4 Experimental Results

The use of experimental results is important for those conductors with an approximate modelling, because of several reasons: non linearity (like in the presence of ferromagnetic materials), variability of properties with frequency (such as magnetic permeability and loss resistance), variability due to other external factors (such as temperature), variability of basic material properties over the entire production (in particular if a specific property doesn't undergo quality sample tests, such as magnetic properties for running rails, often neglected in favor of the more important mechanical characteristics). Rails have been considered by means of some tests and measurements and results have been published in less than a dozen papers in the last 100 years; for measurements taken at power supply frequency, Mariscotti and Pozzobon in [511] report a survey for the various types of rails and supply frequencies adopted in the tests; for higher frequency (i.e. audiofrequency) tests were conducted at large signal and also at small signal, superimposed to an underlying waveform at power supply frequency (either dc or 50 Hz).

Achard, Kennelly and Dana (AKD) tested several rails at 25 and 60 Hz using a loop test circuit with the rail supplied in differential mode with a return conductor and compensating for the external inductance of the so formed loop circuit. Trueblood and Wascheck (TW) ran similar tests including also 40 Hz supply frequency. Hill and Carpenter (HC) again performed similar tests at 50 Hz only, but zeroing out the external inductance using a hollow rail in copper that replaced the rail sample in the test setup, simulating an extreme skin effect. The measurements appearing in the ORE question were made for the first time using a coaxial structure to reduce external inductance to a minimum and also to shield the test setup from external magnetic fields; the tests are done only at 50 Hz on some rail samples, but almost doubling the test current to 2000 A. Mariscotti and Pozzobon (MP) repeated the tests at 50 Hz using a loop structure made of two identical rails at the standard gauge distance (1435 mm on internal profiles of rail head) and at a reduced distance (0.5 m, also to increase the circulating current reducing the test circuit external inductance), attaining a maximum current of about 1200 Arms and compensating for the external inductance by first, calculation and, second, measurement at very high frequency (100 kHz).

Given the relevance of the 50 Hz supply frequency, the results of the most recent publications are summarized in Figure 7.37 after interpolation to 50 Hz for those measured at different values of supply frequency. The same experimental results are compared with theoretical formulation (7.236) for variable relative magnetic permeability μ_r; the spread of values for each μ_r identified by the gray rectangles is given by the two different values for the equivalent radius in terms of surface or perimeter.

As a synthetic picture the experimental values collected from the various publications are treated statistically to quantify the dispersion of the same

Table 7.5 Geometrical data of the tested rail types

Type	Section [mm^2]	Perimeter [mm]	Weight
(AKD) 96A	6320	574	100 lb/yd
(AKD) 100	6320	620	100 lb/yd
(AKD) 163	6320	617	100 lb/yd
(AKD) 175	6320	620	100 lb/yd
(AKD) 161	6320	584	100 lb/yd
(TW) 100	—	625	100 lb/yd
(TW) 130	—	654	130 lb/yd
(HC) 113A	7500	630	100 lb/yd
(MP) UNI60	7679	656	60 kg/m
(ORE) UIC48	—	—	48 kg/m
(ORE) UIC54	—	—	54 kg/m
(ORE) UIC65	—	—	65 kg/m

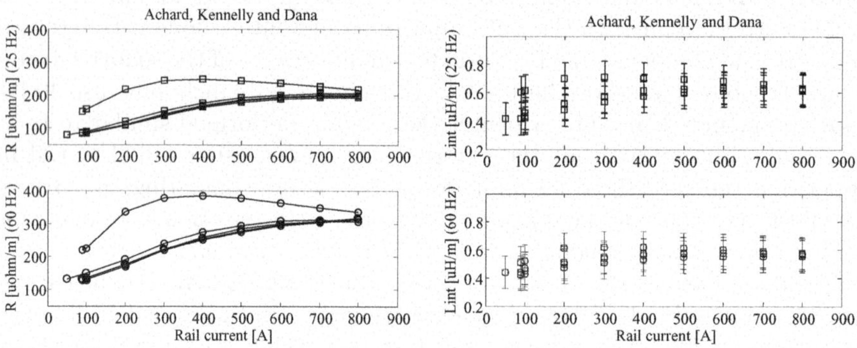

(a) (b)

Fig. 7.34 Measured (a) resistance, (b) internal inductance in AKD (square 25 Hz, circle 60 Hz)

data; the mean value and $\pm 1\sigma$ intervals are shown in Figure 7.38 versus supply current amplitude.

For the high frequency results Mariscotti and Pozzobon adopted a coaxial structure (using an external copper shield) and performed small signal measurements at audiofrequency superimposed on a biasing current at either dc or 50 Hz; the current levels of the biasing current cover the usual supply current levels in real traction systems, while the small signal current ranges from a fraction of to a few Ampere, as does the nominal current of track circuits. Some results are also reported by Hill and Carpenter, yet without a specific description of the measurement setup. The importance of the amplitude of the biasing current is the setting of the operating point of the magnetic circuit

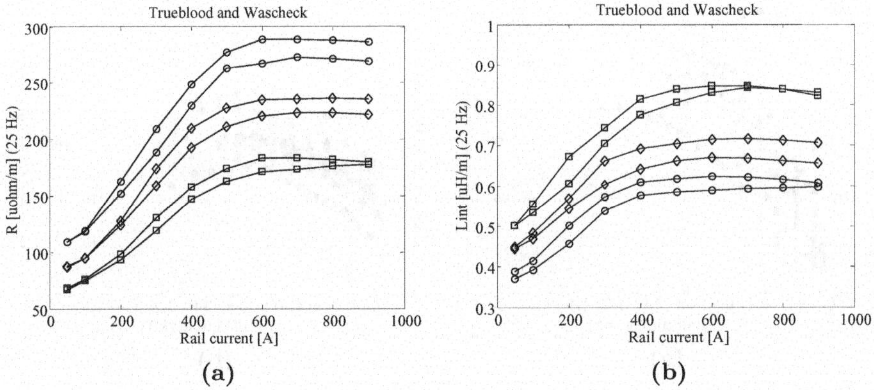

Fig. 7.35 Measured (a) resistance, (b) internal inductance in TW (square 25 Hz, diamond 40 Hz, circle 60 Hz)

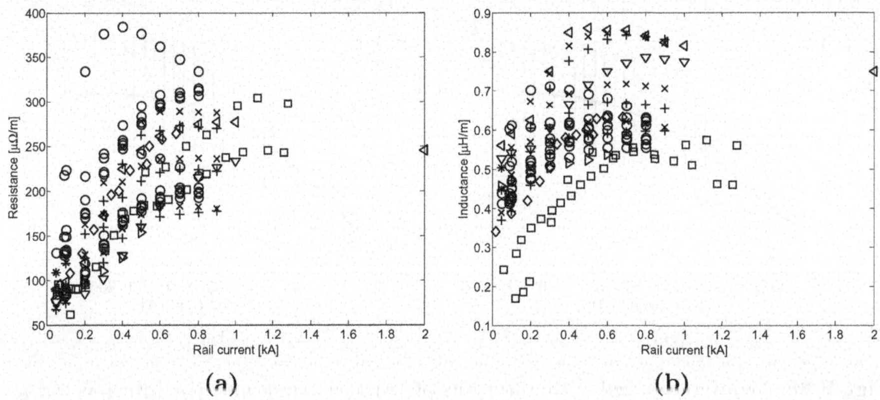

Fig. 7.36 Measured (a) resistance, (b) internal inductance at 50 Hz (square [3], diamond [5 corr. lower curve, 8 upper curve], circle [6,7])

along its main hysteresis cycle, so that the faster hysteresis cycles driven by the high frequency test signal are located in different operating points with different slopes.

The dependency of the resistance and inductance on the instantaneous current was demonstrated in [510, 285]: for the lower audiofrequency values the dependency on the 50 Hz current waveform is clear ([510], Figure 9a), where both the resistance and internal inductance values follow the four quadrants of the 50 Hz waveform (see 7.39 below as an example). In the same publication the harmonic components of the internal inductance are analyzed with the supply frequency (50 Hz) as the fundamental, identifying relevant components at 50, 100 and 200 Hz for test frequencies up to 10 kHz.

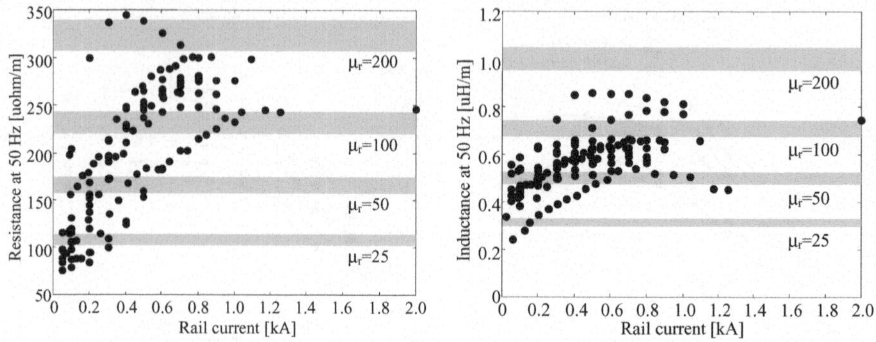

(a) (b)

Fig. 7.37 Comparison of theoretical and experimental data for (a) resistance and (b) internal inductance at 50 Hz for different μ_r.

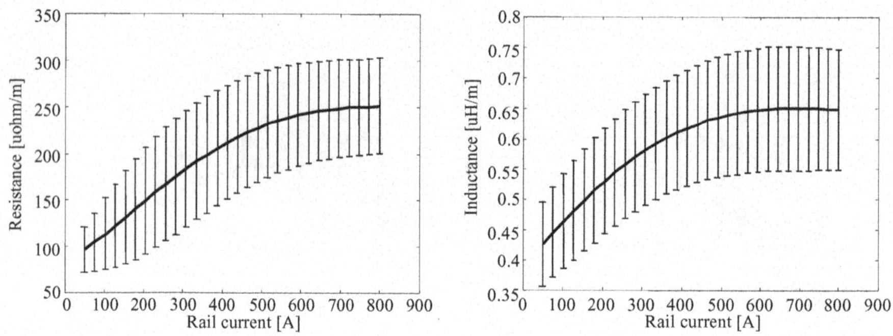

(a) (b)

Fig. 7.38 Mean value and $\pm 1\sigma$ intervals of (a) resistance and (b) internal inductance at 50 Hz

Considering the problem of the temperature dependency of the resistance and inductance discussed so far, in [511] the temperature coefficients determined by several authors are compared and discussed: the average values for various types of rails at supply frequency is 3070 and 2510 ppm/°C for the resistance and inductance respectively.

7.5.5.5 Single Conductor over a Lossy Earth

A solution for the electromagnetic field and current density in the earth for an infinitely long conductor over a homogeneous non-ideal earth was presented by Carson [46]. Carson's solution is based on Maxwell's equation and circuit theory, however due to the use of the circuit theory the validity of the solution

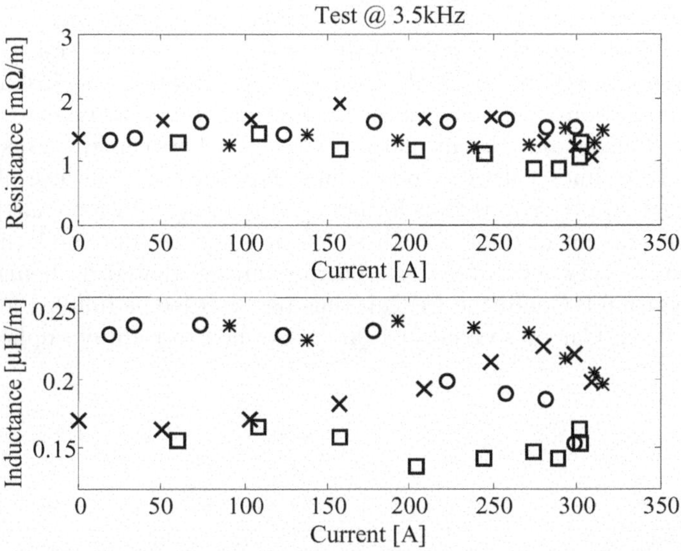

Fig. 7.39 Internal resistance and inductance as a function of 50 Hz instantaneous current value for a 3.5 kHz test frequency: 'circle' 1st quadrant, 'square' 2nd quadrant, 'cross' 3rd quadrant, 'star' 4th quadrant.

is limited to low frequencies. There are four necessary conditions that must be fulfilled for the application of Carson's equation to be valid, these are:

1) The magnitude of the wave number of the earth must be far greater than that of free space;

2) The length of the observation point to the conductor (or its image) should be far less than the wavelength at free space;

3) The radius of the conductor (rail, catenary, signalling cable etc) must be far less than the height of the conductor above ground;

4) The propagation constant of the conductor must be far less than the wave number of the earth.

For railway installations, conditions (1) and (2) are easily fulfilled due to the frequency of operations, 50 Hz (or 16 2/3 Hz as used in some European countries). The issue regarding the application of Carson's equation to railway applications lays in the fulfillment of condition (3) and (4). Condition (3) can not be fulfilled as both rails and signalling/communications cables are in continuous contact with the ground. Similarly, the condition (4) is partially fulfilled as the propagation constant of the rail is of the same order as the wave number of the earth.

Olsen and Pankaskie [592] provided an exact solution of Carson's equation in terms of the approximations that must be made to derive Carson's equation from the exact theory. In [592], two approximate expressions for the propagation constant of wire is given; these expressions apply to a case of a conductor above or at the earth surface. The results obtained from these approximate expressions suggest that the violation (or otherwise) of the Condition (4) depends on the quality of the contact between the conductors and earth. Chang and Wait in [215] also came to the same conclusion. In the railway installations, the insulation between rails and earth is relatively good due to the ballast and the insulating pads used between the rails and the ties on railway lines with concretes slabs. Furthermore, the use of booster transformers (BTs) or autotransformers (ATs) in ac railway, contributes to keeping the current in the rails. These features ensure that the level of leakage current is very low; and thus the rail can be viewed as being slightly above the ground. Condition (4) can thus be regarded as fulfilled and, as an approximation, Carson's equations can be applied to railway applications.

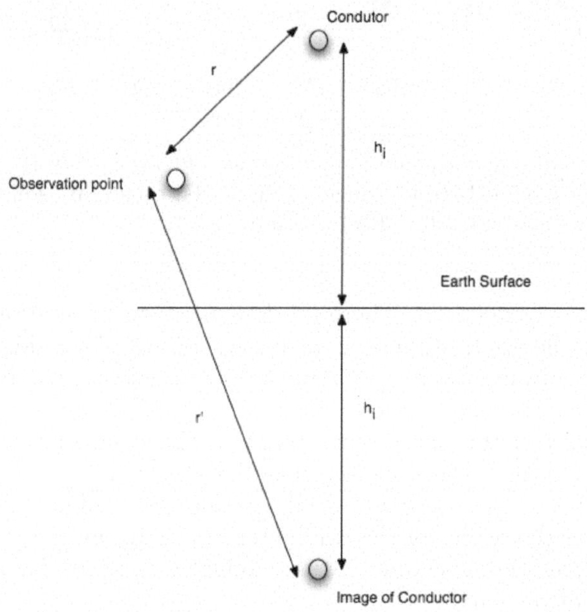

Fig. 7.40 Single conductor over lossy earth

The series impedance for a conductor over a lossy earth (as shown in Figure 7.40) can be expressed by the sum of the impedance over a perfectly conducting earth and a contribution that is due to the non-ideal characteristic of the earth. The later is given by an infinite integral; the solution to the infinite integral can be expressed in terms of two functions P and Q. The function P accounts for the dissipated energy in the earth path, while Q accounts for the magnetic effects of the earth current. The following expression can be written for the self impedance of the conductor:

$$Z_{ii} = R_{int} + jX_{int} + j\frac{\omega\mu_0}{2\pi}\ln\frac{2h_i}{r_i} + \frac{2\omega\mu_0}{2\pi}\left(P_{ii} + Q_{ii}\right) \tag{7.243}$$

Both P and Q can be expressed as infinite series, which are rapidly convergent. For cases where $r = 2h_i\sqrt{\omega\mu\sigma_e} \leq 1/4$, the series can be written as:

$$P_{ii} = \frac{\pi}{8} - \frac{r}{3\sqrt{2}} + \frac{r^2}{16}\left[0.6728 + \ln\frac{2}{r}\right] + \frac{r^3}{45\sqrt{2}} - \frac{\pi r^4}{1536} + \cdots \tag{7.244}$$

$$Q_{ii} = -0.0386 + \frac{1}{2}\ln\frac{2}{r} + \frac{r}{3\sqrt{2}} - \frac{\pi r^2}{64} + \frac{r^3}{45\sqrt{2}} - \frac{r^4}{384} - \frac{r^4}{384}\left[\ln\frac{2}{r} + 1.0895\right] + \cdots \tag{7.245}$$

The leading terms of (7.244) and (7.245) is dependent on the frequency and on the accuracy required. For low frequency problems, an equivalent depth approximation deduced from Carson's equation can be used. The approximation obtained for cases where and only the leading terms are retained with the earth path viewed as an equivalent conductor at a given depth. The depth is dependent on earth resistivity and frequency and is given by:

$$D_e = 660\sqrt{\frac{\rho}{f}} = 1.31\sqrt{\frac{2\rho}{\mu\omega}} \tag{7.246}$$

The imaginary return conductor has an internal resistance that can be determined from the dissipated energy in the earth. A simplified expression for the series impedance of a conductor over a lossy earth is given as:

$$Z_{ii} = R_{int} + R_e + jX_{int} + j\mu_0 f \ln\frac{D_e}{r_i} \tag{7.247}$$

where $R_e = \mu\pi f/4$, a closer look at (7.247) reveals an independence of the conductor's height above the ground. However, the assumption that the conductor height above ground is much larger than the conductor radius must be fulfilled. The analysis presented above assumes that the earth is a homogeneous conducting half space, while in reality railway installations span several kilometres and variations in soil characteristics is to be expected. This variation can result in difficulties in defining a multilayer soil model which adequately describes the whole installation. Equation (7.247) can be extended to deal with horizontally stratified earth with multiple layers. Nakagawa *et al* detailed and provided expressions for a 3 layer case in [559] and a multilayer case in [560].

For buried conductors, if propagation effects are neglected, an expression for the impedance of a conductor at the earth's surface for cases where $|k_e r_i| \leq 0.25$ is given by Sunde [650] as:

$$Z_{ii} = Z_{int} + \frac{j\omega\mu}{2\pi}\left[\frac{-j\pi}{4} + \ln\left(\frac{1.85}{|r_i k_e|}\right)\right] \tag{7.248}$$

where $k_e = j\omega\mu\sigma_e$ and r_i is the radius of the conductor. According to Sunde, the self inductance of an insulated conductor buried at normal depth will be essentially the same as the self inductance of a conductor at the earth's surface. Consequently, Carson's equations can be used for buried conductors as long as the leakage currents are small. For situations where the propagation effects are non-negligible, i.e. large leakage currents, the expressions for the impedance will be dependent on the propagation constant, γ, of the buried conductor and is given as:

$$Z_{ii}(\gamma) = Z_{int} + \frac{j\omega\mu}{2\pi} \ln\left(\frac{1.85}{r_i\sqrt{k_e^2+\gamma^2}}\right) \tag{7.249}$$

Since the development and the symbology are very similar, Carson's expressions in terms of infinite series are treated in the next subsection for both self and mutual terms altogether.

7.5.5.6 Finite Element Methods

In this section Finite Element Methods (FEM) that can be and have been used to analyse specific problems related to the scope of railway traction system are briefly reviewed. FEM are appropriate whenever the problem geometry is too complex for a closed form formulation; they are also used as a subsidiary tool for the validation of results obtained by other means (e.g. by approximate closed form formulas or as measurement results) and may be adopted also as the primary tool for simulation and theoretical calculations, since they are in generally well trusted and their validity is not questioned (provided that they are fed with correct geometries and parameters and they are used for electromagnetic phenomena that can be modelled and solved by the internal engine).

Several different methods are normally conveyed under the name "FEM", such as Transmission Line Method (TLM), Method of Moments (MoM) and Finite Difference Time Domain (FDTD).

Going from the particular to the general in the following a few examples are briefly reviewed, based on the published results and on the authors' direct experience.

FEM have been used in many cases to model specific elements and components of a railway system, in particular where the phenomena are non-linear and there is not a common agreement on the values that characterize them:

- running rails (considered in the previous section) are known as a non-linear component, whose p.u.l. electrical parameters are influenced by the magnetic properties of the material, and thus by the instantaneous current flowing, and by the frequency; several authors [245, 311, 244, 306] have presented results obtained by applying FEM techniques, that have been already discussed and compared with other theoretical values and measurement results;

- also cables and other more regular conductors may benefit of FEM modeling, if coupling at short distance and proximity effect in cable ways are of interest, as well as magnetic diffusion and other phenomena related to a more accurate evaluation of transfer impedance [496];
- the modelling of the whole track, composed of the running rails, sleepers and ballast, and taking into account rail and sleeper shape as well as the properties of the used materials was considered in [307]; the results appear in sec. 7.5.8 and 7.5.7;
- electromagnetic emissions from the vehicle body and from an entire section of the traction line were evaluated by means of TLM-2D method [38] and the results there, as well as the cited references, represent an attempt to use meshing and FEM techniques to a wide domain, with very demanding extension with respect to the mesh size and the wavelength for the analysed frequency range; a more classical approach based on the modeling of emissions of current elements [231] is more suitable for a lower frequency range, limited to a few MHz, where significant deviations from measurement results begin to appear.

7.5.6 *Mutual Impedance of Conductors*

The mutual impedance between two conductors is that established between two independent circuits formed by each of the two conductors and their ground return paths. In the absence of internal components that establish a mutual path, the real part (the mutual resistance) is zero. The determination of the mutual impedance for long parallel conductors may be derived from expressions of the electromagnetic fields generated by electric dipoles (Carson's formulation), or from the complex penetration depth method. In cases, where the condition of parallelism between the conductors is not fulfilled, the segmentation method, discussed in [419], is often applied where the non-paralleled conductor is sub-divided into segments and each resulting region is modelled as being parallel to the transmission line at a distance equal to the average equivalent separation from the transmission line.

7.5.6.1 Two Conductors with Earth Return

An expression for the mutual impedance for two conductors over a perfectly conducting earth (as shown in Figure (7.41)) and a term representing the non-ideal characteristic of the earth was provided by Carson [46].

$$Z_{ji} = \frac{j\omega\mu_0}{2\pi} \ln\left[\frac{D'_{ji}}{D_{ji}}\right] + \frac{2\omega\mu_0}{2\pi}\left(P_{ji} + Q_{ji}\right) \tag{7.250}$$

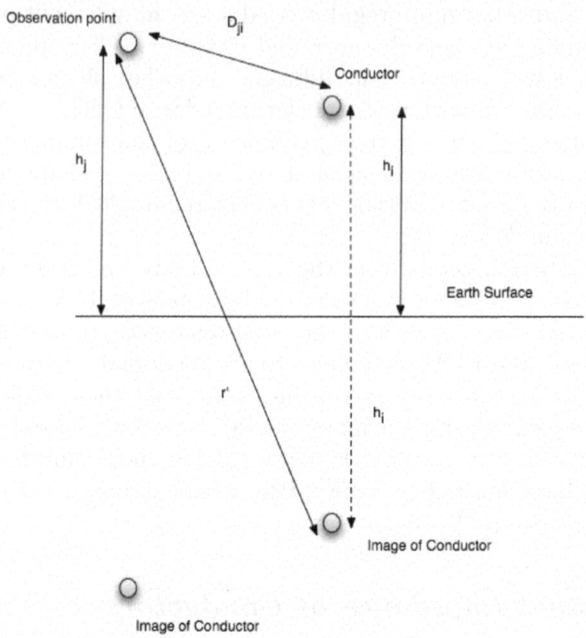

Fig. 7.41 Two Conductors over a perfectly conducting earth

For $r \leq 1/4$, the infinite series P_{ji} and Q_{ji} are given by:

$$P_{ji} = \frac{\pi}{8} - \frac{r\cos\theta}{3\sqrt{2}} + \frac{r^2}{16}\cos 2\theta\left(0.6728 + \log\frac{2}{r}\right) + \frac{r^2\theta\sin 2\theta}{16} \tag{7.251}$$

$$Q_{ji} = -0.0386 + \frac{1}{2}\log\frac{2}{r} + \frac{r\cos\theta}{3\sqrt{2}} \tag{7.252}$$

where $r = kD'_{ji} = \sqrt{\omega\mu_e\sigma_e}D_{ji}$ and $\theta = \sin^{-1}\left|\frac{D_{ji}}{D'_{ji}}\right|$

Using the concept of equivalent depth of return for this system leads to the following expression for the mutual impedance between the two conductors:

$$Z_{ji} = R_e + j\mu_0 f\ln\left|\frac{D_e}{D_{ji}}\right| \tag{7.253}$$

This equation applies to cases where the earth is supposed to be homogeneous. A simplified expression based on Carson and Pollaczek [603] for mutual coupling of multiple conductors referenced in [579] is given as:

$$M_{ij} = \frac{\mu_0}{2\pi}\left|-j\alpha\frac{\pi}{4} + \ln\frac{D_e}{D_{ij}}\right| \tag{7.254}$$

7.5.6.2 Carson's Formulation

In this section we provide a general treatment of Carson's equation and its application to electrified railways. Carson's equations are widespread expressions for the calculation of the external part of the self and mutual impedance of conductors. In the original integral form they appeared in [46], but their mostly used and referenced formulation is the one reported about sixty years later in differential form in CCITT publications [419]. These expressions are based on some general assumptions about conductors arrangement and surrounding medium: the conductors are of infinite length, they are parallel and the return ground is homogeneous of constant resistivity, they assume a constant magnetic permeability and neglect any non-linearity effect. An important feature of this formulation is its treatment of the effect of the image current circulating in the reference ground and thus adding corrective terms for the effect of ground, this is particularly important when the conductors are close to the ground, such as the running rails and the third rail, not to mention the BGC (that being buried below the ground level needs an even more accurate approach).

The external self and mutual impedance expressions are

$$Z_{ii,ext} = j\omega \frac{\mu}{2\pi} \ln \frac{2h_i}{r_i} + 2(\Delta R_{ii} + j\Delta X_{ii}) \tag{7.255}$$

$$Z_{ij,ext} = j\omega \frac{\mu}{2\pi} \ln \frac{D_{ij}}{d_{ij}} + 2(\Delta R_{ij} + j\Delta X_{ij}) \tag{7.256}$$

where

d_{ij} is the distance between conductors i and j;

D_{ij} is the distance between conductor i and image of conductor j, and vice-versa;

h_i is the height above ground of conductor i;

r_i is the radius of conductor i;

ΔR_{ii}, ΔX_{ii}, ΔR_{ij}, ΔX_{ij} are the self and mutual, resistive and inductive correction terms respectively.

The first term on the right hand side of each equation assumes that the ground is a perfect conductor with infinite conductivity, the correction terms for the finite conductivity ground are defined as infinite series in [418, 419], to approximate the exact relationships.

The infinite series may be truncated to the desired accuracy, observing that under all sensible conductors properties and frequency values only a few terms are needed to keep the approximation error below some hundreds ppm.

The infinite series have been calculated for some sets of conditions on the parameter p, with $p = 2\alpha h$ for the self impedance expression and $p = \alpha D$ for the mutual impedance expression, where $\alpha = \sqrt{\frac{\mu\omega}{\rho}}$, and on the parameter θ, such as for a pair of conductors i and j $\cos\theta = (h_i + h_j)/D_{ij})$.

For $p \leq 4$:

$$\Delta R = 2\omega 10^{-7} \left\{ \frac{\pi}{8} - b_1 p\cos\theta + b_2 p^2 \left[(c_2 - \ln p)\cos 2\theta + \theta\sin 2\theta \right] + \right. \\ +b_3 p^3 \cos 3\theta - d_4 p^4 \cos 4\theta - b_5 p^5 \cos 5\theta + \\ b_6 p^6 \left[(c_6 - \ln p)\cos 6\theta + \theta\sin 6\theta \right] + \\ \left. +b_7 p^7 \cos 7\theta - d_8 p^8 \cos 8\theta - b_9 p^9 \cos 9\theta + ... \right\} \tag{7.257}$$

$$\Delta X = 2\omega 10^{-7} \left\{ \frac{1}{2}(0.6159315 - \ln p) - b_1 p\cos\theta - d_2 p^2 \cos 2\theta + \right. \\ +b_3 p^3 \cos 3\theta - b_4 p^4 \left[(c_4 - \ln p)\cos 4\theta + \theta\sin 4\theta \right] + \\ +b_5 p^5 \cos 5\theta - d_6 p^6 \cos 6\theta + b_7 p^7 \cos 7\theta + \\ \left. -b_8 p^8 \left[(c_8 - \ln p)\cos 8\theta + \theta\sin 8\theta \right] + b_9 p^9 \cos 9\theta + ... \right\} \tag{7.258}$$

where the b_i, c_i and d_i coefficients are defined by the following recursive expressions:

$$b_i = |b_{i-2}| \frac{s}{i(i+2)} \qquad s = \begin{cases} +1 & \text{if } i = 1,2,3,4,9,10,11,12 \\ -1 & \text{if } i = 5,6,7,8,13,14,15,16 \end{cases} \tag{7.259}$$

$$b_1 = \frac{\sqrt{2}}{6} \qquad b_2 = \frac{1}{16} \tag{7.260}$$

$$c_i = c_{i-2} + \frac{1}{i} + \frac{1}{i+2} \qquad c_2 = 1.3659315 \tag{7.261}$$

$$d_i = \frac{\pi}{4} b_i \tag{7.262}$$

For $p > 4$ and $\theta \leq \frac{\pi}{2}$:

$$\Delta R = 2\omega 10^{-7} \left\{ \frac{\sqrt{2}}{2} \left[\frac{\cos\theta}{p} - \frac{\sqrt{2}\cos 2\theta}{p^2} + \frac{\cos 3\theta}{p^3} + \frac{3\cos 5\theta}{p^5} - \frac{45\cos 7\theta}{p^7} \right] \right\} \tag{7.263}$$

$$\Delta X = 2\omega 10^{-7}\left\{\frac{\sqrt{2}}{2}\left[\frac{\cos\theta}{p} - \frac{\cos 3\theta}{p^3} + \frac{3\cos 5\theta}{p^5} + \frac{45\cos 7\theta}{p^7}\right]\right\} \quad (7.264)$$

For $p > 4$ and $\theta > \frac{\pi}{2}$:

the same expressions above for $\theta \leq \frac{\pi}{2}$ hold, but an additional corrective impedance term is introduced.

$$\Delta Z = 2\omega 10^{-7}\left\{\frac{\sqrt{\pi}}{2}\frac{\exp(2w)}{w\sqrt{w}}\left[1 - \frac{3}{16w} - \frac{15}{512w^2} - \frac{105}{8192w^3} - \ldots\right]\right\} \quad (7.265)$$

where

$$w = \frac{1}{2}p\exp(j(\theta + 3\pi/4)) \quad (7.266)$$

The number of terms in the expressions above needed to attain a given accuracy has been verified numerically for different conductors geometries (horizontal and vertical distance between conductors pairs and with respect to the ground), soil properties (ground resistivity) and frequency. Percent changes have been calculated by adding progressively new terms of higher order: for small distances (up to some tens of meters) and low/medium frequency (approximately lower than 50 kHz) the calculations are already accurate including the first ten terms; at larger distances and/or frequency values the error gets negligible for twice as many terms.

A known simplification of the Carson's equations, that holds for small distance and the most common geometries (when $x = \alpha r_i$ and $x = \alpha d_{ij}$ defined for the self and mutual impedance terms respectively are both smaller than 0.25), is that of the Carson-Clem expressions:

$$\widetilde{Z}_{ii,ext} = 10^{-6}\left[0.987f + j\omega 2\ln\frac{D}{r_i}\right] \quad (7.267)$$

$$\widetilde{Z}_{ij,ext} = 10^{-6}\left[0.990f + j\omega 2\ln\frac{D}{d_{ij}}\right] \quad (7.268)$$

where D_e is the complex earth depth, defined as an equivalent depth of a fictitious return path of the earth current, is dependent on earth resistivity and frequency.The parameter p used so far is a complex distance, that accounting for the frequency, the earth resistivity and the permeability of the medium, indicates also the depth below the ground level of the conductive plane through which the earth current ideally returns and that represents the reference for the application of an image method. This is the base for the derivation of the Dubanton expressions

$$Z_{ii,ext} = j\omega \frac{\mu}{2\pi} \ln \frac{2(h_i + p)}{r_i} \tag{7.269}$$

$$Z_{ij,ext} = j\omega \frac{\mu}{2\pi} \ln \frac{\sqrt{(h_i + h_j + 2p)^2 + d_{ij}^2}}{\sqrt{(h_i - h_j)^2 + d_{ij}^2}} \tag{7.270}$$

where p becomes $p = \dfrac{1}{\sqrt{j\alpha}} = \sqrt{\dfrac{\rho}{j\omega\mu}}$.

An extreme simplification in the form of a simple polynomial expression is the mutual coupling coefficient M_{ij}, obtained by curve fitting of Carson's equations, as proposed by CCITT [418, 419] for the calculation of induced voltage. Starting from the complex depth x, defined as

$$x = \begin{cases} \alpha r & \text{for self terms} \\ \alpha d & \text{for mutual terms} \end{cases} \tag{7.271}$$

the CCITT proposes the following simplifications (the expressions are in the original formulation, so that M_{ij} is expressed in µH/km):

For $x \leq 10$:

$$|M_{ij}| = 142.5 + 45.96x - 1.413x^2 - 198.4 \ln x \tag{7.272}$$

$$\begin{aligned} \mathrm{Re}\{M_{ij}\} &= 123.36 - 1.69x + 23.937x^2 - 4.9614x^3 + \\ &+ 0.44212x^4 - 0.01526x^5 + 0.001215 \exp(x) - 200 \ln x \end{aligned} \tag{7.273}$$

$$\begin{aligned} \mathrm{Im}\{M_{ij}\} &= -339 + 193.67x - 49.77x^2 + 69.79x^3 - 0.5243x^4 + \\ &+ 0.01672x^5 + 180.42 \exp(-x) + 0.00146 \exp(x) - 0.274 \ln x \end{aligned} \tag{7.274}$$

For $x > 10$:

$$|M_{ij}| = \frac{400}{x^2} \tag{7.275}$$

$$\mathrm{Re}\{M_{ij}\} = 0 \qquad \mathrm{Im}\{M_{ij}\} = \frac{400}{x^2} \tag{7.276}$$

The mutual impedance is then derived as $Z_{ij,ext} = j\omega M_{ij}$.

In [248] the degree of approximation of these expressions has been analysed and a slightly better approximation for the second interval is proposed to reduced the observed difference of about 6-8%, whence for $x > 6$ (and not 10) the new expression is

$$|M_{ij}| = \frac{425}{x^2} \tag{7.277}$$

that ensures a continuity between the two intervals within 0.1%.

7.5.6.3 The Complex Penetration Method - Dubanton's Method

Carson-Clem expressions given in (7.267) and (7.268) are not valid for high frequencies analysis, where h is not negligible compared with D_e. For the calculation of the self and mutual impedance at any frequency, Dubanton [248] provided expressions which are in very good agreement with Carson's integral over a wide range of frequencies. Dubanton expressions are based on the assumption that the earth return currents are concentrated in an imaginary flat plane in depth p under the earth surface; p being a complex quantity:

$$p = (1 - j)\,\frac{\delta}{2} = \frac{1}{\sqrt{j\omega\mu_0\sigma}} \tag{7.278}$$

The calculation of impedances therefore follows considering the image conductors with respect to the defined imaginary plane as given in (7.269) and (7.270).

7.5.7 *Shunt Capacitance*

With similar assumptions used to derive the Carson's expressions (the conductors are of infinite length, they are parallel and the ground is considered a perfect conductor), it is possible to apply the Charge Image Method (CIM), derived from electrostatics [609, 512]. The reference system is that of conductors i and j, above a ground plane, and the determination of the self and mutual capacitive coefficients is obtained by first, defining the self and mutual potential coefficients p_{ii} and p_{ij} (that form the matrix $\mathbf{P}$ of the same size as the number of conductors),

$$p_{ii} = \frac{1}{2\pi\varepsilon}\ln\frac{2h_i}{r_i} \qquad p_{ij} = p_{ji} = \frac{1}{2\pi\varepsilon}\ln\frac{D_{ij}}{d_{ij}} \tag{7.279}$$

$$\mathbf{P} = \begin{bmatrix} \ddots & \vdots & & \vdots & \\ \cdots & p_{ii} & & p_{ij} & \cdots \\ & & \ddots & & \\ \cdots & p_{ji} & & p_{jj} & \cdots \\ & \vdots & & \vdots & \ddots \end{bmatrix} \tag{7.280}$$

and then, calculating the inverse of the obtained matrix $\mathbf{C} = \mathbf{P}^{-1}$

The conductors assumed above are circular; it is known yet that the capacitance depends on the facing surface, so that non-circular conductors, such as the rails, may feature a larger effective capacitance, especially if the distance with respect to the other conductors is not much larger than their physical dimension.

The medium is assumed homogeneous, and conductors close to the ground feature a medium made of half air and half soil, so that doubts may be raised for accuracy, as well as when the conductor position is very close to the ground surface, comparable to its height.

So, the method needs to be integrated by more accurate estimations coming from either numerical simulations or experimental data:

- finite element methods (see sec. 7.5.5.6) may be used for any problem geometry, by assigning the correct electrical properties to materials; at low frequency or dc only the permittivities of the various media are needed, but it is known that sleepers, ballast and soil may have widely varying values, in particular at low frequency (see sec. **7.5.7.1**) and this adds some uncertainty to the computed values;
- measurement results may be given in terms of single material properties (such as the permittivity) or in terms of capacitance values between two conductors or two groups of conductors (each brought to the same potential) of a real system, such as a line section.

7.5.7.1 Measurement Results

In order to compare the theoretical capacitive terms c_{ij} with measured results, it is important to understand what is really measured by an experimental test setup.

Let's consider an admittance matrix $\mathbf{Y}^{ph}$ of the physical shunt parameters (including shunt conductance and capacitance); on the contrary the admittance matrix $\mathbf{Y}$ does not contain physical parameters. The physical self and mutual admittance terms are obtained by manipulation of the $\mathbf{Y}$ matrix parameters according to the following relationships:

$$Y_{ii}^{ph.} = Y_{ii} - \sum_{i \neq j} (i \neq j) Y_{ij} \qquad Y_{ij}^{ph.} = -Y_{ij} \tag{7.281}$$

As pointed out by Hill *et al* [307] the $\mathbf{Y}$ responds to the exigency of mathematical consistency, while $\mathbf{Y}^{ph}$ contains positive quantities and is preferable for the derivation of any equivalent circuit. Yet, even $\mathbf{Y}^{ph}$ is different from the measurable quantities, that may be identified by the $\mathbf{Y}^{meas}$ matrix, and that for brevity in a two conductors setup may be exemplified as follows.

Without loss of generality let's assume that the measurements are carried on by volt-amperometric methods, so that we can think clearly of a voltage applied between two conductors and a current flowing. The most reliable measurement is a differential measurement that doesn't involve the ground: this measurement gives the rail-rail admittance Y_{ij}^{meas}, that in our two-conductor system takes the form of $Y_{12}^{meas} (= Y_{21}^{meas})$. For a common mode measurement that involves one of the two conductors (let's say conductor 1) and the earth, the measured rail-earth admittance is Y_{ii}^{meas}, that in our two-conductor system takes the form of Y_{10}^{meas}; it is underlined that in a real case it is not ensured that the same common mode measurement on the other conductor gives exactly the same result, so in general $Y_{10}^{meas} \neq Y_{20}^{meas}$. Referring back to $\mathbf{Y}^{ph}$ and to the equivalent circuit of the two conductors setup, measured and physical quantities may be put in relationship, observing that:

- when the rail-rail measurement is taken, not only the physical rail-rail term is measured, but also the two self terms insisting on the two conductors 1 and 2 in parallel; if the transversal impedance of the earth between the two points of reference for admittance terms may be neglected, then

$$Y_{12}^{meas} = Y_{12}^{ph} + \frac{Y_{11}^{ph} Y_{22}^{ph}}{Y_{11}^{ph} + Y_{22}^{ph}} = Y_{12}^{ph} + \frac{Y_{11}^{ph}}{2} \tag{7.282}$$

where the further simplification is possible if the two physical self terms are equal (as it is in reality for a symmetrical problem).

- when the rail-earth measurement is taken (let's say on conductor 1), not only the physical rail-earth term Y_{11}^{ph} is measured, but also the contribution of the other self term Y_{22}^{ph} through the physical mutual term Y_{12}^{ph} and the transversal impedance of the earth (that is again neglected):

$$Y_{11}^{meas} = Y_{11}^{ph} + \frac{Y_{12}^{ph} Y_{22}^{ph}}{Y_{12}^{ph} + Y_{22}^{ph}} = Y_{11}^{ph} + \frac{Y_{12}^{ph} Y_{11}^{ph}}{Y_{12}^{ph} + Y_{11}^{ph}} = Y_{11}^{ph} \left(1 + \frac{2Y_{12}^{ph}}{Y_{12}^{ph} + Y_{11}^{ph}} \right) \tag{7.283}$$

When the physical self term is much larger than the mutual term, $Y_{11}^{meas} \cong Y_{11}^{ph}$; on the contrary, when the mutual term is the largest one, $Y_{12}^{meas} \cong Y_{12}^{ph}$. The first condition is more realistic if both the real part (the conductance) and the imaginary part (the susceptance) of the admittance are considered; we will see in the next section that the second condition may be applied only to the conductance for wood sleepers and in general for track in bad conditions, where the poor isolation properties increase the transversal conductance.

In general a differential mode measurement is always preferable, since the electrical contact with an ideal ground return circuit is rarely available, since the ground return circuit in a real railway system is made of different conductors and conductive parts, some accessible and some connected to the other parts of the ground circuit with unknown resistive and inductive components. It is underlined that the series impedance of the same conductors under measurement becomes relevant at increasing frequency and increasing length of the line section under measurement, so that there is always a trade-off between the extension of the explored frequency interval and the length of the line section, that possesses a self resonance frequency, close or above which the use of measurement results is very difficult.

7.5.7.2 Combination of Calculated and Measured Values

When measured data are available the adopted methods are thus either that of recalculating the theoretical terms by solution of the inverse problem using (7.282) and (7.283), in case the measured data refer simply to the shunt admittance of the said conductors, or the use of a simulation of the whole line section to link the theoretical terms to the measurable variables. The latter is a more practical approach, that may be applied to any parameter that is not observable, and it thus falls within the category of indirect methods. The test setup and the measurement conditions (such as terminal impedances, test frequency, sensitivity of instruments and test configurations) must be conceived in order to maximize the effectiveness of the method. For example, given a pair of parallel conductors, such as two rails, the most effective configuration for a volt-amperometric measurement applied to one of the two ends is that of an open circuit on the other end, and the use of a large enough voltage so that the small current at low frequency can be measured with enough accuracy; the capacitive reactance under measurement is in parallel to any rail-to-ground and rail-to-rail conductance term; increasing the frequency improves the sensitivity since the capacitive reactance gets smaller and the circulating current increases, but series parameters are no longer negligible and first, phase rotation and longitudinal voltage drops, and then resonance effects come into play. From this the need of a model driven measurement technique, that accounts for the contemporary effect of all the electrical parameters as it is in reality, and that necessitates a post-processing activity, based on curve fitting and optimization techniques.

For the application of Finite Element Methods several properties of the materials and elements down to the physical level are needed and some assumptions and approximations need to be adopted: FEM analysis is to be used to derive valuable estimations of the capacitive terms, where the charge image method is not reliable any longer, so when one or some of the assumptions no longer hold. In particular, the running rails are conductors close to the ground and with a far from circular shape. They represent also a very important problem since the rail-to-rail capacitance influences directly the transmission line properties of the track circuit and the tuning and behaviour of connected signalling equipment.

The measured capacitance term between the overhead contact wire and the rails (named in brief “loop capacitance”) reported in [512] is very close to the values obtained by the charge image method within a few % up to about 100 kHz, then the error gets slightly larger approaching the test line resonance. The value is approximately 12 pF/m, that is sensible for a single track line and is confirmed by the values reported by CCITT [420]:

- for a single track line the calculated value ranges from 10.8 and 11.8 pF/m;
- a calculation made by SNCF for a dc railway line reports 12.5 and 19.7 pF/m for single track and double track configurations;
- practical elements can increase the loop capacitance, such as line masts (about 10%), the presence of coaches between the overhead contact wires and the rails (up to 60%) and the effect of the capacitance of feeding cables at substations and transformers or on the trains roof;
- the values applied by railway operators are 11/18 pF/m and 14/23 pF/m for DB and SNCF and for single and double track lines respectively, the other railways adopting values between these extremes.

On the contrary for the rail-to-rail capacitance the values calculated by CIM show a significant spread by about 4:1, produced by the sensitivity analysis with respect to the variation of soil properties; the measured values fall in the upper part of the sensitivity interval, with a peculiar behaviour at low (10 Hz) and high (above 10 kHz) frequency, where it is possible that the measurement error gets larger due to the practical issues identified above (small and noisy input current and resonance influence). The FEM results reported in [307] are somewhat influenced by the necessary assumption on sleeper-ballast properties (as it is for the rail conductance calculations): the rail-earth and rail-rail capacitance terms calculated at an unknown frequency value (but supposed around 100 Hz) vary between 30 and 80 and between 30 and 50 pF/m respectively, with the largest values given by the lowest track conductivity values (sleeper and ballast conductivity is below 1 mS/m); the rail-rail capacitance values all show an inverse dependency on frequency and are detailed in 7.6 below.

Table 7.6 FEM calculated rail-rail capacitance term versus frequency for sleeper/ballast conductivity $\sigma_{sb} = 0.5/0.2\,\mathrm{S/km}$ and variable moisture content $m = 0.1, 0.5, 1, 5\,\%$ (the shown values are the minimum and maximum over the simulated m values for each frequency/conductivity combination) [307]

Frequency [kHz]	$\sigma_{sb} = 0.5\,\mathrm{S/km}$	$\sigma_{sb} = 0.2\,\mathrm{S/km}$
0.1	40 - 180	20 - 80
1	1.3 - 3.5	0.5 - 2.0
10	0.45 - 0.90	0.4 - 0.65
50	0.35 - 0.50	0.3 - 0.35

The experimental values reported in [307] have the same behaviour versus frequency and correspond to the largest values for $\sigma_{sb} = 0.5\,\mathrm{S/km}$ above. The spread of these values brackets both the sensitivity intervals and the measured results reported in [512].

7.5.8 Shunt Conductance

The shunt conductance is a widely variable term, depending on the considered conductor: it may be assumed null for aerial conductors and may reach relatively large values for conductors in contact with or buried in the soil (such as the running rails and the BGC respectively). Considering the running rails, the supporting sleepers and the rest of the mechanical elements for electrical insulation and mechanical fastening, represents various conductive paths, so that the rail conductance is the synthesis of a complex conduction mechanism that takes place between and along the different surfaces and also through the various elements, thus distinguishing surface and volume conductivity.

For non-earthed running rails the desired behaviour is that of an overall rail-to-earth conductance below a certain limit value, posed mainly by signalling requirements. The various conduction paths combine with the various elements and with the environmental conditions and agents (temperature, moisture, dirt, dust and metallic particles, vegetation) and make the rail-to-earth conductance a highly variable parameter, also with frequency. The variability with frequency is due to some of the conduction interfaces that involve ionic species and also to the concomitance of capacitance terms in parallel, that cannot be separated during measurements. It is thus evident that there is no theoretical formulation nor calculation tool that allows the a-priori determination or estimation of the conductance terms for the running rails. If the BGC is considered, by recalling its use as a horizontal earth electrode, the formulas reported in [359] for the determination of the resistance to earth may be used: the resistance to earth of a buried counterpoise of circular section is given by

$$R_e = \frac{\rho}{\pi L_h} \ln\left(\frac{2L_h}{d}\right) \tag{7.284}$$

where

d is the conductor diameter in m;

L_h is the horizontal length of the conductor in m;

ρ is the resistivity of earth assumed homogeneous in Ωm.

This oversimplified formula doesn't take into account the conductor depth that is normally set to 0.5-1 m. Yet, the formula correctly indicates an asymptotic behaviour for increasing L_h, since the longitudinal resistance of the conductor prohibits that the resistance to earth R_e gets arbitrarily small (simulated values for $\rho = 50\,\Omega$m gives a limit of about $0.5\,\Omega$). For an infinite BGC the conductance to earth is the reciprocal of the resistance to earth limit value.

For the running rails the conductance terms are to be determined by measurements. The application of the differential signal to the two running rails captures not only the rail-to-rail conductance term, but also the common mode rail-to-earth conductance term, that in reality encompasses the direct conduction term to the sleepers beneath the rails themselves, the transversal conduction of the sleeper (both superficial and volumic) and the further conduction terms to the ballast and earth. Since these terms may be hardly measured or evaluated separately in a real environment, a tradeoff between accuracy and simplicity is represented by the representation shown in Figure 7.42 the simplified equivalent circuit that implements the admittance formulae (7.282) and (7.283). The measured differential conductance is indicated by $G_{rr,meas}$ and is related to the physical self- and mutual conductance terms $G_{s,phys}$ and $G_{m,phys}$ by a simple series/parallel circuit.

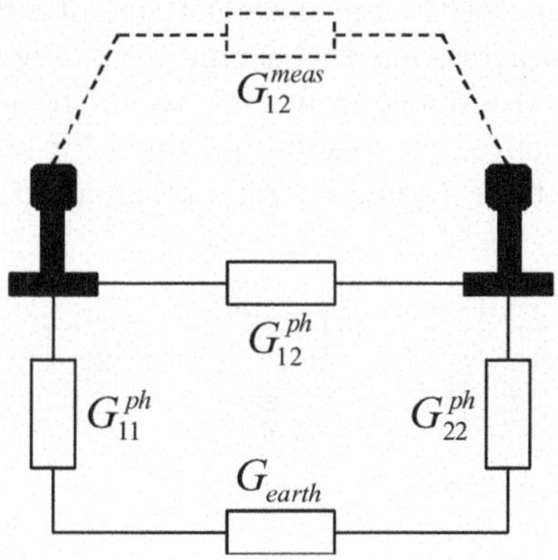

Fig. 7.42 Simplified representation of measured and physical conductance terms

Under the symmetry assumption already discussed above in sec. 7.5.7. The relationship between measured and physical terms is simply

$$G_{rr,meas} = \frac{1}{2} G_{s,phys} + G_{m,phys} \tag{7.285}$$

Normally $G_{m,phys}$ is several times smaller than $G_{s,phys}$ because of the unfavourable ratio of distances, represented by the the horizontal gauge and a fraction of the sleeper thickness, even if the ratio of the G terms heavily depends on the sleeper surface conditions and on the weather conditions. If $G_{s,phys} = 10 \div 20\, G_{m,phys}$ we may assume $G_{s,phys} = 1.6 \div 2\, G_{rr,meas}$.

In general, based on the results shown in [307], it is possible to conclude that:

1) for small sleeper and ballast conductivity (below about 0.05 S/km) the self-conductance term $G_{s,phys}$ is larger (even by an order of magnitude) than the mutual conductance term $G_{m,phys}$;

2) for medium sleeper and ballast conductivity (around 0.1-1 S/km) the self- and mutual conductance terms are comparable, with oscillation between $G_{s,phys} < G_{m,phys}$ and $G_{s,phys} > G_{m,phys}$ for small and high earth conductivity respectively;

3) for large sleeper and ballast conductivity (above 1 S/km) the self- conductance term is smaller than the mutual conductance term (even by an order of magnitude) for earth conductivity between 0.1 and 10 mS/m.

It is underlined that the earth conductivity considered here is a global parameter, that takes somehow into account also the contribution of the ballast and of impermeabilization and isolation of the concrete bed below the ballast, so that the applicable earth conductivity values are very likely low conductivity values, around 0.1 to 1 mS/m.

The said measurements are the most profitable if carried on real lines, isolating a line section of adequate length: if the line is too short, the equivalent resistance of the conductance is so large, so that several secondary effects may reduce the accuracy of the measurement results; if the line is too long, the series impedance starts soon to become a significant longitudinal voltage drop and also line resonance limits the useful frequency range. In [513] there are reported measurement results obtained for test sections of slightly different length, so that the frequency intervals extend up to 10 to 30 kHz.

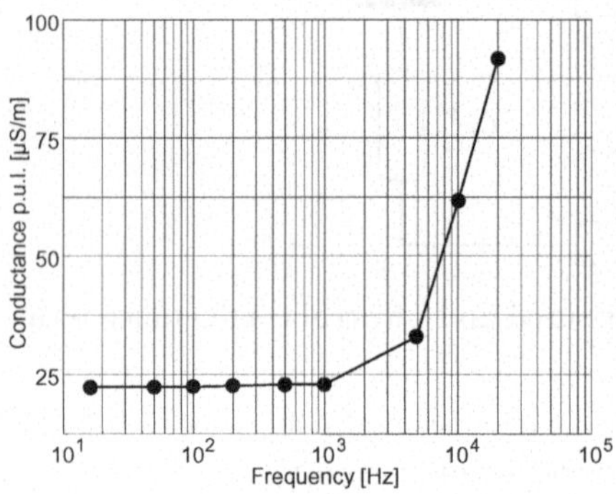

(a) Spezia (dc system at station)

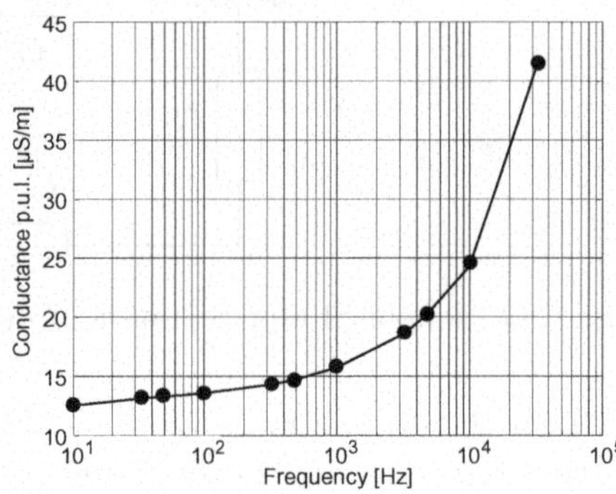

(b) Rigutino (dc system open line)

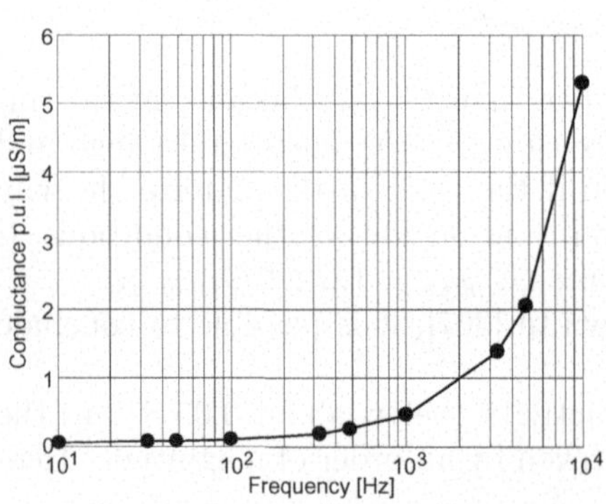

(c) Anagni (new ac system open line)

Fig. 7.43 Rail-to-rail measured conductance vs. frequency

In [513] the FEM results shown in [307] are compared with the measured data shown above in Figure 7.43. FEM results are in agreement with the measured data for larger conductance systems, while the measured values taken on the new Italian High Speed Line at Anagni represent a lower bound typical of very well insulated systems.

There exist however intervals of values of general reference reported in [418] and shown here in Table 7.7.

Table 7.7 Rail to earth conductance values in S/m as per [418]

	Minimum	Average	Maximum
Non-isolated rail	$5\,10^{-5} \div 1\,10^{-4}$	$1\,10^{-3} \div 2\,10^{-3}$	$4\,10^{-3} \div 5\,10^{-3}$
Isolated rail	$2\,10^{-5} \div 5\,10^{-5}$	$5\,10^{-4} \div 1\,10^{-3}$	$1\,10^{-3} \div 2\,10^{-3}$

7.6 Traction Line Modeling

The frequency behavior of a traction line can be readily be studied by putting to work all the expressions and parameters concerning transmission line theory, multiconductor transmission lines, self and mutual impedance of conductors and calculation and estimation of parameters, including practical values and the electrical properties of materials. A quick theoretical introduction can be found in [314, 507], where the basic transmission line theory is applied to a reference case of a train at an intermediate position between two substations, that define the terminal conditions for the traction line.

7.6.1 Pantograph Impedance for a Simple Straight Line

The simple straight line configuration considered in [314, 507] is governed by simple equations, based on transmission line theory, for the prediction of the line impedance seen at locomotive pantograph, the so called pantograph impedance Z. At the locomotive pantograph the current injected by the on-board power converters and the line impedance Z interact, causing pantograph voltage distortion; the locomotive pantograph is supplied by two opposite substations through two line sections with length x and $L - x$ respectively, as shown in Figure 7.44.

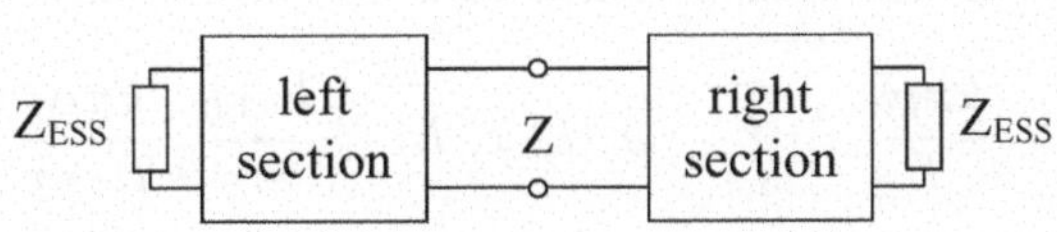

Fig. 7.44 Reference configuration for the simplified case of one straight line and two substations

Considering a simplified traction line made of two conductors and taking zero the two substation impedances, so that the line is short circuited at its ends, the pantograph impedance is given by

$$Z = Z_c \frac{\tanh(\gamma x)\, \tanh(\gamma(L - x))}{\tanh(\gamma x) + \tanh(\gamma(L - x))} \qquad (7.286)$$

where the symbols Z_c and γ stand for characteristic impedance and propagation constant respectively.

Given that the propagation constant may be expressed as $\gamma = \alpha + j\beta$, with α and β indicating the attenuation and phase constant, in [507] there are proposed several simplified expressions, for different frequency ranges, and a sensitivity analysis of Z is carried on with respect to line geometrical and electrical parameters and terminal conditions.

The pantograph impedance is subject to a resonance (maximum) and anti-resonance (minimum) phenomenon depending on frequency and on train position, as shown in Figure 7.45.

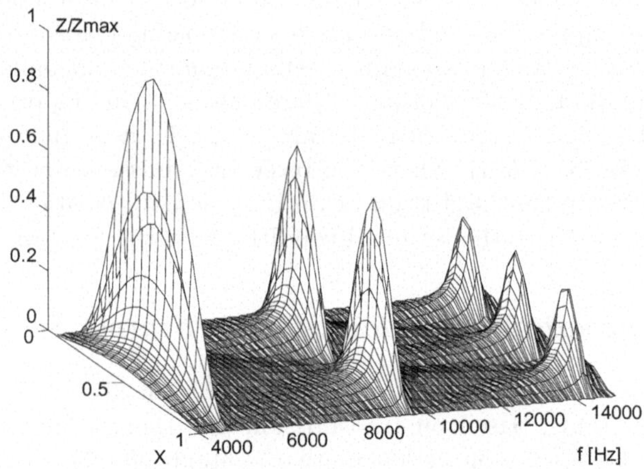

Fig. 7.45 Normalized pantograph impedance versus locomotive position (normalized to line length L) and frequency

Resonance and anti-resonance frequencies are defined as the poles and zeros of the impedance expression (7.286):

$$\omega_r(k, L) = \frac{k\pi v_0}{L} \qquad \omega_a(k, x, L) = \frac{k\pi v_0}{x}, \frac{k\pi v_0}{L - x} \qquad (7.287)$$

The corresponding impedance values at these frequencies are

$$Z_r = Z(\omega_r) = \frac{Z_c}{2}\frac{1-\cos(2\pi x/L)}{\alpha L}$$
$$Z_r = Z(\omega_r) = \begin{cases} Z_c\alpha x \\ Z_c\alpha(L-x) \end{cases} \tag{7.288}$$

A traction line differs from an ideal transmission line in that the characteristic impedance is not constant and is not a real quantity. In [507] Z_c is evaluated at four different frequencies for the line p.u.l. parameters r, l and c distributed uniformly around their average values; the reasons for such a variability are many: the resistance depends for example on the running rails losses (in sec. 7.5.5 it is shown that they depend on the frequency and current amplitude), analogously for the inductance of the running rails there is a similar dependence on frequency and current amplitude, while the inductance of the aerial conductors depends on their geometrical position; the capacitance also depends on the aerial conductors position, on the type of track (tunnel, viaduct, etc.), on the presence of pieces of high voltage cable.

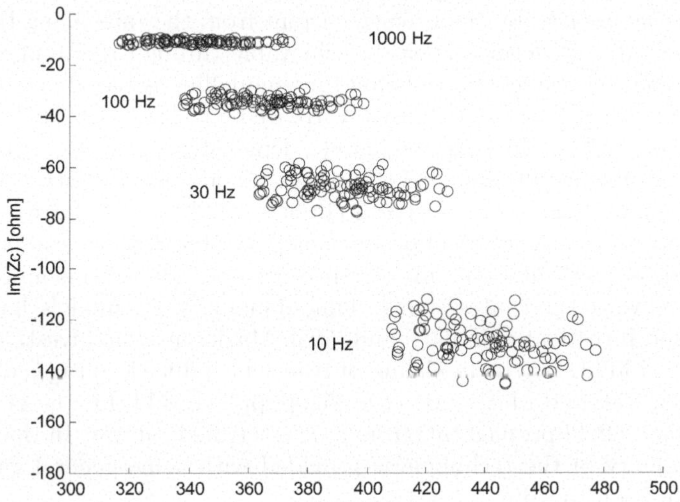

Fig. 7.46 Scatter plots of Re$\{Z_c\}$ and Im$\{Z_c\}$ at 10, 30, 100 and 1000 Hz (r, l and c uniformly distributed)

The line characteristic impedance has a relevant imaginary part at low frequency, progressively reducing for increasing frequency, as long as the resistive part (even if increasing slightly for the skin effect and some additional losses in the rails) becomes a smaller fraction of the inductive reactance; also the real part of the characteristic impedance reduces slightly, following the reduction of the rail internal inductance and the capacitance to ground (the soil has a much larger dielectric constant at low frequency).

For more complex traction line arrangements and when a more complete analysis is needed, that goes beyond the prediction of the main supply line resonances and propagation modes and that evaluates various coupling terms and transfer functions, a multiconductor transmission line approach is often used [508, 418, 419, 599] or alternatively a lumped parameters mesh network with adequate discretization is sometimes considered.

7.6.2 Multiconductor Transmission Line Modeling

The railway is often modelled as a multiple conductor transmission line due to that fact that it consist of more than two parallel conductors that run for long lengths. This is particularly important for the evaluation of the inductive and conductive coupling from the railway line to neigbouring (buried) metallic structures. A key parameter derived from the evaluation is the current distribution between the different conductors of the return path, forming the multiple conductor transmission line, i.e. the rails, earth and possible return conductors. For conductive coupling, the necessity to determine the earth potential created by the leakage current from the rails along the route of buried metallic structures is easily achievable through the application of MTL modeling. Multiple Transmission Line modelling is based on the transmission line theory that is extensively treated in [418, 419, 599] and it's application to railway lines is adequately demonstrated in literature. Multiconductor transmission-line (MTL) models are applied to both frequency- and time-domain analysis of electrical circuits. MTL models have been successfully applied to railway traction systems [233, 508, 284], to evaluate EMC aspects and, in particular, the interference between subsystems and components and toward external systems. Time-domain MTL models have been proposed and investigated in [504] and [595]. Mazloom *et al*, investigated the application of MTL for the modeling of transient induced voltages along the railway [533, 534] and Mariscotti *et al* [520] applied MTL for the evaluation of the magnetic field propagating from a dc electrified railway. In this section a brief summary of the technique is provided with some general guidelines on its application to railway lines.

The MTL formulation requires the following fundamental assumptions to be applied:

- the line is uniform with no significant variations along the axial direction of the physical and geometrical characteristics of the conductors and of the medium surrounding them;
- the electromagnetic fields have TEM structure lying in the cross section and propagating along the line axis;

- the algebraic sum of the line currents at any cross section is zero; the net return current of the other n conductors flows in the reference conductor ($n+1$-th conductor);
- the medium surrounding the conductors must be linear, homogeneous, and isotropic.

For a generic MTL consisting of $n+1$ conductor, the governing transmission line equations are a coupled set of $2n$ first order partial differential equations

$$\frac{\partial}{\partial z}\mathbf{V}(z,t) = -\mathbf{R}\mathbf{I}(z,t) - \mathbf{L}\frac{\partial}{\partial t}\mathbf{I}(z,t)$$
$$\frac{\partial}{\partial z}\mathbf{I}(z,t) = -\mathbf{G}\mathbf{V}(z,t) - \mathbf{C}\frac{\partial}{\partial t}\mathbf{V}(z,t) \tag{7.289}$$

where $\mathbf{V}$ is a $n \times 1$ column vector, containing the n line voltage with respect to the reference conductor; $\mathbf{I}$ is a $n \times 1$ column vector containing the n line currents; $\mathbf{R}$,$\mathbf{L}$,$\mathbf{G}$,and $\mathbf{C}$ denote the per unit length parameters matrices of size $n \times n$. The solution of the general phasor MTL equations for a line of length L can be written in the chain parameter form as:

$$\begin{bmatrix}\mathbf{V}(L)\\ \mathbf{I}(L)\end{bmatrix} = \begin{bmatrix}\mathbf{\Phi}_{11}(L) & \mathbf{\Phi}_{12}(L)\\ \mathbf{\Phi}_{21}(L) & \mathbf{\Phi}_{22}(L)\end{bmatrix}\begin{bmatrix}\mathbf{V}(0)\\ \mathbf{I}(0)\end{bmatrix} \tag{7.290}$$

where $\mathbf{V}(0), \mathbf{I}(0)$ are the current at the line near end; $\mathbf{V}(L), \mathbf{I}(L)$ are the current at the line far end; $\mathbf{\Phi}_{ij}$ are the chain parameters sub-matrices and are related to the line electrical parameters by

$$\mathbf{\Phi}_{11} = \cosh\left(\sqrt{\mathbf{ZY}}L\right)$$
$$\Phi_{12} = -\mathbf{Z}_c \sinh\left(\sqrt{\mathbf{YZ}}L\right)$$
$$\mathbf{\Phi}_{21} = -\mathbf{Z}_c^{-1} \sinh\left(\sqrt{\mathbf{ZY}}L\right)$$
$$\mathbf{\Phi}_{22} = \cosh\left(\sqrt{\mathbf{YZ}}L\right) \tag{7.291}$$

The constant Z_C is the characteristics impedance matrix defined as

$$\mathbf{Z}_C = \mathbf{Y}^{-1}\sqrt{\mathbf{YZ}} \tag{7.292}$$

where

$$\mathbf{Z} = \mathbf{R} + j\omega\mathbf{L}$$
$$\mathbf{Y} = \mathbf{G} + j\omega\mathbf{C} \tag{7.293}$$

The phasor MTL equations are solved adding $2n$ additional constraints, provided by the terminal conditions. The $2n$ terminal conditions derived from the terminal networks are represented using the generalised mixed representation, applying the Kirchhoff voltage and current laws

$$\mathbf{Y}_0 V(0) + \mathbf{Z}_0 I(0) = \mathbf{P}_0$$
$$\mathbf{Y}_L \mathbf{V}(L) + \mathbf{Z}_L \mathbf{I}(L) = \mathbf{P}_L \tag{7.294}$$

where $\mathbf{Y}_0, \mathbf{Y}_L, \mathbf{Z}_0$ and $\mathbf{Z}_L$ are the result of the branch elements and many controlled sources in the terminal networks; and $\mathbf{P}_0$ and $\mathbf{P}_L$ are the voltage and current excitation sources. The R and L matrices are computed using Carson's equation taking due cognisance of the frequency dependent parameters such as rail resistance. The C matrix can determined, as a first approximation, with the method of images assuming soil and ballast as perfect grounds; a more precise formulation would take into consideration the electrical properties of soil and ballast.

Chapter 8
Earthing, Bonding and Cable Layout

In this chapter the concepts related to earthing and bonding as well as to cable to cable coupling and segregation are briefly reviewed, with extensive reference to Chap. 7 for basic theory and formulas. It is however underlined that earthing, bonding and cable layout in a railway system are to be specified for different subsystems and different types and categories of equipment, while taking into account possible variable system conditions and coping with the needs and requirements of electrical safety, equipment protection, signal reference for correct equipment operation, and interference mitigation. Reference is made to applicable standards with respect to the involved equipment and subsystems, aspect of the railway installation or specific phenomenon. For such standard reference, we may categories as follows:

- standards that specify earthing methods and provisions for railway applications with specific attention to the traction circuit and the infrastructure;
- standards addressing cabling methods and practices, in particular for information technology, automation and telecommunication applications, targeted to residential and industrial environments;
- standards for telecommunication applications, that take into account a variety of installations indoor and outdoor, with additional requirements and guidelines;
- standards specifically addressing lightning phenomena and related provisions;
- standards specifically addressing EMC and interference phenomena and related provisions.

Lightning is a complex subject that is not addressed in this book, but lightning provisions and design have a significant impact on grounding and cable layout, and are thus briefly reviewed in this section. Since lightning effects have an impact on people, equipment and installations safety, they usually take precedence over EMC aspects, also because they are considered at an earlier stage of the project.

A. Ogunsola and A. Mariscotti: Electromagn. Compat. in Railways, LNEE 168, pp. 429–482.
springerlink.com

Some definitions are provided and will be used for the following discussion.

- Bonding network (BN). Set of interconnected conductive structures that provides an "electromagnetic shield" for electronic systems and personnel at frequencies from direct current (DC) to low radio frequency (RF). The term "electromagnetic shield" denotes any structure used to divert, block or impede the passage of electromagnetic energy.
- Common bonding network (CBN). It is the set of metallic components that are intentionally or incidentally interconnected to form the principal BN in a building. These components include: structural steel or reinforcing rods, metallic plumbing, alternating current power conduit, protective conductors (PE), cables racks and bonding conductors. The CBN always has a mesh topology and is connected to the earthing network.
- Earth electrode. Conductive part or group of conductive parts in intimate contact with and providing an electrical connection with earth.
- Earthing network. Part of an earthing installation which is restricted to the earth electrodes and their interconnections.
- Equipotential bonding. Electrical connection putting various exposed conductive parts and extraneous conductive parts at a substantially equal potential.
- Parallel earthing conductor (PEC). Earthing conductor that is parallel to the cable.
- PEN conductor. Earthed conductor combining the functions of both protective conductor and neutral conductor.

There are several reasons for earthing and bonding in railway installations, these include:

- System earthing path for traction current;
- Low impedance return path for fault currents;
- Equipment protection against adverse voltages; and
- Attaining electromagnetic compatibility between different sub-systems.

The reasons stated above are no different for the implementation of an earthing system in any installation, however the dual nature of rails forming part of the signalling circuits as well as acting as protective earth, makes the implementation of an earthing system in the railway special. The requirement for a low impedance return path requires the rails to be insulated from earth, however this requirement aligns with that of the signalling systems where leakage current from the rails may results in mal-operation of the said signalling systems as well as the equipment, of other systems, to prevent the flow of return current flowing through them. Rails are also used as protective earth - they represent earth electrodes to which wayside equipment is bonded. To fulfill the requirement of touch voltages, low impedance path from the rails to ground is required and this is the paradox of earthing and bonding in railways.

8.1 Basic Concept of Earthing and Bonding

In simple terms, earthing implies (1) a physical connection to earth, and (2) an equipotential bond between or to metallic structures. The objective of earthing is the following:

- to provide a path to earth for the discharge of lightning strikes in a manner that protects a structure, its occupant and equipment within the structure
- to restrict touch voltages, in areas accessible to people, to below hazardous levels even under lightning discharge or power faults conditions
- to decrease noise in signal and control circuits by minimising voltage differential between different signal reference sub-system

8.1.1 Protective Requirement

The primary objective of an earthing system is to assure people's safety and to protect installations against damage. The earthing system is usually designed to protect against lightning and power system faults, as these phenomena can result in large circulating currents which may create hazardous voltages in the railway. The earthing system design goal is to be the path to earth for fault currents, while maintaining voltage differences between any two point of an installation as low as possible. The earthing system is generally characterised by its earthing resistance, a parameter that is deemed adequate at railway power frequencies.The earthing resistance is however considered inadequate when dealing with high frequency disturbance where earthing impedance is more appropriate.

8.1.2 Electromagnetic Compatibility

The secondary function of the earthing system is to serve as a common voltage reference and to contribute to the mitigation of all kinds of electromagnetic disturbance regardless of their coupling mode. The earthing system contributes to the mitigation of disturbance by being the path for return currents, between a source of disturbance and a victim equipment. The electromagnetic disturbance is generally treated in Chap. 7 and covered in IEC 61000-2-5 and EN 50121 (IEC 62236) series of standards. Irrespective of the coupling modes, it is possible to determine a transfer between the source of electromagnetic disturbance and the victim equipment. The transfer function can be an impedance, an admittance or a dimensionless parameter, depending on the nature of the measurand and the outcome of the coupling - i.e. in terms of current or voltage. In all cases, the electromagnetic interaction between a source and victim can be modelled by a two port network, represented in Figure 8.1 in its simplest form, by a T network.

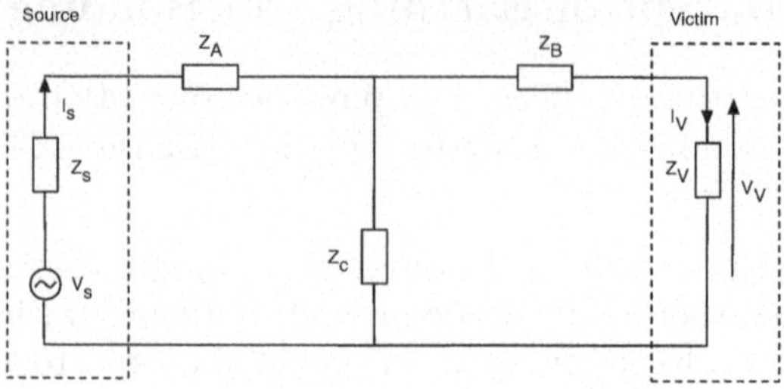

Fig. 8.1 A two port network representative of the interference model

This circuit model, in which the return conductor is usually the ground suggest two main strategies for mitigating electromagnetic disturbance: Short circuit and Open circuit strategies. If Z_C is zero, the no energy from the source is able to reach the victim; similarly if either Z_A and Z_B.is a infinite in magnitude (i.e. open-circuit) no energy from the source is able to reach the victim. The sort circuit strategy is aimed at reducing the impedance of all grounding conductors and in particular their inductance which usually dominates above 1 kHz. The open-circuit strategy is aimed at isolating the victim circuit from the source either by increasing the distance or by inserting physical barriers (reducing the coupling factor), or by making an isolated bonding network with "single point" connections to earth[1].

8.1.3 Earthing and Equipotential Bonding

Quite often this terms are mis-representated and thus it is beneficial to distinguish between the terms. Earthing implies that a structure is connected to earth electrodes such as rods, wires or other metallic objects buried in the soil whereas equipotential bonding is the means of connection of different exposed conductive parts or extraneous conductive parts to put them at a substantial equal potential in order not to get high voltages between different parts of a installation.

8.1.4 Bonding Configuration

Three main bonding configurations exist:

- The Isolating Bonding Network;
- The Isolating Bonding Network with Start or Mesh configuration; and
- Multiple point with Mesh configuration

[1] We state the limitation of a single point grounding topology later in this chapter.

8.1.4.1 Isolating Bonding Network

The principle of the Isolating Bonding Network (IBN) is to connect each piece of equipment to the earth network at one point only by single and direct links which may be shared only by equipment belonging to the same circuit or system. The idea is to prevent, as far as possible, common impedance coupling due to current flowing in the earth network. This is the simplest method of low-resistance grounding. Since there is only one ground source, it provides lower ground fault current than with multiple point grounding. The implementation of this network is longer encouraged, at least for new installations.

8.1.4.2 Mesh Bonding Network

A Mesh Bonding Network (Mesh-BN) is composed of a great number of highly interconnected conductors. No effort is made to avoid ground loops, but efforts is made to ensure that the ground loops area is kept small in order to reduce the impedance. With this bonding network, the mesh is made tighter or finer depending on the sensitivity of the victim circuit. One of the advantages of this bonding network, in addition to its effectiveness, is the ease it which it can be built up, it does not require any controls or checks, and is self developing.

8.1.4.3 Hybrid Network

A hybrid network is a bonding network that combines the topology of a MESH-BN and IBN. It is usually implemented with the MESH-BN being at the local level - for instance each floor or each room of a building, with the IBN as the only connection to earth. This earthing topology has the advantage in that it enables the separation of sensitive circuits from major noise generators; however this advantage comes at a price, in that it is complex and difficult to maintain, and it usually includes the insertion of an isolation transformer between the safety bonding networking and the EMC bonding network.

8.2 Earthing Methods and Provisions for Railway Applications

8.2.1 Traction Circuit

As mentioned previously, the primary function of an earthing system is to provide a path for the return current; how this is achieved in railway installation, particular when dealing with system earth depends largely on the form of electrification (i.e. ac or dc electrification). The main standard reference is EN 50122-1 [85] that presents earthing and bonding as a means to ensure electrical safety in railway applications both in ac and dc. For ac electrified railway the preferred earthing method is the connection of all metallic parts to the "traction earth". The standard says "if it is determined that the running rails and the metallic parts cannot be directly connected to the traction earth, then a voltage limiting device shall be used"; the referred case is that of the presence of track circuits, for which the commonly implemented solution is the installation of impedance bonds for the earthing of the running rails, while the other metallic parts are however earthed to the traction earth. The use of a voltage limiting device for running rails is a common solution to ensure on one side that the rail potential doesn't reach dangerous values and on the other side that the running rails are normally kept isolated from, or at least not solidly bonded to earth. This is the case of dc electrified railway, for which direct earthing of running rails is not desirable in order to minimize stray currents [85, 86]. The limits for the rail potential are specified in EN 50122-1 for ac and dc systems, with decreasing values for increasing time duration, down to the limit for permanent voltage at 60 V and 120 V respectively.

The earthing of the traction circuit is harmonized with the earthing of the necessary elements of the traction line, such as the masts supporting the overhead conductors. The bonding of the other wayside elements may be accomplished by means of a buried through-earth collector (either dedicated or part of the traction circuit), a suspended overhead earth wire or in some cases by the running rails themselves. While the latter is a solution implemented only in low traffic systems (typically old railway networks) or in unelectrified systems, the overhead earth wire may be used as a cheaper bonding means that allows easier inspection and maintenance; its position relative to the live overhead traction conductors makes it effect in reducing the return current circulating in the running rails, because of the favourable mutual inductance (current exchange between the rails and this wire is possible of course only at the first bonding element, either a straight connection through the same mast or an impedance bond). The through-earth collector is added to increase the total cross section without increasing the weight of the wire suspended to masts, to ensure an additional distributed earthing along the traction line, and to improve the geometry of the external earthed conductors for the screening of the magnetic field emissions and consequently the reduction of the induced voltage in particular onto external installations and metallic structures. Moreover, the use of a through-earth collector, insulated from the surrounding earth (being located for example in a sand filled trough and protected by a non-conductive sheath), is highly advisable to keep the traction current away from tunnel structures: at both ends of the tunnel the running rails are interrupted by insulating joints and the return current flows in the dedicated collector; in this case tunnel installations are kept separate and earthed independently.

To summarize, equipotential bonding between the running rails (through which the return current from the rolling stock flows) and the earthed conductors (overhead and through-earth) is accomplished in various ways:

- straight connection at each mast if track circuits are no connected to the the running rails , but the signalling is implemented by axle counters, balises or radio (the mast acts as the equipotential bond between the through-earth collector and the overhead wire or in the absence of the former, as the connecting element to reach the running rails);
- use of impedance bonds for continuously welded running rails with the center terminal bonded at the through-earth collector or a mast, the impedance bonds located at a convenient separation that may range from 0.5 to 2 km approximately;
- use of impedance bonds with insulating mechanical joints, that break the electrical continuity of running rails, with the center terminal again bonded at the through-earth collector or a mast with a similar spatial separation.

8.2.1.1 Earthing Methods

Earthing is obtained by a combination of different methods, provided that the same earthing arrangement must be effective not only for the potentials developed by the traction return current, but also in case of various type of faults and lightning strokes.

Vertical Electrodes: Supporting Masts, Earthing Rods. The easiest way is the use of the masts as earthing electrodes: recalling the formula for the calculation of the resistance to earth of vertical rods, possibly encased in concrete, given by the IEEE Std. 80 [359], the earthing resistance ranges from a few ohm to a tens of ohm for various diameters and buried depths; the equipotential bonding of the masts located every 50 to 60 m, one row for each track, ensures a total earthing conductance of some S/km.

If a single vertical rod is considered, that is the case of a pole, the following expressions are applicable, with and without the presence of a concrete encasing; concrete is hygroscopic and attracts moisture, so it helps in lowering the resistance to remote earth of buried metallic rods. For a bare metallic rod, the resistance R_{v1} is

$$R_{v1} = \frac{\rho}{2\pi L_r}[\ln(8L_r/d) - 1] \tag{8.1}$$

where L_r is the length of the grounding rod in m; d is the diameter of the grounding rod in m; ρ is the resistivity of soil, assumed uniform, in Ωm.

If the concrete casing is added, the following expression is obtained:

$$R_{v2} = \frac{1}{2\pi L_r}\{\rho\,[\ln(8L_r/D_c) - 1] + \rho_c\,[\ln(8L_r/d) - 1] - \rho_c\,[\ln(8L_r/D_c) - 1]\} \tag{8.2}$$

where D_c is the diameter of the concrete encasing in m; ρ_c is the resistivity of concrete, assumed uniform, in Ωm.

A series of curves of the resistance to earth provided by the first formula (8.1) are shown below for variable L_r and d and soil resistivity $\rho = 100\,\Omega$m in Figure 8.2.

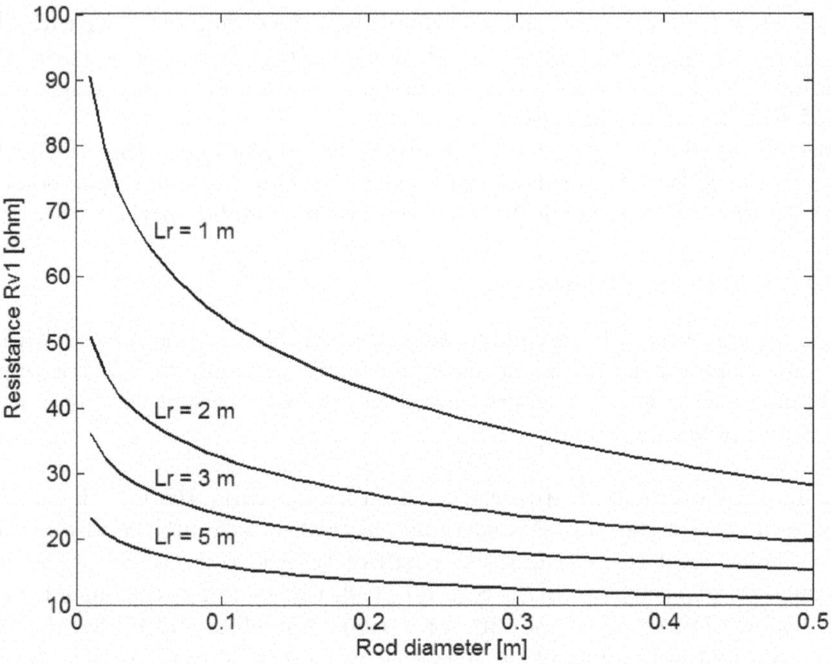

Fig. 8.2 Resistance to earth of vertical conductor as per IEEE formula (8.1)

Additional earthing rods may be included in the design, to drain a significant percentage of the current away from the masts, thus preserving them from severe corrosion. For their sizing the same IEEE formula may be used to calculate the earth resistance and the minimum cross section for carrying fault current shall be met, as dictated by the adopted standard (IEEE Std. 80 itself or EN 62305 for lightning phenomena).

Finally, it is observed that the formulas above have one soil resistivity value, constant throughout the different soil layers that the vertical rod may encounter along its length; if geological prospections and soil resistivity measurements have identified a layer with favourable resistivity, the rod length may be adapted always within practical limits to reach that layer, and that portion of the rod towards its buried tip will carry an even larger percentage of the total earth current entering the rod at the top terminal.

Horizontal Electrodes: Through-earth Collector. The through-earth collector is also a horizontal earthing electrode, that is normally sized for carrying the short circuit current and for mechanical robustness, also considering possible corrosion due to the contact with soil. Yet, there is no general agreement on a closed form expression for the calculation of its resistance to earth, taking into account its physical size, its length, the burial depth and the resistivity of soil: the IEEE Std. 80 suggests the use of an equivalent length and the solved circuit is a ladder network of longitudinal and shunt resistive elements, standing for the ohmic and leaking resistance respectively; the total resistance to earth for a long collector may be also determined by the set of curves appearing in CEI 11-1 [49].

The resistance to remote earth for a long buried horizontal conductor of circular section is given by the simplified formulation

$$R_{h1} = \frac{\rho}{\pi L_h} \ln(2L_h/d) \tag{8.3}$$

where L_h is the length of the horizontal element in m; d is the diameter of the element in m.

The results of the calculation of R_{h1} for variable L_h and d are shown in Figure 8.3.

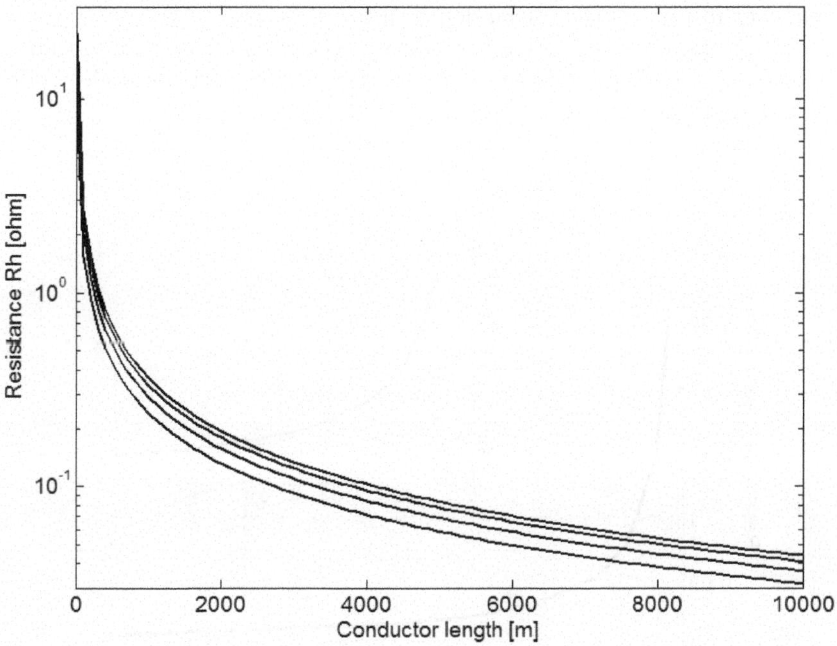

Fig. 8.3 Resistance to earth of horizontal conductor as per IEEE simplified formula (8.3)

This formula does not take account of the effect of h, the burial depth, normally set from 0.5 to 1.0 m for trackside through-earth collectors. Moreover, for long conductors the longitudinal reactance becomes an additional limiting factor for the value of the earthing impedance; in this case a flat strap may be chosen instead of a circular conductor, so to reduce the inductive reactance by approximately 10 to 15%.

The formula indicates an asymptotic behavior for increasing L_h, since the longitudinal resistance prohibits that the resistance to earth R_h gets arbitrarily small. Practical values for this limit at low frequency (thus neglecting the longitudinal reactance) are around 0.5 Ω for $\rho = 50$ Ωm, as shown below in Figure 8.5.

CCITT proposes a different formula, that accounts for length and depth as well, together with the conductor diameter; depth and diameter have a weak influence on the final value of the resistance to earth, with the log of the square root, as it is commonly observed in practice, provided that both are above a minimum advisable value.

$$R_{h2} = \frac{\rho}{\pi L_h} \ln(L_h/1.36\sqrt{hd}) \tag{8.4}$$

where L_h is the length of the horizontal element in m; d is the diameter of the element in m; h is the burial depth in m.

This formula optimistically gives values of the resistance to earth that can be arbitrarily small, provided that the length of the horizontal electrode is increased, as shown in Figure 8.4.

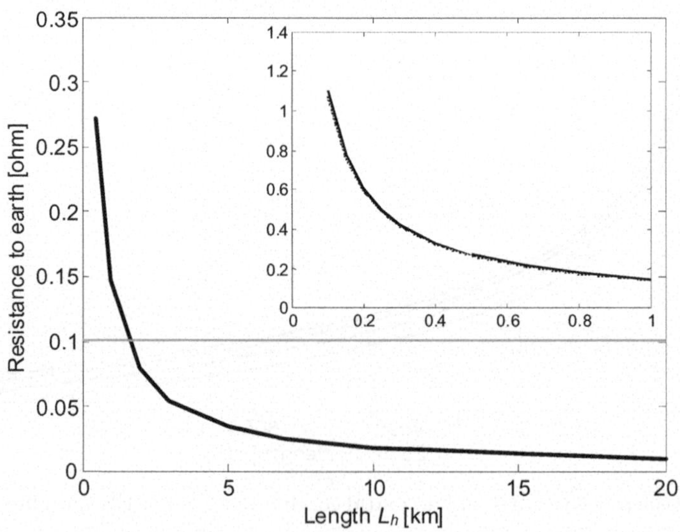

Fig. 8.4 Resistance to earth for a horizontal conductor as per CCITT formulation (8.4)

As already mentioned above, a practical reference may be found in CEI 11-1 [49] (equivalent to other national standards, all derived from the same European harmonized document, HD 637 S1), that is shown in Figure 8.5.

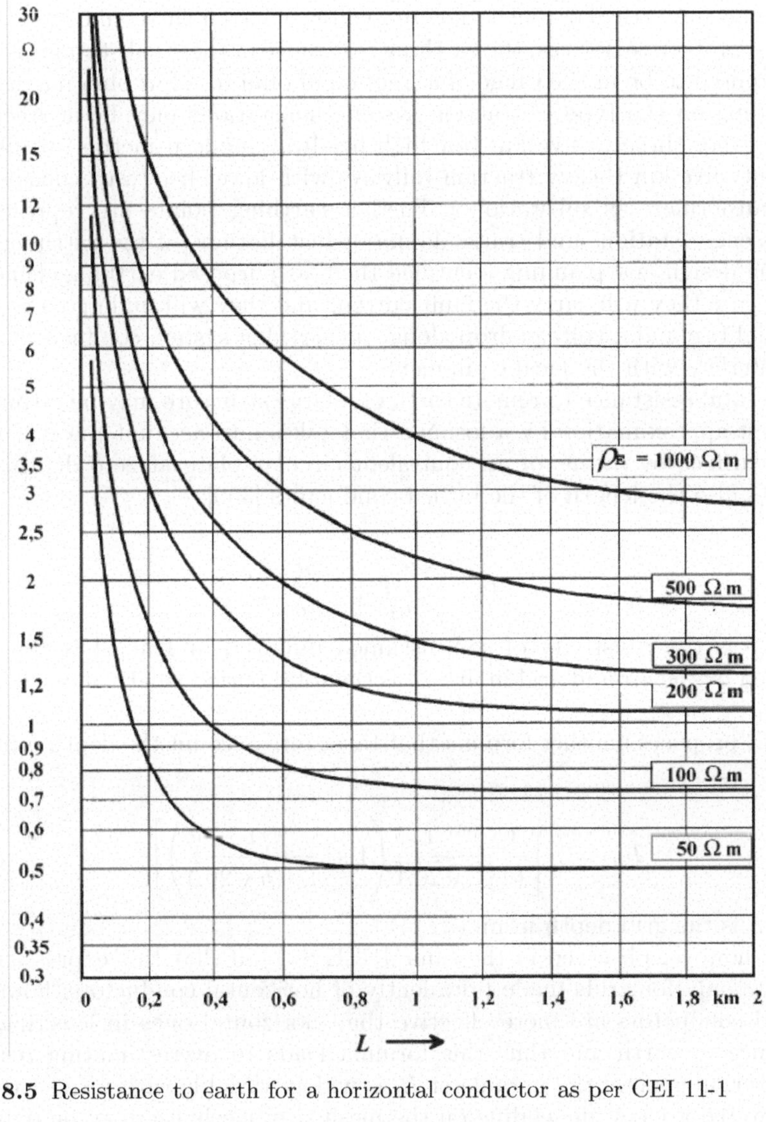

Fig. 8.5 Resistance to earth for a horizontal conductor as per CEI 11-1

Finally, the location of the horizontal electrode running trackside must be selected accurately, since in any new construction the first meters around the installation (in this case, the railway line and in particular the track) are contaminated by gravel and stones, poured during construction activities and also from the same ballast and this may both increase the soil resistivity and worsen the electrical contact at the electrode surface.

Earthing Grids. Finally, the traction supply substations represent additional earthing points sized for the full fault current with respect to electrical safety and in particular touch and step voltages; given the common clearing times of protective devices, the earthing resistance of these substations is usually required to be in the range of a fraction of ohm to a few ohm maximum. Depending on the type of traction system, substations may be as frequent as one every three to five km (for high headway modern metros) to fifteen to twenty-five km (heavy traction railways with lower headway, considering autotransformers as substations). Further earthing points are represented by passenger stations and trackside major installations, if the earthing and bonding design and planning identified them as purposed earthing elements: in this case they will carry the fault current and they will undergo a ground potential rise and a voltage drop along the earthing system conductors, that may interfere with the local equipment.

The total resistance to remote earth of a large structure may be estimated at a first approximation by a formula that takes into account the extension of the surface by means of an equivalent circular plate at zero depth, and includes also the length of the buried conductors [359]:

$$R_{g1} = \frac{\rho}{4}\sqrt{\frac{\pi}{A}} + \frac{\rho}{L_t} \tag{8.5}$$

where ρ is the resistivity of soil, assumed uniform, in Ωm; A is the area occupied by the ground grid in m^2; L_t is the total buried length of conductors in m.

IEEE proposes another formula that takes into account the depth h in the soil:

$$R_{g2} = \rho\left[\frac{1}{L_t} + \frac{1}{\sqrt{20A}}\left(1 + \frac{1}{1 + h\sqrt{20A}}\right)\right] \tag{8.6}$$

where h is the grid depth in m.

It is however observed in the same IEEE Std. 80 that this expressions is accurate only for grids made prevalently of horizontal conductors; however, vertical conductors are more effective than horizontal ones in lowering the resistance to earth and thus this formula leads to overestimating results. Moreover, in general the parameter L_t is widely variable with respect to the grid covered area A, depending on the mesh size: mesh size may be reduced down to 1 m or slightly less in favour of the limitation of step voltage, on the other side any further reduction with a corresponding increase of L_t does not cause in reality a corresponding decrease in earthing resistance, because all these formulas do not take into account proximity effects between nearby electrodes.

When the earthing system must withstand also lightning currents, several prescriptions need to be revised and improved (sizing of conductors, resistance to earth, bonding resistance, separation distance, clamping and welding techniques), taking into account the additional requirements brought along by the adopted standards for lightning protection, such as the IEC/EN62305 [210, 211, 212, 213].

8.2.1.2 Coordination of Leakage to Earth Prescriptions, Bonding and Earthing

In particular for dc systems, leakage to earth of the running rails is a complex subject, that needs to take into due consideration the control of the stray current phenomenon and the protection of metal and concrete structures from corrosion, both within and outside the traction system premises [85, 86, 98]. To simplify, it could be stated that the control of the stray current is achieved by minimizing the amount of current that leaves the running rails and couples to external victim structures, and this is equivalent to minimizing the impressed potential on the same structure, that in turn determines the speed of the reduction chemical reaction and the corrosion rate. It is underlined that there are no fixed commonly valid thresholds or limits for this potential, since the corrosion rate highly depends on the chemical species that participate in the reaction, the soil composition, the level of moisture, the type of victim metal and its electrochemical potential, the type of victim concrete and possible surface treatment, and temperature.

When driving the current away from victims, attention should be paid to the possible corrosion of track and line elements, by observing the current density through the various bottlenecks: rails fastening featuring small mating surfaces, masts foundations, parts of partially earthed track devices (such as point machines and internal motor parts), piles and pillars of bridges and viaducts. In particular for modern metros and light railways, featuring large return currents at dc and routed through dense urban areas, the subject is taken in high consideration:

- for traction system without ballast, a supporting concrete mat (also called "concrete slab") is usually provided with the running rails no longer electrically insulated from it, thus reducing the cost of the fastenings and avoiding high current densities through them;
- the concrete mat is then insulated from the rest of the structure by means of asphalt or polymeric/plastic layers, with a positive effect also for the reduction of externally conducted vibrations;
- the concrete mat may be arranged in different sections of various lengths galvanically separated in the longitudinal direction, in order to avoid the uncontrolled circulation of the return current along the line the and to keep the insulation with respect to lightning phenomena of exposed areas;
- the equipotential bonding of each concrete mat section is ensured by means of a trackside conductor, acting as the through-earth collector, draining

the return current from the the concrete and rebars; the collector of different sections may be bonded together, or at a single point only to avoid stray current circulation, especially for tunnels and viaducts, when also external structures may be involved by means of unwanted transversal leakage terms; the same collector sections may be used for equipotentialization of trackside installations and/or lightning protection purposes, if the length, and as a consequence the inductive reactance, is kept below a convenient limit and each collector is correctly terminated onto a low resistance earthing point; to avoid unwanted leakage to earth of the concrete mat current, two separate collectors may be provided, one for equipotentialization and traction current draining of the concrete mat itself, the other one for equipotenzialization and earthing of trackside installations.

Earthing in tunnel sections highly depends on whether the tunnel walls are insulated from the surrounding soil or not, the possibility of water leakage, the presence of in-out conductive paths represented by the running rails, cable screens, metallic pipes, etc. EN 50122-1 [85] identifies different earthing arrangements and distinguishes between AC systems (for which the traction earth may be used as the reference earth for electrical safety, so that all exposed conductive parts shall be connected to it) and dc systems (for which, because of the stray current phenomenon, the traction circuit and the earth are kept separate, and earthing of exposed conductive parts is operated only on the system earth, leaving the traction circuit without direct earthing). The prescription for the isolation of the return circuit with respect to earth is given in terms of the conductance to earth, and in tunnels the EN 50122-2 [86] reduces the limit for the track conductance to earth to 0.1 S/km, so five times less than open air sections. The reason is the higher risk that corrosion takes place, with the stray current leaving the track collected by the concrete structure of the tunnel.

8.2.2 Buildings, Stations and Substations

For large fixed installations, occupied by large power equipment and possibly open to public or workers, the earthing philosophy is the one dictated by standards for electrical safety. To simplify, touch and step voltages are kept under control by means of an earthing grid with a suitable mesh hosted in the concrete slab of the building or installation, with the same grid brought to earth by means of additional earthing electrodes located on its periphery. IEEE Std. 80 provides some formulas for the estimation of the earthing resistance for various grid arrangements and for single earthing elements; in several cases more detailed calculations by means of finite element methods are needed, not only to obtain more accurate and less overestimating results, but also to confirm the design prescriptions by an independent method: there are several commercial software tools that allow the calculation of the earthing resistance of arbitrary conductors arrangements, together with the profile of the touch and step voltages for various injection points of the fault current.

The same electrodes may be used also for lightning protection as earthing points, which the down conductors from the air terminals are connected to, provided the recommendations and prescriptions in the IEC/EN 62305 are met. The calculation of the lightning induced overvoltages is not handy and even commercial tools and simulation programs need to be used carefully; it is in general preferable to follow all the prescriptions in terms of welding of rebars for solid connection under lightning stroke current, terminals clamping, conductor sizing, clearance and separation distances to avoid spark over, and so on.

At electric substations where the incoming MV or HV electric energy is transformed (and possibly rectified) for the supply of loads, including the traction line, the incoming lines bring their own philosophy for earthing arrangement. The relevant factors are two: the status of the neutral wire (earthed or not earthed, and if not, with which insulation impedance, including the parasitic capacitance of the incoming line), that determines the response of the system in case of faults in terms of intensity of the flowing fault current, earth potential rise and neutral voltage versus earth); and the earthing of the cable shields (that is normally done at both ends in all cases, and thus sets up a connections between the earthing systems of two remote substations of the same traction system through the in/out connections). The latter factor indicates that the earthing systems of two substations are thus normally connected and that a specific prescription is needed to keep them separate. Concerning the separation of two earthing systems, ; another it is further considered consideration in this direction is that the earth electrodes of near systems influence each other and thus the respective earthing systems are not really isolated; (in urban areas this leads to the concept of "global earth", that is the combination of all the earth electrodes, earthing systems, buried metallic parts and installations, and so on). Bringing the concept at the single building or substation level leads to the natural choice of a Common Bonding Network (CBN) at stations and substations: the step voltages, as well as any potential difference, are kept under control by an adequate mesh size of the mesh or grid of electrodes that is located at each floor (with the known exigency of equipotential bonding between the various grid rods). Similarly, for vertical walls rebars and steel columns (or dedicated conductors) are interconnected with the horizontal grids to ensure the lowest bonding impedance as possible, with particular attention to prescriptions for lightning protection; for such a structure, the ensemble of horizontal and vertical mesh not only represents a high performance bonding network, able to withstand the lightning current intensity that normally flows through the down conductors, but also serves as an electromagnetic shield against the electromagnetic field produced by lightning strokes occurring at close distance. Another element that is part of the CBN is the protective earth conductor of the LV distribution system, whose interconnection is possible if in agreement with provisions and requirements of the type of adopted LV system (TT, TN, TN-S, etc.).

On the other hand, the exigency of providing a reference earth for connected electrical and electronic equipment must be met, where there may be a single equipment, a rack of them, or an entire equipment room, that may require a different earthing solution to address the necessity of avoiding earth loops, providing a single point of grounding. For electronic equipment located outdoor a standalone earthing system and earthing electrode are provided, as it is discussed below in sec. 8.2.3 and 8.4. The problem is even more complex if the possibility of cables entering the installations are considered, whose shields may put in connection the two earthing systems and may be subject to potential fluctuations due to external sources, such as again lightning.

The ITU-T K.27 [431] represents a valuable reference for the problem of earthing of electronic equipment inside buildings: the bonding network featuring a single connection to preserve electronic equipment from earth loops is called Isolated Bonding Network (IBN) and the possibilities of connection to the building Common Bonding Network (CBN) are shown in Figure 8.6. There it may be seen that the single equipment may be connected independently to a single point on the CBN of the building in a pure star architecture, or the equipment of an entire room or set of racks may be locally meshed (so featuring a local CBN) and then driven to the building CBN again at a single point of connection ("Mesh-IBN"). In case the electrical safety of persons and equipment is the main concern, and earth loops and reference earth are of no relevance, a complete CBN approach may be followed, where the bonding connections are several, and not only they follow the rule of the shortest path, but also they become part of the CBN, contributing in reducing the overall impedance.

8.2.3 Trackside Installations

Trackside installations include a variety of elements that for various reasons need an earthing connection, that may be provided either by means of a trackside collector or dedicated earthing points. For the rest of the metallic objects that are located wayside, for the sole electrical safety, when they do not support or contain any electrical equipment, earthing is not necessary, if they are of small dimensions: signal posts, level crossing posts, warning labels, single poles, short fences (EN 50122-1). This was decided to reduce costs and earthing complexity and in favour of the minimization of the stray current phenomenon. For exceptional conditions, such as a high keraunic factor and lightning strokes of particular intensity, earthing might be required for all conductive objects that extend above a given height, or with a particular shape, or far from taller objects. Examples of the most common trackside installations are:

- cabinets for signalling, telecommunication and other electrical or electronic apparatus, located longitudinally, where the earthing connection is needed for functional earth, electrical safety and lightning protection;
- signals, poles and other metallic structures mainly to be protected for electrical safety and lightning.

The earthing may be ensured by bonding to a dedicated trackside collector, earthed with a conveniently low earthing resistance. Alternatively, for very long lines with low headway, long block sections and few installations

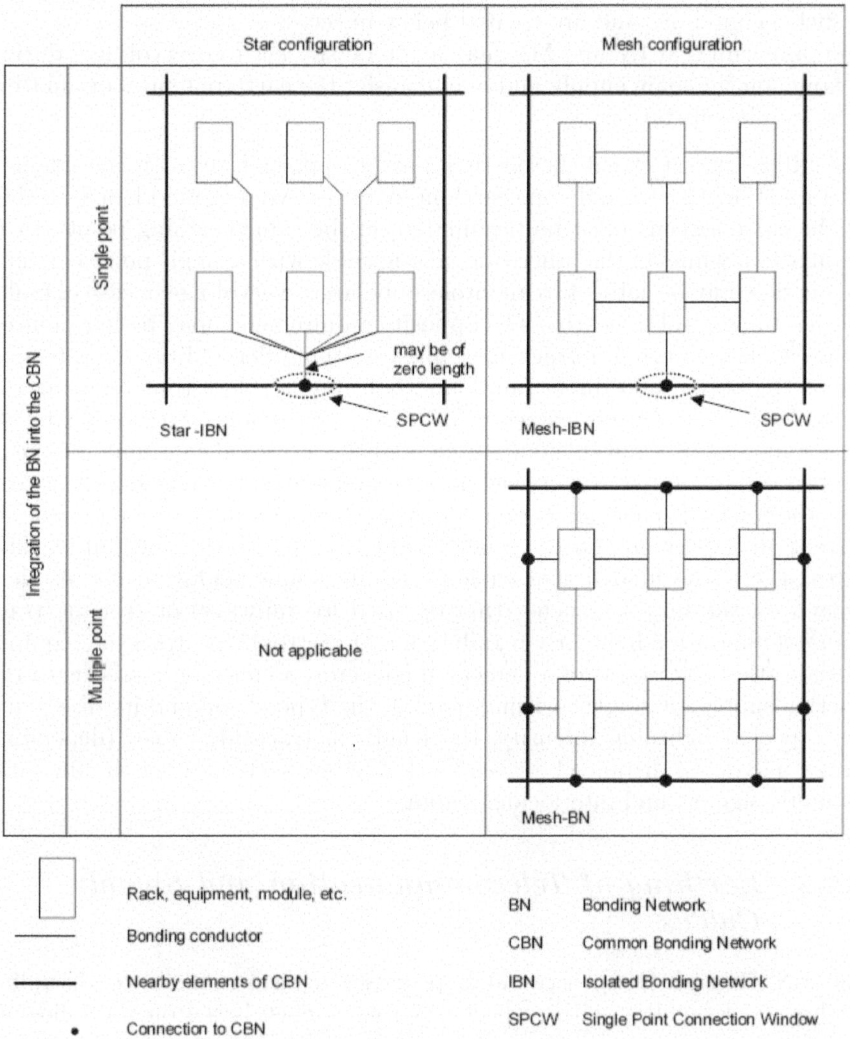

Fig. 8.6 Connection of CBN and IBN (ITU-T K.27 [431])

between stations, earthing may be accomplished locally by means of autonomous earthing points, e.g. vertical rods as earthing electrodes or more articulated earthing grids, for large installations. In all cases the major issues that are to be considered for a correct coordination of earthing and the control of procured overvoltages on the earth terminals are:

- signal or auxiliary supply cables entering the installation may reclose ground loops, thus procuring interference and unwanted current circulation, and may bring dangerous potentials inside the protected installation, originating from faults and lightning strokes, either direct or induced; direct strokes are usually in relationships with the most exposed elements such as antennas and are treated below in sec. 8.4;
- supply cables at LV and MV may again be subject to overvoltages during faults on the main supply and may transfer the earth potential originating at supply substations.

The criteria used to treat these issues may be opposing, since on one side it is desirable that there is only one earthing point to avoid ground loops, so that all the cable screens must be terminated on the same earthing point at the installation, while on the other side, a long cable with a single point earthing on one side may be subject to an intolerably large overvoltage in case of faults and lightning strokes nearby. The opposite requirements may be harmonized if the cables are hosted in trackside cableways that offer additional protection against induction and direct faults and lightning strokes, thus reducing the overvoltages to tolerable values in all cases, or if cables with one external screen and one internal shield are used, and the electrical continuity of one of the two is interrupted at a critical maximum distance terminating the screen onto the trackside collector at one side only (thus avoiding any common mode current on the screen and the related induction due to the not null transfer impedance). The final design choices and the implementation details then depend on the use of trenches (mostly used for railways) or troughs/trays (more common for light urban railways and metros), the trackside equipotentialization (by means of a dedicated collector, a collector shared with the traction supply, or single earthing points), the type of soil and its resistivity, the expected lightning intensity, the length of trackside cables (depending on the maximum distance between track devices, such as track circuits, axle counters, signals, and interlocking cabins).

8.2.4 Earthing of Telecommunication and Signal Cables

The EN 50174-2 [106], sec. 6.3.2, proposes a clear scheme to discipline earthing and bonding with respect to cable shields, based on the following considerations:

a) screen not bonded to equipment: not recommended;

b) screen bonded at both ends to equipment (i.e. connected with the chassis of the terminal equipment): reduces electromagnetic radiation by the principle of the Faraday cage;

c) screen earthed on one end: provides protection against electrical fields;

d) screen earthed on both ends: provides protection against electrical fields and, where problems with high screen currents exist, partially compensates the interfering magnetic field;

e) screen in all cases a) to d): virtually no effect against very low frequency magnetic fields (e.g. 50 Hz), unless special materials are used (μ-metal, Permalloy, etc.).

According to the above considerations the standard gives the following prescriptions:

1) the cable screen shall be continuous from the transmitter to the receiver; in any case a cable screen shall be connected at both ends to terminals or sockets;

2) the cable screen shall have a low transfer impedance according to EN 50173;

3) special attention shall be paid to the assembly of connection elements; the screen contact should be applied over 360 degrees according to the principle of a Faraday cage; the screening connection should be of a low impedance design;

4) the cable screen should totally surround the cable along its entire length; a screening contact applied only through the drain wire has little effect at high frequencies;

5) the screening should continue through an adequate screen connection; normal pin contacts shall not be used;

6) avoid (even small) discontinuities in the screening: e.g. holes in the screen, pigtails, loops; discontinuity dimensions of the order of 1% to 5% of the wavelength can decrease the overall effectiveness of screening.

If everybody agrees with the prescriptions on screen integrity, connector terminations and low impedance path, there are however some critics or considerations that can be formulated on the absolute correctness of some of the prescriptions above.

- The earth termination of cable screen at both ends opens the door to the circulation of the so called loop currents due to earth potential differences and induction and occurring mostly at the power supply frequency; the consequence is the possible overheating of cable screen, if not properly sized, and the coupling through the transfer impedance of noise at supply frequency onto the internal conductors. Countermeasures that can be adopted are: the use of single point earthing instead; the use of double screening with single point earthing on opposite ends or combinations of single and double point earthing; the use of a supplemental earthed conductor in parallel to the cable bundle to drain loop currents; the use of a selective single/double earthing scheme, terminating the cable shield to earth on either sides with a SPD (that triggers to double point earthing only in case of dangerous overvoltage) or a parallel RC network (to increase the impedance at the supply fundamental and reduce loop current to a negligible value); the use of a parallel PE conductor earthed at both ends is highly recommended also by the IEEE Std. 575 [376]. In any case the EN 50174-3 [107] prescribes isolation of the cable shield and no earthing in case of routing within "hot zones" featuring a significant earth potential rise.

- Normal pin contacts are not to be used for earthing, but they are widely used for reference common, that is earthed then at the terminal equipment through the functional earth; the use of a reference conductor close to each signal wire is normally used to reduce interference and to improve the transmission performance, in particular for unbalanced telecommunication lines.

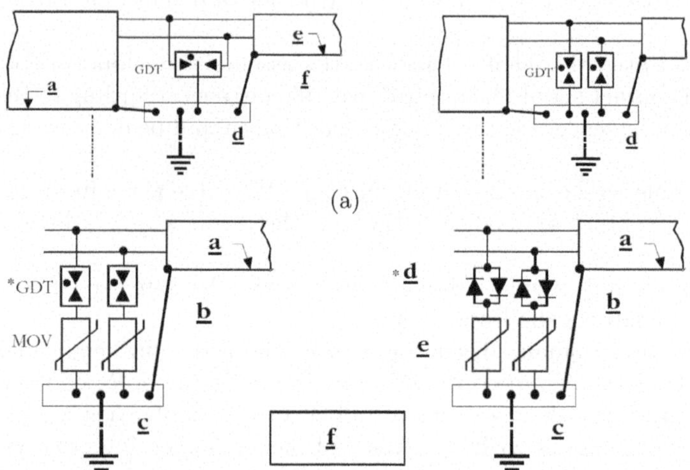

Fig. 8.7 Protection of cables entering the installation against overvoltages: (a) telecommunication cables, (b) power supply cables

8.3 Cabling Methods and Practices

While in general we could say that cabling practices and cable layout must follow workmanship rules, that are known to designers, installers, and in general stakeholders, and are coded in standards and manuals, more specifically cables are routed and laid down according to procedures based upon:

- protection of cableways from fire and in case of accidents;
- location convenience with respect to trackside installations, passages, etc.;
- use and exploitation of the provided cableways, designed also considering the installation practices of the specific country, past experience of the subcontractor, costs, harmonization with civil works, and so on;
- cable segregation criteria, often specified in companies internal documents, possibly set out in consortium documentation and following a few standards of reference.

8.3.1 Cableways Categories

The protection of cableways from fire and in case of accident takes precedence over "subtler" elements concerning interference and EMC. It is quite common that cables, as well as the location of other vital elements for the safety of the system and of the circulation, are arranged on either sides of the track, protected by concrete or metallic housing, at a convenient height above or below rails level, following unwritten rules for the harmonization with the civil and mechanical works. Cableways may be broadly classified in:

- trenches, often excavated trackside in open-air long distance track sections, such as for railways and light railways, but provided with concrete compartments or plastic pipes for cable protection;
- cable troughs, located in an existing concrete mat (in particular in modern metros and light railways) or setup trackside (for railways and in particular old installations);
- cable trays, secured to walls of internal installations (such as power stations, automation and control rooms, etc.) and trackside in tunnels (at various heights, in parallel to other cables and services).

Segregation criteria are mainly based on the classification of the emission and susceptibility properties of each cable category (because of the connected equipment and its characteristics) and a set of additional properties aimed at the reduction of the coupling between different cable classes (depending on shielding characteristics of materials, distance between cables, cable type, use of grounded wires for additional shielding, and so on). The used cableway may represent an additional shielding with respect to cables alone, and is to be considered in the definition of the minimum segregation distance for cable classes. Cable segregation criteria and categories are detailed below in sec. 8.6.

8.3.1.1 Electromagnetic Properties

Cable trays are preferable since their shielding properties are better than the other solutions and they allow a more accurate cable routing and separation among cables sharing the same cable-way. However, for outdoor installations cable trenches and cable troughs ensure mechanical robustness to prevent damages, also in case of falling conductors and direct or nearby lightning strokes. In order to ensure and preserve the shielding properties of metallic cable trays, those with solid walls are preferred for EMC purposes, not for ventilation and weight (see Figure 8.8).

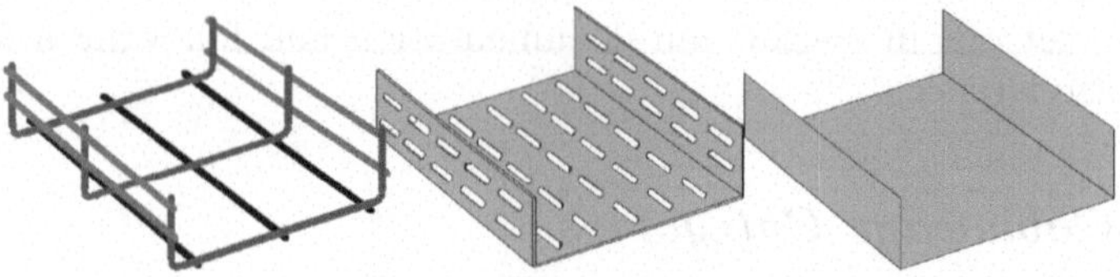

Fig. 8.8 Ranking of different metallic cable trays: no shielding effect (left), moderate effect, but adequate (center), preferred (right)

For U shaped cable trays the magnetic field is attenuated by the tray walls with a maximum attenuation near the bottom part and in particular the two corners, so deep sections are preferred as shown in Figure 8.9. For deep sections the inductive part M of the transfer impedance of a test wire in the middle lower part of the cable tray is reduced: starting from a flat metal plate as reference, for which the height/width ratio is 0, with M_0 value, when the ratio is brought to 1.0 (the walls are the same size as the width) $M_{1.0} = 0.5M_0$, and when the ratio becomes 2.0 (the walls are twice the size of the width) $M_{2.0} = 0.1M_0$. Moreover, even if aluminum or copper have a dc resistance lower than steel, the latter due to a higher magnetic permeability as a more pronounced skin effect that starts from a lower frequency, so that steel is the preferred material [322] (also for mechanical robustness and cost, of course). A metallic cover is always advisable for U shaped cable trays and in this case many electrical contacts along the longitudinal dimension are preferred.

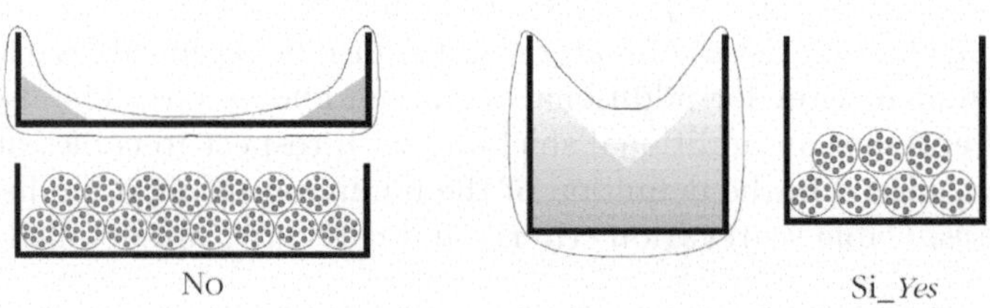

Fig. 8.9 Magnetic field attenuation in a cable tray and best cable allocation

There are cases where metallic construction elements may serve as cable trays, providing similar shielding performance, as shown in Figure 8.10.

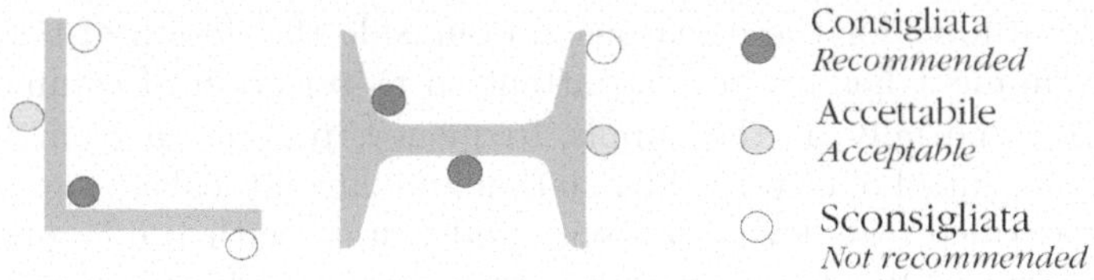

Fig. 8.10 Location of cables inside metallic construction elements

8.3.1.2 Bonding of Cableways

The good or excellent shielding effect of a cable tray, in particular for solid trays, may be compromised at medium and high frequencies if separate sections are bonded together improperly, as shown in Figure 8.11, where a gap may be present, and bonding may be made by wire sections, or preferably with flat straps; solid bonding between adjacent sections, by using specific joining elements, is the preferred solution.

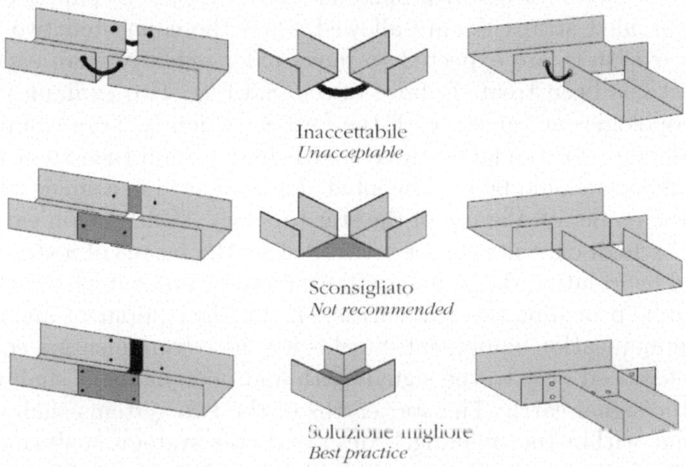

Fig. 8.11 Bonding of cable trays

8.4 Telecommunication Products and Installations, Indoor and Outdoor

While telecommunication equipment may be in principle considered similar to other electrical equipment, it distinguishes for some factors: in-band susceptibility is a major concern, in particular for analog telephones and digital data networks; signal ports and telecommunication protocols operate on weak signals, with respect to external disturbances, and may be more exposed to interference and electrical damage; the same signal ports are often connected to very long cables, that cross several electromagnetic environments and may be located outdoor; for these reasons, on one side the design of telecommunication equipment has given more attention to rejection of common mode disturbance, especially at the supply frequency, protection from transient overvoltages, control of electrical insulation and current leakage; on the other side the prescribed tests for interference and voltage withstanding capability have been historically the most severe. Moreover, the design guidelines and suggested practices have been verified on-site for many years with a significant sample of equipment and installations in the most variable electromagnetic conditions.

The bonding and earthing arrangements for communication facilities inside industrial plants, so in the presence of a potentially harsh electromagnetic environment, reported in IEC/EN 61918 [205] are the already considered mesh and star topologies. Basic prescriptions are given, concerning i) the minimum size of the equipotentialization (called "equalisation" in the standard) or earthing conductor to reach a total 1 Ω resistance and so depending on its length, ii) bonding straps cross-section, iii) use of protective surface treatment for bonding plates, iv) bonding of metallic pathways "shall be connected by using rigid metal straps with maximum coverage of the gap; the connection with braided straps is only allowed when the connected two parts of the pathway system are expected to move independently from each other" (this aspect has been treated above in sec. 8.3.1.2). Two examples of mesh and star topologies are given, with the former evidently more complete and articulated, while for the latter the reader is in doubt on the extent at which the star topology should be implemented. The standard in a single paragraph instructs the reader on the use of the star topology: "Currents in earth paths generated by high currents can be controlled by the means of a star earthing system and by isolating the signal earth from the equipment earth. This is accomplished by providing two star earths, one for the equipment and a second for the communication equipment. Shields for the communication equipment shall be referenced only to the signal earth and no equipment shall be referenced to the signal earth. The star earths of the two systems shall converge to one point within the building." Mesh and star systems may cohabit the same plant, where the mesh system is used for the less susceptible equipment (such as power equipment, motors, etc.) and for structural parts, to provide a minimum resistance bonding to the earthing point, while the star system is used for communication, automation, signaling subsystems, where interference and common mode coupling are also of concern. While the prescriptions for single earthing can be met locally, it is normally challenging to preserve earth segregation over long cable runs and across the most different electrical and electronic equipment, thus ensuring a unique earthing point for the star topology:

- several equipment come with earthing provided by their PE connection in the plug together with the live conductors (this is the case of IT equipment, such as computers, printers, and so on);

- there may be no distinction between protective earth and functional earth with the equipment providing a unique screw for earthing;
- cable shields, if earthed on both sides, are thus setting a bonding path between remote functional earths and may bond also protective earths; the immediate effect is the increase of the circulating current at the supply frequency due to unavoidable potential differences; if one of the two ends of the cable shield is earthed through a capacitor or parallel RC network (high frequency selective earthing), this occurs only at high frequency and it is much more difficult to be located.

Even if functional earths shall be bonded to the protective earth at the "source of supply", that is at the main rack or cabinet, with sub-racks and remote units receiving the earth reference by means of cable shields and/or dedicated signal reference conductors within cables, the IEC/EN 61918 [205], App. E, considers the case of earth connection reclosing through shield clamps, enclosures and protective earth. It observes that with TN-C power supply system, using a unique conductor for neutral and PE, in the presence of switching supplies and static converters, the current unbalance on live wires occurring at high frequency recloses on the PE connection and thus possibly on all connections in parallel, such as cable shields. It is thus recommended that TN-S power supply systems are used, with neutral and PE as separate conductors (the so called "5 wires" system).

EN 50173-1 [104], Table 3, identifies the characteristics of the electromagnetic environments E1, E2 and E3 defined within the MICE framework (ISO/IEC 24702 [416]): E1 corresponds to the immunity tests and levels specified by EN 55024, E2 to the residential e.m. environment and E3 to the industrial environment. The same standard subdivides the cables into two categories, unscreened cables and screened cables, which two different performance factors for cable-to-cable coupling are made corresponding to: the TCL (Transverse conversion loss) and the ELTCTL (Equal level transverse conversion transfer loss) are applicable to unscreened and screened cables respectively, and defined over various cable categories[2], that are made correspond to different frequency ranges. For unscreened cables at lower frequency the specified TCL limit is constant, identical for the three electromagnetic environments E1, E2 and E3: for category A, up to 100 kHz the required coupling loss is 30 dB; for category B, up to 100 kHz the required coupling loss becomes 45 dB, but 20 dB are required at 1 MHz. For higher cable categories the required coupling loss is given by the formula $A - 15 \ln f$, where A stands for 53, 63 and 73 dB for E1, E2 and E3 environments (for the [1, 30] MHz frequency range) and by the formula $A - 20 \ln f$, where A stands

[2] The EN 50173-1 [104] standard specifies the cable categories based on the operating frequency interval: a) Class A: specified up to 100 kHz; b) Class B: specified up to 1 MHz; c) Class C: specified up to 16 MHz; d) Class D: specified up to 100 MHz; e) Class E: specified up to 250 MHz; f) Class F: specified up to 600 MHz; g) Class CCCB: specified up to 0.1 MHz; h) Class BCT-B specified up to 1000 MHz.

for 60.4, 70.4 and 80.4 dB for E1, E2 and E3 environments (for the [30, 1000] MHz frequency range); all the formulas are limited to a maximum of 40 dB, For screened cables of category D or above, the prescribed coupling loss is given by the formula $A - 20 \ln f$, where A stands for 80, 90 and 100 dB for E1, E2 and E3 environments, with a maximum of 40, 50 and 60 dB respectively, that represents the constant value prescribed for category D. This explains the performance that is common to Cat. 5 and 6 cables of a generic "screening factor" of more than 50 dB, often around 54 to 56, that make them suitable for all residential and information technology environments, but strictly speaking non for industrial ones. In reality, if exceptional circumstances can be excluded, it is likely that a "more than 50 dB coupling loss" is suitable for the vast majority of industrial applications; this is the case of a specific aspect that must be upheld by a company cable segregation policy with indisputable benefits also in terms of costs. It is worth underlining that the prescribed coupling loss values above are referred to the respective electromagnetic environments, characterized in terms of the parameters of several electromagnetic phenomena, such as electromagnetic field intensity, but missing to fully consider the reactive electromagnetic field caused by cable-to-cable coupling: conducted emissions along a cable connected to a disturbing equipment are likely to produce an equivalent electromagnetic field intensity higher than the assumed far field values, if minimum separation distance is not prescribed. So, the coupling loss values are a useful support and turn out to be a valid reference for evaluation of minimum distance criteria.

8.5 Provisions against Lightning

In this section lightning provisions are only considered for their influence on the cable arrangements, shield earthing and bonding techniques, that stem from their level of exposure with respect to lightning phenomena and to the chosen location, layout and housing. The reference standards for definitions, prescriptions and guidelines are the EN 62305 [210, 211, 212, 213]. The assignment of a cable to a defined Lightning Protection Zone (LPZ) and related Lightning Protection Level (LPL) automatically reflects as a set of requirements related to the cable insulation level, sparkover protection and need of surge protective devices (SPDs), in relationship to the voltage withstanding capability and susceptibility of connected equipment. Cable insulation level and protection of cable by itself against sparkover may increase its cost and pose a serious issue on its availability on market: cables of various categories and for various applications are commonly available with a given insulation level, suited for a broad range of cable uses and applications. In case of particularly exposed areas (because for example of a particularly high iso-keraunic level, for the severity of lightning strokes, or for elevated structures, such as viaducts, or towers and masts) the lightning threat must be reduced as soon as possible with the selection and use of the correct cable housing and additional shielding, if necessary. It is not uncommon that for cables that are inevitably exposed, such as while routing to a signal on a mast or to a tower top antenna, an additional protection is provided, such as a solidly bonded conduit.

On the other side, for the connected equipment the surge phenomena picked up and transmitted along the cable are attenuated by the use of SPDs, SPDs combinations and SPDs and filters combinations. All these solutions rely on a low impedance earth connection, that is not only ensured by means of a low resistance to earth, but also with a short path to the earthing termination, where the used conductor is of adequate size and features in particular an optimized path with respect to self inductance and to the possible induction to nearby circuits. It must be remembered that the lightning current is characterized by a very large peak value (in the order of one or two hundreds of kA) and a fast rise time (between one and a tens of microseconds), thus exhibiting the characteristics of a high energy pulse (the time duration ranges from fifty to several hundreds microseconds) with a fast leading edge typical of fast transients (featuring an extended bandwidth, up to some MHz). The combination of fast rise time and large current produces large overvoltages in the same conductor carrying the lightning current and nearby circuits by induction, so that both the conductor cross section and its routing (turns, total length, proximity to nearby conductors) are of concern. Several metallic parts and conductors at earth potential cannot be part of the lightning protection circuit because of the limited current capability with respect to self heating and sparkover: this is the case of rebars (in particular if they are not brazed between each other with an adequate mating surface), bonding terminals and not properly insulated down conductors along masts and towers.

For cables of considerable length we have said that the limiting factor is the associated inductance and thus the impedance with respect to the earthing point, that may exhibit a very low resistance to earth in the range of a fraction of ohm. For trackside cables as well as cables crossing a depot or an open area exposed to the most severe lightning phenomena, protection and control of overvoltages is thus reached by means of both interposition of screening surfaces, such as the covers of the cable trays, troughs and trenches, and reduction of the lightning stroke reaching the victim, by installing a lightning protection system employing an adequate number of air terminals. When it is possible to show that the victim cable is subject to a residual lightning current of amplitude not exceeding a given threshold, then it will be protected against spark-over and it won't transmit to connected equipment overvoltages more severe than those specified in the assumed electromagnetic environments. When a set of electrical and electronic equipment is purchased with a given voltage withstanding capability and surge immunity, this determines the required level of protection for the cable and the required SPDs to meet the prescribed levels with a given safety margin.

8.6 Cable Segregation Guidelines

Cable segregation is based on the classification of the cable classes in terms of emissions and susceptibility of the connected equipment. Cable segregation guidelines and policies are normally company specific and there are a few standards to be used as references: the old IEEE Std. 518 [374], dating back to 1981 and thus limited in particular in terms of an appropriate and up-to-date description of sources and victims, and the group of standards of CENELEC EN 50173 and 50174 [104, 105, 106], that focus on residential and light industrial environments, neglecting the characterization of heavy industrial sources: large power converters, switching of inductive loads, loads subject to highly variable operating conditions, such as the traction loads, radiofrequency transmitting devices, etc.

Equipment for railway applications, such as signalling circuits and related cables, have no specific standard for cable segregation, but the required minimum distances from other cables classes, again common in railway applications, are mostly decided within a company or as a requirement specification of a specific project (such as the construction of a new metro or a high speed line). In the absence of a standard reference, the minimum separation is estimated by careful calculation of transfer functions, by worst case assumptions on cables and cable trays parameters and on the behavior of connected sources. This approach is not so uncommon even in the case an applicable standard exists, because the approach followed for some railway systems has applied safety concepts to EMC, so to ensure that no interference occurs under a degraded configuration (e.g. with a first undetected failure, such as loss of cable insulation) and in case of an exceptional event (such as a short circuit on the traction line supply caused by the catenary falling on the running rails).

8.6.1 Classification of Emissions and Susceptibility Levels

The approach followed in general and in particular by the IEEE Std. 518 is that of classifying emission and susceptibility into a manageable number of levels, depending on broad and evident characteristics of the connected equipment, not to require expensive and complicated measurements and analysis. The usual levels are H(ighly), M(edium), L(ow), N(egligible) applied to E(missions) and S(usceptibility), and all the possible combinations of one E level with one S level are considered, related to cables that share the same cable-way or that run in parallel for a significant length: the correct combination is that of extreme categories, so it is preferred that a HE is run close to NS cables or to LS with increased distance; MS are considered only if the distance is increased further with other levels of segregation or barriers, and so on.

Already the assignment of a source or victim to the proper E or S category bears a significant level of approximation, since emissions in particular highly depend on the operating conditions of the connected emitting equipment (think for example of a thyristor converter, that may operate under various firing delay angle values and overlap angles, depending also on the load characteristics). Moreover, the intensity of emissions is highly variable over the frequency axis, depending also on the use of filters and on the characteristics of the feeding line. Finally, the severity of the emissions is in strong relationship with the susceptibility of victim equipment: at a first glance, this statement could be not sensible and opposite to the purpose, however we must remember that the classification of emissions is not a quantitative ranking, but rather a classification that combines the intensity of emissions in an absolute scale with the likeliness of interference onto a given set of susceptible victims, maybe based also on past experience of similar systems.

To better describe the variability of the emissions characteristics and associated classification some examples of typical sources are considered.

8.6.1.1 Transient Emissions from Contactors with Inductive Loads

This example is one of the most popular and was already considered in the IEEE Std. 518. The emission mechanism is that of the switching of an inductive load and the overvoltage associated to the magnetic flux variation: the overvoltage reached during the initial disconnection of the inductive load is able to ionize the air and to re-ignite the electric arc between the two electric contacts that are moving further away, increasing their separation; at shorter contacts separation the overvoltage level is smaller, and increases while one of the contacts is moving away. The obtained effect is that of a series of re-ignitions triggered by steep spikes, the so called "burst of spikes", or "burst of transients", or "shower effect". It is so common and its effect so devastating that one of the mandatory and most demanding immunity tests is the one called "Electrical Fast Transients Burst", where bursts of spikes with 5 ns rise time and 1-2 kV amplitude each are applied to supply and I/O ports; since the radiated coupling of these spikes onto victim cables occurs for a significant parallelism with the "dirty" emitting line the test for I/O lines is mandatory only for cables longer than a prescribed length, that may vary depending on the electromagnetic environment and it characterization by the standardization committee.

Of course there are countermeasures that mitigate the phenomenon at the origin and that are now well known and deemed necessary: the overvoltage at the origin of electric arc re-ignition and causing repetitive spikes is attenuated by using a series RC damping network (called "snubber" in the world of semiconductor converters) connected across the ac terminals of the contactor. Its sizing is straightforward, where the capacitance C is selected not to drain any significant current at the supply fundamental when the contacts are open and the resistance R adjust the RC time constant to be in the order of the ns: C is thus in the range of a fraction of to a few nF (with a sub-mA leakage current for all LV nominal voltages) and R between a few and a ten of Ω.

Even if a snubber circuit is foreseen and installed, the residual overvoltage may be of relevant amplitude, and this justifies in any case the adoption of the said immunity test. Moreover, it is not the occasional operation of a contactor for emergency or protection that represents a relevant source of emissions, rather the frequent connection and disconnection of an inductive load such as a fan or compressor for ventilation or air-conditioning. For this reason this class of contactors is always classified of the HE type.

8.6.1.2 Harmonic and Radiofrequency Emissions from Static Converters

The term "static converter" covers a very wide range of equipment, different for their architecture and operating principle, for their rated power, voltage and current, for the switching frequency. It is not in our scope to detail the characteristics of the various type of static converters, but only to briefly identify them as per the classification task.

All large power converters are a relevant source of low order harmonics, that may reach current intensity of several tens or even hundreds of Amperes; this is the case of diode and thyristor rectifiers, cycloconverters, square wave inverters, etc. At higher frequency their conducted emissions are broadband and of lower intensity and because of the use of relatively slow semiconductors, they are considered relevant by experience only up to some MHz, where some resonance of internal cabling and of stray parameters can occur.

Smaller converters used for example as battery chargers (rectifiers), power supplies for dc voltage regulation hosted in cabinets or racks (dc/dc converters), Uninterruptible Power Supplies are characterized by a much diverse spectrum of conducted emissions; the harmonic phenomena are located further, around the switching frequency of the adopted modulation (mostly PWM) in the range of some or tens of kHz (or even a hundred kHz for extremely fast, light and performing switching power supplies) and they harmonic in the sense that narrow-band components are visible with patterns peculiar of the adopted modulation. Since faster semiconductors are used, it is extremely likely that almost narrow-band phenomena appear also in the MHz range, due to for instance the ringing of a MOS or IGBT against is turn-off circuit; this extends the band of relevant conducted emissions to several tens of MHz or even a hundred MHz. It is easy to see that correspondingly the evaluation of the coupling of such emissions and related countermeasures may be extremely challenging.

Almost all static converters have an input EMI filter that drastically reduces the radiofrequency conducted emissions, whose performances depend on the prescribed limits on one side and on the rated current and thus its feasibility on the other side. Also low frequency harmonic phenomena are attenuated by low pass and tuned resonant LC filters, again subject to constraints on sizing with respect to rated current and to possible resonances with the supply line. So, in general conducted emissions are much lower than the theoretical values and may be assumed of the same order of magnitude of the applied limits, allowing for some margin de to difference termination impedance and line resonance. In extreme cases, when there is no applicable limit or there are margins of agreement with the customer or the utility, emissions are much larger and they may represent a real threat. Examples are represented by the substation rectifiers polluting the MV feeding line or the output of large power UPS of category C3 and output current above 100 A (see Chap. 6, sec. 6.2.3.4). The latter are surely classified as HE, while an adequately filtered converter, meeting industrial or residential environment limits on conducted emissions may be classified as ME.

8.6.1.3 Susceptibility of Data Network Equipment

Data networks operate with base-band signalling and, if the reduced amplitude of the exchanged signals and the increased data rates are considered, they are characterized by a very large bandwidth and a medium susceptibility. The Ethernet protocol is widely used for data communication and for industrial automation: the former uses normally the 100 Mbps data rate format, while the latter operates with the more consolidated 10 Mbps format or even at lower rates (31.25 kHz), as prescribed by EN 61158-2 [201]. The signal format used by the physical layer is Manchester coded (with "return to zero" feature), so that the spectrum of the transmitted signal is located around the data rate fundamental frequency and has a negligible low frequency content. The susceptibility is a result of the deformation and corruption of the transmitted signal for the presence of the external interference; without going into the details of the various methods for detection, decoding and regeneration of the received signal, it is however evident that, if the corruption affects several successive symbols (bits, in the case of the Ethernet format) and the deformation is extended to a significant part of each symbol with respect to the symbol interval (corresponding to the reciprocal of the data rate for the Ethernet format), then the level of interference increases, even beyond any possibility of correction and recovery. Thus, taking a constant amplitude of the applied disturbance, the susceptibility is maximum when its spectrum overlaps with that of the normal transmission. Ethernet rejects quite efficiently low frequency disturbance and is on the contrary susceptible to in band noise; noise at a much higher frequency is however rejected by the input filters that define its operating bandwidth.

8.6.1.4 Medium Voltage and Low Voltage Cables Emissions

These cables are a source of magnetic field emissions in the low frequency range caused by the supply current at the fundamental supply frequency and harmonics, related to the presence of distorting loads, such as the static converters considered above. The most relevant harmonic currents occur for dc systems on the MV cables supplying substation rectifiers (see Chap. 7, sec. 3.3) and it is remembered that the induced voltage onto nearby victim cables increases proportionally with the harmonic order. In case of a short circuit the amplitude of the current increases above the nominal one by approximately a factor of several tens to a hundred and so the induced voltage; harmonic components have a different behavior and do not increase by the same amount. For this reason supply cables are classified as HE and in some cases LV cables with limited harmonic distortion may be classified as ME. An example of the latter is a LV cable line dedicated to the supply of non-distorting loads, such as an induction motor, of limited power and operated seldom or in almost steady state conditions.

8.6.1.5 Earthed Conductors

This class of conductors is normally assumed not noisy and it is often advised to route signal cables along, in order to increase their protection against external interference. However, in particular for down conductors of a lightning system or major protective earth connections, these earthed conductors must be considered as noisy conductors of the highest class like power cables, since they could carry intense fault and lightning currents, and they collect to earth also all the potentially noisy return currents from all connected equipment.

8.6.2 Levels of Segregation

In an easy and straightforward way a series of countermeasures related to the reduction of disturbance pick-up are grouped under the name "level of segregation" or "barrier", if they are able to reduce the received disturbance by about two orders of magnitude or, in other words, if the associated insertion loss is about 40 dB. This is of course a simplifying approach for many reasons:

- different countermeasures may offer widely different insertion losses, some below or nearly equal an order of magnitude, other well above this reference level;
- the same countermeasure, such as for example shielding of a victim cable, may have different performance depending on the used materials (e.g. braided shield, tin or silver covered braid, double braid, aluminum foil, etc.), the quality of the assembly (e.g. flexible or rigid cable, optical coverage and symmetry) and the final installation (e.g. electrical continuity through the terminal connectors, grounding connection, etc.);
- depending on the type of countermeasure (such as twisting, shielding, etc.) and on the installation aspects (such as the type and length of grounding connections) the offered insertion loss is variable with frequency, and its value cannot be retain a constant.

Stating the equivalence between different countermeasures is of course only approximate, but it's a good starting point for a preliminary discussion. Assuming a reference value of 20 dB of attenuation in order to define one level of segregation, the most popular countermeasures are briefly reviewed.

1) Separation by distance is the most intuitive countermeasure, that reduces the coupling between the source and the victim; under the assumption of high frequency electromagnetic coupling occurring in far-field a 40 dB attenuation corresponds to a hundred times larger separation; for mixed magnetic-electric coupling, as it occurs in practice given the frequency interval of relevant emissions and the separation of cables in cableways, the dependency with distance is a combination of linear and quadratic terms, so the expected required separation is somewhat larger than ten times.

2) Twisting of pair conductors is a very effective countermeasure that ensures a reduction of two orders of magnitude for magnetically coupled disturbance, provided that the separation between the source and the victim cables is in the order of the twist pitch or larger. It is known that twisting is ineffective with respect to electric field coupling.

3) The use of shielded cables is on the contrary quite effective against electric field coupling, and possibly also with respect to magnetic field coupling, provided that the shield is terminated to ground on both sides. Even if accurate estimations based on the real values of the transfer, termination and grounding impedances should be carried on, an approximate estimation based on typical values given by manufacturers says that a two order of magnitude reduction is attainable, for a broad frequency interval. Double shielded cables cannot be assigned simply two levels of segregation for several reasons: first, the two shields normally complement each other, operating on different frequency ranges, or for slightly different coupling mechanism; second, the proximity of the two shields prohibits from applying the same simplifying assumption concerning the wave impedance and the direction of the impinging electromagnetic wave, so that the effect is closer to having simply a thicker shield; third, if the two shields are in reality an internal shield and an external armor, physically separated by an insulating layer, then it is likely that the external armor is there for mechanical protection mainly, and that may be used for protection against heavy magnetic induction and electrical safety.

4) Additional shielding on the contrary may be achieved if the victim cable, or a set of cables compatible from the segregation viewpoint, are placed in a conduit, a solid metallic pipe, properly grounded, with a 100% optical coverage, that ensures one more level of segregation up to very high frequency.

5) When a more economic solution is sought, a metallic partition may be placed inside a cable tray to keep two cable categories physically distant and to ensure additional shielding; if the additional shielding corresponds to one level of segregation or only part of it must be properly addressed, taking into account materials and geometries as well as the electrical continuity of the added element with respect to the tray walls.

So, the correct approach to define the adopted barriers and their effectiveness is that of preparing a section of a document on cable segregation criteria dedicated to the evaluation of the performances of the adopted countermeasures:

- the cable datasheets report a general evaluation of cable performance in terms of per-unit-length parameters and/or characteristic impedance, transfer impedance of the shield(s), if present, screening coefficient (identifying the common mode to differential mode transformation);
- the performance of the grounding termination depends on the resistance to ground offered by the grounding system and on the grounding connection in terms of length, shape and routing;
- the electrical properties of the materials used for cableways and an analysis of the offered equivalent shielding may be very complex, since many non-idealities and simplifications come into play: concrete performance may vary depending on the water content, the cement type and the rebar arrangement; metallic trays have different thickness and geometry, cut slots begin to represent a relevant element only at higher frequency, the bonding technique between different tray sections is a critical aspect;
- grounded conductors inside cableways offer additional shielding factor provided that the coupling geometry with respect to source and victim cables is favorable and their effectiveness depends on the frequency.

8.6.3 Segregation Criteria

Once a list of relevant emissions and susceptibility related to installed equipment is provided and the characteristics of the used cables and cable-ways are known (as far as the assigned level of segregation), the cable categories may be grouped and assigned to the respective cable ways, provided that:

- the available cable-ways are filled up to the maximum percentage indicated by the adopted electric codes;
- the requirement on minimum separation distance is fulfilled for all categories, that is if the distinct cable-ways are sufficient to accommodate for the cable classification categories.

The necessary number of cable-ways need to be discussed with the responsibles of civil works and track works already at the beginning of the project and the number of emission and susceptibility levels adapted accordingly. Cable trenches and troughs, even if of large size and able to host a large number of cables, cannot keep them physically separated unless specific procedures for cable lay-down are adopted (such as filling first with HE cables on bottom, or grouping HE cables together on flanks), and in any case the separation distance is not absolutely ensured.

Examples of combinations of separation distance and use of screened cables, as recommended in the standards are reported in the following.

Guidelines common to all applications and standards are: power and signal cables, that in our classification are likely to belong to extreme categories of HE/ME and MS/HS respectively, shall never be in the same bundle; cable crossings shall be at right angles; signal cables shall be at least 0.1-0.2 m from lumped sources, such as fluorescent lamps.

8.6.3.1 Power and Information Technology Cables inside Buildings

The EN 50174-2 [106] makes reference to the old EN 50081 and EN 50082 standards, that defined a residential and a light industrial environment with respect to emissions and immunity, with a set of immunity tests and levels slightly more benign than the more recent EN 61000-6 standards, as we have discussed in Chap. 6. Within such a scope the standard specifies minimum distances between IT cables and power cables for various cases, including the use of a metallic divider (also called "metallic partition" in other standards). The standard says that for other electromagnetic environments the distance may need to be increased, but doesn't specify the amount. The separation prescription is applied in case of screened cables for a parallelism longer than 35 m, excluding however the last 15 m to the outlet.

Table 8.1 Power and IT cables distance in mm for screened/unscreened cables and with/without metallic divider (EN 50174-2)

	Without divider or non-metallic divider	Aluminum divider	Steel divider
Unscreened power cable and unscreened IT cable	200	100	50
Unscreened power cable and screened IT cable	50	20	5
Screened power cable and unscreened IT cable	30	10	2
Screened power cable and screened IT cable	0	0	0

The IEC/EN 61918, Table 16, provides additional recommendations for the minimum distance of communication cables with respect to various types of power sources. In general it is not specified if the cables are screened and reference is to be made to common practice and workmanship: LV ac power lines are not normally screened, digital ac and dc I/O may be screened, analog and low power digital lines, communication lines, process signals are always

Table 8.2 Power and communication cables distance in mm depending on power circuit type (IEC/EN 61918)

	Distance for routing outside enclosure	Distance for routing inside enclosure or metallic cable-way
AC power lines of greater than 100 kVA High-power digital a.c. I/O High-power digital d.c. I/O Power connections from drives to motors	600	300
Analog I/O lines and analog circuits Low-power digital a.c./d.c. I/O lines Communication cables for control AC power lines of $\geq$20 A, up to 100 kVA	300	150
Low-voltage d.c. power lines Communication's cables to connect between system components within the same enclosure Process signals (25 V) Unscreened d.c. voltages (60 V) Unscreened a.c. voltages (25 V) Conductors of less than 20 A	150	80
Electric light and power	Minimum distance: 80 0 V – 100 V: 80 101 V – 200 V: 110 201 V – 300 V: 130 301 V – 400 V: 160	

screened, power connections from drives (i.e. static converters) to motors are normally screened, in order to control the flow of high frequency common mode currents and the related radiated emissions.

Since it is known that some devices may be a significant source of disturbance to cables, in its Appendix B the IEC/EN 61918 identifies the reference distance with respect to cabling and the electromagnetic environment conditions due to the operation of such devices; the electromagnetic environment conditions are identified there with reference to the MICE classification (see sec. 8.4): E1 and E2 correspond to the Residential Environment, while E3 corresponds to the Industrial Environment. The reference distance is only used to identify the corresponding electromagnetic environment without a clear identification of the minimum recommended distance, even if the same reference distance may be used, thus ensuring the adoption of the most benign environment.

Table 8.3 Relationship between electromagnetic disturbance generating devices and electromagnetic environment classification: environments for shorter and longer distances are separated by a "/" [205]

Electromagnetic disturbance generating device	Ref. dist. from cabling [m]	E.m. env. classification
Contactor relay	<0.5 >0.5	E2 E1
Transmitters (< 1 W)	<0.5 >0.5	E3 E2 or E1
Transmitters (1 W to 3 W)	<1.0 >1.0	E3 E2 or E1
Transmitters (TV, radio, mobile, base station)	<300 >300	E3 E2 or E1
High HP motors	<3.0 >3.0	E3 E3
Motor controllers	<0.5 0.5 to 3.0 >3.0	E3 E2 E1
Induction heating < 8 MW	<0.5 0.5 to 3.0 >3.0	E3 E2 E1
Resistance heating	<0.5 >0.5	E2 E1
Fluorescent lights	<0.15 >0.15	E3 E2 or E1
Thermostatic switches 110 V to 230 V	<0.5 >0.5	E3 or E2 E1

8.6.3.2 Power and Information Technology Cables Outdoor

If for cable segregation inside buildings there are various prescriptions that agree at a certain extent, for outdoor installations there are few references spanning over a long period and making reference to cases and configurations that are much different and are not treated exhaustively.

The EN 50174-3 addresses among the outdoor installations and configurations the use of overhead IT cables, such as in the case of telephone cables in rural or scarcely populated areas, even if the most interesting and widespread are the underground configurations.

The considered interfaces of overhead IT cables are with overhead power lines, in the form of cables or separated conductors (as for medium and high voltage overhead lines), and with roads and railway lines. For overhead power lines minimum distances are specified distinguishing for IT cables and IT poles, so it is evident that it is not interference, but electrical safety, the relevant phenomenon, as confirmed also by the first sentence of the section concerning work in correspondence of crossing of overhead power cables. Distances are specified as fixed values or depending on the nominal voltage of the overhead power line, while it is known that the coupled interference is largely due to inductive coupling and the current intensity in steady and fault conditions. For underground cables crossings between IT cables and power cables shall be at minimum 0.3 m distance, and a protection pipe may be fitted to reduce this distance.

In rural environments the separation prescription is extended to earthed electrodes of power cables, because it is more likely that some earth current couples with the IT cable, being the only conductive element nearby. Even if it is possible to find in the standard literature a generic prescription on earthing electrodes as a possible source of disturbance and on the need as a consequence of keeping cable trenches away, quite interestingly the EN 50174-3 [107] seems the only standard to quantify the required minimum separation. Correctly the separation is done for systems located in a rural environment, both for a practical consideration of reduced clearance in urban environments and - most of all - for the concept of global earthing related to urban environments, that reduces the earth resistivity perceived by the electrode and the gradient of the electric field in the soil. The prescribed distance is related not to the normal operating conditions, but to the abnormal current circulation in case of lightning; since the lightning parameters considered to derive the separation values are not clearly stated, it is possible to assume a harmonized approach by the standardization committee, using the one proposed by the IEC/EN 62305 standards, or at least of the old superseded standard IEC/EN 61024.

Inductive coupling is clearly identified later on, with reference to the ITU-T Standard K.53 [444], concerning limits of induced voltage; a safe distance is indicated, that depends on the length of the victim IT cable and on the

Table 8.4 Minimum distance between information technology earthing systems and earthed electrodes of power systems in rural environment (EN 50174-3)

<table>
<tr><th>Soil resistivity [Ωm]</th><th>LV neutral earthing electrode (TT system)</th><th>LV neutral earthing electrode (TN system and IT system)</th><th>HV earthing system (isolated neutral or arc suppression coil)</th><th>HV earthing system (directly earthed neutral)</th></tr>
<tr><td>< 50</td><td>2 m</td><td rowspan="5">2 m</td><td>5 m</td><td>10 m</td></tr>
<tr><td>> 50 & < 500</td><td>4 m</td><td>10 m</td><td>20 m</td></tr>
<tr><td>> 500 & < 5000</td><td>8 m</td><td>20 m</td><td>50 m</td></tr>
<tr><td>> 5000 & < 10000</td><td>16 m</td><td>20 m</td><td>100 m</td></tr>
<tr><td>> 10000</td><td>16 m</td><td>20 m</td><td>200 m</td></tr>
</table>

soil resistivity, beyond which there is no need of investigation by means of calculations or measurements. In an attempt to simplify, a distinction is made for power lines with nominal voltage below 30 and 25 kV, in normal and fault conditions respectively, for which investigation is not due; it is known that the amount of induced voltage depends on the conductor separation, the current intensity in the source (in this case the rated power and the fault characteristics) and on the length of parallelism, of which only the first one may be put in relationship with the voltage.

Concerning the use of long shielded cables routed outdoor and the adoption of a correct earthing principle, it is possible to identify the occurrence of overvoltages due to faults in nearby circuits and to lightning as the most severe threat, that prohibits the use of single point earthing, operated at the telecommunication building only. For this reason, the use of overvoltage limiters is advisable, in case single point earthing is strongly required by interference and loop current issues: during the occurrence of overvoltages the cable screen is earthed at both ends and a large fault current may flow through, so that the overvoltage protective devices must be adequately selected.

8.6.3.3 Industrial

The IEEE Std. 518 [374] is a thirty-years-old standard that addresses the problem of noise coupling to control equipment in industrial applications, so characterized in the IEC/CENELEC jargon by a light industrial or heavy industrial electromagnetic environment. In Chap. 6 this standard was briefly introduced and the possibility of slightly superseded classifications was already underlined. However, this standard remains one of the few references for cable segregation specifications, first, proposing a classification of emission and susceptibility characteristics of cables and, second, defining separation distances taking into account also of the characteristics of cableways and shields - if present - . In the following paragraphs the Std. 518 classification is reported and commented for applicability to modern railway systems, and then the minimum spacings between cable categories are given, as a function of the shielding characteristics of the same cable-way, divided into trays and conduits; the separation distances are expressed in meters, starting from the Std. 518 values given in inches, assuming 1 inch = 2.5 cm. A 0 m distance means that no minimum distance is prescribed.

Level 1 - High Susceptibility. Analog signals of less than 50 V and digital signals of less than 15 V:

(1) Common returns to high-susceptibility equipment; in this case the return circuit, so the 0 V reference, is treated as a signal circuit;

(2) Control common tie (CCT); since the process control is performed nowadays by means of an extensive use of digital equipment, leaving analog devices only as actuators and transducers, control signals have now a more robust format;

(3) DC power supply buses feeding sensitive analog hardware; an example could be signalling equipment, but it is uncommon that such a DC bus is routed along for long distances, whilst it is usually local to an equipment or an apparatus;
(4) All wiring connected to components associated with sensitive analog hardware (for example, strain gauges, thermocouples, etc);
(5) Operational amplifier signals; this item and the following two are typical of old applications of '60s and '70s, where all the control functions were typically analogue and calculations were not performed digitally;
(6) Power amplifier signals;
(7) Output of isolation amplifiers feeding sensitive analog hardware;
(8) Phone circuits; analogue telephones have been recently more and more often replaced by digital telephones and their application is now limited to emergency call boxes that operate on an analogue connection, redundant with respect to digital networks;
(9) Logic buses feeding sensitive digital hardware; this item and the next one are in relationship with the connection of PLC units at the rack level or between racks, or the extensive use of digital field buses for the connection of intelligent devices and the automation and control equipment;
(10) All signal wires associated with digital hardware.

Level 2 - Medium Susceptibility. Analog signals greater than 50 V and switching circuits.
(1) Common returns to medium-susceptibility equipment;
(2) DC bus feeding digital relays, lights; and input buffers;
(3) All wiring connected to input signal conditioning buffers;
(4) Lights and relays operated by less than 50 V;
(5) Analog tachometer signals.

Level 3 - Low Susceptibility. Switching signals greater than 50 V, analog signals greater than 50 V, regulating signals of 50 V with currents less than 20 A, and ac feeders less than 20 A.
(1) Fused control bus 50-250 V dc;
(2) Indicating lights greater than 50 V;
(3) 50-250 V dc relay and contactor coils;
(4) Circuit breaker coils of less than 20 A;
(5) Machine fields of less than 20 A;
(6) Static master reference power source;
(7) Machine armature voltage feedback;
(8) Machine ground-detection circuits;
(9) Line-shunt signals for induction;
(10) All ac feeders of less than 20A;
(11) Convenience outlets, rear panel lighting;
(12) Recording meter chart drives;
(13) Thyristor field exciter ac power input and dc output of less than 20 A.

In general the signals belonging to this level are: signals driving contactors and relays coils, that are usually fed at 110 V dc or 115/230 V ac; signals for electric machinery monitoring and control; low power LV supply lines even featuring distorting loads, such as thyristor converters. At the time of the Std. 518 switching power supplies were not considered because they were not present on the market: even for rated current less than 20 A the conducted emissions may be relevant up to some MHz, causing a much larger induced voltage than a corresponding low frequency harmonic current; however, it must be remembered that such equipment are normally filtered in order to pass successfully the tests on conducted emissions, even if in some cases limits may be higher than expected, such as for UPS of category C3.

Level 4 - Power. AC and DC buses of 0-1000 V with currents of 20-800 A.

(1) Motor armature circuits;
(2) Generator armature circuits;
(3) Thyristor ac power input and dc outputs;
(4) Primaries and secondaries of transformers above 5 kVA;
(5) Thyristor field exciter ac power input and dc output with currents greater than 20 A;
(6) Static exciter (regulated and unregulated) ac power input and dc output;
(7) 250 V shop bus;
(8) Machine fields over 20 A.

It is evident that this level is not including all types of power supply lines and installations, in particular those installed at substations, so regarding both traction power generation and supply incoming at MV or HV levels.

Class Codes. Within a level, conditions may exist that require specific cables, and regrouping is not allowed. This condition may be identified by a class coding system similar to the following:

A, analog inputs, outputs;
B, pulse inputs;
C, contact and interrupt inputs;
D, decimal switch inputs;
E, output data lines;
F, display outputs, contact outputs;
G, logic input buffers;
S, special handling of special levels may require special spacing of conduits and trays, such as signals from commutating field and line resistors, or signals from line shunts to regulators, or power $>$ 1000 V or $>$ 800 A, or both;
U, high-voltage potential unfused greater than 600 V dc.

Tray-tray Spacing. Table 8.5 indicates the minimum distance in inches between the top of one tray and the bottom of the tray above, or between the sides of adjacent trays. This also applies to the distance between trays and to power equipment of less than 100 kVA.

Table 8.5 Tray-tray spacing as per IEEE Std. 518 (distances are in meters)

Level	1	2	3	3S	4	4S
1	0	See (2)	0.15	0.15	0.65	0.65
2	See (2)	0	0.15	0.15	0.45	0.65
3	0.15	0.15	0	0	See (1)	0.30
3S	0.15	0.15	0	0	0.20	0.45
4	0.65	0.45	See (9)	0.20	0	0
4S	0.65	0.65	0.30	0.45	0	0

Tray-Conduit Spacing. Table 8.6 indicates the minimum distance in inches between trays and conduits. This also applies to the distance between trays or conduits and power equipment of less than 100 kVA.

Table 8.6 Tray-conduit spacing as per IEEE Std. 518 (distances are in meters)

Level	1	2	3	3S	4	4S
1	0	0.025	0.10	0.10	0.45	0.45
2	0.025	0	0.10	0.10	0.30	0.65
3	0.10	0.10	0	0	0	0.20
3S	0.10	0.10	0	0	0.15	0.30
4	0.45	0.30	0	0.15	0	0
4S	0.45	0.30	0.20	0.30	0	0

Conduit-Conduit Spacing. Table 8.7 indicates the minimum distance in inches between the outside surfaces of conduits being run in banks. This also applies to the distance between conduits and power equipment of less than 100 kVA.

Table 8.7 Conduit-conduit spacing as per IEEE Std. 518 (distances are in meters)

Level	1	2	3	3S	4	4S
1	0	0.025	0.075	0.075	0.30	0.30
2	0.025	0	0.075	0.075	0.225	0.30
3	0.075	0.075	0	0	0	0.15
3S	0.075	0.075	0	0	0.15	0.225
4	0.30	0.225	0	0.15	0	0
4S	0.30	0.30	0.15	0.225	0	0

Notes on Tray and Conduit Separation. The most relevant notes of the standard are reported between quotes and commented for suitability and interpretation in case of railway applications.

(1) "Levels 3 and 4 may be run in a common tray, but should be separated by a barrier. Spacing should be level 4. (This barrier does not necessarily have to be grounded.)" This is what is normally done when power supply cables of different voltage levels (e.g. MV and LV cables) are laid down in the same cable trough; the indication that the barrier needs not to be grounded implies that it is required to shield magnetic field components, that are probably the most likely for near field induction of low frequency large current components flowing in the said cables.

(2) "When separate trays are impractical, levels 1 and 2 may be combined in a common tray, provided the levels are separated by a grounded steel barrier. This practice is not as effective as separation, and some rerouting at installation may be required. When levels 1 and 2 are run side by side in trays, a 1 inch (2.5 cm) minimum spacing is recommended." This is sometimes done for the less susceptible classes within level 1, such as well shielded telecommunication cables, with respect to other analog signals and low voltage relays, as for SCADA applications. The possibility that telecommunication cables are also double shielded (as for some Cat. 6 cables) and exhibit a large crosstalk attenuation may reduce the requirement of the steel barrier in some cases, even if it is always advisable for other more susceptible cable classes.

(3) "When unlike signal levels must cross either in trays or in conduits, they should cross at 90° angles at a maximum spacing. Where it is not possible to maintain spacing, a grounded steel barrier should be placed between unlike levels at the crossover point." This is normally done in industrial and on-board ship applications, where cable trays may move in three dimensions, so that crossings of two cableways running in parallel or joining from different routes are done by applying 45° turns in opposite directions to each of them; in railway applications this approach may be followed in the large electric substation environment or even in large equipment rooms, even if it is uncommon, but not for trackside cableways, where the geometry of the available space is clearly two-dimensional and all cableways run in the same direction.

(4) "Trays containing level 1 and 2 wiring should have solid bottoms. Ventilation slots or louvers may be used on other trays. Tray covers on level 1 and 2 trays must be used to provide complete shielding. Cover contact to side rails must be positive and continuous to avoid high-reluctance air gaps which impair shielding. Trays for all levels should be metal, solidly grounded with good ground continuity." The use of solid bottom cable trays was already considered in Figure 8.8, where the solid wall cable tray is the preferred solution. As far as the good electrical contact of the tray cover with the metallic sides, this is nowadays ensured by all good quality products, and only attention must be given to a correct installation and

that a wrong storage hasn't compromised the mating of the two parts. Then, grounding of cable trays is prescribed also by electrical safety rules and is usually provided by bonding and earthing straps and short conductors.

(5) "Trays and conduits containing levels 1, 2, and 3S should not be routed parallel to high-power equipment enclosures of 100 kVA and larger at a spacing of less than 5 ft (1.5 m) for trays and 2.5 ft (0.75 m) for conduit." This is approximately the distance that is usually kept in trackside cableways, providing a cable trough for MV cables and traction voltage feeders, or hanging them on the tunnel wall in some cases, and routing all the other signalling and communication cables in cable trays. By direct computation of the expected induced voltage under the most common situations (in terms of nominal current and harmonic distortion), it is however observed that these distances represent a conservative prescription.

(6) Where practical for levels 4 and 4S wiring, the complete power circuit between equipment should be routed in the same path, that is, the same tray or conduit. This practice will minimize the possibility of power and control encircling each other.

(7) "When entering terminal equipment and the spacings listed above are difficult to maintain, parallel runs should not exceed 5 ft (1.5 m) in the overall run." This of course follows the common sense that the crosstalk and the coupled disturbance is proportional to the length of parallelism and thus closer spacings are allowed for short cable runs; the 1.5 m prescription should be adequate for most equipments, but for larger apparatus this could be slightly adapted to contingency.

(8) "All spacing given in Table 8.6 assumes that the level 1 and 2 trays will not be left uncovered; otherwise Table 8.5 spacing must be used." While cable trays should be covered all the time, this notes leaves the impression that the tray-tray distance is prescribed for uncovered trays, while it is believed that the correct interpretation is that for the case of tray-conduit spacing, the distance reduction that could be gained by the improved shielding characteristics of the conduit is lost if the tray is left uncovered, and the tray-tray spacing must be used instead. It is however a little misleading note, since one then wanders what happens if also the trays in Table 8.5 are left uncovered.

(9) "Within the confines of pull boxes and junction boxes, levels should be kept separate, and grounded barriers should be used to assist level spacing." Again, as for the case of entering equipment and apparatus, spacing may be reduced; because of the reduced length of parallelism at pull and junction boxes, it is believed that an earthed barrier needs to be used only to separate cables particularly noisy or coming from harsh electromagnetic environments (such as the outdoor environment), where the chance of picking up high energy pulse and transients, due for example to lightning phenomena and to the commutation of inductive loads, is high.

(10) "Tray-to-conduit transition spacings and separations are potential sources of noise. Care should be taken to cross unlike levels at right angles and maintain required separations. Transition areas should be protected in accordance with the level recommendations." While we can interpret this note in a broad sense, thus encompassing low and high frequency phenomena, featuring all types of coupling, and in particular high frequency radiated electromagnetic field from slots and discontinuities in the metallic surfaces at transitions, the most relevant aspect at transitions is that the electrical continuity is ensured as for a continuous tray or conduit; geometrical factors (such as slots and misalignments and discontinuities) may be relevant in case of large near field magnetic components at a low to medium frequency range, because they modify the field distribution causing field gradients (see Figure 8.9).

8.7 Stray Current

Stray current is defined as currents flowing into electrolytic environments (i.e. soil and water) from electric circuits inadequately insulated from the environment. Apart from the rare cases such as those when stray currents are formed in the earth's external crust as a result of natural phenomena connected with solar activity (the so called telluric currents or geo-currents) stray currents are largely the effects of human activity. From a corrosion point of view, dc industrial equipment remains the most prominent source of stray currents for metallic structures. Any metallic structure, such a pipeline, buried in soil represents a low resistance current path and is therefore fundamentally vulnerable to the effects of stray currents. In a dc electrified railway, leakage of stray currents take places from the running rails, which are an element of the return circuit for traction currents. Current pulled by the Rail Vehicle returned to the Traction Power Substation (TPS) is returned to the TPS through the current return circuit – running rails and return conductors (if one is installed). Part of the return current leaks out of the running rail and returns to the TPS through the ground and any underground metallic structure in close proximity to the railway. The magnitude of the leakage from the running rails depends largely on the voltage drop in the rails (as a function current flowing in the rails and the resistance rails) and resistance of the rails with respect to ground. For typical dc electrified railway, it is generally thought that 95 % of the return current returns to the TPS through the running rail (the intended return path), while the remaining 5% flow to ground. Each electrical discontinuity of the return circuit increases the leakage of stray current (even up to 50% of the return current) and their instantaneous intensity in the ground can result in stray current amplitudes of up to hundreds of Amperes. Traction stray currents are dynamic in character, their amplitude and direction of flow continuously changes, depending on the traction load, and position of the Rail Vehicle in relation to the TPS.

Dynamic potential changes of underground structures and their random periods of anodic (corrosion) and cathodic polarization (lack of corrosion) are a consequence of this.

8.7.1 Stray Current Corrosion

Corrosion processes occur on external surfaces of metallic underground structures, which due to the character of the environment, are referred to as earth corrosion. Stray currents flow in the field in the direction of negative pole of the current source (TPS); such current may flow through nearby metallic structures, such as underground pipelines and cables. The point at which the current flows out of the metallic structure to the electrolytic will experience corrosion – this type of corrosion is referred to as electrolytic corrosion. Electrolytic corrosion is the most severe form of corrosion because the metallic structure is forced to become an anode and the amount of current will be translated directly into metal loss. The corrosion occurs at the point where the current discharges. Through the current circuit, stray current will cause severe corrosion on buried metallic structure, as well as on facilities belonging to the transit system itself. At the point where the stray current leaves the metallic structures, an anodic corrosion reaction takes place, resulting in oxidation (dissolution) of the metal. At the point where the stray current enters the metallic structure, the cathodic part of the corrosion reaction occurs. In the anodic area, the reaction is iron dissolution: $2Fe \to 2Fe^{2+} + 4e^-$; for the cathodic part the reaction is oxygen evolution: $O_2 + 2H_2O + 4e^- \to 4OH$.

The amplitude of stray current that is generated by a rail transit system depends largely on the resistance of the rail to earth. Insulated rail fasteners, ungrounded traction power substations and clean, well-maintained ballasts are used to control stray current at the source

8.7.2 Stray Current Protection Design

The control of the hazard of stray currents can be achieved either by:

- preventive actions implemented at the current source – this is mainly by limitation of leakage of current to ground;
- application of electrochemical protection to metallic structures at risk;
- sectioning of long structures – this is mainly by insulating short sections from each other;
- the application of coating to relevant metallic structures.

8.7.2.1 Control of Stray Current Sources

In general, the following actions are undertaken as part of the prophylactic measures to limit the leakage of current from the return current circuit:

- welding of rails, which decrease the resistance to the flow of return current;
- application of insulation between the rails and the sleepers (or viaduct) – increases the rail to ground resistance;
- maintenance of a 3 cm (minimum) separation distance between the flanges of the rails and the ballast of the track – this increases the rail to ground resistance;
- installation of rail bonds this decrease the resistance of rail network and maintains an equal potential between rails.

It is not unusual for a number of the above mentioned measures to be implemented as part of the design and construction of the railway to control stray current.

8.7.2.2 Cathodic Protection System

There are three basic types of corrosion control:

- the sacrificial anode system;
- the impressed current system; and
- the traction current drainage system.

The Figures 8.12 and 8.13 illustrate impressed current and sacrificial systems respectively. Both of these systems introduce direct current into the soil, a component of which can flow through nearby structures and cause corrosion.

Figure 8.14 illustrates a drainage system as a deliberate conducting path for the return of stray current to the railway line in a way that will

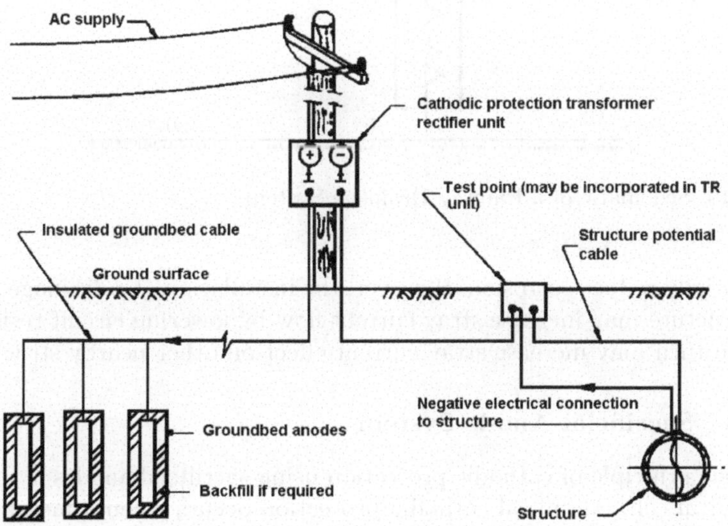

Fig. 8.12 Schematic of a Cathodic Protection with Impressed Current System

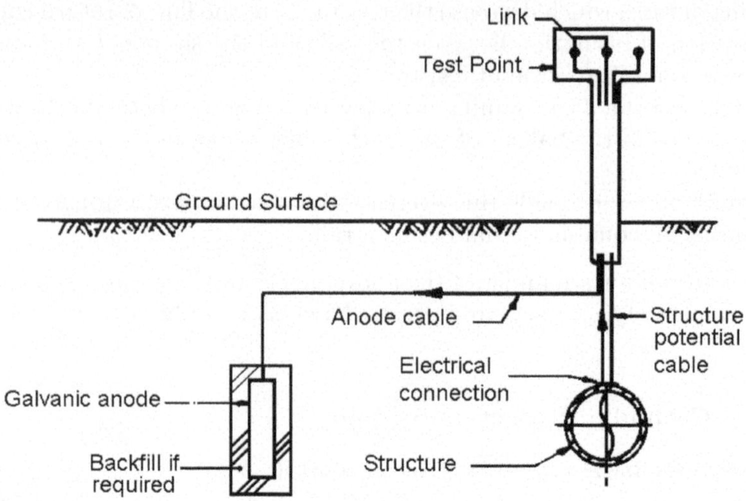

Fig. 8.13 Schematic of a Cathodic Protection with Galvanic Anode

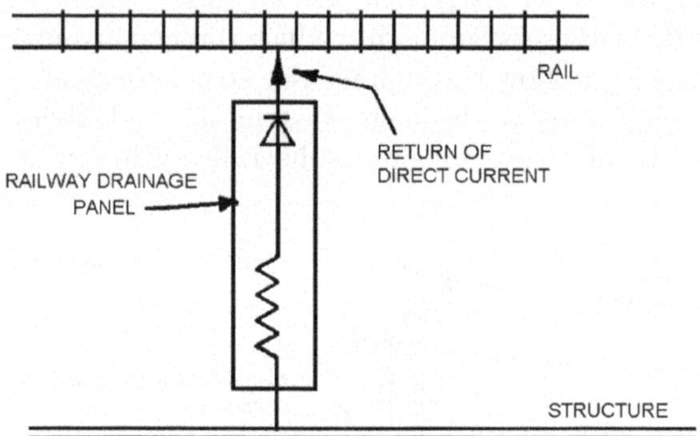

Fig. 8.14 Schematic of a Railway Drainage System

substantially reduce corrosion. However, the installation of a drainage system on a structure may increase stray current flow by lowering circuit resistance, which in turn may increase stray current effect on other nearby structures.

8.7.2.3 Sacrificial Anode System

The basic principle of cathodic protection using sacrificial anodes is the electrochemical cell. Sacrificial cathodic protection occurs when a metal is coupled to a more reactive (anodic) metal. This connection is referred to as a galvanic couple. High energy (potential) electrons are forced to flow from the anode to the structure to be protected, in order to effectively transfer corrosion from the metal structure, the anode material must have a large enough natural voltage difference to produce an electrical current flow. The structure-to-electrolyte potentials required for protection are identical to those for impressed current cathodic protection systems. The high potential electrons are generated through the corrosion of an active metal such as magnesium or zinc (In this type of system, the anode material is consumed, or sacrificed in the process, and the anodes must be periodically replaced in order to obtain continued protection. In order to minimize periodic anode replacement, sufficient anode material is normally provided so that the anode replacement interval is a desired number of years. Common practice for buried systems is to design the system for a 10- to 15-year anode life. For submerged systems, or for buried systems where anode replacement is difficult, longer (20- to 30-year) anode life is often used as a design criteria

An advantage of sacrificial anode systems is the flexibility in application. Anodes can be installed in a variety of applications and configurations. No outside power is required for cathodic protection to be effective. Another advantage is the minimal maintenance required for these systems to function.

Disadvantages of sacrificial anode systems include the limited protection current available and limited life. Sacrificial anodes are subject to rapid corrosion (consumption) and require replacement on a regular basis. Typical design life of a pipeline system anode is five to ten years.

8.7.2.4 Impressed Current System

An impressed current cathodic protection involves the application of an external dc through long lasting anodes. For effectiveness, the impressed current anodes must be designed for long life at high current output. This requires the selection of materials with very low corrosion rates; the typical design life for a impressed current anode is twenty years. Anodes are normally installed in grouped configurations in the electrolyte; these groupings (both horizontal and vertical) in an underground application are known as ground-beds. The ground-beds are connected to the power by a positive cable to the power source whilst a negative cable connects the power source to the structure (pipeline). Advantages of impressed current systems include the unlimited current opportunities and longer life. Impressed current systems are typically installed where the structure to be protected is large, requiring higher levels of current. Disadvantages include the requirement for an outside power source and higher maintenance requirements. Outside power might come from sources such as commercial ac converted to dc through a rectifier, thermoelectric generator, or solar panels. A significantly higher monitoring and maintenance effort is required by comparison to sacrificial anode systems.

8.7.2.5 Traction Current Drainage System

A drainage system is a deliberate conducting path to return stray current to the railway line in a way that will substantially reduce corrosion. However, the installation of a drainage system on a structure may increase stray current flow by lowering circuit resistance which in turn may increase stray current effect on other nearby structures.

8.7.3 Stray Current Detection

The best means of assessing stray current interference is to determine the change in current density on the buried pipeline. It is not an easy task, due to the complexity and uncertainty of stray current. The potential of the pipeline soil is a fluctuating parameter as a function of current flows. Therefore, it is possible to estimate the magnitude of the stray current via the changes of the pipeline to soil potential. The potential can be measured through the use of a reference electrode, e.g. a saturated copper-copper-sulphate electrode ($Cu/CuSO_4$) placed in the soil near the pipeline. When electrically influenced by stray current, the potential of the buried metallic structure (i.e. pipeline) shifts in the positive or in the negative direction where the current leaves or enters the metal surface. The pipe potential value and the magnitude of its variation are most frequently proposed as the electrolytic corrosion hazard criterion in standards and regulations of many countries. The positive potential shifts of the affected structure and the potential gradient of soil is to be considered in the presence of stray current.

8.7.4 Stray Current Prediction

Numerous methods exist for the prediction of stray current due to an electrified railway. These methods can broadly be divided into two analytical approaches: the field approach and the electric circuit approach. The field approach can be used to model the electrochemical polarization that occurs on the buried metallic structure [501]. In the electric circuit approach, based on the earth-return circuit theory, which is more suitable for modeling extended structures (pipelines, cables, etc.), the polarization phenomenon is not taken into account. A much more detailed categorisation of the methods applied is given below:

1. direct discretisation into circuit model;
2. resistor network model;
3. cylinder layer model;
4. 3- or 2- D modeling.

An advantage of the direct discretisation method is the ability to derive a simple model, however the characterisation of the circuit elements based on the key parameters of the railway being modeled can be quite problematic.

The resistor network model is equally relatively simple, however the accuracy depends on the application of appropriate boundary conditions, furthermore the model becomes complex for scenarios where multi return paths exist. the next two models are basically field models in either 2 or 3 dimensions. The cylinder layer model can be linked to 3-D models, as implemented by Hill and Brillante in [308]; such models lends themselves to large structures but suffer when multi-layer soil, multi tracks with signalling assets are present. 3 dimensional models exist however they suffer from the effort required to set up such model, as well as the discrepancy between transversal and longitudinal distance of interest in a railway.

8.7.5 Pipeline Assessment Criteria

Methods and criteria for the assessment for stray current corrosion are defined in a number of international standards such as [63, 64, 413, 39, 98]. Clause 6 of the EN 50162 [98] specifies criteria for structures without (Clause 6.1.1) and with cathodic protection (Clause 6.1.2). Clause 6.1.2 states that a pipeline is deemed to be exposed to an unacceptable stray current if the IR free potential is outside the protective potential range specified in EN 12954. The protection potential of most common metals are listed in Table 1 of EN 12954. Generally, corrosion potential and corrosion current density are used to characterize the active dissolution ability of materials, while passivation current density and passivation potential are used to characterize the passivation ability of materials[538]. Without knowing the type of metal the pipeline is constructed off, it is difficult for the railway designer to determine the applicable criteria from Table 1 of EN 12954 and thus relies on the utility owner providing the relevant data; furthermore the protection criteria and thus the safe corrosion current limit varies from application to application.

While a standardised method exist for the assessment of stray current danger, usually by measuring the positive deviations ΔV of the total (including the IR component) potential V of a pipe from its steady-state potential V_{st}, i.e., $\Delta V = V - V_{st}$ in a stray current field, there appears to be no common permissible limit with respect to the positive value of $\triangle V$. In practice, the values of V_{st} and V are usually measured with respect to a copper-sulfate reference electrode placed on the ground, with V_{st} measured in the absence of railway traffic (for example at night) and V measured during peak traffic. From a theoretical view point, it is possible for the railway engineer to estimate the permissible level of the anodic deviations of the potential of the stray current field, however to do so, as is shown below, requires a consideration of the dependence of $\triangle V$ on the size and shape of the through defects in the insulating coating of the pipe, the coating thickness, soil resistivity etc which are not necessarily available to the railway engineer and thus the application of this theoretical approach requires some simplification. The through defects in the insulating coating is generally referred to as coating holiday and this implies a damaged or missing coating on a pipeline that exposes the pipeline metal.

For a pipe with an insulation thickness t, it is assumed that the axis of a through defect channel in the insulation is normal to the metal surface, and its cross-section is circular having a diameter d. The anodic shift $\triangle V$, of the potential, V, is the sum of the polarization $\triangle E$, and the ohmic potential drop:

$$\triangle V = \triangle E + \triangle V_{\Omega} = \triangle E + IR \quad (8.7)$$

where R is the sum of the ohmic resistance of soil in the defect channel and the ohmic resistance to the current flowing from the channel mouth. For a soil of resistivity, ρ,

$$R = \rho \left[(2d)^{-1} + \left(\frac{t}{0.785d^2} \right) \right] \quad (8.8)$$

Then,

$$\triangle V = \triangle E + J\rho\,(0.39d + t) \quad (8.9)$$

where J, is the stray current density. The anodic curve of active steel usually has a Tafel slope $\beta_a \approx 40 \pm 10\,mV$, whereas the cathodic range can be approximated by an exponent with a quasi Tafel slope $\beta_c \approx 350 - 800\,mV$[410]. Usually, in the presence of stray current, the $\triangle E$ values are small compared with $\triangle V_{\Omega}$; however, irrespective of this, the limiting stray current density value taken below is so small that the following Stern-Geary equation[645] for linear polarization is fulfilled,

$$J = I_{corr} \triangle E \times 2.3 \frac{(\beta_a + \beta_c)}{\beta_a \beta_c} = 2.3 I_{corr} B \triangle E \quad (8.10)$$

where I_{corr} is the free corrosion rate and $\triangle V = V - V_{corr}$. Hence, (8.10) can therefore be written as:

$$J = \frac{\triangle V}{[r_p + \rho\,(0.39d + t)]} \quad (8.11)$$

The maximum yearly average corrosion rate (mm/year) due to a permissible stray current is given as:

$$K_{max} = \frac{2.43 \times 10^{-2} \triangle V_{max} \gamma T_{int}}{[r_p + \rho\,(0.39d + t)]} \quad (8.12)$$

where r_p, is the polarization resistance and is determined from the slopes of the potential-current plots measured by the linear polarization curve in the range of ± 10 mV about the open-circuit potential, γ is the stray current asymmetry coefficient and T_{int} is the duration of peak operation of the LRT, with unit of hrs/day. It is therefore possible to express $\triangle V_{max}$ in the form:

$$\triangle V_{max} = \frac{\left(5.2B^{-1} + 10.25K_{corr}\rho d_{ef}\right)}{\gamma T_{int}} \tag{8.13}$$

where $K_{corr}\rho d_{ef}$ is the spread resistance (otherwise referred to as the leakage resistance in CEN/TS 15280[60]) related to a circular coating defect (as shown in Fig. 8.15) with area A and diameter d embedded in soil with a soil resistivity ρ. The spread resistance plays a very important role in pipeline corrosion and cathodic protection aspects. Along with the dc current density, it determines the level the IR drop measured in dc potentials. It equally determines the level of ac current exchanged between the soil and the pipe through the coating holiday at a given pipe-to-soil ac voltage.

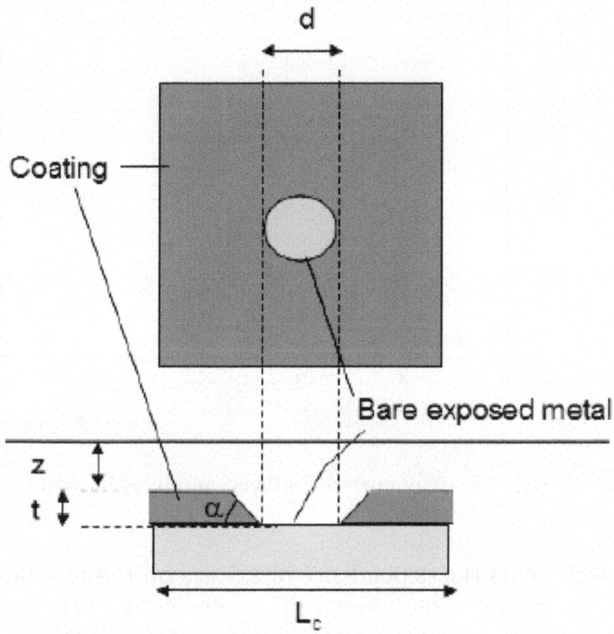

Fig. 8.15 Schematic of a pipe showing coating defect

The total spread resistance R_s of a coating holiday as shown in Fig. 8.15 is given by the following[584]:

$$R_s = K_{corr}\rho d_{ef} = \frac{R_{s1}R_{s2}}{R_{s1} + R_{s2}} \tag{8.14}$$

For a circular defect:

$$R_{s1} = \rho_{soil}\frac{1}{2d}\left(1 - \frac{d}{5L_c}\right) + \rho_{pore}\frac{4t}{\pi d^2} \tag{8.15}$$

$$R_{s2} = \left[\rho_{soil}\frac{1}{2\pi t}\left(1 - \frac{d}{5L_c}\right) + \rho_{pore}\frac{2}{\pi d}\right]\frac{1}{1 - \left(\frac{\alpha}{90}\right)^{3/2}} \tag{8.16}$$

Equation 8.13 can therefore be used to estimate the maximum safe stray current induced anodic shift of the pipe potential calculated for the most dangerous defects, as shown in Fig. 8.16.

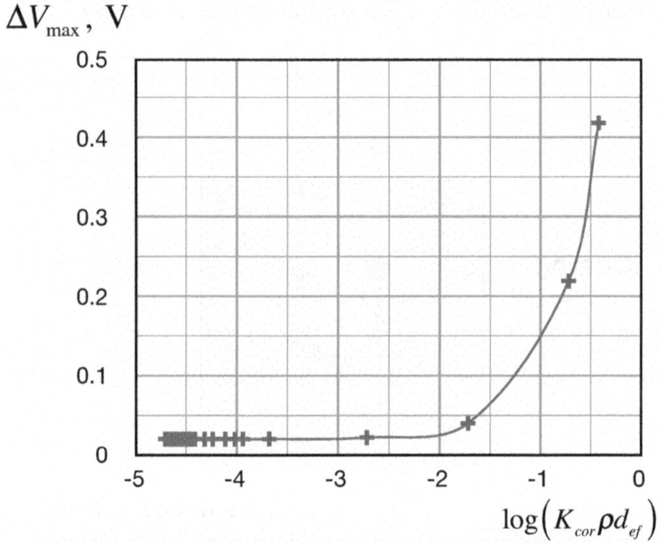

Fig. 8.16 Estimated safe stray current induced anodic shift of pipe potential.

Fig. 8.16 represents the dependence of $\triangle V_{max}$ on the product of $K_{corr}d_{ef}$ varied deliberately over wide limits. In the calculation of the above, the pipe is assumed to be a low pressure gas pipeline having a wall thickness of approximately $t = 4.5\,mm$, and a specified operational life of 40 years. Depending on $K_{corr}\rho d_{ef}$ for a given T_{int}, the tolerable safe value of $\triangle V_{max}$ can vary in wide limits, that is from $\approx 20 - 930\,mV$ and the values $\triangle V_{max} \leq 20\,mV$ may be considered as safe in practice under given conditions.

References

[1] Akagi, H.: Large Static Converters for Industry and Utility Applications. Proceedings of the IEEE 89(6), 976–983 (2001)

[2] ALSTOM, Type B vital relays - Operation and maintenance manual, p. 1457 (2010)

[3] ALSTOM, Solid-State Vane Relay - Operation and maintenance manual, p. 2522 (2009)

[4] Ametani, A.: A highly efficient method for calculating transmission line transients. IEEE Transactions on Power Apparatus and Systems 95(5), 1545–1551 (1976)

[5] Ametani, A., Fuse, I.: Approximate method for calculating the impedances of multiconductor with cross sections of arbitrary shapes. Electrical Engineering Review of Japan 111(2), 117–123 (1992)

[6] Ametani, A., et al.: Wave propagation characteristics of iron conductors in an intelligent building. Transactions of IEE Japan 120-B(1), 31–37 (2000)

[7] Angelidis, G., Semlyen, A.: Direct Phase-domain calculation of transmission line transients using two-sided recursions. IEEE Transactions on Power Delivery 10(2), 941–949 (1995)

[8] Ansaldo Signal, TWT - Train-Wayside transmission system (undated)

[9] Antonini, G., Ferri, G.: A new approach for closed form transient analysis of multiconductor transmission lines. IEEE Transactions on Electromagnetic Compatibility 46(4) (November 2004)

[10] Apollonskii, S.M.: Estimation of the electromagnetic environment on objects of the railway electrified on direct current. In: Procedings of the IEEE EUROCON, St. Petersburg, Russia, May 18-23, pp. 1545–1555 (2009)

[11] APTA SS-E-010-98, Standard for the development of an electromagnetic compatibility plan, 2nd edn. APTA PRESS Task Force, APTA, pp. 1–8 (2006)

[12] APTA SS-E-005-98, Standard for Grounding and Bonding, 2nd edn. APTA PRESS Task Force (1998)

[13] AREMA C&S Manual 11.4.5, Recommended Instructions for Protecting Electronic Equipment Modules and Sensitive Components from Electrostatic Discharge (ESD) for Field Locations (2008)

[14] AREMA Communication and Signals Manual 11.5.1, Recommended Environmental Requirements for Electrical and Electronic Railroad Signal System Equipment (2009)

[15] AREMA Communication and Signals Manual 11.5.2, Recommended Electromagnetic Interference Immunity and Emissions Test Methods for Signaling Equipment (2009)
[16] AREMA Communication and Signals Manual 20.1.4, Recommended Principles and Practices for Inductive Coordination of Railway Electrical Supply Facilities and the Commercial Communication Facilities (2007)
[17] Armstrong, D.S.: EMC aspects of electrified railways at high frequencies. In: IEE Colloquium on EMC in Large Systems, pp. 7/1–7/4 (1994)
[18] Armstrong, K.: Why EMC Immunity Testing is Inadequate for Functional Safety. In: IEEE International Symposium on EMC, Santa Clara, CA, USA, vol. 1, pp. 145–149 (2004)
[19] Armstrong, K.: EMC for the functional safety of automobiles why EMC testing is insufficient, and what is necessary. In: IEEE International Symposium on Electromagnetic Compatibility, EMC 2008, pp. 1–6 (2008)
[20] Arrillaga, J., Watson, N.R.: Power System Harmonics, 2nd edn. John Wiley & Sons (2003)
[21] Audone, B., Amisano, F.: A new approach to immunity testing. In: International Symposium on Electromagnetic Compatibility, EMC, Europe (2008)
[22] Azar, R., et al.: Advanced SPICE modeling of large power IGBT modules. IEEE Transactions on Industry Applications 40(3), 710–716 (2004)
[23] Babinet, M.: Memories d'optique metrologique. Comptes Rendus de l'Academie de Science 4, 638–648 (1837)
[24] Baghzouz, Y., et al.: Time-varying harmonics: Part II – Harmonics summation and propagation. IEEE Transactions on Power Delivery 17(1), 279–285 (2002)
[25] Balanis, C.A.: Antenna Theory - Analysis and Design, 2nd edn. John Wiley & Sons (1997)
[26] Batista, E., Dienot, J.M.: EMC Characterization for Switching Noise Investigation on Power Transistors. In: Proceedings of the IEEE International Symposium on Electromagnetic Compatibility, pp. 1–7 (2008)
[27] Batista, E., Dienot, J.M., Mermet-Guyennet, M., Castellazzi, A., Ciappa, M., Fichtner, W.: Accurate Mixed Electrical and Electromagnetic Model of a 6.5 kV IGBT Module. In: Proceedings of the 7th IEEE Internatonal Conference on Power Electronics (ICPE), pp. 1151–1155 (2007)
[28] Bedford, T., Cooke, R.: Probabilistic Risk Analysis – Foundations and Methods. Cambridge Press (2001)
[29] Ben Dhia, S., Ndoye, A.C., Boyer, A., Guillot, L., Vrignon, B.: IC emission spectrum drifts after burn-in cycles. In: Asia-Pacific Symposium on Electromagnetic Compatibility APEMC 2008 and 19th International Zurich Symposium on Electromagnetic Compatibility, pp. 255–258 (2008)
[30] Berleze, S.L.M., Robert, R.: Skin and Proximity Effects in Nonmagnetic Conductors. IEEE Transanctions on Education 46(3), 368–372 (2003)
[31] Bethe, H.A.: Theory of diffraction by small holes. Physical Review 66(7-8), 163–182 (1944)
[32] Bhargava, B.: Benefits of a Low Frequency, Low Voltage Railway Electrification System. In: Proc. of the 1996 ASME/IEEE Joint Railroad Conference, pp. 177–184 (1996)
[33] Blanchard, B.S.: System Engineering Management. John Wiley & Sons, New York (1998)

[34] Bolljahn, J.T.: Antennas near conducting sheets of finite size. University of California (1950)
[35] Booker, H.G.: Slot aerial and their relation to complementary wire aerials (Babinet's Principle). Journal IEE 93(pt. III-A), 620–626 (1946)
[36] Booker, H.G.: Slot Aerials and Their Relation to Complementary Wire Aerials. Journal of the Institution of Electrical Engineers - Part IIIA: Radiolocation 93(4), 620–626 (1946)
[37] Bouwkamp, C.J.: Diffraction theory. Rep. Prog. Phys. 17, 35–100 (1954)
[38] Brillante, S., Mariscotti, A., Pozzobon, P.: Calculation of Electromagnetic Emissions around a Railway Vehicle by TLM-2D. In: International Conference on Modern Electric Traction, Gdansk, Poland, May 31-June 3, pp. pp. 7.1–7.6 (2001)
[39] British Standard BS 7361 Part 1, Cathodic protection – Part 1: Code of practice for land and marine applications – (formerly CP 1021) (1991)
[40] British Standard BS 5760 Part 2, Reliability of systems, equipment and components. Guide to the assessment of reliability, BSI (1994)
[41] Brown, J.C., Allan, J., Mellit, B.: Calculation and measurement of rail impedances applicable to remote short circuit fault currents. IEE Proc. Part B 139(4) (July 1992)
[42] Burke, J.J., Engel, A.P., Gilligan, S.R., Mincer, N.A.: Increasing the power system capacity of the 50 kV Black Mesa and Lake Powell railroad through harmonic filtering and series compensation. IEEE Transactions on Power Apparatus and Systems 98(4), 1268–1274 (1979)
[43] Busatto, G., Abbate, C., Iannuzzo, F., Fratelli, L., Cascone, B., Giannini, G.: EMI Characterisation of high power IGBT modules for Traction Application. In: Proceedings of the 36th IEEE Power Electronics Specialists Conference (PESC), pp. 2180–2186 (2005)
[44] Camp, M., Gerth, H., Garbe, H., Haase, H.: Predicting the breakdown behavior of microcontrollers under EMP/UWB impact using a statistical analysis. IEEE Transactions on Electromagnetic Compatibility 46(3), 368–379 (2004)
[45] Carpenter, D.C., Hill, R..J.: A continuous-function ground conductivity model for the determination of electric railway earth conductance. IEEE Trans. Geosci. Remote Sens. 31(5), 953–960 (1993)
[46] Carson, J.R.: Wave propagation in overhead wires with ground return. Bell Systems Technical Journal 5, 539–554 (1926)
[47] CEI 9-20-1, Guida d'applicazione della normativa di sicurezza per gli impianti fissi di trazione a corrente continua in presenza di strutture metalliche od in cemento armato – Parte 1: Sistemi con tensione nominale di linea sino a 1500 V (Italian, Application guide for the safety standardisation in railway d.c. fixed installations facing metallic structures or reinforced concrete – Part 1: Systems with line nominal voltage up to 1500 V) (March 2002)
[48] CEI 9-34, Effetti delle interferenze elettromagnetiche provocate da linee di trazione ferroviarie elettrificate in corrente alternata su tubazioni (Italian, Effects of electromagnetic interference from ac railway traction lines on pipes) (October 1997)
[49] CEI 11-1, Impianti elettrici con tensione superiore a 1 kV in corrente alternata (Italian, Power installations exceeding 1 kV a.c.) (January 1999)
[50] CEI 103-1/12, Impianti telefonici interni – Parte 12: Protezione degli impianti telefonici interni (Italian, Private telephone exchanges – Part 12: Protection of the private telephone exchanges) (June 2001)

[51] CEI 103-6, Protezione delle linee di telecomunicazione dagli effetti dell'induzione elettromagnetica provocata dalle linee elettriche vicine in caso di guasto (Italian, Protection of telecommunication lines against harmful effects produced by electrical power lines in fault conditions) (December 1997)
[52] CEI 103-10, Protezione delle linee di telecomunicazioni dagli effetti dell'induzione elettromagnetica provocata da linee ferroviarie elettrificate in corrente alternata (Italian, Protection of telecommunication lines against harmful effects produced by electrified a.c. railway lines) (December 1998)
[53] CEI 304-1, Interferenza elettromagnetica prodotta da linee elettriche su tubazioni metalliche – Identificazione dei rischi e limiti di interferenza (Italian, Electromagnetic interference produced by electric power lines on metallic pipelines – Identification of hazards and interference limits) (November 2005)
[54] CEI 304-2, Prove di immunità condotta alle frequenze fondamentali della rete elettrica su apparecchiature di telecomunicazione (Italian, Conducted immunity test method and level at fundamental frequencies for telecommunication equipment) (November 2005)
[55] CEI 306-2, Guida per il cablaggio per telecomunicazioni e distribuzione multimediale negli edifici residenziali (Italian, Cabling criteria for telecommunications and multimedia signals distribution in residential buildings) (January 2003)
[56] CEI 306-10, Sistemi di cablaggio strutturato – Guida alla realizzazione e alle Norme tecniche (Italian, Generic cabling systems – Guide to the implementation and the standards) (September 2006)
[57] CEI 306-11, Tecnologia dell'informazione – Guida al cablaggio degli access point wireless (Italian, Information technology – Customer premises cabling for wireless access points) (September 2006)
[58] Cella, R., Giangaspero, G., Mariscotti, A., Montepagano, A., Pozzobon, P., Ruscelli, M., Vanti, M.: Measurement of AT Electric Railway System currents and validation of a Multiconductor Transmission Line model. IEEE Transactions on Power Delivery 21(3), 1721–1726 (2006)
[59] Celozzi, S., Araneo, R., Lovat, G.: Electromagnetic Shielding. John Wiley & Sons (2008)
[60] CEN/TS 15280, Evaluation of AC corrosion likelihood of buried pipelines – application to cathodically protected pipelines (2000)
[61] CENELEC EN 12015, Electromagnetic compatibility – Product family standard for lifts, escalators and moving walks – Emission (2004)
[62] CENELEC EN 12016, Electromagnetic compatibility – Product family standard for lifts, escalators and moving walks – Immunity (2004)
[63] CENELEC EN 12954, Cathodic protection of buried or immersed metallic structures – General principles and application for pipelines, pp. 1–34 (2001)
[64] CENELEC EN 13509, Cathodic protection measurement techniques, pp. 1–42 (2003)
[65] CENELEC EN 13146-5, Railway applications – Track – Test methods for fastening systems – Part 5: Determination of electrical resistance (November 2002)
[66] CENELEC prEN 13509, Cathodic protection measurement techniques (September 2002)
[67] CENELEC EN 50081-1, Electromagnetic compatibility – Generic emission standard – Residential, commercial and light industry (1992)

[68] CENELEC EN 50081-2, Electromagnetic compatibility – Generic emission standard – Industrial environment (1993)
[69] CENELEC EN 50082-1, Electromagnetic compatibility – Generic immunity standard – Residential, commercial and light industry (1992)
[70] CENELEC EN 50082-2, Electromagnetic compatibility – Generic immunity standard – Industrial environment (1995)
[71] CENELEC EN 50083-8, Cable networks for television signals, sound signals and interactive services – Part 8: Electromagnetic compatibility for networks) (May 2002)
[72] CENELEC EN CLC/TR 50083-10-1, Cable networks for television signals, sound signals and interactive services – Part 10-1: Guidelines for the implementation of return paths in cable networks (February 2009)
[73] CENELEC EN 50121-1 (IEC 62236-1), Railway applications – Electromagnetic compatibility – Part 1: General (September 2000)
[74] CENELEC EN 50121-1 (IEC 62236-1), Railway applications – Electromagnetic compatibility – Part 1: General (July 2006)
[75] CENELEC EN 50121-2 (IEC 62236-2), Railway applications – Electromagnetic compatibility – Part 2: Emission of the whole railway system to the outside world (September 2000)
[76] CENELEC EN 50121-2 (IEC 62236-2), Railway applications – Electromagnetic compatibility – Part 2: Emission of the whole railway system to the outside world (July 2006)
[77] CENELEC EN 50121-3-1 (IEC 62236-3-1), Railway applications – Electromagnetic compatibility – Rolling stock – Part 3.1: Train and complete vehicle (September 2000)
[78] CENELEC EN 50121-3-1 (IEC 62236-3-1), Railway applications – Electromagnetic compatibility – Rolling stock – Part 3.1: Train and complete vehicle (July 2006)
[79] CENELEC EN 50121-3-2 (IEC 62236-3-2), Railway applications – Electromagnetic compatibility – Rolling stock – Part 3.2: Rolling Stock Apparatus (September 2000)
[80] CENELEC EN 50121-3-2 (IEC 62236-3-1), Railway applications – Electromagnetic compatibility – Rolling stock – Part 3.2: Rolling Stock Apparatus (July 2006)
[81] CENELEC EN 50121-4 (IEC 62236-4), Railway applications – Electromagnetic compatibility – Part 4: Emission and immunity of the Signalling and Telecommunications Apparatus (July 2000)
[82] CENELEC EN 50121-4 (IEC 62236-4), Railway applications – Electromagnetic compatibility – Part 4: Emission and immunity of the Signalling and Telecommunications Apparatus (July 2006)
[83] CENELEC EN 50121-5 (IEC 62236-5), Railway applications – Electromagnetic compatibility – Part 5: Emission and immunity of railway fixed power supply installations (September 2000)
[84] CENELEC EN 50121-5 (IEC 62236-5), Railway applications – Electromagnetic compatibility – Part 5: Emission and immunity of railway fixed power supply installations (July 2006)
[85] CENELEC EN 50122-1, Railway applications – Fixed installations – Part 1: Protective provisions relating to electrical safety and earthing (June 1997)
[86] CENELEC EN 50122-2, Railway applications – Fixed installations – Part 2: Protective provisions against the effects of stray currents caused by d.c. traction systems (April 1998)

[87] CENELEC EN 50124-2, Railway application – Insulation coordination – Part 2: Overvoltages and related protection (March 2001)
[88] CENELEC EN 50125-1, Railway applications – Environmental conditions for equipment – Part 1: Equipment on board of rolling stock (August 1999)
[89] CENELEC EN 50125-2, Railway applications – Environmental conditions for equipment – Part 2: Fixed electrical installations (December 2002)
[90] CENELEC EN 50125-3, Railway applications – Environmental conditions for equipment – Part 3: Equipment for signalling and telecommunications (January 2003)
[91] CENELEC EN 50130-4, Alarm systems – Part 4: Electromagnetic compatibility – Product family standard: Immunity requirements for components of fire, intruder and social alarm system (December 1995)
[92] CENELEC EN 50155, Railway applications – Electronic equipment used on rolling stock (August 2001)
[93] CENELEC EN 50155, Railway applications – Electronic equipment used on rolling stock (July 2007)
[94] CENELEC EN 50159-1, Railway applications – Communication, signalling and processing systems – Safety-related communication in closed transmission systems (March 2001)
[95] CENELEC EN 50159-2, Railway applications – Communication, signalling and processing systems – Safety-related communication in open transmission systems (March 2001)
[96] CENELEC prEN 50160, Voltage characteristics of electricity supplied by public distribution networks (2006)
[97] CENELEC EN 50160, Voltage characteristics of electricity supplied by public distribution networks (July 2010)
[98] CENELEC EN 50162, Protection against corrosion by stray current from direct current systems (August 2004)
[99] CENELEC EN 50163, Railway applications – Supply voltages of traction systems (November 2004)
[100] CENELEC EN 50163/A1, Railway applications – Supply voltages of traction systems (July 2007)
[101] CENELEC EN 50164-1, Lightning Protection Components (LPC) – Part 1: Requirements for connection components (August 2008)
[102] CENELEC EN 50164-2, Lightning protection components (LPC) – Part 2: Requirements for conductors and earth electrodes (August 2008)
[103] CENELEC EN 50164-3, Lightning Protection Components (LPC) – Part 3: Requirements for isolating spark gaps (August 2006)
[104] CENELEC EN 50173-1, Information technology – Generic cabling systems – Part 1: General requirements (May 2007)
[105] CENELEC EN 50173-3, Information technology – Generic cabling systems – Part 3: Industrial premises (September 2007)
[106] CENELEC EN 50174-2, Information technology – Cabling installation – Part 2: Installation planning and practices inside buildings (August 2000)
[107] CENELEC EN 50174-3, Information technology – Cabling installation – Part 3: Installation planning and practices outside buildings (November 2003)
[108] CENELEC EN 50206-1, Railway applications – Rolling stock – Pantographs: Characteristics and tests Part 1: Pantographs for main line vehicles (March 1998)

[109] CENELEC EN 50206-2, Railway applications – Rolling stock – Pantographs: Characteristics and tests Part 2: Pantographs for metros and light rail vehicles (February 1999)
[110] CENELEC EN 50215, Railway application – Testing of rolling stock after completion of construction and before entry into service (April 1999)
[111] CENELEC EN 50215, Railway applications – Rolling stock – Testing of rolling stock on completion of construction and before entry into service (October 2009)
[112] CENELEC EN 50238, Railway applications – Compatibility between rolling stock and train detection systems (February 2003)
[113] CENELEC TR 50238-2 (Project PR16697), Compatibility between Rolling Stock and Track Circuits (April 2006)
[114] CENELEC FprEN 50238-2, Railway applications – Compatibility between rolling stock and train detection systems – Part 2: Compatibility with Track Circuits (June 2008)
[115] CENELEC CLC/TS 50238-2, Railway applications – Compatibility between rolling stock and train detection systems – Part 2: Compatibility with track circuits (July 2010)
[116] CENELEC FprEN 50238-3, Railway applications – Compatibility between rolling stock and train detection systems – Part 3: Compatibility with Axle Counters (June 2008)
[117] CENELEC EN 50239, Railway applications. Radio remote control system of traction vehicle for freight traffic (February 2000)
[118] CENELEC EN 50263, Electromagnetic compatibility (EMC). Product standard for measuring relays and protection equipment (May 2000)
[119] CENELEC EN 50289-1-6, Communication cables – Specifications for test methods – Part 1.6: Electrical test methods – Electromagnetic performance (March 2002)
[120] CENELEC EN 50289-1-8, Communication cables – Specifications for test methods – Part 1.8: Electrical test methods – Attenuation (June 2001)
[121] CENELEC EN 50310, Application of equipotential bonding and earthing in buildings with information technology equipment (March 2006)
[122] CENELEC EN 50317, Railway applications – Current collection systems – Requirements for and validation of measurements of the dynamic interaction between pantograph and overhead contact line (July 2002)
[123] CENELEC EN 50367, Railway applications – Current collection systems – Technical criteria for the interaction between pantograph and overhead line (to achieve free access) (April 2006)
[124] CENELEC EN 50388, Railway applications – Power supply and rolling stock – Technical criteria for the coordination between power supply (substation) and rolling stock to achieve interoperability (August 2005)
[125] CENELEC EN 50463, Railway applications. Energy measurement on board trains (January 2008)
[126] CENELEC CLC/TS 50467, Railway applications – Rolling stock – Electrical connectors, requirements and test methods (April 2008)
[127] CENELEC EN 50468, Resistibility requirements to over-voltages and overcurrents due to lightning for equipment having telecommunication ports (April 2009)
[128] CENELEC EN 50500, Measurement procedures of magnetic field levels generated by electronic and electrical apparatus in the railway environment with respect to human exposure (August 2008)

[129] CENELEC CLC/TS 50502, Railway applications – Rolling stock – Electric equipment in trolley buses - Safety requirements and connection systems (July 2008)
[130] CENELEC CLC/TR 50507, Railway applications – Interference limits of existing track circuits used on European railways (May 2007)
[131] CENELEC CLC/TS 50511, Railway applications – Communications, signalling and processing systems – ERTMS/ETCS - External signalling for lines equipped with ERTMS/ETCS Level 2 (August 2007)
[132] CENELEC EN 55011 (CISPR 11), Industrial, scientific and medical (ISM) radio-frequency equipment – Radio disturbance characteristics – Limits and methods of measurement (May 1998)
[133] CENELEC EN 55011 (CISPR 11), Industrial, scientific and medical (ISM) radio-frequency equipment – Radio disturbance characteristics – Limits and methods of measurement (March 2007)
[134] CENELEC EN 55016-1-1 (CISPR 16-1-1), Specification for radio disturbance and immunity measuring apparatus and methods – Part 1-1: Radio disturbance and immunity measuring apparatus – Measuring apparatus (February 2008)
[135] CENELEC EN 55016-1-2 (CISPR 16-1-2), Specification for radio disturbance and immunity measuring apparatus and methods – Part 1-2: Radio disturbance and immunity measuring apparatus – Ancillary equipment – Conducted disturbances (October 2004)
[136] CENELEC EN 55016-1-3 (CISPR 16-1-3), Specification for radio disturbance and immunity measuring apparatus and methods – Part 1-3: Radio disturbance and immunity measuring apparatus – Ancillary equipment – Disturbance power (June 2004)
[137] CENELEC EN 55016-1-4 (CISPR 16-1-4), Specification for radio disturbance and immunity measuring apparatus and methods – Part 1-4: Radio disturbance and immunity measuring apparatus – Ancillary equipment – Radiated disturbance (February 2007)
[138] CENELEC EN 55016-1-5 (CISPR 16-1-5), Specification for radio disturbance and immunity measuring apparatus and methods – Part 1-5: Radio disturbance and immunity measuring apparatus – Antenna calibration test sites for 30 MHz to 1 000 MHz (November 2003)
[139] CENELEC EN 55016-2-1 (CISPR 16-2-1), Specification for radio disturbance and immunity measuring apparatus and methods – Part 2-1: Methods of measurement of disturbances and immunity – Conducted disturbance measurements (October 2008)
[140] CENELEC EN 55016-2-2 (CISPR 16-2-2), Specification for radio disturbance and immunity measuring apparatus and methods – Part 2-2: Methods of measurement of disturbances and immunity – Measurement of disturbance power (November 2003)
[141] CENELEC EN 55016-2-3 (CISPR 16-2-3), Specification for radio disturbance and immunity measuring apparatus and methods – Part 2-3: Methods of measurement of disturbances and immunity – Radiated disturbance measurements (July 2006)
[142] CENELEC EN 55016-2-4 (CISPR 16-2-4), Specification for radio disturbance and immunity measuring apparatus and methods – Part 2-4: Methods of measurement of disturbances and immunity – Immunity measurements (November 2003)

[143] CENELEC EN 55016-4-2 (CISPR 16-4-2), Specification for radio disturbance and immunity measuring apparatus and methods – Part 4-2: Uncertainties, statistics and limit modelling – Uncertainty in EMC measurements (November 2003)
[144] CENELEC EN 55022 (CISPR 22), Information technology equipment – Radio disturbance characteristics – Limits and methods of measurement (September 1998)
[145] CENELEC EN 55022 (CISPR 22), Information technology equipment – Radio disturbance characteristics – Limits and methods of measurement (October 2007)
[146] CENELEC EN 55024 (CISPR 24), Information technology equipment – Immunity characteristics – Limits and methods of measurement (September 1998)
[147] CENELEC EN 55024/A1, Information technology equipment – Immunity characteristics – Limits and methods of measurement (October 2001)
[148] CENELEC EN 55024/A2, Information technology equipment – Immunity characteristics – Limits and methods of measurement (January 2003)
[149] CENELEC EN 55024 (CISPR 24), Information technology equipment – Immunity characteristics – Limits and methods of measurement (November 2010)
[150] CENELEC EN 55103-1, Electromagnetic compatibility – Product family standard for audio, video, audio-visual and entertainment lighting control apparatus for professional use Part 1: Emission (November 1996)
[151] CENELEC EN 55103-2, Electromagnetic compatibility – Product family standard for audio, video, audio-visual and entertainment lighting control apparatus for professional use Part 2: Immunity (November 1996)
[152] CENELEC EN/IEC 60146-1-1, Semiconductor convertors – General requirements and line commutated convertors – Part 1-1: Specifications of basic requirements (February 1993)
[153] CENELEC EN/IEC 60255-22-1, Measuring relays and protection equipment - Part 22-1: Electrical disturbance tests - 1 MHz burst immunity tests, 3rd edn. (October 2007)
[154] CENELEC EN/IEC 60255-22-2, Measuring relays and protection equipment - Part 22-2: Electrical disturbance tests - Electrostatic discharge tests, 3rd edn. (April 2008)
[155] CENELEC EN/IEC 60255-22-3, Measuring relays and protection equipment - Part 22-3: Electrical disturbance tests - Radiated electromagnetic field immunity, 3rd edn. (July 2007)
[156] CENELEC EN/IEC 60255-22-4, Measuring relays and protection equipment - Part 22-4: Electrical disturbance tests - Electrical fast transient/burst immunity test, 3rd edn. (April 2008)
[157] CENELEC EN/IEC 60255-22-5, Measuring relays and protection equipment - Part 22-5: Electrical disturbance tests - Surge immunity test, 2nd edn. (November 2008)
[158] CENELEC EN/IEC 60255-22-6, Electrical relays - Part 22-6: Electrical disturbance tests for measuring relays and protection equipment - Immunity to conducted disturbances induced by radio frequency fields, 1st edn. (April 2001)

[159] CENELEC EN/IEC 60255-22-7, Electrical relays - Part 22-7: Electrical disturbance tests for measuring relays and protection equipment - Power frequency immunity tests, 1st edn. (April 2003)
[160] CENELEC EN/IEC 60255-22-8, Electrical relays - Part 8: Thermal electrical relays, 2nd edn. (October 1990)
[161] CENELEC EN/IEC 60349-2, Railway applications – Rotating electrical machines for rail and road vehicles – Part 2: Electronic converter-fed alternating current motors (October 2001)
[162] CENELEC EN/IEC 60694, Common specifications for high-voltage switchgear and controlgear standards (January 2002)
[163] IEC 60812, Analysis techniques for system reliability. Procedure for failure mode and effects analysis (FMEA) (2006)
[164] CENELEC EN/IEC 60870-2-1, Telecontrol equipment and systems – Part 2: Operating conditions – Section 1: Power supply and electromagnetic compatibility (January 1996)
[165] CENELEC EN/IEC 60898, Circuit-breakers for overcurrent protection for household and similar installations, various amendments between 1991 and 1998
[166] CENELEC EN/IEC 60909-0, Short-circuit currents in three-phase a.c. systems – Part 0: Calculation of currents (2001)
[167] CENELEC EN/IEC 60950-1, Information technology equipment – Safety Part 1: General requirements (April 2006)
[168] CENELEC EN/IEC 61000-2-2, Electromagnetic compatibility – Part 2.2: Environment – Compatibility levels for low-frequency conducted disturbances and signalling in public low-voltage power supply systems (June 2002)
[169] CENELEC EN/IEC 61000-2-4, Electromagnetic compatibility – Part 2.4: Environment – Compatibility levels in industrial plant for low-frequency conducted disturbances (September 2002)
[170] CENELEC EN/IEC 61000-2-12, Electromagnetic compatibility – Part 2.12: Environment – Compatibility levels for low-frequency conducted disturbances and signalling in public medium-voltage power supply systems (July 2003)
[171] CENELEC EN/IEC 61000-3-4, Electromagnetic compatibility (EMC) – Part 3-4: Limits – Limitation of emission of harmonic currents in low-voltage power supply systems for equipment with rated current greater than 16 A (October 1998)
[172] CENELEC EN/IEC 61000-3-6, Electromagnetic compatibility (EMC) – Part 3-6: Limits – Assessment of emission limits for distorting loads in MV and HV power systems (February 2008)
[173] CENELEC EN/IEC 61000-3-12, Electromagnetic compatibility (EMC) – Part 3-12: Limits – Limits for harmonic currents produced by equipment connected to public low-voltage systems with input current >16 A and <=75 A per phase (November 2004)
[174] CENELEC EN/IEC 61000-4-2, Electromagnetic compatibility – Part 4: Testing and measurement techniques – Section 2: Electrostatic discharge immunity test (March 2009)
[175] CENELEC EN/IEC 61000-4-3, Electromagnetic compatibility (EMC) - Part 4-3: Testing and measurement techniques – Radiated, radio-frequency, electromagnetic field immunity test (May 2006)
[176] CENELEC EN/IEC 61000-4-4, Electromagnetic compatibility – Part 4: Testing and measurement techniques – Section 4: Electrical fast transient/burst immunity test (December 2004)

[177] CENELEC EN/IEC 61000-4-5, Electromagnetic compatibility – Part 4: Testing and measurement techniques – Section 5: Surge immunity test (November 2006)

[178] CENELEC EN/IEC 61000-4-6, Electromagnetic compatibility (EMC) - Part 4-6: Testing and measurement techniques - Immunity to conducted disturbances, induced by radio-frequency fields (August 2007)

[179] CENELEC IEC/EN 61000-4-8, Electromagnetic compatibility – Part 4: Testing and measurement techniques – Section 8: Power frequency magnetic field immunity test (September 1993)

[180] CENELEC EN/IEC 61000-4-9, Electromagnetic compatibility – Part 4: Testing and measurement techniques – Section 9: Pulse magnetic field immunity test – Basic EMC Publication (September 1993)

[181] CENELEC EN/IEC 61000-4-10, Electromagnetic compatibility – Part 4: Testing and measurement techniques – Section 10: Damped oscillatory magnetic field immunity test – Basic EMC Publication (September 1993)

[182] CENELEC EN/IEC 61000-4-11, Electromagnetic compatibility – Part 4: Testing and measurement techniques – Section 11: Voltage dips, short interruptions and voltage variations immunity tests (August 2004)

[183] CENELEC EN/IEC 61000-4-12, Electromagnetic compatibility – Part 4: Testing and measurement techniques – Section 12: Ring wave immunity tests (December 2006)

[184] CENELEC EN/IEC 61000-4-13, Electromagnetic compatibility – Part 4: Testing and measurement techniques – Section 13: Harmonics and interharmonics including mains signalling at a.c. power port, low frequency immunity tests (June 2002)

[185] CENELEC EN/IEC 61000-4-14, Electromagnetic compatibility – Part 4: Testing and measurement techniques – Section 14: Voltage fluctuations immunity test (April 1999)

[186] CENELEC EN/IEC 61000-4-15, Electromagnetic compatibility – Part 4: Testing and measurement techniques – Section 15: Flickermeter – Functional and design specifications, 2nd edn. 2010 (April 1998)

[187] CENELEC EN/IEC 61000-4-16, Electromagnetic compatibility – Part 4: Testing and measurement techniques – Section 16: Test for immunity to conducted, common mode disturbances in the frequency range 0 Hz to 150 kHz (October 1998)

[188] CENELEC EN/IEC 61000-4-17, Electromagnetic compatibility – Part 4: Testing and measurement techniques – Section 17: Ripple on d.c. input power port immunity test (August 1999)

[189] CENELEC EN/IEC 61000-4-18, Electromagnetic compatibility – Part 4: Testing and measurement techniques –Section 18: Oscillatory wave immunity test (September 2007)

[190] CENELEC EN/IEC 61000-4-27, Electromagnetic compatibility – Part 4: Testing and measurement techniques – Section 27: Unbalance immunity test (November 2000)

[191] CENELEC EN/IEC 61000-4-28, Electromagnetic compatibility – Part 4: Testing and measurement techniques – Section 28: Variation of power frequency, immunity test (March 2000)

[192] CENELEC EN/IEC 61000-4-29, Electromagnetic compatibility – Part 4: Testing and measurement techniques – Section 29: Voltage dips, short interruptions and voltage variations on d.c. input power ports, immunity tests (November 2000)
[193] CENELEC EN/IEC 61000-4-34, Electromagnetic compatibility – Part 4: Testing and measurement techniques – Section 34: Voltage dips, short interruptions and voltage variations immunity tests for equipment with input current more than 16 A per phase (May 2007)
[194] CENELEC EN/IEC 61000-6-1, Electromagnetic compatibility (EMC) – Part 6-1: Generic standards – Immunity for residential, commercial and light-industrial environments (January 2007)
[195] CENELEC EN/IEC 61000-6-2, Electromagnetic compatibility (EMC) – Part 6-2: Generic standards – Immunity for industrial environments (November 2005)
[196] CENELEC EN/IEC 61000-6-3, Electromagnetic compatibility (EMC) – Part 6-3: Generic standards – Emission for residential, commercial and light-industrial environments (January 2007)
[197] CENELEC EN/IEC 61000-6-4, Electromagnetic compatibility (EMC) – Part 6-4: Generic standards – Emission for industrial environments (January 2007)
[198] CENELEC EN/IEC 61008-1, Residual current operated circuit-breakers without integral overcurrent protection for household and similar uses (RCCBs) – Part 1: General rules (September 2004)
[199] CENELEC EN/IEC 61009-1, Residual current operated circuit-breakers with integral overcurrent protection for household and similar uses (RCCOs) – Part 1: General rules (July 2006)
[200] CENELEC EN/IEC 61204-3, Low voltage power supplies, d.c. output – Part 3: Electromagnetic compatibility (EMC) (December 2000)
[201] CENELEC EN/IEC 61158-2, Fieldbus standard for use in industrial control systems – Part 2: Physical layer specification and service definition (May 2003)
[202] CENELEC EN/IEC 61543, Residual current-operated protective devices (RCDs) for household and similar use – Electromagnetic compatibility (July 1995)
[203] CENELEC EN/IEC 61557-4, Electrical safety in low voltage distribution systems up to 1000 V a.c. and 1 500 V d.c. – Equipment for testing, measuring or monitoring of protective measures – Part 4: Resistance of earth connection and equipotential bonding (March 2007)
[204] CENELEC EN/IEC 61557-5, Electrical safety in low voltage distribution systems up to 1000 V a.c. and 1 500 V d.c. – Equipment for testing, measuring or monitoring of protective measures – Part 5: Resistance to earth (March 2007)
[205] CENELEC EN/IEC 61918, Industrial communication networks – Installation of communication networks in industrial premises (May 2008)
[206] CENELEC EN/IEC 61935-1, Specification for the testing of balanced and coaxial information technology cabling – Part 1: Installed balanced cabling as specified in the standards series EN 50173 (December 2009)
[207] CENELEC EN/IEC 61935-3, Testing of balanced and coaxial information technology cabling – Part 3: Installed cabling as specified in EN 50173-4 and related standards (December 2009)

[208] CENELEC EN 62040-2, Uninterruptible power systems (UPS) – Part 2: Electromagnetic compatibility (EMC) requirements (March 2006)

[209] CENELEC EN 62040-3, Uninterruptible power systems (UPS) – Part 3: Method of specifying the performance and test requirements (June 2001)

[210] CENELEC EN 62305-1, Protection against lightning – Part 1: General principles (February 2006)

[211] CENELEC EN 62305-2, Protection against lightning – Part 2: Risk management (February 2006)

[212] CENELEC EN 62305-3, Protection against lightning – Part 3: Physical damage to structures and life hazards (February 2006)

[213] CENELEC EN 62305-4, Protection against lightning – Part 4: Electrical and electronic systems within structures (February 2006)

[214] CENELEC HD 472 S1, Nominal voltages for low voltage public electricity supply systems (1989)

[215] Chang, D.C., Wait, J.R.: Extremely low frequency (ELF) propagation along a horizontal wire located above or buried in the earth. IEEE Transactions on Communications 22(4), 421–427 (1974)

[216] Chen, S.-L., Li, R.-J., Hsi, P.-H.: Traction System Unbalance Problem - Analysis Methodologies. IEEE Transactions on Power Delivery 19(4), 1877–1883 (2004)

[217] Chen, T.-H.: Simplified Models of Electric Railway Substations for Three-phase Power-Flow Studies. In: Proc. of the 1994 IEEE Industry Applications Society Annual Meeting, Denver, CO, USA, October 2-6, pp. 2245–2248 (1994)

[218] Cheng, D.K.: Field and wave electromagnetic, 2nd edn. Addison-Wesley Publishing Company (1991)

[219] Chibante, R.F.M., Araújo, A.L.S., da Silva Carvalho, A.: A new physics based SPICE sub-circuit model for insulated gate bipolar transistors (IGBTs). In: Proceedings of the 10th EPE Conference, Toulouse, France (2003)

[220] Chibante, R., Araújo, A., Carvalho, A.: A new physics based SPICE model for NPT IGBTs 2, 1156–1161 (2003)

[221] Chipman, R.A.: Theory and problems of transmission lines. McGraw Hill, New York (1968)

[222] Chu, M., Ho, R.: EMC management methodology and lifecycle adopted by the West Rail project in Hong Kong. In: IEEE International Symposium on Electromagnetic Compatibility, EMC 2002, vol. 2, pp. 791–794 (2002)

[223] Ciurlo, S., Mariscotti, A., Viacava, A.: A Helmholtz coil for high frequency high field intensity applications. Metrology and Measurement Systems XVI(1), M-0254 (2009)

[224] Ciurlo, S., Mariscotti, A.: Track compensation with Audiofrequency Track Circuits. In: Proc. of IEEE Conference on Electrical Systems for Aircraft, Railway and Ship Propulsion, ESARS 2010, Bologna, Italy, October 21-23, pp. 1–6 (2010)

[225] Clark, M.R., Gillespie, R., Kemp, T., McCann, D.M., Forde, M.C.: Electromagnetic properties of railway ballast. NDT&E International 34(5), 305–311 (2001)

[226] Clark, M.R., Gordon, M.O., Giannopoulos, A., Forde, M.C.: Experimental and Computer Modelling of GPR to Characterize Trackbed Ballast. In: Proceedings of the World Congress on Railway Research, Scotland, 16 pages (2003)

[227] Cole, A.H., et al.: Development of power supply and supervisory equipment for British Railways 50 Hz ac electrification. Proceedings of IEE 119, 189–199 (1972)

[228] Coles, P.C., Fracchia, M., Hill, R.J., Pozzobon, P., Szelag, A.: Identification of catenary harmonics in 3kV DC railway traction systems. In: Proceedings of the Mediterranean Electrotechnical Conference, MELECON 1994, pp. 825–828 (1994)

[229] Costamagna, E., Gamba, P., Pellegrini, P.F.: Railway cab signalling by spread spectrum modulations and compatibility with narrow band track circuits. In: Proceedings of the 8th Mediterranean Electrotechnical Conference, MELECON 1996, vol. 3, pp. 1666–1669 (1996)

[230] Costamagna, E., Favalli, L., Gamba, P., Savazzi, P.: Compatibility of railway cab signalling by track circuits and spread spectrum modulations with four-quadrant controller AC drives. In: Proceedings of the IEEE 49th Vehicular Technology Conference, vol. 2, pp. 1182–1186 (1999)

[231] Cozza, A., Démoulin, B.: On the Modeling of Electric Railway Lines for the Assessment of Infrastructure Impact in Radiated Emission Tests of Rolling Stock. IEEE Transactions on Electromagnetic Compatibility 50(3), 566–576 (2008)

[232] D'Addio, G., Ferrari, P., Mariscotti, A., Pozzobon, P.: Integrated modelling of audiofrequency track circuits. In: IEE EMC York 1999, York, UK, July 12-13 (1999)

[233] D'Addio, G., Fracchia, M., Mariscotti, A., Pozzobon, P.: Sensitivity analysis of railway line impedance to variations of electrical and geometrical parameters. In: Proc. of the World Congress on Railway Research, Tokyo, Japan, October 19-23 (1999)

[234] D'Amore, M., Sarto, M.S.: Simulation models of a dissipative transmission line above a lossy ground for a wide-frequency range – Part I: Single conductor configuration. IEEE Transactions on Electromagnetic Compatibility 38(2), 127–138 (1996)

[235] D'Amore, M., Sarto, M.S.: Simulation models of a dissipative transmission line above a lossy ground for a wide-frequency range – Part II: Multi conductor configuration. IEEE Transactions on Electromagnetic Compatibility 38(2), 139–149 (1996)

[236] Dan, S., Yougang, G., Yuanmao, S.: Determination of shielding effectiveness of multilayer shield by making use of transmission line theory. In: 7th International Symposium on Electromagnetic Compatibility and Electromagnetic Ecology, pp. 95–97 (2007)

[237] Davidson, W.: Radiated emissions from contact wire. In: IEE Colloquium on EMC in Electric Traction and Signalling, pp. 3/1– 3/6

[238] de Arizon, P., Dommel, H.W.: Computation of cable impedances based on sub-division of conductors. IEEE Transactions on Power Delivery 2(1), 21–27 (1987)

[239] DEF STAN 59-41, Part 7, Issue 1, Electromagnetic Compatibility - Part 7: Code of practice for HM ships - Installation Guidelines (November 10, 1995)

[240] de Leon, F., Semlyen, A.: Reduced order model for transformer transients. IEEE Transactions on Power Delivery 7(1), 361–369 (1992)

[241] de Leon, F., Semlyen, A.: Reduced order model for transformer transients. IEEE Transactions on Power Delivery 9(1), 231–239 (1994)

[242] Depenbrock, M.: Direct Self-Control (DSC) of Inverter-Fed Induction Machine. IEEE Transactions on Power Electronics 3(4), 420–429 (1988)

[243] Dickson, R.: Orex upgrade targets more capacity. In: Railway Gazzette International, June 3 (2007)

[244] Di Rienzo, L., Zhang, Z., Pignari, S.A.: Boundary-Element Computation of Per-Unit-Length Series Parameters of Railway Lines. IEEE Transactions on Electromagnetic Compatibility 51(3), 825–832 (2009)

[245] Dolara, A., Leva, S.: Calculation of Rail Internal Impedance by Using Finite Elements Methods and ComplexMagnetic Permeability. International Journal of Vehicular Technology 2009, Article ID 505246

[246] Dolara, A., Gualdoni, M., Leva, S.: Impact of High Voltage Primary Supply Lines in the 2×25 kV – 50 Hz Railway System on the Equivalent Impedance at Pantograph Terminals. IEEE Transactions on Power Delivery 27(1), 164–175 (2012)

[247] Dounavis, A., Ramachandra, A., Nikhla, M.: Efficient passive circuit models for distributed networks with frequency-dependent parameters. IEEE Transactions on Advanced Packaging 3(3) (August 2000)

[248] Dubanton, C.: Calcul approché des paramètres primaires et secondaires d'une ligne de transport. EDF Bulletin de la Direction des Études et Recherches (1), 53–62 (1969)

[249] Dwight, H.B.: Skin effect in tubular and flat conductors. Trans. A.I.E.E., 1379–1403 (1918)

[250] Dwight, H.B.: Proximity effect in wires and thin tubes. Trans. A.I.E.E., 850–859 (June 1923)

[251] Ehreler, U., Ketteler, K.H.: Interference current at 3 kV DC traction application. In: Proc. of EPE 1997, Trondheim, Norway, September 8-10, pp. 439–446 (1997)

[252] Electronic Communications Committee (ECC) within the European Conference of Postal and Telecommunications Administrations (CEPT), ECC REPORT 98 - Studying the Compatibility Issues of the UIC Euroloop System with Other Systems in the Frequency Band 9.5 to 17.5 MHz, Bern (February 2007)

[253] Elfadel, I., Huang, H.M., Ruehli, A., Dounavis, A., Nakhla, M.: A comparative study of two transient analysis algorithms for lossy transmission lines with frequency dependent data. IEEE Transactions on Advanced Packaging 25(2) (May 2002)

[254] ETSI EN 300 253, Environmental Engineering (EE); Earthing and bonding configuration inside telecommunications centres (V2.1.0) (December 2001)

[255] ETSI EG 201 147, Equipment Engineering (EE); Interworking between Direct Current/Isolated (DC/I) and Direct Current/Common (DC/C) electrical power systems (V1.1.2) (February 1998)

[256] ETSI R0BT-002/EG 201 212, Electrical safety; Classification of interfaces for equipment to be connected to telecommunication networks (V1.2.1) (November 1998)

[257] ETSI EN 300 220-1, Electromagnetic compatibility and Radio spectrum Matters (ERM); Short Range Devices (SRD); Radio equipment to be used in the 25 MHz to 1 000 MHz frequency range with power levels ranging up to 500 mW; Part 1: Technical characteristics and test methods (V2.3.1) (February 2010)

[258] ETSI EN 300 330-1, Electromagnetic compatibility and Radio spectrum Matters (ERM); Short Range Devices (SRD); Radio equipment in the frequency range 9 kHz to 25 MHz and inductive loop systems in the frequency range 9 kHz to 30 MHz – Part 1: Technical characteristics and test methods (V1.7.1) (February 2010)

[259] ETSI EN 300 440-1, Electromagnetic compatibility and Radio spectrum Matters (ERM); Short Range Devices (SRD); Radio equipment to be used in the 1 GHz to 40 GHz frequency range – Part 1: Technical characteristics and test methods (V1.6.1) (August 2010)

[260] ETSI EN 300 328-1, Electromagnetic compatibility and Radio Spectrum Matters (ERM); Wideband Transmission systems; Data transmission equipment operating in the 2.4 GHz ISM band and using spread spectrum modulation techniques – Part 1: Technical characteristics and test conditions (V1.7.1.) (May 2006)

[261] ETSI EN 300 652, Broadband Radio Access Networks (BRAN); HIgh PErformance Radio Local Area Network (HIPERLAN) Type 1; Functional specification (V1.2.1) (July 1998)

[262] ETSI EN 300 683, Electromagnetic compatibility and Radio spectrum Matters (ERM); ElectroMagnetic Compatibility (EMC) standard for Short Range Devices (SRD) operating on frequencies between 9 kHz and 25 GHz (V1.2.1) (July 1998)

[263] ETSI EN 301 893, Broadband Radio Access Networks (BRAN); HIPERLAN Type 2; Harmonized EN covering essential requirements of article 3.2 of the R&TTE Directive (V1.5.1) (December 2008)

[264] ETSI EN 301 489-1, Electromagnetic compatibility and Radio spectrum Matters (ERM); ElectroMagnetic Compatibility (EMC) standard for radio equipment and services – Part 1: Common technical requirements (V1.8.1) (April 2008)

[265] ETSI EN 301 489-3, Electromagnetic compatibility and Radio spectrum Matters (ERM); ElectroMagnetic Compatibility (EMC) standard for radio equipment and services – Part 3: Specific conditions for Short-Range Devices (SRD) operating on frequencies between 9 kHz and 40 GHz (V1.4.1) (August 2002)

[266] ETSI EN 301 489-7, Electromagnetic compatibility and Radio spectrum Matters (ERM); ElectroMagnetic Compatibility (EMC) standard for radio equipment and services – Part 7: Specific conditions for mobile and portable radio and ancillary equipment of digital cellular radio telecommunications systems (GSM and DCS) (V1.3.1) (November 2005)

[267] ETSI EN 301 489-17, Electromagnetic compatibility and Radio spectrum Matters (ERM); ElectroMagnetic Compatibility (EMC) standard for radio equipment and services – Part 17: Specific conditions for 2.4 GHz wideband transmission systems, 5 GHz high performance RLAN equipment and 5.8 GHz Broadband Data Transmitting Systems (V2.1.1) (May 2009)

[268] ETSI EN 301 489-18, Electromagnetic compatibility and Radio spectrum Matters (ERM); ElectroMagnetic Compatibility (EMC) standard for radio equipment and services – Part 18: Specific conditions for Terrestrial Trunked Radio (TETRA) equipment (V1.3.1) (August 2002)

[269] ETSI EN 301 489-19, Electromagnetic compatibility and Radio spectrum Matters (ERM); ElectroMagnetic Compatibility (EMC) standard for radio equipment and services – Part 19: Specific conditions for Receive Only Mobile Earth Stations (ROMES) operating in the 1.5 GHz band providing data communications (V1.2.1) (November 2002)
[270] ETSI EN 301 489-33, Electromagnetic compatibility and Radio spectrum Matters (ERM); ElectroMagnetic Compatibility (EMC) standard for radio equipment and services – Part 33: Specific conditions for Ultra Wide Band (UWB) communications devices (V1.1.1) (February 2009)
[271] ETSI ETR 108, European digital cellular telecommunications system (Phase 2); GSM Electro Magnetic Compatibility (EMC) considerations (GSM 05.90) (August 1993)
[272] ETSI ETR 108 A1, European digital cellular telecommunications system (Phase 2); GSM Electro Magnetic Compatibility (EMC) considerations (GSM 05.90) (August 1993)
[273] ETSI ETR 357, Digital cellular telecommunications system; GSM Electro Magnetic Compatibility (EMC) considerations (GSM 05.90 version 5.0.0) (January 1997)
[274] ETSI ETS 300 573, Digital cellular telecommunications system (Phase 2); Physical layer on the radio path General description (GSM 05.01) (May 1996)
[275] ETSI ETS 300 577, Digital cellular telecommunications system (Phase 2); Radio transmission and reception (GSM 05.05) (March 1997)
[276] ETSI ETS 300 910, Digital cellular telecommunications system (Phase 2+); Radio transmission and reception (GSM 05.05) (March 1997)
[277] European Union Directive 89/336/EC, On the approximation of laws of the Member States relating to electromagnetic compatibility (May 03,1989)
[278] European Union Directive 2004/104/EC, Adapting to technical progress Council Directive 72/245/EEC relating to the radio interference (electromagnetic compatibility) of vehicles and amending Directive 70/156/EEC on the approximation of the laws of the Member States relating to the type-approval of motor vehicles and their trailers (October 31, 1995)
[279] European Union Directive 2004/104/EC, Adapting to technical progress Council Directive 72/245/EEC on the approximation of the laws of the Member States relating to the suppression of radio interference produced by spark-ignition engines fitted to motor vehicles and amending Directive 70/156/EEC on the approximation of the laws of the Member States relating to the type-approval of motor vehicles and their trailers (November 13, 2004)
[280] European Union Directive 2004/108/EC, On the approximation of the laws of the Member States relating to electromagnetic compatibility and repealing Directive 89/336/EEC (December 05, 2004)
[281] Falvo, M.C., Fedeli, E., Stellin, M., Vannini, E.: Signalling system interferences due to railway electric substation failures on Italian conventional railway networks. In: International Symposium on Electronics, Electrical Drives, Automation and Motion, SPEEDAM 2006, pp. 1111–1114 (2006)
[282] European Radiocommunication Committee ERC/DEC/(96)04, "ERC Decision of 7 March 1996 on the frequency bands for the introduction of the Trans European Trunked Radio System (TETRA)" (December 1996)
[283] Federal Code of Regulation (CFR), Title 47, FCC Part 15 (47 CFR 15), Radio Frequency Devices

[284] Ferrari, P., Mariscotti, A., Pozzobon, P.: Reference curves of the pantograph impedance in DC railway systems. In: Proceedings of the International Symposium on Circuits and Systems, ISCAS 2000, Geneva, Switzerland, May 28-31, vol. 1, pp. 555–558 (2000)

[285] Filippone, F., Mariscotti, A., Pozzobon, P.: The Internal Impedance of Traction Rails for DC Railways in the 1-100 kHz Frequency Range. IEEE Transactions on Instrumentation and Measurement 55(5), 1616–1619 (2006)

[286] Frugier, D., Ladoux, P.: Voltage Disturbances on 25kV-50 Hz Railway Lines - Modelling Method and Analysis. In: International Symposium on Power Electronics, Electrical Drives, Automation and Motion, SPEEDAM 2010, Pisa, Italy, June 14-16, pp. 1080–1085 (2010)

[287] Gelman, V.: Thyristor rectifier for traction: practical at last. In: Proc. of the Commuter Rail/Transit Conference, Baltimore, MD, USA, June 9-13 (2002)

[288] Gerber, M., Drabek, E., Mueller, R.: Locomotive 2000 - Series 460 - of the Swiss federal Railways. Schweizer Eisenbahn Review, 3–47 (October 1991)

[289] Glazar, A.J.: A Software Implementation of TL Field-to-Cable Coupling Equations. IEEE Practical Paper, `http://www.ieee.org/organizations/pubs/newsletters/emcs/fall00/a_software.htm`

[290] Glover, J.D., Kusko, A., Peeran, S.M.: Train Voltage Analysis for AC Railroad Electrification. IEEE Trans. on Industry Applications 20(4), 925–934 (1984)

[291] Gustavsen, B., Semlyen, A.: Simulation of transmission line transients using vector fitting and modal decomposition. IEEE Transactions on Power Delivery 13(2), 605–614 (1998)

[292] Gustavsen, B., Semlyen, A.: Calculation of transmission line transients using polar decomposition. IEEE Transactions on Power Delivery 13(3), 855–862 (1998)

[293] Gustavsen, B., Sletback, J., Henriksen, T.: Calculation of electromagnetic transients in transmission cables and lines taking frequency dependent effects accurately into account. IEEE Transactions on Power Delivery 10(2), 1076–1084 (1995)

[294] Hallén, E.: On antenna impedances. Transactions of the Royal Institute of Technology (13) (1947)

[295] Hallén, E.: Admittance diagrams for antennas and the relation between antenna theories. Cruft Lab. Technical Report, vol. 46. Harvard University (1948)

[296] Hallén, E.: Theoretical investigations into the transmitting and receiving qualities of antennae. Nova Acta Regiae Soc. Sci. Upsaliensis, Ser. IV, vol. 11(4), pp. 1–44 (1938)

[297] Hansen, R.C.: Fundamental limitations in antennas. Proceedings of the IEEE 69(2), 170–182 (1981)

[298] Hapeman, M.J., Long, J., Plette, D.L.: Diesel Electric Locomotive Propulsion Systems - A Look into the Future. IEEE Transactions on Industry Applications 22(3), 495–501 (1986)

[299] Hartal, O.: Electromagnetic Compatibility by Design, 4th edn., West Conshohocken PA. R & B Enterprizes (1996)

[300] Hemmer, B., Mariscotti, A., Wuergler, D.: Recommendations for the calculation of the total disturbing return current from electric traction vehicles. IEEE Transactions on Power Delivery 19(2), 1190–1197 (2004)

[301] Hill, P.N., Green, H.E.: In situ measurement of soil permittivity and permeability. Journal of Electrical and Electronic Engineers Australia 2(4), 202–208 (1982)

[302] Hill, J.R.: Optimal Construction of Synchronizable Coding for Railwav Track Circuit Data Transmission. IEEE Transactions on Vehicular Technology 39(4), 390–399 (1990)

[303] Hill, R.J.: Electric railway traction. VI. Electromagnetic compatibility disturbance-sources and equipment susceptibility. Power Engineering Journal 11(1), 31–39 (1997)

[304] Hill, R.J., Carpenter, D.C.: Determination of rail internal impedance for electric railway traction system simulation. IEE Proc. Part B 138(6) (November 1991)

[305] Hill, R.J., Brillante, S., De Souza, C.R., Leonard, P.J.: Electrical material data for railway track transmission line parameter studies. IEE Proc. Electric Power Applications 146(1), 60–68 (1999)

[306] Hill, R.J., Brillante, S., Leonard, P.J.: Railway track transmission line parameters from finite element field modelling: series impedance. IEE Proc. Electric Power Applications 146(6), 647–660 (1999)

[307] Hill, R.J., Brillante, S., Leonard, P.J.: Railway track transmission line parameters from finite element field modelling: shunt admittance. IEE Proc. Electric Power Applications 147(3), 227–238 (2000)

[308] Hill, R.J., Brillante, S., Leonard, P.J.: Modelling rail track electrical behaviour using two-dimensional finite elements. In: Proceedings of the 1996 ASME/IEEE Joint Railroad Conference, April 30-May 2, pp. 101–109 (1996)

[309] Hill, J.R., Weedon, D.N.: Safety & Reliability of Synchronizable Digital Coding in Railway Track-Circuits. IEEE Transactions on Reliability 39(5), 581–591 (1990)

[310] Hill, R.J., Cevik, I.H.: On-line simulation of voltage regulation in Autotransformer-fed AC electric railroad traction networks. IEEE Trans. on Vehicular Technology 42(3), 365–372 (1993)

[311] Hill, R.J., Carpenter, D.C.: Determination of rail internal impedance for electric railway traction system simulation. IEE Proc. 138(6), pt. B (November 1991)

[312] Ho, T.K., Chi, Y.L., Wang, J., Leung, K.K., Siu, L.K., Tse, C.T.: Probabilistic load flow in AC electrified railways. IEE Proc. Electric Power Applications 152(4), 1003–1013 (2005)

[313] Holmstrom, F.R., Long, L.E., Gagnon, R.: Assuring compatibility of rapid transit propulsion and signalling systems. In: Proceedings of the 32nd IEEE Vehicular Technology Conference, vol. 32, pp. 445–453 (1982)

[314] Holtz, J., Klein, H.J.: The propagation of harmonic currents generated by inverter fed locomotives in the distributed overhead supply systems. IEEE Transactions on Power Electronics 4(2), 168–174 (1989)

[315] Holtz, J., Stadtfeld, S., Lammert, P.: An economic very high power PWM inverter for induction motor drives. In: Proc. of. Europ. Conf. on Power Electronics, Brussels (1985)

[316] Humpage, W.D., Wong, K.P., Nguyen, T.T., Sutanto, D.: Z-transform electromagnetic transient analysis in power system. IEE Proceedings on Power Transmission 127(6), pt. C, 370–378 (1980)

[317] IEC Multilingual Dictionary, Electricity, Electronics and Telecommunications (2005)

[318] IEC 61000-1-1, Electromagnetic compatibility (EMC) Part 1: General Section 1: Application and interpretation of fundamental definitions and terms, 1st edn. (April 1992)
[319] IEC/TS 61000-1-2, Electromagnetic compatibility (EMC) - Part 1-2: General - Methodology for the achievement of functional safety of electrical and electronic systems including equipment with regard to electromagnetic phenomena, 1st edn. (2001)
[320] IEC/TS 61000-1-2, Electromagnetic compatibility (EMC) - Part 1-2: General - Methodology for the achievement of functional safety of electrical and electronic systems including equipment with regard to electromagnetic phenomena, 2nd edn. (November 2008)
[321] IEC 61000-5-1, Electromagnetic compatibility (EMC) – Part 5: Installation and mitigation guidelines – Section 1: General considerations - Basic EMC publication (December 1996)
[322] IEC 61000-5-2, Electromagnetic compatibility (EMC) – Part 5: Installation and mitigation guidelines – Section 2: Earthing and cabling (November 1997)
[323] IEC 61000-6-5, Electromagnetic compatibility (EMC) – Part 6-5: Generic standards – Immunity for power station and substation environments (July 2001)
[324] IEC 77/389/CD, IEC 61000-6-7, Electromagnetic compatibility (EMC) - Part 6-7: Generic standards - Immunity requirements for safety-related systems and equipment intended to perform functions in a safety-related system (functional safety) in industrial environments, 1st edn. (2011)
[325] IEC 61025, Reliability of Systems, Equipment and Components. Guide to Fault Tree Analysis, IEC, Geneva (2006)
[326] IEC 61508 Part 1, Functional safety of electrical/electronic/programmable electronic safety-related systems - Part 1: General requirements, 2nd edn. (April 2010)
[327] IEC 61508 Part 2, Functional safety of electrical/electronic/programmable electronic safety-related systems - Part 2: Requirements for electrical/electronic/programmable electronic safety-related systems, 2nd edn. (April 2010)
[328] IEC 61508 Part 3, Functional safety of electrical/electronic/programmable electronic safety-related systems - Part 3: Software requirements, 2nd edn. (April 2010)
[329] IEC 61508 Part 4, Functional safety of electrical/electronic/programmable electronic safety-related systems - Part 4: Definitions and abbreviations, 2nd edn. (April 2010)
[330] IEC 61508 Part 5, Functional safety of electrical/electronic/programmable electronic safety-related systems - Part 5: Examples of methods for the determination of safety integrity levels, 2nd edn. (April 2010)
[331] IEC 61508 Part 6, Functional safety of electrical/electronic/programmable electronic safety-related systems - Part 6: Guidelines on the application of IEC 61508-2 and IEC 61508-3, 2nd edn. (April 2010)
[332] IEC 61508 Part 7, Functional safety of electrical/electronic/programmable electronic safety-related systems - Part 7: Overview of techniques and measures, 2nd edn. (April 2010)
[333] IEC 61508 Part 0, Functional safety of electrical/electronic/programmable electronic safety-related systems - Part 0: Functional safety and IEC 61508, 2nd edn. (April 2010)

[334] ANSI/IEEE Std. C37.010, IEEE Application Guide for AC High_Voltage Circuit Breakers Rated on a Symmetrical Current Basis (1999)
[335] ANSI/IEEE C63.022, American National Standard for Limits and Methods of Measurement of Radio Disturbance Characteristics of Information Technology Equipment (1996)
[336] ANSI/IEEE C63.022, American National Standard Guide for Automated Electromagnetic Interference Measurements (2004)
[337] ANSI/IEEE C63.2, American National Standard for Electromagnetic Noise and Field Strength Instrumentation, 10 Hz to 40 GHz – Specifications (1996)
[338] ANSI C63.4, American National Standard for Methods of Measurement of Radio-Noise Emissions from Low-Voltage Electrical and Electronic Equipment in the Range of 9 kHz to 40 GHz (1992)
[339] ANSI C63.5, American National Standard for Electromagnetic Compatibility – Radiated Emission Measurements in Electromagnetic Interference (EMI) Control – Calibration of Antennas (9 kHz to 40 GHz) (2004)
[340] ANSI C63.6, American National Standard Guide for the Computation of Errors in Open-Area Test Site Measurements (1996)
[341] ANSI C63.7, American National Standard Guide for Construction of Open-Area Test Sites for Performing Radiated Emission Measurements (2005)
[342] ANSI C63.9, American National Standard for RF Immunity of Audio Office Equipment to General Use Transmitting Devices with Transmitter Power Levels up to 8 Watts (2008)
[343] ANSI C63.10, American National Standard for Testing Unlicensed Wireless Devices (2009)
[344] ANSI C63.11 Draft, American National Standard for Inter-lab Comparison EMC Testing
[345] ANSI C63.12, American National Standard for Electromagnetic Compatibility Limits – Recommended Practice (2007)
[346] ANSI C 63.13, American National Standard Guide on the Application and Evaluation of EMI Power-Line Filters for Commercial Use (1991)
[347] ANSI C63.14, American National Standard Dictionary for Technologies of Electromagnetic Compatibility (EMC), Electromagnetic Pulse (EMP), and Electrostatic Discharge (ESD) (1998)
[348] ANSI C63.15, American National Standard Recommended Practice for the Immunity Measurement of Electrical and Electronic Equipment (2010)
[349] ANSI C63.16, American National Standard Guide for Electrostatic Discharge Test Methodologies and Criteria for Electronic Equipment (1993)
[350] ANSI C63.17, American National Standard for Methods of Measurement of the Electromagnetic and Operational Compatibility of Unlicensed Personal Communications Services (UPCS) Devices (1998)
[351] ANSI C63.18, On-site, Ad-Hoc Test Method for Estimating Radiated Electromagnetic Immunity of Medical Devices to Specific Radio Frequency Transmitters (1997)
[352] ANSI C63.19, American National Standard for Methods of Measurement of Compatibility between Wireless Communications Devices and Hearing Aids (2006)
[353] ANSI C63.20 Draft, American National Standard for EMC Immunity Qualification of Instrumentation & Control Equipment and Systems intended for use in Nuclear Power Stations

[354] ANSI C63.22, American National Standard Guide for Automated Electromagnetic Interference Measurements (2004)
[355] ANSI C63.23 Draft, Guide for Computations and Treatment of Measurement Uncertainty
[356] ANSI C63.24 Draft, American National Standard Recommended Practice for In-Situ RF Immunity Evaluation of Products, Instrumentation, and Control Systems in High Reliability Installations
[357] ANSI C63.25 Draft, Test Site validation time domain
[358] ANSI C63.26 Draft, American National Standard of procedures for compliance testing of licensed transmitters
[359] IEEE Std. 80, IEEE Guide for Safety in AC Substation Grounding (2000)
[360] IEEE Std. 81, IEEE Guide for Measuring Earth Resistivity, Ground Impedance, and Earth Surface Potentials of a Ground System (1983)
[361] IEEE Std. 140, IEEE Recommended Practice for Minimization of Interference from Radio-Frequency Heating Equipment (1990)
[362] IEEE Std. 141, IEEE Recommended Practice for Electric Power Distribution for Industrial Plants (1993)
[363] IEEE Std. 142, IEEE Recommended Practice for Grounding of Industrial and Commercial Power Systems (2007)
[364] IEEE Std. 211, IEEE Standard Definitions of Terms for Radio Wave Propagation (1997)
[365] IEEE Std. 291, IEEE Standard Methods for Measuring Electromagnetic Field Strength of Sinusoidal Continuous Waves, 30 Hz to 30 GHz (1991)
[366] IEEE Std. 299, IEEE Standard Method for Measuring the Effectiveness of Electromagnetic Shielding Enclosures (1997)
[367] IEEE Std. 302, IEEE Standard Methods for Measuring Electromagnetic Field Strength for Frequencies Below 1000 MHz in Radio Wave Propagation (1969)
[368] IEEE Std. 323, IEEE Standard for Qualifying Class E Equipment for Nuclear Power Generating Stations (2003)
[369] IEEE Std. 356, IEEE Guide for Measurements of Electromagnetic Properties of Earth Media (2001)
[370] ANSI/IEEE Std. 356, IEEE Standard for the Measurement of Impulse Strength and Impulse Bandwidth (1975)
[371] ANSI/IEEE Std. 430, IEEE Standard Procedures for the Measurement of Radio Noise from Overhead Power Lines and Substations (1986)
[372] IEEE Std. 473, IEEE Recommended Practice for an Electromagnetic Site Survey (10 kHz to 10 GHz) (1985)
[373] IEEE Std. 487, IEEE Recommended Practice for the Protection of Wire-Line Communication Facilities Serving Electric Supply Locations (2007)
[374] IEEE Std. 518, IEEE Guide for the Installation of Electrical Equipment to minimize Electrical Noise Inputs to Controllers from External Sources (1982)
[375] IEEE Std. 519, IEEE Recommended Practices and Requirements for Harmonic Control in Electrical Power Systems (1992)
[376] ANSI/IEEE Std. 575, IEEE Guide for the Application of Sheath-Bonding Methods for Single-Conductor Cables and the Calculation of Induced Voltages and Currents in Cable Sheaths (1988)
[377] IEEE Std. 644, IEEE Standard Procedure for Measurement of Power Frequency Electrric and Magnetic Fields from AC Power Lines (1994)
[378] IEEE Std. 665, IEEE Guide for Generating Station Grounding (R.2001) (1995)

[379] IEEE Std. 776, IEEE Guide for Inductive Coordination of Electric Supply and Communication Lines (1992)

[380] ANSI/IEEE Std. 802.11, Information technology – Telecommunications and information exchange between systems – Local and metropolitan area networks – Specific requirements – Part 11: Wireless LAN Medium Access Control (MAC) and Physical Layer (PHY) Specifications (1999)

[381] ANSI/IEEE Std. 802.11g, Information technology – Telecommunications and information exchange between systems – Local and metropolitan area networks – Specific requirements – Part 11: Wireless LAN Medium Access Control (MAC) and Physical Layer (PHY) Specifications – Amendment 4: Further Higher Data Rate Extension in the 2.4 GHz Band (2003)

[382] IEEE Std. 802.11h, Information technology – Telecommunications and information exchange between systems – Local and metropolitan area networks – Specific requirements – Part 11: Wireless LAN Medium Access Control (MAC) and Physical Layer (PHY) Specifications – Amendment 5: Spectrum and Transmit Power Management Extensions in the 5 GHz band in Europe (2006)

[383] IEEE Std. 1048, IEEE Guide for Protective Grounding of Power Lines (2003)

[384] IEEE Std. 1050, IEEE Guide for Instrumentation and Control Equipment Grounding in Generating Stations (2004)

[385] IEEE Std. 1100, IEEE Recommended Practice for Powering and Grounding Electronic Equipment (2005)

[386] IEEE Std. 1137, IEEE Guide for the Implementation of Inductive Coordination Mitigation Techniques and Application (1991)

[387] IEEE Std. 1143, IEEE Guide on Shielding Practice for Low Voltage Cables (1994)

[388] ANSI/IEEE Std. 1299, IEEE Guide for the Connection of Surge Arresters to protect Insulated, Shielded Electric Power Cable Systems (1996)

[389] IEEE Std. 1346, IEEE Recommended Practice for Evaluating Electric Power System Compatibility with Electronic Process Equipment (1998)

[390] IEEE Std. 1473, IEEE Standard for Communications Protocol Aboard Trains (1999)

[391] IEEE Std. 1474.1, IEEE Standard for Communications- Based Train Control (CBTC) Performance and Functional Requirements (2004)

[392] IEEE Std. 1474.2, IEEE Standard for User Interface Requirements in Communications-Based Train Control (CBTC) Systems (2003)

[393] IEEE Std. 1474.3, IEEE Recommended Practice for Communications-Based Train Control (CBTC) System Design and Functional Allocations (2008)

[394] IEEE Std. 1475, IEEE Standard for the Functioning of and Interfaces Among Propulsion, Friction Brake, and Train-borne Master Control on Rail Rapid Transit Vehicles (1999)

[395] IEEE Std. 1476, IEEE Standard for Passenger Train Auxiliary Power Systems Interfaces, R.2008 (2000)

[396] IEEE Std. 1477, IEEE Standard for Passenger Information System for Rail Transit Vehicles (1998)

[397] IEEE Std. 1478, IEEE Standard for Passenger Information System for Rail Transit Vehicles (2001)

[398] IEEE Std. 1488, IEEE Trial-Use Standard for Message Set Template for Intelligent Transportation Systems (2000)

[399] IEEE Std. 1489, 'IEEE Standard for Data Dictionaries for Intelligent Transportation Systems (1999)

[400] IEEE Std. 1531, IEEE Guide for Application and Specification of Harmonic Filters (2003)

[401] IEEE Std. 1570, IEEE Standard for the Interface Between the Rail Subsystem and the Highway Subsystem at a Highway Rail Intersection (2002)

[402] IEEE Std. P1582, Draft Standard for Environmental Requirements for Rail Transit Automatic Train Control Systems Wayside Equipment (2002)

[403] IEEE Std. 1597.1, IEEE Standard for Validation of Computational Electromagnetics Computer Modeling and Simulations (2008)

[404] ANSI/IEEE Std. C62.47, IEEE Guide on Electrostatic Discharge (ESD): Characterization of the ESD Environment (1992)

[405] ANSI Std. C63.2, American National Standard for Electromagnetic Noise and Field Strength Instrumentation, 10 Hz to 40 GHz – Specifications (1996)

[406] ANSI Std. C63.4, American National Standard for Methods of Measurement of Radio-Noise Emissions from Low-Voltage Electrical and Electronic Equipment in the Range of 9 kHz to 40 GHz (1992)

[407] Institution of Railway Signal Engineers, Railway signalling: a treatise on the recent practice of British Railways, Nock, O.S. (ed.) (1980)

[408] Invensys, Phase Shift Overlay 4000 Track Circuit, Datasheet 5A-4 rev. 1.0, `www.invensysrail.com`

[409] IS 402, Prove di Tipo e di Accettazione per le apparecchiature elettroniche ed elettromeccaniche destinate agli Impianti di Sicurezza e Segnalamento; RFI - Divisione Infrastruttura, rev. A (July 18, 2000)

[410] Ismail, A., Sanad, S., El-Meligi, A.: Inhibiting effect of indole and some of its derivatives on corrosion of C-steel in HCl. Journal of materials science and Technology 16(4), 397–400 (2000)

[411] ISO/IEC Directives, Part 1, Procedures for the technical work, 8th edn. (2011)

[412] ISO/IEC Guide 2, Standardisation and Related Activity – General Vocabulary (2004)

[413] ISO 8044, Corrosion of metals and alloys - Basic terms and definitions, British Standards Institute, pp. 1–56 (2000)

[414] ISO/ENV 13005, Guide to the expression of uncertainty in measurement (May 1999)

[415] ISO/IEC TR 14763-2, Information technology – Implementation and operation of customer premises cabling – Part 2: Planning and installation (July 2000)

[416] ISO/IEC 24702, Information technology – Generic cabling – Industrial premises (October 2006)

[417] CCITT (International Telegraph and Telephone Consultative Committee), Directives concerning the protection of telecommunication lines against harmful effects from electric power and electrified railway lines, vol. I, Geneva, Switzerland (1989) ISBN 92-61-03941-3

[418] CCITT (International Telegraph and Telephone Consultative Committee), Directives concerning the protection of telecommunication lines against harmful effects from electric power and electrified railway lines, vol. II, Geneva, Switzerland (1989) ISBN 92-61-08001-4

[419] CCITT (International Telegraph and Telephone Consultative Committee), Directives concerning the protection of telecommunication lines against harmful effects from electric power and electrified railway lines, vol. III, Geneva, Switzerland (1989) ISBN 92-61-04041-1
[420] CCITT (International Telegraph and Telephone Consultative Committee), Directives concerning the protection of telecommunication lines against harmful effects from electric power and electrified railway lines, vol. IV, Geneva, Switzerland (1989) ISBN 92-61-04051-9
[421] CCITT (International Telegraph and Telephone Consultative Committee), Directives concerning the protection of telecommunication lines against harmful effects from electric power and electrified railway lines, vol. V, Geneva, Switzerland (1989) ISBN 92-61-04061-6
[422] CCITT (International Telegraph and Telephone Consultative Committee), Directives concerning the protection of telecommunication lines against harmful effects from electric power and electrified railway lines, vol. VI, Geneva, Switzerland (1989) ISBN 92-61-04071-3
[423] ITU-T K.9, Series K: Protection against interference – Protection of telecommunication staff and plant against a large earth potential due to a neighbouring electric traction line (1993)
[424] ITU-T K.10, Series K: Protection against interference – Low frequency interference due to unbalance about earth of telecommunication equipment (October 1996)
[425] ITU-T K.11, Series K: Protection against interference – Principles of protection against over-voltages and overcurrents (January 2009)
[426] ITU-T K.13, Series K: Protection against interference – Induced voltage in cables with plastic-insulated conductors (1993)
[427] ITU-T K.18, Series K: Protection against interference – Calculation of voltage induced into telecommunication lines from radio station broadcasts and methods of reducing interference (1993)
[428] ITU-T K.20, Series K: Protection against interference – Resistibility of telecommunication equipment installed in a telecommunications centre to over-voltages and overcurrents (April 2008)
[429] ITU-T K.21, Series K: Protection against interference – Resistibility of telecommunication equipment installed in customer premises to over-voltages and overcurrents (April 2008)
[430] ITU-T K.26, Series K: Protection against interference – Protection of telecommunication lines against harmful effects from electric power and electrified railway lines (April 2008)
[431] ITU-T K.27, Series K: Protection against interference – Bonding configurations and earthing inside a telecommunication building (May 1996)
[432] ITU-T K.33, Series K: Protection against interference – Limits for people safety related to coupling into telecommunications system from ac electric power and ac electrified railway installations in fault conditions (October 1996)
[433] ITU-T K.34, Series K: Protection against interference – Classification of electromagnetic environmental conditions for telecommunication equipment – Basic EMC Recommendation (July 2003)
[434] ITU-T K.35, Series K: Protection against interference – Bonding configurations and earthing at remote electronic sites (May 1996)

[435] ITU-T K.37, Series K: Protection against interference – Low and high frequency EMC mitigation techniques for telecommunication installations and systems – Basic EMC Recommendation (February 1999)

[436] ITU-T K.39, Series K: Protection against interference – Risk assessment of damages to telecommunication sites due to lightning discharges (October 1996)

[437] ITU-T K.40, Series K: Protection against interference – Protection against LEMP in telecommunication centers (October 1996)

[438] ITU-T K.43, Series K: Protection against interference – Immunity requirements for telecommunication network equipment (July 2009)

[439] ITU-T K.44, Series K: Protection against interference – Resistibility tests for telecommunication equipment exposed to over-voltages and overcurrents – Basic Recommendation (April 2008)

[440] ITU-T K.45, Series K: Protection against interference – Resistibility of telecommunication equipment installed in the access and trunk networks to over-voltages and overcurrents (April 2008)

[441] ITU-T K.46, Series K: Protection against interference – Protection of telecommunication lines using metallic symmetric conductors against lightning-induced surges (April 2008)

[442] ITU-T K.47, Series K: Protection against interference – Protection of telecommunication lines using metallic conductors against direct lightning discharges (April 2008)

[443] ITU-T K.51, Series K: Protection against interference – Safety criteria for telecommunication equipment (July 2009)

[444] ITU-T K.53, Series K: Protection against interference – Values of induced voltages on telecommunication installations to establish telecom and a.c. power and railway operators responsibilities (February 2000)

[445] ITU-T K.56, Series K: Protection against interference – Protection of radio base stations against lightning discharges (January 2010)

[446] ITU-T K.58, Series K: Protection against interference – EMC, resistibility and safety requirements and guidance for determining responsibility under co-located telecommunication installations (April 2008)

[447] ITU-T K.66, Series K: Protection against interference – Protection of customer premises from over-voltages (December 2004)

[448] ITU-T K.67, Series K: Protection against interference – Expected surges on telecommunications and signalling networks due to lightning (February 2006)

[449] ITU-T K.68, Series K: Protection against interference – Management of electromagnetic interference on telecommunication systems due to power systems (February 2006)

[450] ITU-T K.71, Series K: Protection against interference – Protection of customer antenna installations (July 2007)

[451] ITU-T K.72, Series K: Protection against interference – Protection of telecommunication lines using metallic conductors against lightning – Risk management (April 2008)

[452] ITU-T K.73, Series K: Protection against interference – Shielding and bonding for cables between buildings (April 2008)

[453] ITU-T K.79, Series K: Protection against interference – Electromagnetic characterization of the radiated environment in the 2.4 GHz ISM band (June 2009)

[454] ITU-T K.80, Series K: Protection against interference – EMC requirements for telecommunication network equipment (1 GHz-6 GHz) (July 2009)
[455] ITU-T O.41, Series O: Specifications for measuring equipment – Equipment for the measurement of analogue parameters – Psophometer for use on telephone-type circuits (October 1994)
[456] ITU-T P.68, Series P: Telephone transmission quality, telephone installations, local line networks – Objective electro-acoustical measurements – Determination of sensitivity/frequency characteristics of local telephone systems (November 2007)
[457] Jaekel, B.: Considerations on immunity test levels and methods with regard to functional safety. In: 18th International Wroclaw Symposium and Exhibition on Electromagnetic Compatibility, Wroclaw, Poland (2006)
[458] Jahns, T.M., Blasko, V.: Recent Advances in Power Electronics Technology for Industrial and Traction Machine Drives. Proceedings of the IEEE 89(6), 963–975 (2001)
[459] Jia, K., Thottappillil, R.: EMC assessment of the railway traction system by using PSpice. In: Proceedings of the IEEE Asia-Pacific Symposium on Electromagnetic Compatibility (APEMC), pp. 598–601 (2010)
[460] Kaden, H.: Wirbelstrome und Schirmung in der Nachrichtentechnik. Springer, Berlin (1959)
[461] Kaplan, S., Garrick, G.J.: On the quantitative definition of risk. Risk Anal. 1(1), 11–27 (1981)
[462] Károlyi, K.: An interactive code to solve multipoint boundary value problems. In: Conference EQUADIFF 1991, Barcelona, pp. 621–626 (1991)
[463] Kastell, K., Bug, S., Nazarov, A., Jakoby, R.: Improvements in Railway Communications via GSM-R. In: Proceedings of the IEEE 63rd Vehicular Technology Conference, vol. 6, pp. 3026–3030 (2006)
[464] Kennelly, E., Achard, F.H., Dana, S.: Experimental researches on the skin effect in steel rails. Journal of the Franklin Institute (August 1916)
[465] Kehrmann, H., Lienau, W., Nill, K.: Vierquadrantentsteller – Eine netzfreundliche Einspeisung fuer Triebfahrzeuge mit Drehstromantrieb. Elektrirche Bahnen 45(6), 135–142 (1974)
[466] Kim, M.-S., Lee, J.-W., Ko, J.-S.: A Study on the Compensation Capacitor in Ballast Track Circuit. In: Proc. of IEEE International Conference on Electrical Machines and Systems, ICEMS 2008, October 17-20, pp. 4182–4187 (2008)
[467] Kim, M.-S., Bae, Y.-B., Lee, S.-Y., Lee, J.-W.: A Study on Electrical Separation Joint in the Slab Track. In: Proc. of IEEE International Conference on Electrical Machines and Systems, ICEMS 2009, November 15-18, pp. 1–6 (2009)
[468] Kimata, M., Koyama, M., Uchida, R., Ikeda, M., Kawamura, T., Okada, T.: Smart IGBT model and its application for power converter design. In: Proceedings of the IEEE Industry Applications Society Annual Meeting, pp. 1168–1173 (1994)
[469] Kimbark, E.W.: Direct Current Transmission, vol. I. John Wiley & Sons, New York (1971)
[470] King, D.D.: The measured impedance of cylindrical dipoles. Journal of Applied Physics 17, 844–852 (1947)
[471] King, R., Harrison Jr., C.W.: The distribution of current along a symmetrical center-driven antenna. Proceedings of the IRE 31, 548–567 (1943)

[472] Kirchhoff, G.R.: Zur Theorie der Lichstrahlen. Ann. Phys. (Leipzig) 18, 663–695 (1883)
[473] Klapas, D., Hackam, R., Benson, F.A.: Electric arc power collection for high-speed trains. Proceedings of IEEE 64, 1699–1715 (1976)
[474] Kneschke, T.: Electrical power supply configurations on 10,000 miles of U.S. railroads, Electrack, Inc. for the Department of Transportation, Contract DTRS-57-80-C-00042 (April 1981)
[475] Kneschke, T.: Simple Method for Determination of Substation Spacing for AC and DC Electrification Systems. IEEE Transactions on Industry Applications 22(4), 763–780 (1986)
[476] Kneschke, T.: Control of Utility System Unbalance Caused by Single-Phase Electric Traction. IEEE Transactions on Industry Applications 21(6), 1559–1570 (1985)
[477] Kneschke, T., Hong, J., Natarajan, R., Naqvi, W.: Impedance calculations for SEPTA's rail power distribution system. In: Proc. of the 1995 IEEE/ASME Joint Railroad Conference, April 4-6, pp. 79–85 (1995)
[478] Kneschke, T., Mbika, P.: Determination of traction power distribution system impedances and susceptances for ac railroad electrification systems. In: Proc. of the 2004 ASME/IEEE Joint Rail Conference, Baltimore, Maryland, USA, April 6-8, pp. 27–34 (2004)
[479] Knockaert, J., Pissoort, D., Vanhee, F.: Comparing EMC-signatures by FSV as a Quality Assessment Tool. In: Progress In Electromagnetics Research Symposium Proceedings, pp. 1099–1103 (April 2011)
[480] Krakowski, M.: Currents and potentials along extensive underground conductors. IEE Proceedings 115(9), 1299–1304 (1968)
[481] Krause, J.D., Marhefka, R.J.: Antennas, 3rd edn. McGraw Hill, New York (2002)
[482] Kraus, R., Turkes, P., Sigg, J.: Physics-based models of power semiconductor devices for the circuit simulator SPICE. In: Proceedings of the 29th Annual IEEE Power Electronics Specialists Conference (PESC), vol. 2, pp. 1726–1731 (1998)
[483] Kuppa, V., Wolfs, P.: An Elegant Solution Using Hybrid Power Filter to Improve the Line Current Spectrum of Multiphase PWM Locomotive Rectifiers with Load Unbalance. In: Proc. of the 2006 Australasian Universities Power Engineering Conference, AUPEC 2006, Melbourne, Australia, December 10-13 (2006)
[484] Kurokawa, S., Pissolato, J., Tavares, M.C., Portela, C.M., Prado, A.J.: A New Procedure to Derive Transmission-Line Parameters: Applications and Restrictions. IEEE Transactions on Power Delivery 21(1), 492–498 (2006)
[485] Lafon, F., de Daran, F., Caves, L., Ramdani, M., Drissi, M.: Influence of aging and environment conditions on EMC performances of electronic equipment. Influence of passive Vs active components. In: 9th International Symposium on EMC Joint With 20th International Wroclaw Symposium on EMC, EMC Europe 2010, Wroclaw, pp. 215–220 (2010)
[486] Langella, R., Sollazzo, A., Testa, A.: Modeling Waveform Distortion Produced by DC Locomotive Conversion System - Part 1: Locomotive Model. In: Proc. of the IEEE 11th International Conference on Harmonics and Quality of Power, Lake Placid, NY, USA, September 12-15, pp. 477–482 (2004)
[487] Laport, E.A.: Long-wire antennas. In: Johnson, R.C. (ed.) Antenna Engineering Handbook, 3rd edn. McGraw Hill, New York (1993)

[488] Lee, H.-M., Lee, C.-M., Jang, G., Kwon, S.-H.: Harmonic Analysis of the Korean High-Speed Railway Using the Eight-Port Representation Model. IEEE Transactions on Power Delivery 21(2), 979–986 (2006)

[489] Leva, S., Morando, A.P., Colombaioni, P.: Dynamic Analysis of a High Speed Train. IEEE Transactions on Vehicular Technology 57(1), 107–119 (2008)

[490] Lewinski, D.A.: A new objective for message circuit noise. Bell System Technical Journal 43, 719–740 (1964)

[491] Li, B., Boyer, A., Ben Dhia, S., Lemoine, C.: Ageing effect on immunity of a mixed signal IC. In: Proceedings of Asia-Pacific International Symposium on Electromagnetic Compatibility, Beijing, China, pp. 1024–1027 (2010)

[492] Li, B., Boyer, A., Bendhia, S., Lemoine, C.: Ageing effect on electromagnetic susceptibility of a phase locked loop. Microelectronics Reliability 50(9), 1304–1308 (2010)

[493] S&CSE-ST0062-A3, Maximum allowable level of electromagnetic interference in safety signalling equipment. Signalling and Control Engineering, London Underground

[494] 1-222, London Underground Category 1 EMC Standard, London Underground Limited

[495] G-222, London Underground Electromagnetic Compatibility Manual of Good Practice, London Underground Limited

[496] Lucas, R., Talukdar, S.: Advances in Finite Element Techniques for calculating cable resistance and inductance. IEEE Transactions on Power Apparatus and Systems PAS-97(3), 875–883 (1978)

[497] Lucca, G.: Per Unit Length Parameters of a Multiconductor Line with Earth Return Inside a Tunnel. IEEE Transactions on Electromagnetic Compatibility 42(1), 6–15 (2000)

[498] Lucca, G.: Electromagnetic fields produced by sliding contact between pantograph and contact wire of an electrified traction line. In: International Symposium on Electromagnetic Compatibility, EMC, Europe (2000)

[499] Machczynski, W.: Currents and potentials in earth return circuits exposed to alternating current electric railways. IEE Proceedings 129(5), pt. B, 279–288 (1982)

[500] Machczynski, W.: Evaluation of conductive effects on earth return circuits. In: EMC, Zurich, pp. 117–120 (1993)

[501] Machczynski, W.: A calculation model for cathodic protection of underground extensive structures using impressed current cable anodes. Archiv für Elektrotechnik 84(1), 33–39 (2002)

[502] Manneback, C.: An integral equation for skin effect in parallel conductors. J. Mathematics Phys. 1, 123–146 (1922)

[503] Mao, J.F., Kuh, E.S.: Fast Simulation and Sensitivity Analysis of Lossy Transmission Lines by the Method of Characteristics. IEEE Transactions on Circuits and Systems – Part I: Fundamental Theory and Applications 44(5), 391–401 (1997)

[504] Mardare, D., LoVetri, J.: The finite-difference time-domain solution of lossy MTL networks with nonlinear junctions. IEEE Transactions on Electromagnetic Compatibility 37(2), 252–259 (1995)

[505] Mariscotti, A., Pozzobon, P.: Measurement of the internal impedance of traction rails at 50 Hz. IEEE Transactions on Instrumentation and Measurement 49(2), 294–299 (2000)

[506] Mariscotti, A.: Analysis of the dc link current spectrum in voltage source inverters. IEEE Transactions on Circuits and Systems - Part I 49(4), 484–491 (2002)
[507] Mariscotti, A., Pozzobon, P.: Synthesis of line impedance expressions for railway traction systems. IEEE Transactions on Vehicular Technology 52(2), 420–430 (2003)
[508] Mariscotti, A.: Distribution of the traction return current in AC and DC electric railway systems. IEEE Transactions on Power Delivery 18(4), 1422–1432 (2003)
[509] Mariscotti, A.: Discussion of Series expansion for line series impedances considering different specific resistances, magnetic permeabilities and dielectric permettivities of conductors, air and ground. IEEE Transactions on Power Delivery 18(4), 1592 (2003)
[510] Mariscotti, A., Pozzobon, P.: Measurement of the Internal Impedance of Traction Rails at Audiofrequency. IEEE Transactions on Instrumentation and Measurement 53(3), 792–797 (2004)
[511] Mariscotti, A., Pozzobon, P.: Resistance and Internal Inductance of traction rails: a survey. IEEE Transactions on Vehicular Technology 53(4), 1069–1075 (2004)
[512] Mariscotti, A., Pozzobon, P.: Determination of the Electrical Parameters of Railway Traction Lines: Calculation, Measurement and Reference Data. IEEE Transactions on Power Delivery 19(4), 1538–1546 (2004)
[513] Mariscotti, A., Pozzobon, P.: Experimental results on low rail-to-rail conductance values. IEEE Transactions on Vehicular Technology 54(3), 1219–1222 (2005)
[514] Mariscotti, A., Pozzobon, P., Vanti, M.: Distribution of the traction return current in AT electric railway systems. IEEE Transactions on Power Delivery 20(3), 2119–2128 (2005)
[515] Mariscotti, A., Pozzobon, P., Vanti, M.: Simplified modelling of 2x25 kV AT Railway System for the solution of low frequency and large scale problems. IEEE Transactions on Power Delivery 22(1), 296–301 (2007)
[516] Mariscotti, A.: Methods for Ripple Index evaluation in DC Low Voltage Distribution Networks. In: IMTC 2007, Warsaw, Poland, May 2-4 (2007)
[517] Mariscotti, A.: Measurement Procedures and Uncertainty Evaluation for Electromagnetic Radiated Emissions from Large Power Electrical Machinery. IEEE Transactions on Instrumentation and Measurement 56(6), 2452–2463 (2007)
[518] Mariscotti, A., Ruscelli, M., Vanti, M.: Modeling of Audio frequency Track Circuits for validation, tuning and conducted interference prediction. IEEE Transactions on Intelligent Transportation Systems 11(1), 52–60 (2010)
[519] Mariscotti, A.: Statistical evaluation of measured voltage spectra in DC railways. In: IMEKO TC4 2010, Kosice, SK, September 08-10 (2010)
[520] Mariscotti, A., Ogunsola, A.: Prediction and mitigation of magnetic fields propagated from DC light rail transit systems. In: Proceedings of the International Conference on Electrical Systems for Aircraft, Railway and Ship Propulsion, ESARS 2010, Bologna, Italy, pp. 1–6 (2010)
[521] Mariscotti, A.: Statistical evaluation of measured voltage spectra in DC railways. In: IMEKO TC4 2010, Kosice, SK, September 08-10 (2010)
[522] Mariscotti, A., Slepicka, D.: Analysis of frequency stability of 16.7 Hz railways. In: I2MTC 2011, Hangzhou, China, May 10-12 (2011)

[523] Mariscotti, A.: DC railway line voltage ripple for periodic and aperiodic phenomena. In: IMEKO 2011, Natal, RN, Brazil, September 27-30 (2011)
[524] Mariscotti, A.: Characterization of Power Quality transient phenomena of DC railway traction supply. ACTA Imeko 1 (in print, 2012)
[525] Mariscotti, A., Slepička, D.: The Frequency Stability of the 50 Hz French Railway. In: Proceedings of the IEEE International Instrumentation and Measurement Technical Conference, I2MTC, Graz, Austria (2012)
[526] Mariscotti, A.: Direct Measurement of Power Quality over Railway Networks with Results of a 16.7 Hz Network. IEEE Transactions on Instrumentation and Measurement 60(5), 1604–1612 (2011)
[527] Mariscotti, A.: Induced Voltage Calculation in Electric Traction Systems: Simplified Methods, Screening Factors and Accuracy. IEEE Transactions on Intelligent Transportation Systems 12(1), 201–210 (2011)
[528] Marti, J.R.: Accurate modelling of frequency-dependent transmission lines in electromagnetic transient simulations. IEEE Transactions on Power Apparatus and Systems 101(1), 147–155 (1982)
[529] Matsumoto, M.: The Revolution of Train Control System in Japan. In: Proc. of IEEE Conference on Autonomous Decentralized Systems, ISADS 2005, April 4-8, pp. 599–606 (2005)
[530] Mauriello, A.J., Clarke, J.M.: Measurement and analysis of radiated electromagnetic emissions from rail-transit vehicles. IEEE Transactions on Electromagnetic Compatibility 25(4), 405–411 (1983)
[531] Maxim, A., Andreu, D., Boucher, J.: High accuracy SPICE behavioral macromodeling of insulated gate bipolar transistor (IGBT). In: Proceedings of the 13th Annual IEEE Applied Power Electronics Conference and Exposition, APEC, vol. 2, pp. 749–755 (1998)
[532] Mayhan, R.J., Bailey, R.E.: An indirect measurement of the effective dielectric constant and loss tangent of typical concrete roadways. IEEE Transactions on Antennas and Propagation 23(4), 565–569 (1975)
[533] Mazloom, Z., Theethayi, N., Thottappillil, R.: A Method for Interfacing Lumped-Circuit Models and Transmission-Line System Models With Application to Railways. IEEE Transactions on Electromagnetic Compatibility 51(3), 833–841 (2009)
[534] Mazloom, Z., Theethayi, N., Thottappillil, R.: Indirect Lightning-Induced Voltages Along a Railway Catenary-Track Multiconductor Transmission-Line System With Lumped Components. IEEE Transactions on Electromagnetic Compatibility 53(2), 537–539 (2011)
[535] McDonald, W.: Controlling the Alice Springs to Darwin railway - A case study in appropriate technology. In: IRSE Technical Conference, Adelaide, November 03 (2006)
[536] McSparran, L.W.: E60C Electric Locomotives for the Black Mesa and Lake Powell Railroad. IEEE Transactions on Industry Applications 10(3), 385–390 (1974)
[537] Mellitt, B., Allan, J., Shao, Z.Y., Johnston, W.B., Goodman, C.J.: Computer-based methods for induced-voltage calculations in AC railways. IEE Proceedings 137(I), pt. B, 59–72 (1990)
[538] Meng, G.Z., Li, Y., Wang, F.H.: The Corrosion Behavior of Fe-10Cr Nanocrystalline Coating. Electrochimica Acta 51(20), 4277–4284 (2006)

[539] Meyer, M., Petersen, A.J.: Handling Large Railway Supply Systems - A Challenge for System Modelling and a Need to Guarantee Rail Vehicle's System Compatibility. In: Proc. of the 8th International Conference on Harmonics and Quality of Power, Athens, Greece, October 14-16, vol. 6, pp. 630–635 (1998)
[540] Meyer, M., Schöning, J.: Netzstabilitat in Grossen Bahnnetzen. Eisenbahn Review 7-8, 312–317 (1999)
[541] Meyer, M., Stadelmann, M., Kernen, R.: Auswirkung der Netzresonanzen im Lötschberg-Basistunnel. Elektrische Zugförderung, 77–83
[542] Meyer, M., Aeberhard, M., Germann, K., Suter, R., Dähler, P.: Messung des Frequenzgangs von Triebfahrzeugen. Elektrische Zugförderung, 71–76
[543] Meyer, M., van Alphen, G.J.: Netzresonanzmessungen auf HSL Zuid und Betuweroute. Eisenbahn Review 7-8, 610–611 (2006)
[544] Midya, S., Bormann, D., Schutte, T., Thottappillil, R.: DC Component From Pantograph Arcing in AC Traction System – Influencing Parameters, Impact, and Mitigation Techniques. IEEE Transactions on Power Delivery 53(1), 18–27 (2011)
[545] Midya, S., Bormann, D., Schutte, T., Thottappillil, R.: Pantograph Arcing in Electrified Railways – Mechanism and Influence of Various Parameters – Part I: With DC Traction Power Supply. IEEE Transactions on Power Delivery 24(4), 1931–1939 (2009)
[546] Midya, S., Bormann, D., Schutte, T., Thottappillil, R.: Pantograph Arcing in Electrified Railways – Mechanism and Influence of Various Parameters – Part II: With AC Traction Power Supply. IEEE Transactions on Power Delivery 24(4), 1940–1950 (2009)
[547] Mihalic, F., Jezernik, K., Krischan, K., Rentmeister, M.: IGBT SPICE model. IEEE Transactions on Industrial Electronics 42(1), 98–105 (1995)
[548] MIL STD 461E, Requirements for the control of Electromagnetic Interference Characteristics of Subsystems and Equipment (August 1999)
[549] MIL-STD-1541A, Military Standard - Electromagnetic Compatibility Requirements for Space Systems, Department of Air Force, USA (1987)
[550] Min, S.H., Swaminathan, M.: Construction of Broadband Passive Macromodels From Frequency Data for Simulation of Distributed Interconnect Networks. IEEE Transactions on Electromagnetic Compatibility 46(4), 544–558 (2004)
[551] Mitter, C.S., Hefner, A.R., Chen, D.Y., Lee, F.C.: Insulated gate bipolar transistor (IGBT) modeling using IG-SPICE. IEEE Transactions on Industry Applications 30(1), 24–33 (1994)
[552] Modarres, M.: What Every Engineer Should Know about Reliability and Risk Analysis. Marcel Dekker, New York (1993)
[553] Möllerstedt, E., Bernhardsson, B.: Out of control because of harmonics - An analysis of the harmonic response of an inverter locomotive. IEEE Control Systems Magazine 20(4), 70–81 (2000)
[554] Mohan, N., Undeland, T., Robbins, W.: Power Electronics: Converters, Applications and Design, 3rd edn. John Wiley & Sons (2002)
[555] Molak, V.: Fundamentals of Risk Analysis and Risk Management. CRC Press, Lewis Publishers, Boca Raton (1997)

[556] Morched, A., Gustavsen, B., Tartibi, M.: A universal model for accurate calculation of electromagnetic transients on overhead lines and underground cables. IEEE Transactions on Power Delivery 14(3), 1032–1038 (1999)

[557] Mowete, A.I., Ogunsola, A.: An analytical model for the sheilding effectiveness of planar multi-layer shield. In: EMC, Europe (2006)

[558] Nabae, A., Takahashi, I., Akagi, H.: A New Neutral-Point-Clamped PWM Inverter. IEEE Trans. on Industrial Applications 17(5), 518–523 (1981)

[559] Nakagawa, M., Ametani, A., Iwamoto, K.: Further studies on wave propagation in overhead lines with earth return: impedance of stratified earth. Proceedings of the IEE 120(12), 1521–1528 (1973)

[560] Nakagawa, M., Iwamoto, K.: Earth return impedance for the multilayer case. IEEE Transaction on Power Apparatus and Systems 95(2), 671–676 (1976)

[561] NASA Contractor Report 4759, System Guidelines for EMC Safety-Critical Circuits: Design, Selection and Margin Demonstration, NASA (October 1996)

[562] Network Rail NR/L1/SIG/30040, EMC Strategy for Network Rail (2008)

[563] Network Rail NR/L1/SIG/30041, EMC Assurance Process for Network Rail

[564] Network Rail NR/SP/SIG/50002, Methodology for the demonstration of compliance with single rail reed track circuit on the ac railway, Network Rail, United Kingdom (2007)

[565] Network Rail NR/SP/SIG/50003, Methodology for the demonstration of compliance with double rail reed track circuits on the dc railway, Network Rail, United Kingdom (2007)

[566] Network Rail NR/SP/SIG/50004, Methodology for the demonstration of electrical compatibility with dc (ac immune) track circuit, Network Rail, United Kingdom (2006)

[567] Network Rail NR/GN/SIG/50005, Methodology for the demonstration of compatibility with 50 Hz single rail track circuits, Network Rail, United Kingdom (2003)

[568] Network Rail NR/SP/SIG/50006, Methodology for the demonstration of compatibility with 50 Hz double rail track circuits, Network Rail, United Kingdom (2006)

[569] Network Rail NR/GN/SIG/50007, Methodology for the demonstration of compatibility with HVI track circuits, Network Rail, United Kingdom (2003)

[570] Network Rail NR/GN/SIG/50008, Methodology for the demonstration of compatibility with TI 21 track circuits, Network Rail, United Kingdom (2003)

[571] Network Rail NR/GN/SIG/50009, Methodology for the demonstration of compatibility with FS 2600 track circuits on the dc railway, Network Rail, United Kingdom (2003)

[572] Network Rail NR/L2/SIG/50010, Methodology for the demonstration of electrical compatibility with train detection systems in use on non-electrified lines, Network Rail, United Kingdom (2008)

[573] Network Rail NR/SP/SIG/50011, Methodology for the demonstration of electrical compatibility with axle couters, Network Rail, United Kingdom (2006)

[574] Network Rail NR/SP/SIG/50012, Methodology for the demonstration of compatibility with TPWS track Sub-system, Network Rail, United Kingdom (2006)

[575] Network Rail NR/SP/SIG/50013, Methodology for the demonstration of compatibility with interlockings, Network Rail, United Kingdom (2003)

[576] Network Rail NR/GN/SIG/50014, Methodology for the demonstration of compatibility with lineside equipment, Network Rail, United Kingdom (2008)
[577] Network Rail NR/SP/SIG/50015, Methodology for the demonstration of compliance with REED FDM Systems on ac and dc railways, Network Rail, United Kingdom (2007)
[578] Network Rail NR/SP/TEL/50016, Methodology for the demonstration of compatibility with telecommunication systems, Network Rail, United Kingdom (2006)
[579] Network Rail NR/GN/SIG/50018, Methodology for the determination of interaction with neighbouring railways, Network Rail, United Kingdom (2003)
[580] Network Rail NR/GN/TEL/31106, Overview of electromagnetic coupling between traction systems and telecommunications cables, Network Rail, United Kingdom (2009)
[581] Nitsch, D., Camp, M., Sabath, F., ter Haseborg, J., Garbe, H.: Susceptibility of some electronic equipment to HPEM threats. IEEE Transactions on Electromagnetic Compatibility 46(3), 380–389 (2004)
[582] Noda, T., Nagakoda, N., Ametani, A.: Phase domain modelling of frequency-dependent transmission lines by means of an ARMA model. IEEE Transactions on Power Delivery 11(1), 401–411 (1996)
[583] Nedelchev, N.: Jointless track circuit length. IEE Proceedings on Electric Power Applications 146(1), 69–74 (1999)
[584] Nielsen, L.V., Petersen, M.B., Bortels, L., Parlongue, J.: Effect of Coating Defect Size, Coating Defect Geometry, and Cathodic Polarization on Spread Resistance: - Consequences in relation to AC Corrosion Monitoring. In: Proceedings of Ceocor, Brugge, pp. 1–14 (2010)
[585] Nikolov, N., Nedelchev, N.: Study on centre-fed boundless track circuits. IEE Proceedings on Electric Power Applications 152(5), 1049–1054 (2005)
[586] Normative Française NF C-13 000, Installations électriques de tensions nominales supérieurs à 1 kV en courant alternatif (French, Fixed installations with nominal voltage greater than 1 kV ac) (April 2003)
[587] NF 07-011, "Installations fixes et matériel roulant ferroviaires – Perturbations électromagnétiques en électrification 25 kV, 50 Hz" (French, "Fixed installations and rolling stock – Electromagnetic interference under the 25 kV 50 Hz supply system") (October 1993)
[588] Ogunsola, A.: EMC assurance and management. In: The IET Colloquium on EMC Assurance in a Railway Environment (January 2003)
[589] Ogunsola, A., Pomeroy, S.: EMC Assurance and Safety Critical Apparatus in a Railway Environment. In: IEEE International Symposium on Electromagnetic Compatibility, vol. 1, pp. 429–432 (2003)
[590] Ogunsola, A., Reggiani, U., Sandrolini, L.: Modelling shielding properties of concrete. In: 17th International Zurich Symposium on Electromagnetic Compatibility, EMC, Zurich, pp. 34–37 (2006)
[591] Ogunsola, A., Reggiani, U., Sandrolini, L.: Demonstrating signalling compatibility between two train control systems. In: 18th International Zurich Symposium on Electromagnetic Compatibility, EMC, Zurich, pp. 361–364 (August 2007)
[592] Olsen, R.G., Pankaskie, T.A.: On the exact Carson and image theories for wire at or above the earth's interface. IEEE Transactions on Power Apparatus and Systems PAS-102(4), 769–778 (1983)

[593] ORE technical document DT 53 (A 122), Bestimmung der elektrischen Gleisparameter in Abhängigkeit von der Frequenz und den Vormagnetisierenden Strömen, Utrecht (November 1976)

[594] ORE question A 171, Calculation methods for determining interference with light current installation – Report n. 2: Parameters of the test lines in Velim, Utrecht (September 1989)

[595] Orlandi, A., Paul, C.R.: FDTD analysis of lossy, multiconductor transmission lines terminated in arbitrary loads. IEEE Transactions on Electromagnetic Compatibility 38(3), 388–399 (1996)

[596] Ouaddi, H., Nottet, G., Baranowski, S., Kone, L., Idir, N.: Determination of the high frequency parameters of the power transformer used in the railway substation. In: Vehicle Power and Propulsion Conference (VPPC), Lille, France, September 01-03 (2010)

[597] Park, K.B., Lee, T.H., Jeon, Y.J., Ryu, Y.T.: A Study on Improvement of High Voltage Impulse Track Circuits at the Railway. In: The International Conference on Electrical Engineering, pp. 1–5 (April 2009)

[598] Patel, H.S., Hoft, R.G.: Generalised techniques of harmonic elimination and voltage control in thyristors inverters: Part I - Harmonic elimination. IEEE Transactions on Industry Applications 9, 310–317 (1973)

[599] Paul, C.R.: Analysis of multiconductor transmission lines. John Wiley & Sons, New York (1994)

[600] Pilo, E., Rouco, L., Fernandez, A.: A reduced representation of 2x25kV electrical systems for high-speed railways. In: Proc. of the 2003 IEEE/ASME Joint Rail Conference, Chicago, Illinois, USA, April 22-24, pp. 199–205 (2003)

[601] Pires, C.L., Nabeta, S.I., Cardoso, J.R.: Second-order model for remote and close-up short-circuit faults currents on DC traction supply. IET Power Electronics 1(3), 348–355 (2008)

[602] Pleym, A.: EMC in railway systems: conductive coupling from track to nearby metallic structure. In: EMC, Zurich, pp. 425–430 (1999)

[603] Pollaczek, F.: Uber das Feld einer unendlich langen wechselstrom durchflossenen Einfachleitung. Electrishe Nactrichten Technik 3(9), 339–39 (1926)

[604] Popovic, B.D., Filipovic, D.N.: Theory of power-frequency proximity effect for strip conductors. Proc. Inst. Elect. Eng. 122(8), 839–842 (1975)

[605] Pozzobon, P.: Transient and Steady-State Short-Circuit Currents in Rectifiers for DC Traction Supply. IEEE Transactions on Vehicular Technology 47(4), 1390–1404 (1998)

[606] Proakis, J.G.: Digital Communications, 3rd edn. Mc Graw-Hill (1995)

[607] Rappaport, T.S.: Wireless communications – Principles and practices, 2nd edn. Prentice Hall, New Jersey (2002)

[608] Strutt Lord Rayleigh, J.W.: On the incidence of aerial and electric wave upon small obstacles in the form of ellipsoids or elliptic cylinders, and on the passage of electric wave through a circular aperture in a conducting screen. Phil. Mag. 44, 28–52 (1897)

[609] Ramo, S., Winnery, J.R., Van Duzer, T.: Fields and waves in communication electronics, pp. 291–297. J. Wiley & Sons, New York (1965)

[610] BRB/LUL/RIA 12, General specification for protection of traction & rolling stock electronic equipment from transients & surges in DC control systems. Railway Industry Association, United kingdom

[611] BRB/LUL/RIA 13, General specification for electrical equipment used on traction and rolling stock. Railway Industry Association, United Kingdom

[612] BRB/LUL/RIA 18, Interference testing for electronic equipment used on traction and rolling stock. Railway Industry Association, United Kingdom
[613] BRB/LUL/TG 20, Technical guide on e.m.c. for electronic equipment used on traction and rolling stock. Railway Industry Association, United Kingdom
[614] Robert, A.: Dielectric permittivity of concrete between 50 MHz and 1 GHz and GPR measurements for building materials evaluation. Journal of Applied Geophysics 40, 89–94 (1998)
[615] Rosa, E.B., Grover, F.W.: Formulas and tables for the calculations of mutual and self inductance. Scientific Papers of Bureau of Standards (169) (1948)
[616] Ruehli, A.E.: Inductance calculation in a complex integrated circuit environment. IBM Journal of Research and Development 16(5), 470–481 (1972)
[617] Runge, W.: Control of line harmonics due to four-quadrant-converter in AC tractive stock by means of filter and transformer. In: EPE 1997, Trondheim, Norway, pp. 2–7 (1997)
[618] Safertal, F.J.: GSM-R – a new standard for rail telecommunications, `http://userpage.fu-berlin.de/~jmueller/its/conf/Madrid02/abstracts/Safertal.pdf`
[619] Sandrolini, L., Reggiani, U., Ogunsola, A.: Modelling the electrical properties of concrete for shielding effectiveness prediction. Journal of Physics D: Applied Physics 40, 5366–5372 (2007)
[620] Schelkunoff, S.A.: On diffraction and radiation of electromagnetic waves. Phys. Rev. 56, 308–316 (1939)
[621] Schmidt, S., Wuergler, D., Terwiesch, P., Henning, U.: Electrical System Compatibility for Advanced Rail Vehicles: a survey. In: Proceedings of the 8th International Conference on Harmonics and Quality of Power, vol. 2, pp. 623–629 (1998)
[622] Scott, J.H., Carrol, R.D., Cunningham, D.R.: Dielectric constant and electrical conductivity measurements of moist rock: a new laboratory method. Journal of Geophysical Research 27(20) (October 1967)
[623] Scott Jr., W.R., Smith, G.S.: Measured electrical constitutive parameters of soil as function of frequency and moisture content. IEEE Transactions on Geoscience and Remote Sensing 30(3) (May 1992)
[624] Seborsky, S.: EMC immunity testing of the electronic equipment. In: International Conference on Applied Electronics (2010)
[625] Semlyen, A., Dabuleanu, A.: Fast and accurate switching transient calculations on transmission lines with ground return using recursive convolutions. IEEE Transactions on Power Apparatus and Systems 94, 561–571 (1975)
[626] Semlyen, A., Roth, A.: Calculation of exponential propagation step responses - accurately for three frequencies. IEEE Transactions on Power Apparatus and Systems 96(2), 667–672 (1977)
[627] Shaari, A., Millard, S.G., Bungey, J.H.: Modeling the propagation of a radar signal through concrete as a low pass filter. NDT&E International 37, 237–242 (2004)
[628] Shan, Q., Wen, Y., Li, H.: Study on the Propagation Characteristics of the Radio Noise form Electric Railway in Shortwave Band. In: International Symposium on Electromagnetic Compatibility, EMC 2007, pp. 170–173 (2007)
[629] Shapira, J.: Electromagnetic Compatibility (EMC) for Systems Engineers. Institution of Electromagnetic and Telecommunications Engineers, Technical Review 28, 70–77 (2011)

[630] Shenhar, A.J., Bonen, Z.: The new taxonomy of systems: toward an adaptive systems engineering framework. IEEE Transactions on Systems, Man and Cybernetics, Part A: Systems and Humans 27(2), 137–145 (1997)

[631] Silvester, P.: The accurate calculation of skin effect in conductors of complicated shape. IEEE Transactions on Power Apparatus and Systems 87(3), 735–742 (1968)

[632] Slama, J., Chariag, D.: Measurement and analysis of magnetic field radiated from DC tramway: a case study for Tunis's metro. Journal of Electrical Systems (January 2008)

[633] Slonim, M.A.: Electromagnetic processes in AC/DC power converters: Part I - General equations of AC-DC and/or DC-AC power converters for steady-state and transient processes. IEEE Transactions on Industrial Electronics and Control Instrumentation 24, 50–56 (1977)

[634] Slonim, M.A.: Electromagnetic processes in AC/DC power converters: Part II - Symmetrical transient processes in power converters. IEEE Transactions on Industrial Electronics and Control Instrumentation 25, 65–73 (1978)

[635] Slonim, M.A.: Electromagnetic processes in AC/DC power converters: Part III - Nonsymmetrical transient processes. IEEE Transactions on Industrial Electronics and Control Instrumentation 25, 65–73 (1978)

[636] Snelson, J.K.: Propagation of traveling waves on transmission lines frequency dependent parameters. IEEE Transactions on Power Apparatus and Systems 91, 85–91 (1972)

[637] Sollerkvist, F.J., Varjú, G.: A Comparison of the AT and BT Railway Systems from an Induction Point of View, Using a Multiconductor Simulation. In: International Symposium on EMC, paper B-1, Rome, pp. 34–39 (1998)

[638] Song, W.-S., Smedley, K., Feng, X., Sun, P.: One-Cycle Control of Induction Machine Traction Drive for High Speed Railway – Part I: Multi-pulse Width Modulation Region. In: Proc. of 36th IEEE Annual Conference of the Industrial Electronics Society, IECON 2010, November 7-10, pp. 2346–2351 (2010)

[639] Song, W.-S., Smedley, K., Feng, X.: One-Cycle Control of Induction Machine Traction Drive for High Speed Railway – Part I: Multi-pulse Width Modulation Region. In: Proc. of 26th IEEE Annual Applied Power Electronics Conference and Exposition, APEC 2011, March 6-11, pp. 1003–1009 (2011)

[640] Soutsos, M.N., Bungey, J.H., Millard, S.G., Shaw, M.R., Patterson, A.: Dielectric properties of concrete and their influence on radar testing. NDT&E International 34, 419–425 (2001)

[641] Spenny, C., Kusko, A.: Survey of worldwide railroad electrification. In: 1st Rail Systems Technology and Operations Symposium, Pittsburgh, PA, December 2 (1980)

[642] Sporleder, H.: Continuous automatic train control and cab signalling with the LZB 80. In: International Conference on Main Line Railway Electrification, September 25-28, York, UK, pp. 40–46 (1989)

[643] Steimel, A.: Electric Railway Traction in Europe. IEEE Industry Applications Magazine, 6–17 (November/December 1996)

[644] Steimel, A.: Electric Traction - Motive Power and Energy Supply: Basic and Practical Experience. Oldenbourg Industrieverlag, Munich (2008)

[645] Stern, M., Geary, A.L.: Electrochemical Polarization. Journal of the Electrochemical Society 104(1), 56–63 (1957)

[646] Storer, J.E.: The impedance of an antenna over a large circular screen. Journal of Applied Physics 22, 1058–1066 (1951)
[647] Storer, J.E.: The radiation pattern of an antenna over a circular ground screen. Journal of Applied Physics 23, 588–593 (1952)
[648] Stratton, J.A., Chu, L.J.: Diffraction theory of electromagnetic waves. Phys. Rev. 56, 99–107 (1939)
[649] Suddards, A.D.: Electrification Systems. In: Rail Engineering - The Way Ahead. Inst. Mech. Eng., vol. 1, pp. 279–293 (September 1975)
[650] Sunde, E.D.: Currents and potentials along leaky ground return conductors. Transactions of the American Institute of Electrical Engineer, 1338–1346 (December 1936)
[651] Sunde, E.D.: Earth conduction effects in transmission systems. Van Nostrand, New York (1946)
[652] Sussmann, T.R., O'Hara, K.R., Selig, E.T.: Development of Material Properties for Railway Application of Ground Penetrating Radar. In: Proceedings of the Ninth International Conference on Ground Penetrating Radar, vol. 4758, pp. 42–47 (2002)
[653] Tabbagh, A., Hesse, A., Grard, R.: Determination of electrical properties of the ground at shallow depth with an electrostatic quadrupole: field trials on archaeological sites. Geophysical Prospection 41, 579–597 (1993)
[654] Tachibana, K., Tsuboi, T., Kariya, S.: Harmonic Currents in Catenary Systems from Chopper Control. IEEE Transactions on Industry Applications 8(2), 203–210 (1972)
[655] Taufiq, J.: Step response prediction for traction VSI drive with emphasis on signalling compatibility. In: Proc. of 19th IEEE Power Electronics Specialists Conference, vol. 2, pp. 1271–1275 (March 1988)
[656] Tellini, B., Macucci, M., Giannetti, R., Antonacci, G.A.: Conducted and Radiated Interference Measurements in the Line-Pantograph System. IEEE Transactions on Instrumentation and Measurement 50(6), 1661–1664 (2001)
[657] Trueblood, H.M., Wascheck, G.: Investigation of rail impedances. Electrical Engineering, 898–907 (1933)
[658] Tso, S.K., Lam, F.K., Chan, F.H.Y., Edgely, R.K.: Attenuation effects of undertrack reinforcing on automatic train protection signalling systems. IEE Proc. B 128(2), 92–100 (1981)
[659] Tylavsky, D.J., Trutt, F.C.: Complete Analysis of the Operating Modes of the Resistance-Inductance-Fed Bridge Rectifier. IEEE Transactions on Industrial Electronics 29(4), 287–294 (1982)
[660] Tylavsky, D.J.: Conductor Impedance Approximations for Deep- Underground Mines. IEEE Transactions on Industry Applications 23(4), 723–730 (1987)
[661] Tzeng, Y.-S., Chen, N., Wu, R.-N.: Modes of Operation in Parallel-Connected 12-pulse Uncontrolled Bridge Rectifiers without an Interphase Transformer. IEEE Transactions on Industrial Applications 44(3), 344–355 (1997)
[662] Tzou, Y.Y., Hsu, L.J.: A practical SPICE macro model for the IGBT. In: Proceedings of the IEEE International Conference on Industrial Electronics, Control and Instrumentation (IECON), pp. 762–766 (1993)
[663] Uhlirz, M.: Concept of a GSM-based Communication System for High-Speed Trains. In: Proceedings of the IEEE Vehicular Technology Conference (VTC), pp. 1130–1134 (1994)

[664] UIC 550, Power supply installations for passenger stock, International Union of Railways, 11th edn. (April 2005)
[665] UIC 550-2, Installations pour l'alimentation en énergie des voitures – Essais des types (French, Power supply installations for passenger stock – Type tests (January 1994)
[666] UIC 550-3, Installations pour l'alimentation en énergie des voitures – Influence des équipements électriques a l'extérieur des voitures (French, Power supply installations for passenger stock - Effect on electrical installations outside passenger coaches (April 2005)
[667] UIC 626, Production of electrical power on diesel tractive units for supplying the train cable, International Union of Railways, 3rd edn. (Aprii 2004) (reissue)
[668] U.S. DOT UMTA/TSC, Draft Recommended Practice Rail Transit Intra-System Electromagnetic Compatibility Vehiclular Electrical Power and Track Circuit Signalling Subsystems (July 1980)
[669] UMTA-MA-06-0153-85-6, Conducted Interference in Rapid Transit Signalling Systems, vol. II: Suggested Test Procedures, U.S. Department of Transporation (May 1987)
[670] UMTA-MA-06-0153-85-11, Radiated Interference in Rapid Transit Signalling Systems, vol. II: Suggested Test Procedures, U.S. Department of Transporation (June 1987)
[671] UMTA-MA-06-0153-85-8, Inductive Interference in Rapid Transit Signalling Systems, vol. II: Suggested Test Procedures, U.S. Department of Transporation (March 1987)
[672] Uribe, F.A.: Calculating mutual ground impedances between overhead and buried cables. IEEE Transactions on Electromagnetic Compatibility 50(1), 198–203 (2008)
[673] Union Switch & Signals, M-3, M-23A, M-23B Electric Switch Machines with Permanent Magnet Motor, SM 6263, Rev. 6 (February 2005)
[674] Union Switch & Signals, Audio Frequency Overlay System, SM 6414, Rev. 3 (August 2003)
[675] Union Switch & Signals, MICROTRAX Coded Track Circuit System, SM 6470a (April 1996)
[676] Union Switch & Signals, AF-902/AF-904 Audio Frequency Track Circuit System - Field Maintenance Manual, SM 8051, Rev. 4 (July 2003)
[677] Vance, E.F.: Coupling to Shielded Cable. Wiley-Interscience, New York (1978)
[678] Wang, J.X., Roush, M.L.: What Every Engineer Should Know About Risk Engineering and Management. Marcel Dekker, New York (2000)
[679] Williams, T.: EMC for Product Designers, 3rd edn. Neunes, Oxford (2001)
[680] Wobschall, D.: A theory of the complex dielectric permittivity of soil containing water: the semidisperse model. IEEE Transactions on Geoscience Electronics 15(1), 49–58 (1977)
[681] Yellow Book: Engineering Safety Management, Fundamental and Guidance 1, 2(4) (2005)
[682] Yuen, K.H., Ye, Z.M., Pong, M.H., Lo, W.C.: Modelling of traction harmonic current using statistical method. In: Proceedings of the Conference on Power Electronics and Drive Systems, PEDS 1999, vol. 1, pp. 194–199 (1999)

[683] Zając, W., Szeląg, A.: Harmonic distortion caused by suburban and underground rolling stock with DC motors. In: Proc. of the IREEE International Power Electronics Congress, CIEP 1996, Cuernavaca, Mexico, October 14-17, pp. 200–206 (1996)

[684] Zanobini, A., Ghizzani, M.: Measurement of Some Typical Parameters in the Realization of a Signal Management Plant in Railway Diagnostics. IEEE Transactions on Instrumentation and Measurement 52(5), 1682–1687 (2003)

[685] Zhang, Z., Wu, B., Luo, L., Zhou, L., Liu, F.: A Multi-Purpose Balanced Transformer with Harmonic Eliminating Capability for Railway Traction Applications. In: Proc. of the IEEE Power and Energy Society General Meeting - Conversion and Delivery of Electrical Energy in the 21st Century, Pittsburgh, PA, USA, July 20-24 (2008)

[686] Zhongming, Y., Lo, E., Yuen, K.H., Pong, M.H.: Probabilistic characterization of current harmonics of electrical traction power supply system by analytic method. In: Proceedings of the Industrial Electronics Conference, IECON, vol. 1, pp. 360–366 (1999)

[687] Zynovchenko, A., Xie, J., Jank, S., Klier, F.: Resonance phenomena and propagation of frequency converter harmonics in the catenary of railways with single-phase A.C. In: Proceedings of EPE 2005, Dresden, Germany, September 11-14 (2005)

Index

The manufacturer's authorised representative in the EU is Springer Nature Customer Service Centre GmbH, Europaplatz 3, 69115 Heidelberg, Germany. If you have any concerns regarding our products, please contact ProductSafety@springernature.com

Printed and bound by CPI Group (UK) Ltd, Croydon, CR0 4YY
05/10/2026
02228131-0012